Essentials of Ecology

Essentials of Ecology

FOURTH EDITION

G. TYLER MILLER, JR.

President, Earth Education and Research

CONTRIBUTING EDITOR

SCOTT SPOOLMAN

THOMSON

BROOKS/COLE

Australia • Brazil • Canada • Mexico • Singapore • Spain • United Kingdom • United States

Publisher: *Jack Carey*
Contributing Editor: *Scott Spoolman*
Production Project Manager: *Andy Marinkovich*
Technology Project Manager: *Fiona Chong*
Assistant Editor: *Carol Benedict*
Editorial Assistant: *Kristina Razmara*
Permissions Editor/Photo Researcher: *Abigail Reip*
Marketing Manager: *Kara Kindstrom*
Marketing Assistant/Associate: *Catie Ronquillo*
Advertising Project Manager: *Bryan Vann*
Print/Media Buyer: *Karen Hunt*

Production Management, Copyediting, and Composition: *Thompson Steele, Inc.*
Interior Illustration: *Precision Graphics; Sarah Woodward; Darwin and Vally Hennings; Tasa Graphic Arts, Inc.; Alexander Teshin Associates; John and Judith Walker; Rachel Ciemma; Victor Royer, Electronic Publishing Services, Inc.; J/B Woosley Associates; Kerry Wong; ScEYEnce*
Cover Image: *Alpine sunset, Yosemite National Park* `© *Michael E. Gordon*
Text Printer: *Transcontinental Printing/Interglobe*
Cover Printer: *Transcontinental Printing/Interglobe*
Title Page Photograph: *Salt Marsh in Peru, SuperStock*

Printed in Canada

1 2 3 4 5 6 7 10 09 08 07 06

For more information about our products, contact us at:
Thomson Learning Academic Resource Center
1-800-423-0563
For permission to use material from this text, contact us by:
Phone: 1-800-730-2214
Fax: 1-800-730-2215
Web: http://www.thomsonrights.com

Library of Congress Control Number: 2006930474

ISBN: 0-495-12544-X

Thomson Higher Education
10 Davis Drive
Belmont, CA 94002-3098
USA

Asia (including India)
Thomson Learning
5 Shenton Way #01-01
UIC Building
Singapore 068808

Australia
Nelson Thomson Learning
102 Dodds Street
South Melbourne, Victoria 3205
Australia

Canada
Nelson Thomson Learning
1120 Birchmount Road
Toronto, Ontario M1K 5G4
Canada

Europe/Middle East/Africa
Thomson Learning
High Holborn House
50/51 Bedford Row
London WC1R 4LR
United Kingdom

Latin America
Thomson Learning
Seneca, 53
Colonia Polanco
11560 Mexico D.F.
Mexico

Spain
Paraninfo Thomson Learning
Calle/Magallanes, 25
28015 Madrid
Spain

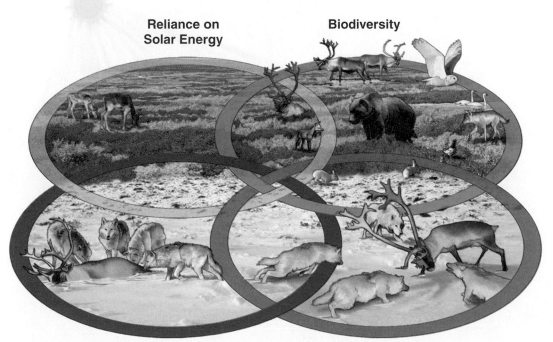

Reliance on Solar Energy

Biodiversity

Nutrient Recycling

Population Control

Four interconnected scientific principles of sustainability, derived from learning how nature has sustained a variety of life on the earth for about 3.7 billion years

Detailed Contents

NASA

SuperStock

Photo 1 Tree plantation

Photo 2 Air pollution from an industrial plant in India

Changes in ice coverage in the northern hemisphere during the past 18,000 years. (Data from the National Oceanic and Atmospheric Administration)

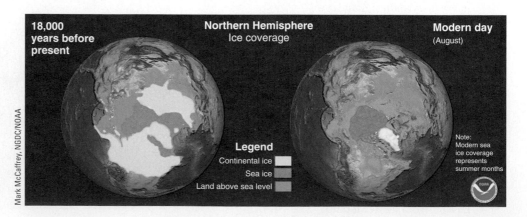

Hartmut Schwartzbach/Peter Arnold, Inc.

Photo 3 Homeless people in Calcutta, India

Ocean hemisphere   Land–ocean hemisphere

The ocean planet, where about 97% of the water is in the interconnected oceans and freshwater systems cover less than 1% of the earth's surface

Photo 5 World's lagest flower (*Rafflesia*) growing in Indonesia

Photo 4 Villager in Mozambique carrying water

Photo 6 Endangered white ukari in a Brazilian tropial forest

SUSTAINING BIODIVERSITY

Photo 7 Scarlet macaw in Brazil's Amazon rainforest

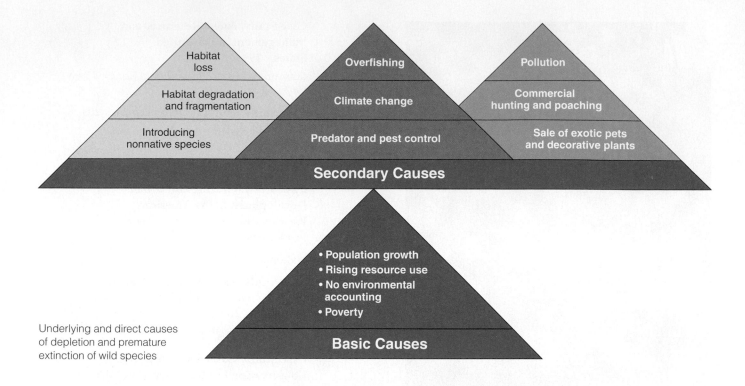

Habitat
loss

Habitat degradation
and fragmentation

Introducing
nonnative species

Overfishing

Climate change

Predator and pest control

Pollution

Commercial
hunting and poaching

Sale of exotic pets
and decorative plants

Secondary Causes

• Population growth
• Rising resource use
• No environmental
  accounting
• Poverty

Underlying and direct causes
of depletion and premature
extinction of wild species

Basic Causes

SuperStock

Photo 8 Endangered ring-tailed lemur on the island of Madagascar

11   Sustaining Biodiversity: The Species Approach   222

Photo 10   Coral reef ecological reserve in Florida Keys (USA)

Photo 9   Temperate deciduous forest, winter, Rhode Island (USA)

12   Sustaining Aquatic Biodiversity   249

Photo 11 Severe gully erosion on cropland in Bolivia

Photo 12 Ship stranded in desert formed by shrinkage of the Aral Sea

SUPPLEMENTS

Photo 13 Air pollution damage to trees in Mount Mitchell State Park, North Carolina (USA)

Photo 14 Creek in Montana (USA) polluted toxic metals by gold mining

How Is This Book Organized?

This book is a brief introduction to basic ecological concepts. It consists of the first twelve chapters of the fifteenth edition of my longer book, *Living in the Environment* (Brooks/Cole, 2007).

Essentials of Ecology, fourth edition, is divided into three major parts. Part I provides an overview of major environmental problems (Chapter 1).

Part II consists of eight chapters covering basic science and principles of ecology.

Part III has three chapters on biodiversity. To get an overview of this book I urge you to look at the Brief Contents (p. v).

Flexibility of use is enhanced by providing 17 *Supplements* (see pp. xiv–xv in the Detailed Contents), which instructors can assign to meet the needs of their specific courses. Examples include maps of biodiversity and ecological footprints (Supplement 4, pp. S8–S15), basic chemistry (Supplement 7, pp. S27–S33), and deforestation and nutrient cycling in an experimental forest (Supplement 9, pp. S36–37).

Case Studies

A *Core Case Study* opens each chapter (see p. 6) and is carried out as a chapter theme (see pp. 7, 11, 12, 14, 16, 18, 20, 24, 26, 27). Each chapter ends by relating the Core Case Study to the four scientific principles of sustainability (see Figure 1-16, p. 24 and pp. 26, 81, and 220). *Thinking About* questions (see pp. 12, 13, 14, 16, 18, 20, 24, 27) found throughout each chapter challenge students to make connections between Core Case Studies and chapter concepts and principles.

In addition to these 12 Core Case Studies, 29 other Case Studies appear throughout the book (see Detailed Contents, pp. vii–xv). These 42 case studies provide a more in-depth look at specific environmental problems and their possible solutions.

Interactive and Visual Learning

Some 74 *Thinking About* exercises (see pp. 98, 119, 145), 11 *How Would You Vote?* exercises (see pp. 165, 212, 259), and questions at the end of captions for 35 figures (see pp. 72, 188, 198) provide an *interactive approach to learning.* These *critical thinking exercises* reinforce information and concepts by asking students to analyze material immediately after it is presented rather than waiting until the end of chapter.

The book's 222 illustrations are designed to present complex ideas in understandable ways and to relate learning to the real world (see pp. 62, 72–73, 131). And there are 94 carefully selected photographs—89 of them new to this edition (see pp.109, 198, 204).

To enhance visual learning and scientific understanding further, 62 *ThomsonNow* interactive animations, referenced to the text (see pp. 65, 72, 180) and diagrams (see pp. 47, 64, 107), are available on the ThomsonNOW website. This learning tool—new to this edition—helps students assess their unique study needs through pretests, posttests, and personalized learning plans. It is FREE with every new copy of the book and can be purchased for used titles. Another feature of this learning tool is *How Do I Prepare?* It allows students to review basic math, chemistry, and other refresher skills.

Currency and Preserving Natural Capital

There are more than 1,200 updates based on information and data published in 2003, 2004, 2005, and 2006. An important example is the 2005 *UN Millennium Ecosystem Assessment* of the health of the world's ecosystems, which emphasized the urgent need to understand and preserve the *natural services* or *natural capital* they provide. Some 93 *Research Frontier* items provide currency by showing students key areas that require more research (see pp. 34, 79, 131).

In keeping with the conclusions of the *Millennium Ecosystem Assessment,* two major subthemes of this book are *natural capital* (see figures on pp. 9, 129, 136, 193) and *natural capital degradation* (see pp. 123, 135, 188, 196) and how protecting natural capital is a key to environmental and economic sustainability (Figure 1-16, p. 24).

A Major Revision

Major changes included in this edition are

- Opening each chapter with a *Core Case Study* and challenging students to connect its material to chapter concepts and principles and to four scientific principes of sustainability

- Adding *flexibility* by the use of 17 optional supplements (see p. xiv–xv in the Detailed Contents)

- Adding 72 *Thinking About* interactive exercises that reinforce learning by asking students to think critically about the implications of various environmental issues immediately after they are discussed in the text. This complements the *How Would You Vote?* interactive feature (see pp. 165, 212, 259)

- Many diagrams with questions that encourage students to think about and evaluate the content of the diagrams (see figures on pp. 72, 129, 188)

- 89 carefully selected new photographs

- 48 new figures and 50 improved or updated figures. This means that 44% of the book's figures are new, improved, or updated.

- More than 1,200 updates based on information and data published in 2003, 2004, 2005, and 2006

- 45 *Research Frontier* items that provide currency by showing students key areas that require more research

- *Environmental* or *Green Careers* indicated in green type in the text (see pp. 96, 152) with in-depth discussions on most careers available on the website

- *Active Graphing* exercises for most chapters, available for assignment at the ThomsonNOW website . These exercises involve students in the graphing and evaluation of data. A new supplement on graphing (see p. S4–S5) was also added.

- More than 50 new or expanded topics including impacts of new affluent consumers in China and India (pp. 15 and 185–187); expanded discussion of four scientific principles of sustainability (pp. 23–25 and Figure 1-16, p. 24); expanded treatment of ecological footprints (Figure 1-7, p. 13, Figure 9-17, p. 188, and Supplement 4, p. S8); 2005 Millennium Ecosystem Assessment (pp. 22, 80, 123, and elsewhere); expanded introduction to biodiversity (pp. 61–63, Figure 3-15, p. 62, and Supplement 4, pp. S8–S15); closer look at photosynthesis (Spotlight, p. 59); geologic processes, climate change, catastrophes and evolution (pp. 87–89); tsunamis (pp. S43–S45 in Supplement 10); hurricanes and New Orleans (Case Study, p. 140); expanded treatment of basic chemistry and nano-technology (Supplement 7, p. S26–S32); deforestation and nutrient cycling in an experimental forest (Supplement 9, p. 35–S36); wolf and moose interactions on Isle Royale, Michigan (Supplement 12, p. S46)

In-Text Study Aids

Each chapter begins with a few general questions to reveal how it is organized and what students will be learning. When a new term is introduced and defined, it is printed in boldface type. A glossary of all key terms is located at the end of the book.

Thinking About exercises reinforce learning by asking students to think critically about the implications of various environmental issues immediately after they are discussed in the text. And the captions of many figures contain questions that involve students in thinking about and evaluating their content.

Each chapter ends with a set of *Critical Thinking* questions to encourage students to think critically and apply what they have learned to their lives. The Companion Website for the book also contains a set of *Overview Questions* covering *all* of the material in the chapter, which can be used as a study guide for students. Some instructors download this from the website to give students a list of questions to answer as a course requirement.

Qualified users of this textbook have free access to the Companion Website for this book at

http://www.thomsonedu.com/biology/miller

At this website they will find the following material for each chapter:

- *Flashcards*, which allow you to test your mastery of the terms and concepts to remember for each chapter

- *Chapter quizzes*, which provide a multiple-choice practice quiz

- *InfoTrac*® *College Edition* articles listed by chapter and section online, as well as by relevant region of the country on InfoTrac® Map. These articles can also support *How Would You Vote?* exercises.

- *Guest Essays* by prominent environmental scientists and leaders

- Information on a variety of environmental careers

- *References*, which list the major books and articles consulted in writing this book

- A brief *What Can You Do?* list addressing key environmental problems

- *Weblinks*, an extensive list of websites with news, research, and images related to individual sections of the chapter.

Qualified adopters of this textbook can request WebCT or Blackboard course cartridges preloaded with ThomsonNOW content and a full array of study tools, including flashcards (with audio), practice quizzes, and web links.

Teachers and students using *new* copies of this textbook also have free and unlimited access to *InfoTrac® College Edition*. This fully searchable online library gives users access to complete environmental articles from several hundred periodicals dating back over the past 24 years.

Other student learning tools include:

- *Audio version for study and review.* Students can listen to the chapters in this book while walking, traveling, sitting in their rooms, or working out. New editions of the book can include pincode access to the audiobook, if requested by the instructor. They can use the pin code to download book chapters free from the Web to any MP3 player.

- Close to 100 *animations and interactions* correlated by chapter available for use by students at ThomsonNOW

- ThomsonNOW also includes *How do I prepare?* tutorials on basic math, chemistry, biology, and study skills.

- *Essential Study Skills for Science Students* by Daniel D. Chiras. This book includes chapters on developing good study habits, sharpening memory, getting the most out of lectures, labs, and reading assignments, improving test-taking abilities, and becoming a critical thinker. Instructors can have this book bundled FREE with the textbook.

- *Laboratory Manual* (fourth edition) by C. Lee Rockett and Kenneth J. Van Dellen. This manual includes a variety of laboratory exercises, workbook exercises, and projects that require a minimum of sophisticated equipment.

Supplements for Instructors

- *PowerLecture CD-ROM.* This CD-ROM—free to adopters—makes it easy to create custom lectures. Each chapter's Microsoft® PowerPoint presentation consolidates all relevant resources—illustrations, photographs, animations, InfoTrac® College Edition references—into one powerful class tool. This program's editing tools allow use of slides from other lectures, modification or removal of figure labels and leaders, insertion of your own slides, saving slides as JPEG images, and preparation of lectures for use on the Web. The CD-ROM also includes Microsoft® Word files of the Instructor's Manual and Test Bank.

- *Transparency Masters and Acetates.* Includes 100 color acetates of line art from the text and over 400 black-and-white master sheets of key diagrams for making overhead transparencies. Free to adopters.

- *CNN™ Today Videos.* These informative news stories, available either on VHS tapes or CD-ROMs while supplies last, contain close to 40 two- to three-minute video clips of current news stories on environmental issues from around the world. Student Workbook Included.

- *ABC News: Environmental Science in the Headlines (2005, 2006)* on DVD. Thomson Brooks/Cole has expanded its portfolio of videos with numerous current segments from ABC News. These two- to three-minute video clips are available on DVD. Student Workbook Included.

- *Online Instructor's Manual with Test Items.* Free to adopters.

- *ExamView.* Allows an instructor to easily create and customize tests, see them on the screen exactly as they will print, and print them out.

- *JoinIn™ on Turning Point®.* Thomson Brooks/Cole offers content for classroom response systems suitable for *Essentials of Ecology.* Transform your classroom and assess your students' progress with instant in-class quizzes and polls on ready-to-use Microsoft PowerPoint slides.

- *Updates Online,* which are organized by chapter allow you to incorporate current environmental news, video clips, and articles into your lectures.

- *ThomsonNOW.* This powerful online course management system saves instructors time through its assignment capabilities, automatic grading, and easy-to-use grade book. Instructors can easily create online assignments for students, using questions from the test bank, Critical Thinking exercises, and other resources. Contact your Thomson Brooks/Cole representative to get set up and for information on integration with WebCT and Blackboard.

- *Outernet* provides 150 customized laboratory experiments that you can choose from for your students.

Other Textbook Options

Instructors wanting a book with a different length and emphasis can use one of my three other books written for various types of environmental science courses: *Living in the Environment,* fifteenth edition (627 pages, Brooks/Cole 2007), *Environmental Science,* eleventh edition (436 pages, Brooks/Cole 2006), and *Sustaining the Earth: An Integrated Approach,* eighth edition (384 pages, Brooks/Cole 2007).

Help Me Improve This Book

Let me know how you think this book can be improved. If you find any errors, bias, or confusing explanations, please e-mail me about them at

mtg89@hotmail.com

Most errors can be corrected in subsequent printings of this edition rather than waiting for a new edition.

Acknowledgments

I wish to thank the many students and teachers who have responded so favorably to the three previous editions of *Essential of Ecology,* the fifteen editions of *Living in the Environment,* the eleven editions of *Environmental Science,* and the eight editions of *Sustaining the Earth,* and who have corrected errors and offered many helpful suggestions for improvement. I am also deeply indebted to the more than 250 reviewers, who pointed out errors and suggested many important improvements in the various editions of these three books. Any errors and deficiencies left are mine.

I am particularly indebted to Scott Spoolman who served as a contributing editor for this new edition and who made numerous suggestions for improving this book. The members of the talented production team, listed on the copyright page, have made vital contributions as well. My thanks also go to production editors Andy Marinkovich at Brooks/Cole and Nicole Barone of Thompson Steele, copy editor Andrea Fincke, layout expert Bonnie Van Slyke, photo researcher Abigail Reip, artist Patrick Lane, Brooks/Cole's hard-working sales staff, and Keli Amann, Fiona Chong, and the other members of the talented team who developed the multimedia, website, and advertising materials associated with this book.

I also thank C. Lee Rockett and Kenneth J. Van Dellen for developing the *Laboratory Manual* to accompany this book; Jane Heinze-Fry for her work on concept mapping; Irene Kokkala for her excellent work on the *Instructor's Manual;* and the people who have translated this book into eighth different languages for use throughout much of the world.

My deepest thanks go to Jack Carey, biology publisher at Brooks/Cole, for his encouragement, help, 40 years of friendship, and superb reviewing system. It helps immensely to work with the best and most experienced editor in college textbook publishing.

I dedicate this book to the earth and to Kathleen Paul Miller, my wife and research associate.

G. Tyler Miller, Jr.

Guest Essayists

Guest essays by the following authors are available online at the website for this book: **M. Kat Anderson,** ethnoecologist with the National Plant Center of the USDA's Natural Resource Conservation Center; **Lester R. Brown,** president, Earth Policy Institute; **Garrett Hardin,** professor emeritus (now deceased) of human ecology, University of California, Santa Barbara; **Paul G. Hawken,** environmental author and business leader; **Jane Heinze-Fry,** environmental educator, **Bobbi S. Low,** professor of resource ecology, University of Michigan; **Lester W. Milbrath,** director of the research program in environment and society, State University of New York, Buffalo; **Peter Montague,** director, Environmental Research Foundation, **Norman Myers,** tropical ecologist and consultant in environment and development; **Donald Worster,** environmental historian and professor of American history, University of Kansas

Essentials of Ecology

Learning Skills

Students who can begin early in their lives to think of things as connected, even if they revise their views every year, have begun the life of learning.

MARK VAN DOREN

Why Is It Important to Study Environmental Science?

Environmental science may be the most important course you will ever take.

Welcome to **environmental science**—an *interdisciplinary* study of how the earth works, how we interact with the earth, and how we can deal with the environmental problems we face.

Environmental issues affect every part of your life. Thus, the concepts, information, and issues discussed in this book and the course you are taking should be useful to you now and throughout your life.

In 1966, I heard Dean Cowie, a physicist with the U.S. Geological Survey, give a lecture on the problems of population growth and pollution. Afterward, I went to him and said, "If even a fraction of what you have said is true, I will feel ethically obligated to give up my research on the corrosion of metals and devote the rest of my life to research and education on environmental problems and solutions. Frankly, I do not want to change my life, and I am going into the literature to try to show that your statements are either untrue or grossly distorted."

After six months of study I was convinced of the seriousness of these and other environmental problems. Since then, I have been studying, teaching, and writing about them. This book summarizes what I have learned in more than four decades of trying to understand environmental principles, problems, connections, and solutions.

Understandably, I am biased. But *I strongly believe that environmental science is the single most important course in your education.* What could be more important than learning how the earth works, how we are affecting its life support system, and how we can reduce our environmental impact?

We live in an incredibly challenging era. There is a growing awareness that during this century we need to make a new cultural transition in which we learn how to live more sustainably by not degrading our life-support system. I hope this book will stimulate you to become involved in this change in the way we view and treat the earth that sustains us, other life, and all economies.

Improving Your Study and Learning Skills

Learning how to learn is life's most important skill.

Maximizing your ability to learn ought to be one of your most important lifetime educational goals. This involves continually trying to *improve your study and learning skills.*

Here are some *general study and learning skills.*

Get organized. Becoming more efficient at studying gives you more time for other interests.

Make daily to-do lists in writing. Put items in order of importance, focus on the most important tasks, and assign a time to work on these items. Because life is full of uncertainties, you will be lucky to accomplish half of the items on your daily list. Shift your schedule as needed to accomplish the most important items.

Set up a study routine in a distraction-free environment. Develop a written daily study schedule and stick to it. Study in a quiet, well-lighted space. Work sitting at a desk or table—not lying down on a couch or bed. Take breaks every hour or so. During each break take several deep breaths and move around to help you stay more alert and focused.

Avoid procrastination—putting work off until another time. Do not fall behind on your reading and other assignments. Accomplish this by setting aside a particular time for studying each day and making it a part of your daily routine.

Do not eat dessert first. Otherwise, you may never get to the main meal (studying). When you have accomplished your study goals then reward yourself with play (dessert).

Make hills out of mountains. It is psychologically difficult to climb a mountain such as reading an entire book, reading a chapter in a book, writing a paper, or cramming to study for a test. Instead, break such large tasks (mountains) down into a series of small tasks (hills). Each day read a few pages of a book or chapter, write a few paragraphs of a paper, and review what you have studied and learned. As American automobile designer and builder Henry Ford put it, "Nothing is particularly hard if you divide it into small jobs."

Look at the big picture first. Get an overview of an assigned reading by looking at the main headings or chapter outline. At the beginning of each chapter in this textbook, I provide a list of the main questions that are the focus of that chapter. Use it as a chapter roadmap.

Ask and answer questions as you read. For example, what is the main point of this section or paragraph? To help you do this I follow the heading of each subsection with a a single sentence (in blue) that describes the main idea of the subsection. You can also use the one-sentence descriptions as a way to review what you have learned. Putting them all together gives you a summary of the chapter's key ideas. I find this running summary to be a more useful learning device than a fairly dense summary at the end of each chapter.

Focus on key terms. Use the glossary in your textbook to look up the meaning of terms or words you do not understand. This book shows all key terms in **boldfaced** type and lesser, but still important, terms in *italicized* type. Flash cards for testing your mastery of key terms for each chapter are also available on the website for this book, or you can make your own by putting a term on one side of a piece of paper and its meaning on the other side.

Interact with what you read. I do this by marking key sentences and paragraphs with a highlighter or pen. I put an asterisk in the margin next to an idea I think is important and double asterisks next to an idea I think is especially important. I write comments in the margins, such as *Beautiful, Confusing, Misleading,* or *Wrong.* I fold down the top corner of pages with highlighted passages and the top and bottom corners of especially important pages. This way, I can flip through a chapter or book and quickly review the key ideas.

Use the audio version of this book for study and review. You can listen to the chapters in this book while walking, traveling, sitting in your room, or working out. You can use the pin code enclosed in any new copy of this book to download book chapters free from the Web to any MP3 player.

Review to reinforce learning. Before each class, review the material you learned in the previous class and read the assigned material.

Become a better note taker. Do not try to take down everything your instructor says. Instead take down main points and key facts using your own shorthand system. Review, fill in, and organize your notes as soon as possible after each class.

Write out answers to questions to focus and reinforce learning. Answer all questions in and at the end of each chapter, or those assigned to you, and the review questions on the website for each chapter. Do this in writing as if you were turning them in for a grade. Save your answers for review and preparation for tests.

Use the buddy system. Study with a friend or become a member of a study group to compare notes, review material, and prepare for tests. Explaining something to someone else is a great way to focus your thoughts and reinforce your learning. Attend any review sessions that might be offered by instructors or teaching assistants.

Learn your instructor's test style. Does your instructor emphasize multiple choice, fill-in-the-blank, true-or-false, factual, thought, or essay questions? How much of the test will come from the textbook and how much from lecture material? Adapt your learning and studying methods to this style. You may disagree with this style and feel that it does not adequately reflect what you know. But the reality is that your instructor is in charge.

Become a better test taker. Avoid cramming. Eat well and get plenty of sleep before a test. Arrive on time or early. Calm yourself and increase your oxygen intake by taking several deep breaths. Do this about every 10–15 minutes. Look over the test and answer the questions you know well first. Then work on the harder ones. Use the process of elimination to narrow down the choices for multiple-choice questions. Getting it down to two choices gives you a 50% chance of guessing the right answer. For essay questions, organize your thoughts before you start writing. If you have no idea what a question means, make an educated guess. You might get some partial credit and avoid a zero. Another strategy for getting some credit is to show your knowledge and reasoning by writing: "If this question means so and so, then my answer is _____."

Develop an optimistic but realistic outlook. Try to be a "glass is half-full" rather than a "glass is half-empty" person. Pessimism, fear, anxiety, and excessive worrying (especially about things you have no control over) are destructive and lead to inaction. Try to keep your energizing feelings of realistic optimism slightly ahead of any immobilizing feelings of pessimism. Then you will always be moving forward.

Take time to enjoy life. Every day take time to laugh and enjoy nature, beauty, and friendship. Becoming an effective and efficient learner is the best way to do this without getting behind and living under a cloud of guilt and anxiety.

Improving Your Critical Thinking Skills: Detecting Baloney

Learning how to think critically is a skill you will need throughout your life.

Critical thinking involves developing skills to help you analyze and evaluate the validity of information and ideas you are exposed to and to make decisions. Critical thinking helps you distinguish between facts and opinions, evaluate evidence and arguments, take and defend informed positions on issues, integrate information and see relationships, and apply your knowledge to dealing with new and different problems. Here are some basic skills for learning how to think more critically.

Question everything and everybody. Be skeptical, as any good scientist is. Do not believe everything you hear or read, including the content of this textbook, without evaluating the information you receive. Seek other sources and opinions. As the famous physicist and philosopher Albert Einstein put it, "The important thing is not to stop questioning."

Identify and evaluate your personal biases and beliefs. Each of us has biases and beliefs taught to us by sources such as parents, teachers, friends, role models, and experience. What are your basic beliefs and biases? Where did they come from? What basic assumptions are they based on? How sure are you that your beliefs and assumptions are right and why? According to the American psychologist and philosopher William James, "A great many people think they are thinking when they are merely rearranging their prejudices."

Be open-minded and flexible. Be open to considering different points of view. Suspend judgment until you gather more evidence, and be capable of changing your mind. Recognize that there may be a number of useful and acceptable solutions to a problem and that very few issues are black or white. One way to evaluate divergent views is to get into another person's head. How do they see or view the world? What are their basic assumptions and beliefs? Is their position logically consistent with their assumptions and beliefs?

Be humble about what you know. Some people are so confident in what they know that they stop thinking and questioning. To paraphrase American writer Mark Twain, "It's not what we don't know that's so bad. It's what we know is true, but just ain't so, that hurts us." Or as philosopher Will Durant put it, "Education is a progressive discovery of our own ignorance."

Evaluate how the information related to an issue was obtained. Are the statements you heard or read based on firsthand knowledge or research or on hearsay? Are unnamed sources used? Is the information based on reproducible and widely accepted scientific studies (*sound* or *consensus science*, p. 32) or preliminary scientific results that may be valid but need further testing (*frontier science*, p. 32)? Is the information based on a few isolated stories or experiences (*anecdotal information*) or on carefully controlled studies? Is it based on unsubstantiated and widely doubted scientific information or beliefs (*junk science* or *pseudoscience*, as discussed on p. 32)?

Question the evidence and conclusions presented. What are the conclusions or claims? What evidence is presented to support them? Does the evidence support them? Is there a need to gather further evidence to test the conclusions? Are there other, more reasonable conclusions?

Try to uncover differences in basic beliefs and assumptions. On the surface most arguments or disagreements involve differences in opinions about the validity or meaning of certain facts or conclusions. Scratch a little deeper and you will find that most disagreements are usually based on different (and often hidden) basic assumptions about how we look at and interpret the world around us. Uncovering these basic differences can allow the parties involved to understand where each is "coming from" and agree to disagree about their basic assumptions or principles.

Try to identify and assess the assumptions and beliefs of those presenting evidence and drawing conclusions. What is their expertise in this area? Do they have any unstated assumptions, beliefs, biases, or values? Do they have a personal agenda? Can they benefit financially or politically from acceptance of their evidence and conclusions? Would investigators with different basic assumptions or beliefs take the same data and come to different conclusions?

Expect and tolerate uncertainty. Recognize that science is an ever-changing adventure that provides only a certain degree of certainty. And the more complex the system or process being investigated, the greater the degree of uncertainty. Scientists can disprove things but they cannot establish absolute proof or certainty.

Do the arguments used involve common logical fallacies or debating tricks? Here are six of many examples. *First,* attack the presenter of an argument rather than the argument itself. *Second,* appeal to emotion rather than facts and logic. *Third,* claim that if one piece of evidence or conclusion is false, then all other pieces of evidence and conclusions are false. *Fourth,* say that a conclusion is false because it has not been scientifically proven (scientists never prove anything absolutely but they can establish degrees of reliability, as discussed on p. 32). *Fifth,* inject irrelevant or misleading information to divert attention from important points. *Sixth,* present only either/or alternatives when there may be a number of alternatives.

Do not believe everything you read on the Internet. The Internet is a wonderful and easily accessible source of information. It is also a useful way to find alternative information and opinions on almost any subject or issue—much of it not available in the mainstream media and scholarly articles. Web logs, or blogs, have become a major source of information, even more important than standard news media for some people. However, because the Internet is so open, anyone can post anything they want to a blog or other website with no editorial control or *peer review*—the method in which scientists and other experts review and comment on an article before it is accepted for publication in a scholarly journal. As a result, evaluating information on the Internet is one of the best ways in which you can put into practice the principles of critical thinking discussed here. Use and enjoy the Internet, but think critically and proceed with caution.

Develop principles or rules for evaluating evidence. Develop a written list of principles, concepts, and rules to serve as guidelines for evaluating evidence and claims and for making decisions. Continually evaluate and modify this list on the basis of your experience.

Become a seeker of wisdom, not a vessel of information. Many people believe that the main goal of education is to learn as much as you can by concentrating on gathering more and more information—much of it useless or misleading. I believe that the primary goal of education is to learn how to sift through mountains of facts and ideas to find the few *nuggets of wisdom* that are the most useful in understanding the world and in making decisions. This book is full of facts and numbers, but they are useful only to the extent that they lead to an understanding of key and useful ideas, scientific laws, concepts, principles, and connections. A major goal of the study of environmental science is to find out how nature works and sustains itself (*environmental wisdom*) and to use *principles of environmental wisdom* to help make our societies and economies more sustainable, more just, and more beneficial and enjoyable for all. As writer Sandra Carey put it, "Never mistake knowledge for wisdom. One helps you make a living; the other helps you make a life."

Critical thinking involves trying to separate useful from useless information. You will find critical thinking questions throughout this book—at the end of each chapter and throughout each chapter in the form of *Thinking About . . . .* There are no right or wrong answers to many of these questions, although experience may show that some answers are better than others. A good way to improve your critical thinking skills is to compare your answers with those of your classmates and discuss how you arrived at your answers.

Know Your Own Learning Style

People learn in different ways, and knowing your own learning style can help you to learn more efficiently and effectively.

People have different ways of learning and it can be helpful to know your own learning style. *Visual learn-*
ers learn best from reading and viewing illustrations and diagrams. They can benefit from using flash cards (available on the website for this book) to memorize key terms and ideas.

Auditory learners learn best by listening and discussing. They might benefit from reading aloud while studying, using a tape recorder in lectures, or using the audio version of this book as a study supplement. *Logical learners* learn best by using concepts and logic to uncover and understand a subject rather than relying mostly on memory.

Part of what determines your learning style is how your brain works. According to the *split-brain hypothesis,* the left hemisphere of your brain is good at logic, analysis, and evaluation and the right half of the brain is good at visualizing, synthesizing, and creating.

The study and critical thinking skills in this book and in most courses largely involve the left brain. However, you can improve these skills by giving your left brain a break and letting your creative side loose. You can do this by brainstorming ideas with classmates with the rule that no left-brain criticism is allowed until the session is over. Other techniques are the following: working backward from where you want to be to where you are, reversing some or all of your assumptions and seeing where this leads, using analogies, visualizing the opposite of the situation, and analyzing a problem from another person's point of view.

When you are trying to solve a problem, rest, meditate, take a walk, exercise, or do something to shut down your controlling left-brain activity and allow the right side of your brain to work on the problem in a less controlled and more creative manner.

Have you ever tried to think of someone's name, and no matter how hard you focus (left-brain activity), you can't think of it? Indeed, the harder you try, the less chance you have because you are "flooding the motor" of your left brain and not allowing your right brain to deal with the problem in an unpressured, more creative way. Then perhaps hours or days later the person's name pops in your head. All this time your right brain was quietly working on the problem in its different way of processing information and ideas. Turning on their right brain is a major way that scientists, artists, novelists, and business leaders come up with creative ideas.

Trade-Offs

There are no simple answers to the environmental problems we face.

There are always *trade-offs* involved in making and implementing environmental decisions. My challenge is to give a fair and balanced presentation of different viewpoints, advantages and disadvantages of various technologies and proposed solutions to environmental problems, and good and bad news about environmental problems without injecting personal bias. My goal is to present a positive vision of our environmental future based on realistic optimism.

Studying a subject as important as environmental science and ending up with no conclusions, opinions, and beliefs means that both the teacher and student have failed. However, such conclusions must be based on sound science and should be reached only through critical thinking to evaluate different ideas and to understand the trade-offs involved.

Help Me Improve This Book

I welcome your help in improving this book.

Researching and writing a book that covers and connects ideas in such a wide variety of disciplines is a challenging and exciting task. Almost every day I learn about some new connection in nature.

In a book this complex, there are bound to be some errors—some typographical mistakes that slip through and some statements that you might question based on your knowledge and research. My goal is to provide you with an interesting, accurate, balanced, and challenging book that furthers your understanding of this vital subject.

I invite you to contact me and point out any remaining bias, correct any errors you find, and suggest ways to improve this book. Over decades of teaching, my students and readers of my textbooks have been some of my best teachers. Please e-mail your suggestions to me at **mtg89@hotmail.com**.

Now start your journey into this fascinating and important study of how the earth works and how we can leave the planet in at least as good a shape as we found it. Have fun.

Study nature love nature, stay close to nature. It will never fail you.
FRANK LLOYD WRIGHT

Environmental Problems, Their Causes, and Sustainability

CORE CASE STUDY
Living in an Exponential Age: Life in the Fast Lane

Two ancient kings enjoyed playing chess; the winner claimed a prize from the loser. After one match, the winning king asked the losing king to pay him by placing one grain of wheat on the first square of the chessboard, two grains on the second square, four on the third, and so on, with the number doubling on each square until all 64 were filled.

The losing king, thinking he was getting off easy, agreed with delight. It was the biggest mistake he ever made. He bankrupted his kingdom because the number of grains of wheat he had promised was probably more than all the wheat ever harvested!

This fictional story illustrates the concept of **exponential growth,** in which a quantity increases at a *fixed percentage* per unit of time, such as 2% per year. Exponential growth starts off slowly, but after only a few doublings, it grows to enormous numbers.

Here is another example. Fold a piece of paper in half to double its thickness. If you could continue doubling the thickness of the paper 42 times, the stack would reach from the earth to the moon, 386,400 kilometers (240,000 miles) away. If you could double it 50 times, the folded paper would almost reach the sun, 149 million kilometers (93 million miles) away!

About 10,000 years ago there were about 5 million humans on the planet. Today there are 6.6 billion. Unless death rates rise sharply, there may be 8–10 billion of us by 2100 (Figure 1-1).

We live in a world of haves and have-nots. Despite a 22-fold increase in economic growth

since 1900, *53% of the people in the world try to survive on a daily income of less than $2 (U.S). And one of every six of the world's people, classified as desperately poor, struggle to survive on less than $1 (U.S.) a day.* Such poverty affects environmental quality because to survive many of the poor must deplete and degrade local forests, grasslands, soils, and wildlife.

Biologists estimate that human activities are causing premature extinction of the earth's life forms, or *species,* at an exponential rate of 0.1–1% per year—an irreversible loss of the earth's incredible variety of life forms and the places or habitats where they live, or *biodiversity* (short for biological diversity). In various parts of the world, forests, grasslands, wetlands, coral reefs, and topsoil from croplands continue to disappear or become degraded as the human ecological footprint gets larger and spreads exponentially across the globe.

There is growing evidence that exponential growth in human activities, such as burning fossil fuels and clearing forests, will play an increasingly significant role in changing the earth's climate during this century. This could ruin some areas for farming, shift water supplies, alter and reduce biodiversity, and disrupt economies in many parts of the world.

Great news. We have potential solutions to these problems that we could implement within a few decades, as you will learn in this book.

This arrow refers back to the **CORE CASE STUDY**

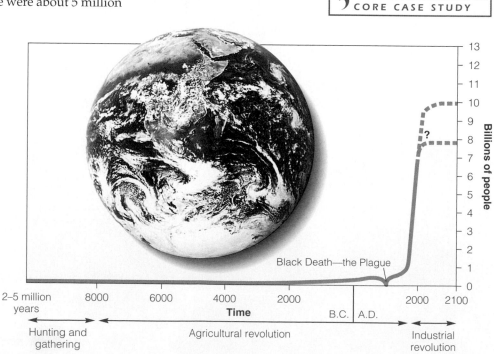

Figure 1-1 Exponential growth: the *J*-shaped curve of past exponential world population growth, with projections to 2100. Exponential growth starts off slowly, but as time passes the curve becomes increasingly steep. Unless death rates rise, the current world population of 6.6 billion people is projected to reach 8–10 billion people sometime this century. (This figure is not to scale.) (Data from the World Bank and United Nations; photo courtesy of NASA)

Alone in space, alone in its life-supporting systems, powered by inconceivable energies, mediating them to us through the most delicate adjustments, wayward, unlikely, unpredictable, but nourishing, enlivening, and enriching in the largest degree—is this not a precious home for all of us? Is it not worth our love?

BARBARA WARD AND RENÉ DUBOS

This chapter presents an overview of environmental problems, their causes, and ways we can live more sustainably. It discusses these questions:

- What are the major themes of this book?

- What keeps us alive? What is an environmentally sustainable society?

- How fast is the human population growing?

- What is the difference between economic growth, economic development, and environmentally sustainable economic development?

- What are the earth's main types of resources? How can they be depleted or degraded?

- What are the principal types of pollution, and what can we do about pollution?

- What are the basic causes of today's environmental problems, and how are these causes connected?

- What are the harmful environmental effects of poverty and affluence?

- What three major human cultural changes have taken place since humans arrived?

- What are four scientific principles of sustainability and how can they help us build more environmentally sustainable and just societies?

LIVING MORE SUSTAINABLY

What Is Environmental Science? (Science)

Environmental science is a study of how the earth works, how we interact with the earth, and how to deal with environmental problems.

Environment is the sum total of all living and nonliving things that affect any living organism. Everything we do affects the environment. Some of our scientific discoveries and actions have led to longer life spans, better health, and increased material wealth for some. At the same time, exponential increases in both the human population (Figure 1-1) and our resource consumption have degraded the air, water, soil, and species in the natural systems that support our lives and economies.* If kept up, such actions can threaten the long-term sustainability of our societies.

This textbook is an introduction to **environmental science:** an interdisciplinary study that integrates information and ideas from the *natural sciences* (such as biology, chemistry, and geology) that study the natural world and the *social sciences* (such as economics, politics, and ethics) that study how humans and their institutions interact with the natural world (Figure 1-2). The goals of environmental science are to learn *how nature works, how the environment affects us, how we affect the environment, and how we can live more sustainably without degrading our life-support system.*

A basic tool used by environmental scientists is **ecology,** a biological science that studies the relationships between living organisms and their environment. See Supplement 3 on p. S6 for a concept map giving an overview of environmental science. To show

---

* The opening Core Case Study is used as a theme to connect and integrate much of the material in each chapter. Curved arrows in the margin point back to the opening of each chapter to indicate these connections.

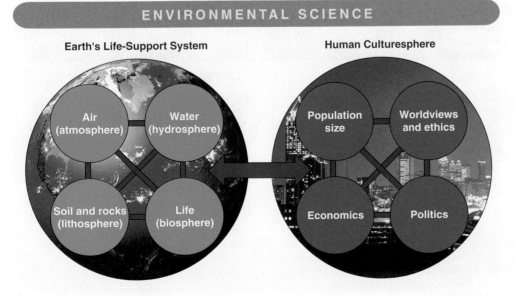

ENVIRONMENTAL SCIENCE

Earth's Life-Support System

- Air (atmosphere)
- Water (hydrosphere)
- Soil and rocks (lithosphere)
- Life (biosphere)

Human Culturesphere

- Population size
- Worldviews and ethics
- Economics
- Politics

Figure 1-2 Environmental science is an interdisciplinary study of connections between the earth's life-support system (left) and the human culturesphere (right).

the integration of natural and social sciences, many subheadings throughout this book are tagged with the terms science, economics, politics, and ethics or some combination of these words.

Environmental science is a fairly young science. As a result it is full of exciting *research frontiers* that can help us understand environmental problems and find balanced and workable solutions to the environmental challenges we face. Many of these frontiers for further study are identified throughout the book. Because environmental science involves a diversity of disciplines, there are many *environmental career* opportunities. A number of these careers are noted throughout the book and more information about most of them is available on the website for this book.

We should not confuse environmental science and ecology with **environmentalism,** a social movement dedicated to protecting the earth's life-support systems for us and other species. Environmentalism is political in nature, involving activities such as working to pass and enforce environmental laws, promoting solutions to environmental problems, and protesting harmful environmental actions.

Sustainability: The Integrative Theme of This Book (Science and Politics)

Sustainability, the central theme of this book, is built on the subthemes of natural capital, natural capital degradation, solutions, trade-offs, and how individuals matter.

Sustainability, or **durability,** is the ability of earth's various systems, including human cultural systems and economies, to survive and adapt to changing environmental conditions indefinitely. It is the central theme that runs through and integrates the material in this book. Figure 1-3 shows the steps along a path to sustainability, based on the five subthemes for this book.

The first step is to understand the components and importance of **natural capital**—the natural resources and natural services that keep us and other species alive and support our economies (Figure 1-4).

We can also think of energy from the sun as **solar capital** that warms the planet and supports photosynthesis, the process that plants use to provide food for themselves and for us and other animals. This direct input of solar energy also produces indirect forms of renewable solar energy such as wind, flowing water, and fuels made from plants and plant residues (biofuels).

Natural capital is not fixed. It has changed over millions of years in response to environmental changes such as global warming and cooling and huge asteroids hitting the earth. Forests have grown and disappeared, as have grasslands and deserts. Species have become extinct because of natural and human causes and new species have appeared. We have transformed many forests and grasslands into croplands—a more simplified form of natural capital created by humans.

The second step toward sustainability is to recognize that many human activities *degrade natural capital* by using normally renewable resources such as forests faster than nature can renew them (Figure 1-3, Step 2). A key variable is the *rate* at which we are transforming parts of the earth to meet our needs and wants. Most natural environmental changes have taken place over thousands to hundreds of thousands of years. Humans are now making major changes in the earth's natural systems within 50 to 100 years. For example, in parts of the world we are clearing many mature forests much faster than nature can re-grow them.

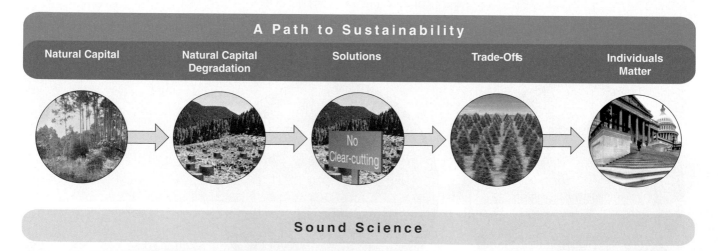

Figure 1-3 A path to sustainability: five subthemes are used throughout this book to illustrate how we can make the transition to more environmentally sustainable or durable societies and economies, based on *sound science*—concepts widely accepted by natural and social scientists in various fields.

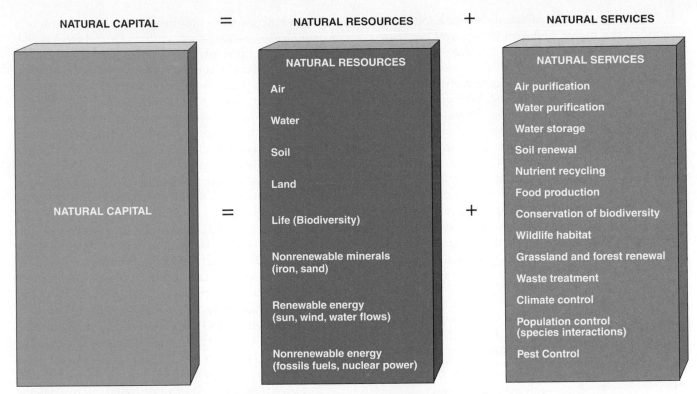

NATURAL CAPITAL = **NATURAL RESOURCES** + **NATURAL SERVICES**

NATURAL CAPITAL

NATURAL RESOURCES

Air

Water

Soil

Land

Life (Biodiversity)

Nonrenewable minerals (iron, sand)

Renewable energy (sun, wind, water flows)

Nonrenewable energy (fossils fuels, nuclear power)

NATURAL SERVICES

Air purification

Water purification

Water storage

Soil renewal

Nutrient recycling

Food production

Conservation of biodiversity

Wildlife habitat

Grassland and forest renewal

Waste treatment

Climate control

Population control (species interactions)

Pest Control

Figure 1-4 Natural capital: the natural resources (center) and natural services (right) that support and sustain the earth's life and economies. For example, *nutrients* or chemicals such as carbon and nitrogen, which plants and animals need as *resources,* are recycled through the air, water, soil, and organisms by the natural process of *nutrient cycling.* And the interactions and competition of different types of plants and animals (species) for *resources* (nutrients) keep any single species from taking over through the natural service *of population control.* Colored wedges are shown at the beginning of most chapters in this book to show the *natural resources* (blue wedges) and *natural services* (orange wedges) discussed in these chapters.

This leads us to search for workable *solutions* to these and other environmental problems (Figure 1-3, Step 3). For example, one solution might be to stop cutting down diverse mature forests.

The search for solutions often involves conflicts, and resolving these conflicts requires us to make *trade-offs,* or compromises (Figure 1-3, Step 4). To provide wood for making paper, for example, we can promote the planting of tree plantations (see photo 1 in the Detailed Contents) in areas that have already been cleared or degraded.

In the search for solutions, *individuals matter,* whether they are working alone or in groups. For example, a scientist might find a way to make paper by using crop residues instead of cutting down trees. Or a group might work together to pass a law banning the clear-cutting of ancient forests while encouraging the planting of tree plantations in areas that have already been cleared or degraded.

The five steps to sustainability must be supported by **sound science**—the concepts and ideas that are widely accepted by experts in a particular field of the natural or social sciences. For example, sound science

tells us that we need to protect and sustain the many natural services provided by diverse mature forests. It also guides us in the design and management of tree plantations and in finding ways to produce paper without using trees.

Environmentally Sustainable Societies: Protecting Natural Capital and Living off Its Income (Science)

An environmentally sustainable society meets the basic resource needs of its people in a just and equitable manner without degrading or depleting the natural capital that supplies these resources.

The ultimate human goal on a path to sustainability or durability is an **environmentally sustainable society**—one that meets the current and future needs of its people for basic resources in a just and equitable manner without compromising the ability of future generations to meet their needs. *Living sustainably* means living off natural income replenished by soils, plants, air, and water and not depleting or degrading the earth's natural capital that supplies this income.

Imagine you win a million dollars in a lottery. If you invest this money and earn 10% interest per year, you will have a sustainable annual income of $100,000 without depleting your capital. If you spend $200,000 per year, your $1 million will be gone early in the seventh year. Even if you spend only $110,000 per year, you will be bankrupt early in the eighteenth year.

The lesson here is an old one: *Protect your capital and live off the income it provides.* Deplete, waste, or squander your capital, and you will move from a sustainable to an unsustainable lifestyle.

The same lesson applies to the earth's natural capital. According to a growing body of scientific evidence, we are living unsustainably by wasting, depleting, and degrading the earth's natural capital at an exponentially accelerating rate (Core Case Study, p. 6).

POPULATION GROWTH, ECONOMIC GROWTH, AND ECONOMIC DEVELOPMENT

Human Population Growth: Slowing but Still Rapid

The rate at which the world's population is growing has slowed, but the population is still increasing rapidly, and it is unequally distributed between rich and poor countries.

Exponential growth of the world's population (Figure 1-1) has slowed but has not ended. Between 1963 and 2006, the exponential rate at which the world's population was growing decreased from 2.2% to 1.23%. This does not seem like a very fast rate but it added about 81 million people (6.6 billion × 0.0123 = 81 million) to the world's population in 2006. This is an average increase of about 222,000 people per day, 9,250 per hour, or 2.6 per second. At this rate it takes only about 2.9 days to replace the 652,000 Americans killed in battle in all U.S. wars and only 1.4 years to replace the 111 million soldiers and civilians killed in all wars fought during the twentieth century. This illustrates the incredible power of exponential growth even at a fairly low rate (Core Case Study, p. 6).

Life in the fast lane can be exhilarating. But it can also be dangerous and reduce the time we have to find solutions to the environmental problems and challenges we face.

Economic Growth and Economic Development (Economics)

Economic growth provides people with more goods and services, and economic development uses economic growth to improve living standards.

Economic growth is an increase in the capacity of a country to provide people with goods and services. Accomplishing this increase requires population growth (more producers and consumers), more production and consumption per person, or both.

Economic growth is usually measured by the percentage change in a country's **gross domestic product (GDP):** the annual market value of all goods and services produced by all firms and organizations, foreign and domestic, operating within a country. Changes in a country's economic growth per person are measured by **per capita GDP:** the GDP divided by the total population at midyear.

In terms of GDP, the world's six largest economies in 2006 were, in order, the United States, Japan, Germany, the United Kingdom, France, and China. To account for differences in purchasing power for basic necessities in different countries, GDPs are also measured in *Purchasing Power Parity* (PPP). In terms of GDP-PPP, the world's six largest economies in 2006 were, in order, the United States, China, Japan, India, Germany, and France.

The current exponential growth (Core Case Study, p. 6) of the global economy is astounding. In 1900, the world's economy grew by billions of dollars; today's annual economic growth is measured in trillions of dollars.

Economic development is the improvement of human living standards by economic growth. The United Nations (UN) classifies the world's countries as economically developed or developing based primarily on their degree of industrialization and their per capita GDP-PPP.

The **developed countries** (with 1.2 billion people) include the United States, Canada, Japan, Australia, New Zealand, and most European countries. Most are highly industrialized and have high average per capita GDP. All other nations (with 5.4 billion people) are classified as **developing countries,** most of them in Africa, Asia, and Latin America. Some are *middle-income, moderately developed countries* such as China, India, Brazil, and Mexico and others are *low-income countries.*

Figure 1-5 compares some key characteristics of developed and developing countries, and Figure 1-6 shows the general distribution of poverty. About 97% of the projected increase in the world's population between 2006 and 2050 is expected to take place in developing countries.

Close to 1 billion people live in developed countries (most in Europe) whose populations are stable or growing slowly. But another billion live in developing countries whose populations are projected to double by 2050. Economically, there is a widening income gap between the world's poor and rich. And socially, there is a wide gap between education and health care in developed and developing countries.

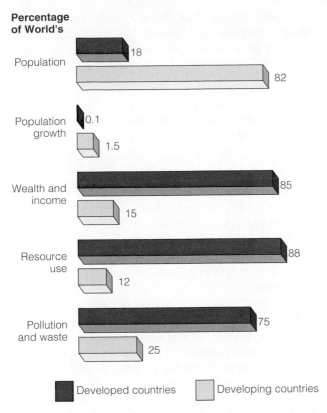

Percentage of World's

Population
18
82

Population growth
0.1
1.5

Wealth and income
85
15

Resource use
88
12

Pollution and waste
75
25

■ Developed countries   □ Developing countries

Figure 1-5 Global outlook: comparison of developed and developing countries, 2006. (Data from the United Nations and the World Bank)

During this century, many analysts call for us to put much greater emphasis on **environmentally sustainable economic development.** Its goal is to use political and economic systems to *encourage* environmentally beneficial and more sustainable forms of economic development and *discourage* environmentally harmful and unsustainable forms of economic growth.

Doubling Time and Exponential Growth: The Rule of 70 (Science)

It is easy to calculate how long it will take for a quantity growing exponentially to double in size.

How long does it take to double the world's population or economic growth at various exponential rates of growth? A quick way to calculate such *doubling times* is the use the *rule of 70*: 70/percentage growth rate = doubling time in years (a simple formula derived from the basic mathematics of exponential growth).

For example, in 2006 the earth's human population grew by 1.23%. If that rate continues, the world's population will double in 57 years (70/1.23 = 57 years). Similarly, if you have money in an investment account growing exponentially at 10% a year and you leave your annual returns in the account, you will double your money in only 10 years (70/7 = 10 years)

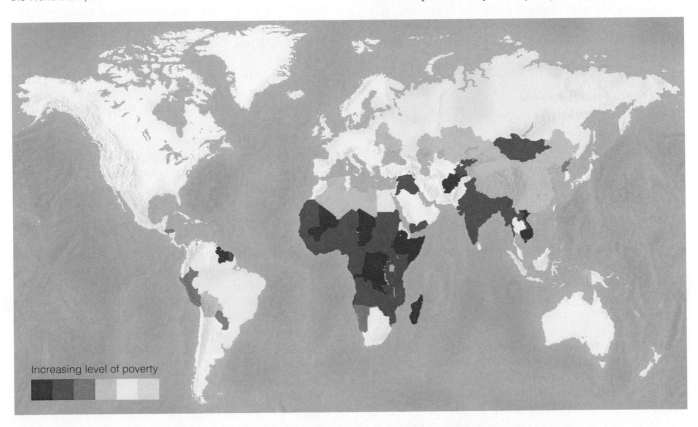

Increasing level of poverty

Figure 1-6 Generalized distribution of poverty. Poverty is found mostly in the southern hemisphere, largely because of unfavorable climates and geological bad luck in terms of fertile soils, minerals, and fossil fuel supplies. (Data from United Nations and World Bank)

THINKING ABOUT DOUBLING TIME Since 1985, China's economy has been growing at an astonishing exponential rate of 9.5% a year. How many years, at this rate, would it take China to double its economic output?

RESOURCES

What Is a Resource? (Science)

We obtain resources from the environment to meet our needs and wants.

From a human standpoint, a **resource** is anything obtained from the environment to meet our needs and wants. Examples include food, water, shelter, and metals used to manufacture goods. On our short human time scale, we classify the material resources we get from the environment as *perpetual, renewable,* or *nonrenewable.*

Some resources, such as solar energy, fresh air, wind, fresh surface water, fertile soil, and wild edible plants, are directly available for use. Other resources, such as petroleum (oil), iron, groundwater (water found underground), and modern crops, are not directly available. They become useful to us only with some effort and technological ingenuity. For example, petroleum was a mysterious fluid until we learned how to find, extract, and convert (refine) it into gasoline, heating oil, and other products that we could sell at affordable prices. In such cases, resources are obtained by an interaction between natural capital and human capital.

Perpetual and Renewable Resources (Science)

Resources renewed by natural processes are sustainable if we do not use them faster than they are replenished.

Solar energy is called a **perpetual resource** because on a human time scale it is renewed continuously. It is expected to last at least 6 billion years as the sun completes its life cycle.

On a human time scale, a **renewable resource** can be replenished fairly rapidly (hours to several decades) through natural processes as long as it is not used up faster than it is replaced. Examples include forests, grasslands, wild animals, fresh water, fresh air, and fertile soil.

Renewable resources can be depleted or degraded. The highest rate at which a renewable resource can be used *indefinitely* without reducing its available supply is called its **sustainable yield.**

When we exceed a resource's natural replacement rate, the available supply begins to shrink, a process known as **environmental degradation.** Examples of such degradation or unsustainable resource use include urbanization of productive land, excessive topsoil erosion, pollution, clearing forests to grow crops, depleting groundwater, and reducing the earth's variety of wildlife (biodiversity) by eliminating habitats and species.

The Tragedy of the Commons (Economics and Politics)

Renewable resources that are freely available to everyone can be degraded.

One cause of environmental degradation of renewable resources is the overuse of **common-property** or **free-access resources.** Individuals do not own these resources, and they are available to users at little or no charge. Examples include clean air, the open ocean and its fish, migratory birds, and gases of the lower atmosphere.

In 1968, biologist Garrett Hardin (1915–2003) called the degradation of renewable free-access resources the **tragedy of the commons.** It happens because each user reasons, "If I do not use this resource, someone else will. The little bit I use or pollute is not enough to matter, and such resources are renewable."

With only a few users, this logic works. Eventually, however, the cumulative effect of many people trying to exploit a free-access resource exhausts or ruins it. Then no one can benefit from it—and that is the tragedy. For example, when we exceed the sustainable catch of an ocean fishery the stock of breeding fish declines. If we fail to reduce catch levels, the fishery collapses.

One solution is to *use free-access resources at rates well below their estimated sustainable yields* by reducing population, regulating access to the resources, or both. Some communities have established rules and traditions to regulate and share their access to common-property resources such as ocean fisheries, grazing lands, and forests. Governments have also enacted laws and international treaties to regulate access to commonly owned resources such as forests, national parks, rangelands, and fisheries in coastal waters.

Another solution is to *convert free-access resources to private ownership.* The reasoning is that if you own something, you are more likely to protect your investment.

That sounds good, but private owners do not always protect natural resources they own when this goal conflicts with protecting their financial capital or increasing their profits. For example, some private forest owners can make more money by clear-cutting timber, selling the degraded land, and investing their profits in other timberlands or businesses. Also, this approach is not practical for global common resources—such as the atmosphere, the open ocean, most wildlife species, and migratory birds—that cannot be divided up and converted to private property.

Our Ecological Footprints (Science)

Supplying each person with renewable resources and absorbing the wastes from such resource use creates a large ecological footprint or environmental impact.

The **ecological footprint** is the amount of biologically productive land and water needed to supply an area with resources and to absorb the wastes and pollution produced by such resource use (Figure 1-7, top left). It is an estimate of the average environmental impact of individuals in a given country or area. The **per capita ecological footprint** is the average ecological footprint of an individual in an area (Figure 1-7, bottom left).

The numbers in Figure 1-7 are estimates, but they show relative differences in resource use and waste production by various countries. See Figures 3 and 4 on pp. S12–S15 in Supplement 4 for maps of the human ecological footprints for the world and the United States. You can use these maps to see the human ecological footprint in the area where you live.

Humanity's ecological footprint exceeds by about 39% the earth's ecological capacity (or biocapacity) to replenish its renewable resources and absorb the resulting waste products and pollution (Figure 1-7, right). When a country's ecological footprint is larger than its ecological capacity, it is using and degrading its cropland, forests, groundwater, and other renewable resources faster than nature can replenish them, and it is exceeding the capacity of its environment to absorb and degrade the resulting wastes and pollution.

When a country depletes its natural capital, it must either suffer the harmful environmental consequences or import food and other resources from other countries and export its pollutants and wastes to the global atmosphere, oceans, and rivers that run through several countries. In effect, such countries are living off of a global ecological credit card instead of using their own renewable resources sustainably.

Currently, the United States, the European Union, China, India, and Japan collectively use about 74% of the earth's ecological capacity, leaving only 26% for the rest of the world's countries and the plants and animals that support all economies (Figure 1-7, top left). The ecological footprints of the United States, the European Union, China, and India are more than twice as large as their domestic ecological capacities and Japan's is 5.6 times its estimated biocapacity.

If these estimates are correct, *it will take the resources of 1.39 planet earths to support indefinitely our current production and consumption of renewable resources!* In other words, we are living unsustainably as more and more of humanity live beyond the earth's biocapacity You can estimate your own ecological footprint by visiting the website **www.myfootprint.org/**. Also see the Guest Essay by Michael Cain on the website for this chapter.

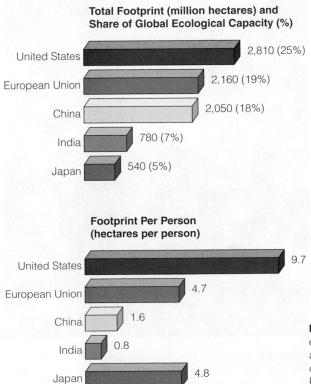

Total Footprint (million hectares) and Share of Global Ecological Capacity (%)

- United States — 2,810 (25%)
- European Union — 2,160 (19%)
- China — 2,050 (18%)
- India — 780 (7%)
- Japan — 540 (5%)

Footprint Per Person (hectares per person)

- United States — 9.7
- European Union — 4.7
- China — 1.6
- India — 0.8
- Japan — 4.8

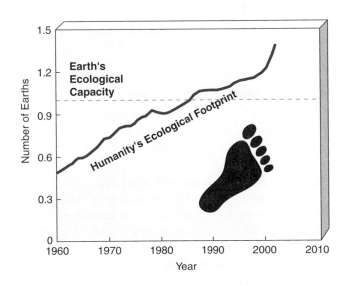

Figure 1-7 Natural capital use and degradation: total and per capita ecological footprints of selected countries in 2002 (left). By 2002, humanity's average ecological footprint was about 39% higher than the earth's ecological capacity (right). (Data from Worldwide Fund for Nature, UN Environment Programme, Global Footprint Network, Worldwatch Institute)

The United States leads the way with an ecological footprint that is twice that of the European Union and Japan, six times that of China, and twelve times that of India (Figure 1-7, bottom left). However, the ecological footprints of China and India are projected to increase rapidly as their economies continue to grow at remarkably high exponential rates (Case Study, p. 15). They will claim increasingly larger shares of the world's resources, and become major polluters of local and global life-support systems.

According to a 2006 study by the Worldwatch Institute, if China and India were to consume resources and emit pollution and wastes at the same rate as the United States currently does, they would require two planet earths. If these estimates are correct the world is on an ecologically and economically unsustainable path. Thus, searching for a more sustainable path is equally important for the United States, the European Union, China, and India.

F RESEARCH FRONTIER Learning more about the ecological footprints for various parts of the earth and evaluating ways to reduce these footprints.

The *size* of the human ecological footprint is not the only concern. A related concept is the *force* of our footprint—how heavily or lightly we walk on the earth through our lifestyles. The three things that have the greatest environmental impact are, in order, agriculture, transportation, and heating and cooling buildings.

[?] THINKING ABOUT OUR ECOLOGICAL FOOTPRINTS AND EXPONENTIAL GROWTH How is the growth of the world's ecological footprint related to exponential growth (Core Case Study, p. 6) of the world's population and economies? What three things would you do to reduce (a) the global ecological footprint and (b) your own ecological footprint?

Nonrenewable Resources (Science)

Nonrenewable resources can be economically depleted to the point where it costs too much to obtain what is left. Some can also be recycled and reused.

Nonrenewable resources exist in a fixed quantity or stock in the earth's crust. On a time scale of millions to billions of years, geological processes can renew such resources. But on the much shorter human time scale of hundreds to thousands of years, these resources can be depleted much faster than they are formed as a result of exponential growth in our resource use (Core Case Study, p. 6).

Such exhaustible resources include *energy resources* (such as coal, oil, and natural gas), *metallic mineral resources* (such as iron, copper, and aluminum), and *nonmetallic mineral resources* (such as salt, clay, sand, and phosphates). Exhaustible resources such as oil and some metals that nature produced over billions of years are being consumed at exponentially growing rates in a single human lifespan.

Although we never completely exhaust a nonrenewable mineral resource, it becomes *economically depleted* when the costs of extracting and using what is left exceed its economic value. At that point, we have five choices: try to find more, recycle or reuse existing supplies (except for nonrenewable energy resources, which cannot be recycled or reused), waste less, use less, or try to develop a substitute.

Some nonrenewable material resources, such as copper and aluminum, can be recycled or reused to extend supplies. **Recycling** involves collecting waste materials, processing them into new materials, and selling these new products. For example, discarded aluminum cans can be crushed and melted to make new aluminum cans or other aluminum items that consumers can buy. **Reuse** is using a resource over and over in the same form. For example, glass bottles can be collected, washed, and refilled many times (Figure 1-8).

Recycling nonrenewable metallic resources takes much less energy, water, and other resources and produces much less pollution and environmental degradation than exploiting virgin metallic resources. Reusing such resources takes even less energy and other resources and produces less pollution and environmental degradation than recycling. In other words, recycling and reuse are important ways to reduce the exponential growth in resource use (Core Case Study, p. 6).

Figure 1-8 Reuse: this child and his family in Katmandu, Nepal, collect beer bottles and sell them for cash to a brewery, where they will be reused.

More than 1 billion super-affluent consumers in the United States, the European Union, and other developed countries are putting immense pressure on the earth's natural capital. And another 1 billion consumers are attaining middle-class, affluent lifestyles in rapidly developing countries such as China, India, Brazil, South Korea, and Mexico. Indeed, the number of middle-class consumers in China and India, which together have 40% of the world's population, already is almost three times larger than the entire U.S. population!

China is now the world's leading consumer of wheat, rice, meat, coal, fertilizers, steel, and cement, and it is the second largest consumer of oil after the United States. China now consumes almost twice as much meat and nearly two and a half times more steel (a basic indicator of industrial development) than the United States.

China also leads the world in the consumption of consumer goods such as television sets, cell phones, and refrigerators and is soon expected to overtake the U.S. in the number of personal computers. By 2020, China is projected to be the world's largest producer and consumer of cars and to have the world's leading economy in terms of GDP-PPP.

If China's economy continues growing exponentially at 8–10% a year, by 2031 the country's income per person will reach that of the United States in 2006. If this happens and China's projected population size reaches 1.47 billion, China will need two-thirds of the world's current grain harvest, twice the world's current paper consumption, and more than the current global production of oil.

According to environmental expert Lester R. Brown:

"The western economic model—the fossil-fuel-based, automobile-centered, throwaway economy—is not going to work for China. Nor will it work for India, which by 2031 is projected to have a population even larger than China's, or for the other 3 billion people in developing countries who are also dreaming the "American dream." And in an increasingly integrated world economy, where all countries are competing for the same oil, grain, and mineral resources, the existing economic model will not work for industrial countries either."

About 22 of the world's current top universities are American. But China and India have top-flight universities that together produce about 500,000 scientists and engineers each year, compared with about 60,000 in the United States. A crucial issue is how much governments will encourage the world's scientists, engineers, and economists to design more sustainable ways to live. For more details on the growing ecological footprint of China, see the Guest Essay by Norman Myers on the website for this chapter.

Critical Thinking

What three things should China and India do to shift towards more sustainable consumption? What three things should the United States, Japan, and the European Union do to shift towards more sustainable consumption?

POLLUTION

Sources and Harmful Effects of Pollutants (Science)

Pollutants are chemicals found at high enough levels in the environment to cause harm to people or other organisms.

Pollution is the presence of chemicals at high enough levels in air, water, soil, or food to threaten the health, survival, or activities of humans or other living organisms. Pollutants can enter the environment naturally, such as from volcanic eruptions or through human activities, such as from burning coal. Most pollution from human activities occurs in or near urban and industrial areas, where pollution sources such as cars and factories are concentrated. Industrialized agriculture also is a major source of air and water pollution. Some pollutants contaminate the areas where they are produced; others are carried by wind or flowing water to other areas.

The pollutants we produce come from two types of sources. **Point sources** of pollutants are single, identifiable sources. Examples are the smokestack of a coal-burning power or industrial plant (Figure 1-9, p. 16, and photo 2 in the Detailed Contents), the drainpipe of a factory, and the exhaust pipe of an automobile. **Nonpoint sources** of pollutants are larger, dispersed, and often difficult to identify. An example is pesticides sprayed into the air or blown by the wind into the atmosphere. Another is the runoff of fertilizers and pesticides from farmlands, golf courses, and suburban lawns and gardens and of eroded sediment from farms and construction sites into streams and lakes. It is much easier and cheaper to identify and control pollution from point sources than from widely dispersed nonpoint sources.

Pollutants can have three types of unwanted effects. *First*, they can disrupt or degrade life-support systems for humans and other species. *Second*, they can damage wildlife, human health, and property. *Third*, they can create nuisances such as noise and unpleasant smells, tastes, and sights.

Figure 1-9 Natural capital degradation: *point-source air pollution* from a pulp mill in New York State (USA).

[?] *THINKING ABOUT POLLUTION AND EXPONENTIAL GROWTH* How is the production of pollution and waste related to exponential growth of the world's population and economies? What three things would you do to reduce the amount of pollution and waste that we produce? List three changes in your lifestyle that would reduce the amount of pollution and wastes you produce.

Solutions: Prevention versus Cleanup (Science and Economics)

We can try to prevent production of pollutants or clean them up after they have been produced.

We use two basic approaches to deal with pollution. One is **pollution prevention,** or **input pollution control,** which reduces or eliminates the production of pollutants. The second is **pollution cleanup,** or **output pollution control,** which involves cleaning up or diluting pollutants after they have been produced.

For example, if smoke is coming out of the stack of a steel mill, we can try to deal with this problem by asking two entirely different questions. One question is "how can we clean up the smoke?" This might involve adding devices to a mill's furnaces or inside its smokestacks to remove some of the pollutants produced by burning coal to produce steel.

An even more important question is "how can we not produce the smoke in the first place?" The answer might be to figure out a way to make steel without burning coal. Germany has done this and is selling the technology in the global marketplace.

Environmental scientists have identified three problems with relying primarily on pollution cleanup. *First,* it is only a temporary bandage as long as population and consumption levels grow without corresponding improvements in pollution control technology. For example, adding catalytic converters to car exhaust systems has reduced some forms of air pollution. At the same time, increases in the number of cars and in the total distance each travels have reduced the effectiveness of this cleanup approach.

Second, cleanup often removes a pollutant from one part of the environment only to cause pollution in another. For example, we can collect garbage, but the garbage is then *burned* (perhaps causing air pollution and leaving toxic ash that must be put somewhere), *dumped* on the land (perhaps causing air and water pollution), or *buried* (perhaps causing soil and groundwater pollution).

Third, once pollutants have entered and become dispersed into the environment at harmful levels, it usually costs too much or is impossible to reduce them to acceptable levels.

Both pollution prevention (front-of-the-pipe) and pollution cleanup (end-of-the-pipe) solutions are needed. Environmental scientists and some economists urge us to put more emphasis on prevention because it works better and is cheaper than cleanup.

[?] *THINKING ABOUT PREVENTING POLLUTION AND EXPONENTIAL GROWTH* Explain how placing much greater emphasis on pollution prevention would help reduce the exponential growth of the human ecological footprint and your own ecological footprint.

ENVIRONMENTAL PROBLEMS: CAUSES AND CONNECTIONS

Key Environmental Problems and Their Basic Causes

The major causes of environmental problems are population growth, wasteful resource use, poverty, poor environmental accounting, and environmental ignorance.

We face a number of interconnected environmental and resource problems mostly as a result of the exponential growth of population and resource use (Figure 1-10). As we run more and more of the earth's natural resources through the global economy, in many parts of the world forests are shrinking, deserts are expanding, soils are eroding, and rangelands are deteriorating. In addition, the atmosphere is warming, glaciers are melting, seas are rising, and storms are becoming more destructive. And in many areas water tables are falling, rivers are running dry, fisheries are collapsing, coral reefs are disappearing, species are be-

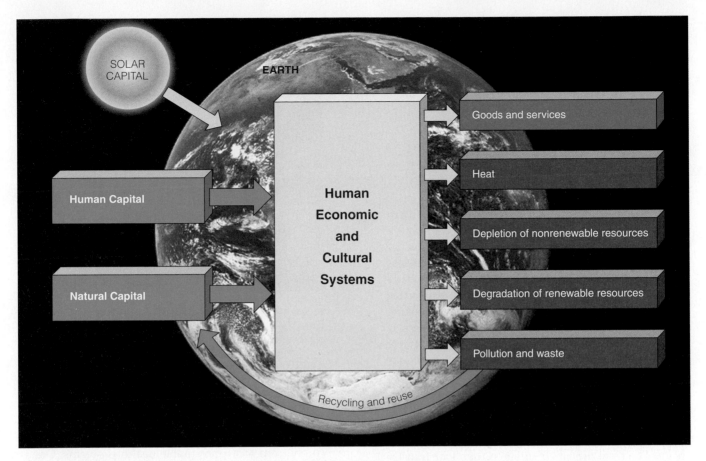

Figure 1-10 Natural capital use, depletion, and degradation: human and natural capital produce an amazing array of goods and services for most of the world's people. But the exponentially increasing flow of material resources through the world's economic systems depletes nonrenewable resources, degrades renewable resources, and adds heat, pollution, and wastes to the environment.

coming extinct, environmental refugees are increasing, and outputs of pollution and wastes are rising.

The first step in dealing with these and other problems is to identify their underlying causes, listed in Figure 1-11. Three other likely causes are global trade policies that can undermine environmental protection, the influence of money in politics, and failure of those concerned about environmental quality to provide inspiring and positive visions of a more sustainable and durable economic and environmental future. One of

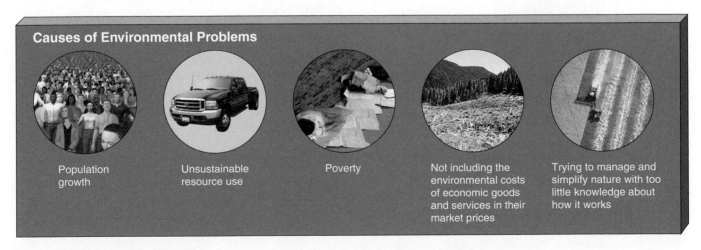

Figure 1-11 Natural capital degradation: five basic causes of the environmental problems we face. QUESTION: *Can you think of any other basic causes?*

the goals of this book is to provide a realistic environmental vision of the future based on energizing hope instead of immobilizing fear and gloom-and-doom.

Poverty and Environmental Problems (Economics and Politics)

Poverty is a major threat to human health and the environment and the world's poorest people suffer the most from pollution and environmental degradation.

Poverty the inability to meet one's basic economic needs and is concentrated mostly in the southern hemisphere (Figure 1-6). Many of the world's poor do not have access to the basic necessities for a healthy and productive life (Figure 1-12). Many are homeless (see photo 3 in the Detailed Contents) and their daily lives are focused on getting enough food, water (see photo 4 in the Detailed Contents), and fuel for cooking and heating to survive. Desperate for land to grow enough food, many of the world's poor people deplete and degrade forests, soil, grasslands, and wildlife for short-term survival. They do not have the luxury of worrying about long-term environmental quality or sustainability.

Poverty also affects population growth. Poor people often have many children as a form of economic security. Their children help them gather fuel (mostly wood and animal dung), haul drinking water, tend crops and livestock, work, and beg in the streets. The children also help their parents survive in their old age before they die, typically in their fifties in the poorest countries.

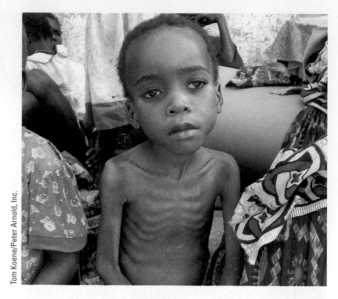

Figure 1-13 Global connections: one in every three children under age 5, such as this child in Lunda, Angola, suffers from severe malnutrition caused by a lack of calories and protein. According to the World Health Organization, each day at least 13,700 children under age 5 die prematurely from malnutrition and infectious diseases, most from drinking contaminated water and being weakened by malnutrition.

Many of the world's desperately poor die prematurely from four preventable health problems. One is *malnutrition* from a lack of protein and other nutrients needed for good health (Figure 1-13). Second is increased susceptibility to normally nonfatal infectious diseases, such as diarrhea and measles, caused by their weakened condition from malnutrition.

Third is lack of access to clean drinking water (see photo 4 in the Detailed Contents). And fourth is severe respiratory disease and premature death from inhaling indoor air pollutants produced by burning wood or coal in open fires or poorly vented stoves for heating and cooking.

According to the World Health Organization, these four factors cause premature death for at least 7 million poor people each year. *This premature death of about 19,200 people per day is equivalent to 48 fully loaded 400-passenger jumbo jet planes crashing every day with no survivors!* Two-thirds of those dying are children younger than age 5. The daily news rarely covers this ongoing human tragedy. The good news is that we have the knowledge and money to solve these problems within two decades if we have the political and ethical will to act.

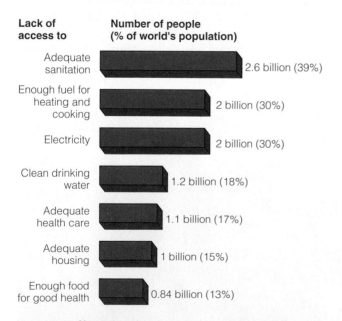

Lack of access to	Number of people (% of world's population)
Adequate sanitation	2.6 billion (39%)
Enough fuel for heating and cooking	2 billion (30%)
Electricity	2 billion (30%)
Clean drinking water	1.2 billion (18%)
Adequate health care	1.1 billion (17%)
Adequate housing	1 billion (15%)
Enough food for good health	0.84 billion (13%)

Figure 1-12 Natural capital degradation: some harmful results of poverty. QUESTION: *Which two of these effects do you believe are the most harmful?* (Data from United Nations, World Bank, and World Health Organization)

[?] THINKING ABOUT THE POOR AND EXPONENTIALLY INCREASING POPULATION GROWTH Some see rapid population growth of the poor in developing countries as the primary cause of our environmental problems. Others say that the much higher resource use per person in developed countries is a more important factor. Which factor do you think is more important? Why?

Resource Consumption and Environmental Problems (Economics and Ethics)

Many consumers in developed countries have become addicted to buying more and more stuff in their search for fulfillment and happiness.

Resource consumption is linked to both poverty and wealth. The poor *underconsume* by not having enough food, water, and other resources to meet their basic needs. Many of the prosperous *overconsume* by using and wasting far more resources than they need.

Affluenza ("af-loo-EN-zuh") is a term used to describe the unsustainable addiction to overconsumption and materialism exhibited in the lifestyles of many affluent consumers in the United States and other developed countries and the rising middle class in countries such as China and India (Case Study, p. 15). It is based on the assumption that buying more things will bring happiness. As humorist Will Rogers said, "Too many people spend money they haven't earned to buy things they don't want, to impress people they don't like."

Affluenza has an enormous environmental impact. Because of the exponential growth (Core Case Study, p. 6) in resource use, it takes about 27 tractor-trailer loads of resources per year to support one American, or 7.9 billion truckloads per year to support the entire U.S. population. Stretched end-to-end, these trucks would more than reach the sun—some 150 million kilometers (93 million miles away).

And using these resources produces large amounts of pollution, environmental degradation, and wastes (Figure 1-10). Globalization and global advertising are now spreading the affluenza virus throughout much of the world.

After a lifetime of studying the growth and decline of the world's human civilizations, historian Arnold Toynbee summarized the true measure of a civilization's growth as the *law of progressive simplification:* "True growth occurs as civilizations transfer an increasing proportion of energy and attention from the material side of life to the nonmaterial side and thereby develop their culture, capacity for compassion, sense of community, and strength of democracy."

Beneficial Effects of Affluence on Environmental Quality (Economics)

Affluent countries have more money for improving environmental quality.

Affluence can lead people to become more concerned about environmental quality. And it provides money for developing technologies to reduce pollution, environmental degradation, and resource waste.

In the United States, the air is cleaner, drinking water is purer, most rivers and lakes are cleaner, and the food supply is more abundant and safer than they were in 1970, and there is less resource waste. Similar advances have been made in most other affluent countries. Affluence financed these improvements in environmental quality.

A downside to wealth is that it allows the affluent to clean up the immediate environment of their homes, cities, and countries by transferring some of their wastes and pollution to more distant locations. It also allows them to obtain the resources they need from almost anywhere in the world without seeing the harmful environmental impacts of their high-consumption life styles (Figure 1-7). In other words, many affluent countries are living beyond their ecological means by running up eventually unsustainable global ecological debts.

Connections between Environmental Problems and Their Causes

Environmental quality is affected by interactions between population size, resource consumption, and technology.

Once we have identified environmental problems and their root causes, the next step is to understand how they are connected to one another. The three-factor model in Figure 1-14 (p. 20) is a starting point.

According to this simple model, the environmental impact (I) of a population on a given area depends on three key factors: the number of people (P), the average resource use per person (affluence, A), and the beneficial and harmful environmental effects of the technologies (T) used to provide and consume each unit of a resource and control or prevent the resulting pollution and environmental degradation.

In developing countries, population size and the resulting degradation of renewable resources (as the poor struggle to stay alive) tend to be the key factors in total environmental impact (Figure 1-14, top). In such countries resource use per person is low.

In developed countries, high rates of resource use per person (affluenza) and the resulting high levels of pollution and environmental degradation per person usually are the key factors determining overall environmental impact (Figure 1-14, bottom) and a country's global ecological footprint and its ecological footprint per person (Figure 1-7).

Consider the United States and India. The U.S. population of 300 million people increases exponentially by roughly 3 million per year, while India's population of 1.1 billion grows by 18.6 million and China's population of 1.3 billion grows by 7.7 million. But the United States has a much larger ecological footprint per person because of its much higher resource use. For example, the average U.S. citizen consumes about 30 times as much as the average citizen of India and 100 times as much as the average person in the world's poorest countries. This means that *poor parents in such developing countries would need 60–200 children to reach*

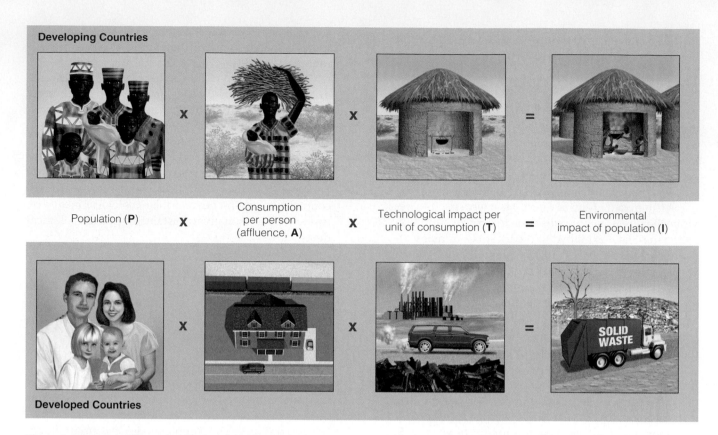

Developing Countries

Population (**P**) **X** Consumption per person (affluence, **A**) **X** Technological impact per unit of consumption (**T**) **=** Environmental impact of population (**I**)

Developed Countries

Figure 1-14 Connections: simplified model of how three factors—number of people, affluence, and technology—affect the environmental impact of the population in developing countries (top) and developed countries (bottom).

the same lifetime family resource consumption level as 2 children in a typical U.S. family.

Some forms of technology, such as polluting factories and motor vehicles and energy-wasting devices, increase environmental impact by raising the T factor in the equation. Other technologies, such as pollution control and prevention, solar cells, and energy-saving devices, lower environmental impact by decreasing the T factor. In other words, some forms of technology are *environmentally harmful* and some are *environmentally beneficial.*

F **RESEARCH FRONTIER** Finding ways to reduce over-consumption

 ? **THINKING ABOUT EXPONENTIAL GROWTH** What role does exponential growth (Core Case Study, p. 6) play in each of the factors in the model in Figure 1-14?

CULTURAL CHANGES AND THE ENVIRONMENT

Human Cultural Changes

Since our hunter–gatherer days, three major cultural changes have increased the human impact on the environment.

Evidence from fossils and studies of ancient cultures suggest that the current form of our species, *Homo sapiens sapiens,* has walked the earth for perhaps 90,000–195,000 years—less than an eye-blink in the earth's 3.7 billion years of life.

Until about 12,000 years ago, we were mostly hunter–gatherers who typically lived in small groups and moved as needed to find enough food for survival. Since then, three major cultural changes have occurred. The *agricultural revolution,* which began 10,000–12,000 years ago, allowed people to settle in villages and raise crops and domesticated animals

Next the *industrial–medical revolution,* which began about 275 years ago, led to a shift from rural villages and animal-powered agriculture to an urban society using fossil fuels for manufacturing material items, agriculture, and transportation. It also involved using science to help us improve sanitation and understand and control disease. The third cultural shift, the *information–globalization revolution,* began about 50 years ago. It is based on using new technologies for gaining rapid access to much more information on a global scale. These technologies include the telephone, radio, television, computers, automated databases, and remote sensing satellites.

Figure 1-15 lists major advantages and disadvantages of the advanced industrial–medical revolution

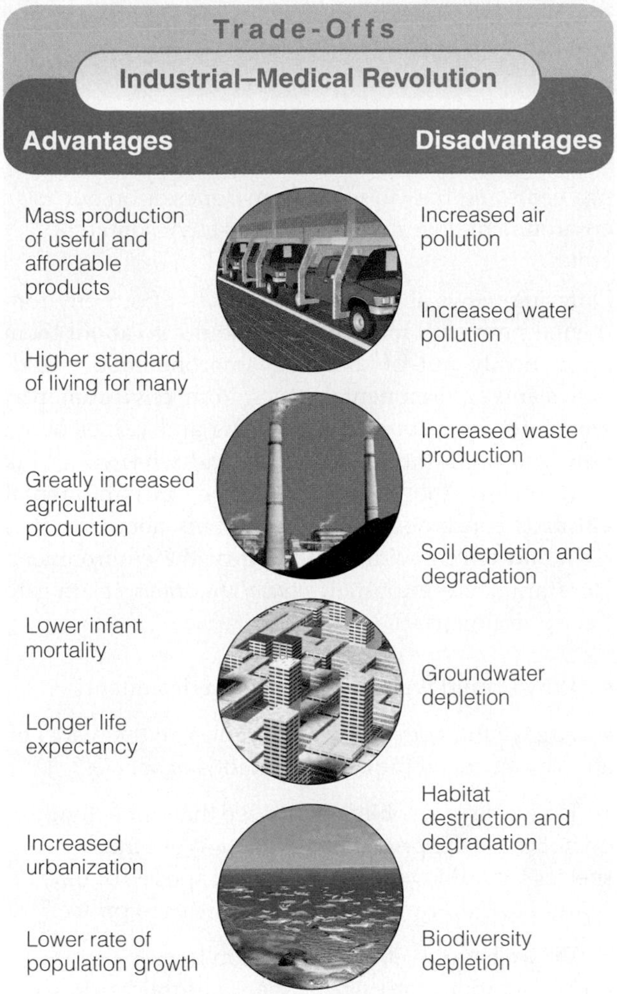

Trade-Offs

Industrial–Medical Revolution

Advantages	Disadvantages
Mass production of useful and affordable products	Increased air pollution
	Increased water pollution
Higher standard of living for many	
Greatly increased agricultural production	Increased waste production
	Soil depletion and degradation
Lower infant mortality	Groundwater depletion
Longer life expectancy	
	Habitat destruction and degradation
Increased urbanization	
Lower rate of population growth	Biodiversity depletion

Figure 1-15 Trade-offs: advantages and disadvantages of the advanced industrial–medical revolution and by extension the information–globalization revolution. QUESTION: *Which single advantage and disadvantage do you think are the most important?*

and by extension the information–globalization revolution. During the last 50 years, living conditions have improved for the majority of the world's population. Most individuals live longer, are better nourished and wealthier, and have the freedom to participate in electing and influencing their leaders. This progress, however, has put an increasing strain on the earth's natural capital.

Eras of Environmental History in the United States

The environmental history of the United States consists of the tribal, frontier, early conservation, and modern environmental eras.

The environmental history of the United States can be divided into four eras. The first was *the tribal era,* during which 5–10 million tribal people (now called Native Americans) occupied North America for at least 10,000 years before European settlers began arriving in the early 1600s.

This was followed by the *frontier era* (1607–1890), when European colonists began settling North America. Faced with a continent containing seemingly inexhaustible forest and wildlife resources and rich soils, the early colonists developed a **frontier environmental worldview.** They viewed most of the continent as having vast resources and as a wilderness to be conquered and managed for human use.

Next came the *early conservation era* (1832–1870), during which some people became alarmed at the scope of resource depletion and degradation in the United States. They urged that part of the unspoiled wilderness on public lands owned jointly by all people (but managed by the government) be protected as a legacy to future generations. Most of these warnings and ideas were not taken seriously.

The early conservation period was followed by an era—lasting from 1870 to the present—with an *increased role of the federal government and private citizens in resource conservation, public health, and environmental protection.* See Supplement 5 on p. S16 for more details and an overview of U.S. environmental history.

SUSTAINABILITY AND ENVIRONMENTAL WORLDVIEWS

Are Things Getting Better or Worse? A Millennium Assessment (Science and Politics)

There is good and bad environmental news.

Experts disagree about how serious our population and environmental problems are and what we should do about them. Some suggest that human ingenuity and technological advances will allow us to clean up pollution to acceptable levels, find substitutes for any scarce resources, and keep expanding the earth's ability to support more humans.

Many leading environmental scientists disagree. They appreciate and applaud the significant environmental and social progress that we have made, but they also cite evidence that we are degrading and disrupting the earth's life-support systems in many parts of the world at an exponentially accelerating rate. They call for much more action to protect the natural capital that supports our economies and all life.

According to environmental expert Lester R. Brown, "We are entering a new world, one where the collisions between our demands and the earth's capacity to satisfy them are becoming daily events. Our global economy is outgrowing the capacity of the earth to support it. No economy, however technologically advanced, can survive the collapse of its environmental support systems."

In 2005, the *UN's Millennium Ecosystem Assessment* was released. According to this four-year study by 1,360 experts from 95 countries, human activities are degrading or using unsustainably about 60% of the world's free natural services (Figure 1-4, right) that sustain life on the earth. In other words, we are living unsustainably.

This pioneering comprehensive examination of the health of the world's life-support systems is also a story of hope. It says we have the tools to preserve the planet's natural capital by 2050 and describes common-sense strategies for doing this.

The most useful answer to the question of whether things are getting better or worse is *both*. Some things are getting better and some are getting worse.

F RESEARCH FRONTIER A crash program to gain better and more comprehensive information about the health of the world's life-support systems

Our challenge is to not get trapped into confusion and inaction by listening primarily to either of two groups of people. *Technological optimists* tend to overstate the situation by telling us to be happy and not to worry, because technological innovations and conventional economic growth and development will lead to a wonder world for everyone. In contrast, *environmental pessimists* overstate the problems to the point where our environmental situation seems hopeless. The noted conservationist Aldo Leopold argued, "I have no hope for a conservation based on fear."

X HOW WOULD YOU VOTE?* Do you believe that the society you live in is on an unsustainable path? Cast your vote online at www.thomsonedu.com/biology/miller.

Many environmental scientists and leaders believe that we must and can make a shift toward a more sustainable economy and civilization during your lifetime. In 2006, Lester R. Brown said, "Sustaining our current global civilization now depends on shifting to a renewable energy-based and a reuse/recycle economy with a diversified transport system, employing a sustainable mix of light rails, buses, bicycles, and cars. Making this transition requires **(1)** restructuring the global economy so that it can sustain civilization, **(2)** an all-out effort to eradicate poverty, stabilize population, and restore hope, and **(3)** a systematic effort to restore natural systems. With each wind farm, rooftop solar panel, paper recycling facility, bicycle path, and

*To cast your vote, go the website for the book and then to the appropriate chapter (in this case, Chapter 1). In most cases, you will be able to compare how you voted with others using this book throughout the United States and the rest of the world.

reforestation program, we move closer to an economy that can sustain economic progress."

Environmental Worldviews and Ethics

The way we view the seriousness of environmental problems and how to solve them depends on our environmental worldview and our environmental ethics.

Differing views about the seriousness of our environmental problems and what we should do about them arise mostly out of differing environmental worldviews and environmental ethics. Your **environmental worldview** is a set of assumptions and values about how you think the world works and what you think your role in the world should be. **Environmental ethics** is concerned with your beliefs about what is right and wrong with how we treat the environment. Here are some important *ethical questions* relating to the environment:

- Why should we care about the environment?

- Are we the most important species on the planet or are we just one of the earth's millions of species?

- Do we have an obligation to see that our activities do not cause the premature extinction of other species? Should we try to protect all species or only some? How do we decide which species to protect?

- Do we have an ethical obligation to pass on to future generations the extraordinary natural world we have inherited in as good condition, if not better, as we inherited?

- Should every person be entitled to equal protection from environmental hazards regardless of race, gender, age, national origin, income, social class, or any other factor? This is the central ethical and political issue for what is known as the *environmental justice* movement. See the Guest Essay by Robert D. Bullard on the website for this book.

? THINKING ABOUT OUR RESPONSIBILITIES How would you answer each of the questions above? Compare your answers with those of your classmates. Record your answers and, at the end of this course, return to these questions to see if your answers have changed.

People with widely differing environmental worldviews and ethical and cultural beliefs can take the same data, be logically consistent, and arrive at quite different conclusions because they start with different assumptions and moral principles or values. Various environmental worldviews are discussed in detail in Chapter 26, but here is a brief introduction.

Some people in today's industrial consumer societies have a **planetary management worldview.** This view holds that we are separate from the nature, that nature exists mainly to meet our needs and increasing wants, and that we can use our ingenuity and technology to manage the earth's life-support systems, mostly for our benefit. It assumes that economic growth is unlimited.

A second environmental worldview, known as the **stewardship worldview,** holds that we can manage the earth for our benefit but that we have an ethical responsibility to be caring and responsible managers, or *stewards,* of the earth. It says we should encourage environmentally beneficial forms of economic growth and discourage environmentally harmful forms.

Another worldview is the **environmental wisdom worldview.** It holds that we are part of and totally dependent on nature and that nature exists for all species, not just for us. It also calls for encouraging earth-sustaining forms of economic growth and development and discouraging earth-degrading forms. According to this view, our success depends on learning how the earth sustains itself and integrating such environmental wisdom into the ways we think and act. Many of the ideas for the stewardship and environmental wisdom worldviews are derived from the writings of Aldo Leopold (Individuals Matter, below).

Four Scientific Principles of Sustainability: Copy Nature (Science)

We can develop more sustainable economies and societies by mimicking the four major ways that nature has adapted and sustained itself for several billion years.

How can we live more sustainably? According to ecologists and environmental scientists, we should find out how life on the earth has survived and adapted for several billion years and use what we learn as guidelines for our lives and economies.

Science reveals that four basic components of the earth's *natural sustainability* are quite simple (Figure 1-16, p 24):

- **Reliance on Solar Energy:** the sun warms the planet and supports photosynthesis used by plants to provide food for us and other animals.

- **Biodiversity:** a great variety of genes, species, ecosystems, and ecological processes have provided

---

INDIVIDUALS MATTER

Aldo Leopold's Environmental Ethics

According to *Aldo Leopold* (Figure 1-A), the role of the human species should be to protect nature, not conquer it.

In 1933, Leopold became a professor at the University of Wisconsin and in 1935 he was one of the founders of the U.S. Wilderness Society. Through his writings and teachings he became one of the leaders of the *conservation* and *environmental movements* of the 20th century. In doing this, he laid important groundwork for the field of environmental ethics.

Leopold's weekends of planting, hiking, and observing nature at his farm in Wisconsin provided material for his most famous book, *A Sand County Almanac,* published in 1949 after his death. Since then more than 2 million copies of this environmental classic have been sold.

The following quotations from his writings reflect Leopold's land ethic, and they form the basis for many of the beliefs of the modern stewardship and environmental wisdom worldviews:

All ethics so far evolved rest upon a single premise: that the individual is a member of a community of interdependent parts.

That land is a community is the basic concept of ecology, but that land is to be loved and respected is an extension of ethics.

The land ethic changes the role of Homo sapiens from conqueror of the land-community to plain member and citizen of it.

We abuse land because we regard it as a commodity belonging to us. When we see land as a community to which we belong, we may begin to use it with love and respect.

Anything is right when it tends to preserve the integrity, stability, and beauty of the biotic community. It is wrong when it tends otherwise.

Courtesy of the University of Wisconsin—Madison Archives

Figure 1-A Individuals matter: *Aldo Leopold* (1887–1948) was a forester, writer, and conservationist. His book *A Sand County Almanac* (published after his death) is considered an environmental classic that inspired the modern environmental and conservation movement.

Critical Thinking

Which of the above quotations do you agree with? Which, if any, of these ethical principles do you put into practice in your own life?

many ways to adapt to changing environmental conditions throughout the 3.7-billion-year history of life on the earth.

- **Population Control:** competition for limited resources among species places a limit on how much any one population can grow. If a population grows beyond those limits, its size decreases from changes in the birth rates and death rates of its members. In nature, no population can grow indefinitely.

- **Nutrient Recycling:** natural processes recycle all chemicals or nutrients that plants and animals need to stay alive and reproduce. In this recycling process, the wastes or dead bodies of all organisms become food or resources for other organisms. There is little waste in nature.

Figure 1-17 summarizes how we can live more sustainably by using these four amazingly simple fundamental lessons from nature (left side) in designing our societies, products, and economies (right side). *Figures 1-16 and 1-17 summarize the sustainability theme central to this book.*

Using the four scientific principles of sustainability to guide our lifestyles and economies could result in an *environmental revolution* during your lifetime. Figure 1-18 lists some of the shifts involved in bringing about this new cultural revolution.

 ? *THINKING ABOUT EXPONENTIAL GROWTH AND SUSTAINABILITY* Is exponential economic growth incompatible with environmental sustainability? Explain.

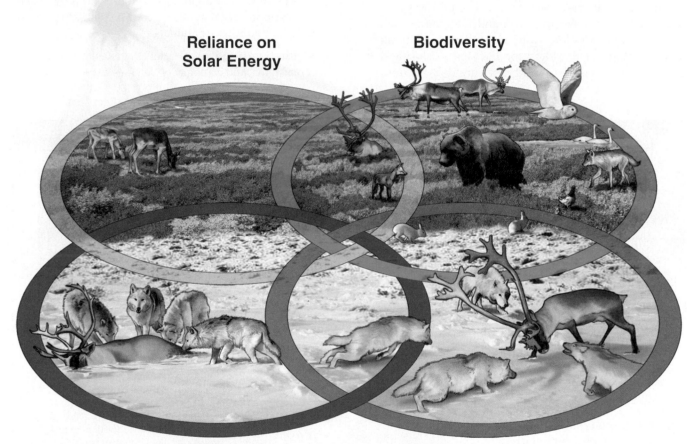

Reliance on Solar Energy

Biodiversity

Nutrient Recycling

Population Control

Figure 1-16 Four scientific principles of sustainability: these four interconnected principles of sustainability are derived from learning how nature has sustained a variety of life on the earth for about 3.7 billion years. The top left oval shows sunlight stimulating the production of vegetation in the Arctic tundra during its brief summer (*solar energy*) and the top right oval shows some of the diversity of species found there during the summer (*biodiversity*). The bottom right oval shows Arctic gray wolves stalking a caribou during the long cold winter (*population control*). The bottom left oval shows Arctic gray wolves feeding on their kill. This plus huge numbers of tiny decomposers that convert dead matter to soil nutrients recycle the nutrients needed to support the plant growth shown in the top left and right ovals (*nutrient recycling*).

How Nature Works	Lessons for Us
Runs on perpetual solar energy.	Rely mostly on direct and indirect solar energy.
Recycles nutrients and wastes. There is little waste in nature.	Prevent and reduce waste and pollution and recycle and reuse resources.
Uses biodiversity to maintain itself and adapt to new environmental conditions.	Preserve biodiversity by protecting ecosystems and preventing premature extinction of species.
Controls population size and resource use of species.	Recognize nature's limits on population size and resource use and learn to live within these limits.

Figure 1-17 Solutions: implications of the four scientific principles of sustainability derived from observing nature (left) for the long-term sustainability of human societies (right).

Scientific evidence indicates that we have perhaps 50 years and no more than 100 years to make such a cultural change. You will witness a historical fork in the road at which point we will choose a path toward sustainability or continue on our current unsustainable course. Everything you do or don't do will play a role in which path we take.

Building Social Capital: Talking and Listening to One Another (Politics and Ethics)

A key to sustaining natural capital is to build social capital by working together to find common ground and implementing an informed and shared vision of a better world based on hope.

Making the shift to more sustainable societies and economies involves building what sociologists call **social capital**. This involves getting people with different views and values to talk and listen to one another, find common ground, and work together to build understanding, trust, and informed shared visions of what their communities, states, nations, and the world

could and should be. This means nurturing openness, communication, cooperation, and hope and discouraging close-mindedness, polarization, confrontation, and fear.

Much of society today has become shrill and less civil. People scream at one another, take strong positions often without investigating the facts, and refuse to listen to those with different ideas. This behavior paralyzes attempts to find workable solutions to common problems and leads to a loss of the most powerful force for change—hope.

The important environmental issues we face are not black and white, but rather all shades of gray because proponents of all sides have some legitimate and useful insights. This means that citizens should strive to build social capital by finding *trade-off solutions*—an important theme of this book—to environmental problems and try to agree on a shared vision of the future they want. Once a shared vision crystallizes, citizens can work together to develop strategies for implementing their vision beginning at the local level, as the citizens of Chattanooga, Tennessee (USA), have done (Case Study, p. 26).

A key to building social capital and implementing solutions to environmental problems is to recognize that most social change results from individual actions and individuals acting together to bring about change

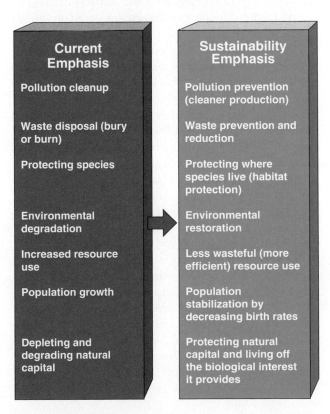

Figure 1-18 Solutions: some shifts involved in bringing about the *environmental* or *sustainability revolution.*

by grassroots action from the bottom up. In other words, *individuals matter*—another important theme of this book. Research by social scientists suggests that it takes only 5–10% of the population of a community, country, or of the world to bring about major social change. Such research also shows that significant social change can occur in a much shorter time than most people think.

Anthropologist Margaret Mead summarized our potential for social change: "Never doubt that a small group of thoughtful, committed citizens can change the world. Indeed, it is the only thing that ever has."

Case Study: Chattanooga, Tennessee (Science, Economics, and Politics)

Local officials, business leaders, and citizens have worked together to transform the U.S. city of Chattanooga, Tennessee, from a highly polluted city to one of the most sustainable and livable cities in the United States.

During the 1960s U.S. government officials rated Chattanooga as having the dirtiest air in the United States. Its air was so polluted by smoke from its coke ovens and steel mills that people sometimes had to turn on their headlights in the middle of the day. The Tennessee River flowing through the city's industrial center bubbled with toxic waste. People and industries fled the downtown area and left a wasteland of abandoned and polluting factories, boarded-up buildings, high unemployment, and crime.

In 1984, the city decided to get serious about improving its environmental quality. Civic leaders started a *Vision 2000* process with a 20-week series of community meetings in which more than 1,700 citizens from all walks of life gathered to build a consensus about what the city could be at the turn of the century. Citizens identified the city's main problems, set goals, and brainstormed thousands of ideas for solutions.

By 1995, Chattanooga had met most of its original goals. The city had encouraged zero-emission industries to locate there and replaced its diesel buses with a fleet of quiet, zero-emission electric buses, made by a new local firm. Downtown car use was reduced by building satellite parking lots and providing free and rapid bus service between the lots and the city center.

Chattanooga also launched an innovative recycling program after environmentally concerned citizens blocked construction of a new garbage incinerator that would have emitted harmful air pollutants. These efforts paid off. Since 1989, the levels of the seven major air pollutants regulated by the EPA in Chattanooga have been lower than those required by federal standards

Another project involved renovating much of the city's existing low-income housing and building new low-income rental units. Chattanooga also built the nation's largest freshwater aquarium, which became the centerpiece for downtown renewal. The city also developed a 35-kilometer-long (22-mile-long) riverfront park along both sides of the Tennessee River running through downtown. The park is filled with shade trees, flowers, fountains, and street musicians, and draws more than 1 million visitors per year. As property values and living conditions have improved, people and businesses have been moving back downtown.

In 1993, the community began the process again in *Revision 2000*. More than 2,600 participants identified additional goals and more than 120 recommendations for further improvements. One goal involves transforming an abandoned and blighted area in South Chattanooga into an environmentally advanced, mixed community of residences, retail stores, and zero-emission industries where employees can live near their workplaces. Most of these goals have been implemented.

Chattanooga's environmental success story, based on people working together to produce a more livable and sustainable city, is a shining example of what other cities can do by building their social capital.

↶ Revisiting Exponential Growth and Sustainability

Making the transition to more sustainable societies and economies challenges us to devise ways to slow down the harmful effects of the powerful force of exponential growth (Core Case Study, p. 6). Accomplishing this vital goal requires better scientific understanding of our interrelated natural and social systems (Figure 1-2), as discussed throughout this book.

We can then use this information to help slow human population growth, sharply reduce poverty, prevent new environmental problems from arising, curb the unsustainable forms of resource use that are eating away at the earth's natural capital, build social capital, and in the process create a better world for ourselves and our children and grandchildren.

Exponential growth is a double-edged sword. It can cause environmental harm. But we can also use it positively to amplify environmentally beneficial changes in our lifestyles and economies based on applying the four scientific principles of sustainability (Figures 1-16 and 1-17). Through our individual and collective actions or inactions we choose which side of the exponential growth sword to use in developing more environmentally sustainable lifestyles and economies. What a challenging and exciting time to be alive!

What's the use of a house if you don't have a decent planet to put it on?

HENRY DAVID THOREAU

CRITICAL THINKING

1. Describe three environmentally beneficial forms of exponential growth.

2. Explain why you agree or disagree with the following propositions:
    a. Stabilizing population is not desirable because without more consumers, economic growth would stop.
    b. The world will never run out of resources because we can use technology to find substitutes and to help us reduce resource waste and pollution.

3. When you read that about 19,200 people die prematurely each day (13 per minute) from preventable malnutrition and infectious disease, do you **(a)** doubt that it is true, **(b)** not want to think about it, **(c)** feel hopeless, **(d)** feel sad, **(e)** feel guilty, or **(f)** want to do something about this problem?

4. How do you feel when you read that the average American consumes about 30 times more resources than the average Indian citizen and human activities are projected to make the earth's climate warmer: **(a)** skeptical about their accuracy, **(b)** indifferent, **(c)** sad, **(d)** helpless, **(e)** guilty, **(f)** concerned, or **(g)** outraged? Which of these feelings help perpetuate such problems, and which can help alleviate them?

5. Which one or more of the four scientific principles of sustainability (Figure 1-16, p. 24) are involved in each of the following actions: **(a)** recycling soda cans; **(b)** using a rake instead of leaf blower; **(c)** choosing to have no more than one child; **(d)** walking to class instead of driving; **(e)** taking your own reusable bags to the grocery store to carry things home in; **(f)** volunteering in a prairie restoration project; and **(g)** lobbying elected officials to require that 20% of your country's electricity be produced by renewable wind power by 2020.

6. Explain why you agree or disagree with each of the following statements: **(a)** humans are superior to other forms of life, **(b)** humans are in charge of the earth, **(c)** all economic growth is good, **(d)** the value of other species depends only on whether they are useful to us, **(e)** because all species eventually become extinct we should not worry about whether our activities cause the premature extinction of a species, **(f)** all species have an inherent right to exist, **(g)** nature has an almost unlimited storehouse of resources for human use, **(h)** technology can solve our environmental problems, **(i)** I do not believe I have any obligation to future generations, and **(j)** I do not believe I have any obligation to other species.

7. What are the basic beliefs of your environmental worldview? Record your answer. Then at the end of this course return to your answer to see if your environmental worldview has changed. Are the beliefs of your environmental worldview consistent with your answers to question 6? Are your environmental actions consistent with your environmental worldview?

8. What three things would you do to model the area where you live after the example of Chattanooga, Tennessee?

9. List two questions that you would like to have answered as a result of reading this chapter.

PROJECTS

1. What are the major resource and environmental problems where you live? Which of these problems affect you directly? Have these problems gotten better or worse during the last 10 years?

2. Write two-page scenarios describing what your life and that of any children you may have might be like 50 years from now if **(a)** we continue on our present path; **(b)** we shift to more sustainable societies throughout most of the world.

3. Make a list of the resources you truly need. Then make another list of the resources you use each day only because you want them. Finally, make a third list of resources you want and hope to use in the future. Compare your lists with those compiled by other members of your class, and relate the overall result to the tragedy of the commons (p. 12).

4. Make a concept map of this chapter's major ideas using the section heads, subheads, and key terms (in boldface type). Look on the website for this book for information about making concept maps.

LEARNING ONLINE

The website for this book contains study aids and many ideas for further reading and research. They include a chapter summary, review questions for the entire chapter, flash cards for key terms and concepts, a multiple-choice practice quiz, interesting Internet sites, references, information about green careers, and a guide for accessing thousands of InfoTrac® College Edition articles. Log into

www.thomsonedu.com/biology/miller

Then choose Chapter 1, and select a learning resource. For access to animations, additional quizzes, chapter outlines and summaries, register and log into

at **www.thomsonedu.com** using the access code card in the front of your book.

Science, Systems, Matter, and Energy

An Environmental Lesson from Easter Island

Easter Island (Rapa Nui) is a small, isolated island in the great expanse of the South Pacific. Polynesians used double-hulled sea-going canoes to colonize this island about 800 years ago.

The settlers found an island paradise with fertile volcanic soil that supported dense and diverse forests and lush grasses. The Polynesians developed a civilization based on two species of the island's trees, giant palms and basswoods (called hauhau). They used the towering palm trees for shelter, as tools, and to build large seagoing canoes used for harpooning fish such as dolphins. They felled the hauhau trees and burned them to cook and keep warm in the island's cool winters and used their fibers to make rope. Forests were also cleared to plant crops.

The islanders had many children and the population reached as many as 15,000. But research in 2006 indicated that almost from the beginning they began using the island's tree and soil resources faster than they could be renewed. As these resources became inadequate to support the growing population, the leaders of the island's different clans appealed to the gods by carving at least 300 divine images from huge stones (Figure 2-1). They directed the people to cut large trees to make platforms for the stone sculptures. They probably placed logs underneath to roll the platforms or had 50 to 500 people use thick ropes to drag the platforms and statues across wooden rails to various locations on the island's coast.

In doing so, they increased the use of the island's precious large trees much faster than they were regenerated—an example of the tragedy of the commons based on living unsustainably. Without large trees, the islanders could not build their traditional seagoing canoes for hunting dolphins and fishing in deeper offshore waters, and no one could escape the island by boat.

Without the once-great forests to absorb and slowly release water, springs and streams dried up, exposed soils eroded, crop yields plummeted, and famine struck. There was no firewood for cooking or keeping warm. The hungry islanders ate all of the island's seabirds and land birds. Then they began raising and eating rats, descendants of hitchhikers on the first canoes to reach Easter Island.

Both the population and the civilization collapsed as rival clans fought one another for dwindling food supplies. Evidence suggests that eventually the islanders began to hunt and eat one another.

When Dutch explorers reached the island on Easter Day, 1722, they found several hundred hungry Polynesians, living in caves on grassland dotted with shrubs.

Like Easter Island at its peak, the earth is an isolated island in the vastness of space with no other suitable planet to migrate to. As on Easter Island, our population and resource consumption are growing and our resources are finite.

Will the humans on Earth Island re-create the tragedy of Easter Island on a grander scale, or will we learn how to live more sustainably on this planet that is our only home?

Scientific knowledge about how the earth works and sustains itself, as discussed in this chapter, is a key to learning how to live more sustainably and thus avoiding the fate of Easter Islanders and numerous other past civilizations (see Supplement 6 on p. S25) that disappeared because they degraded their resource base. In today's ecologically and economically interdependent world the fates of everyone on the earth are intertwined.

Jeremy Woodhouse/MWI/Peter Arnold, Inc.

Figure 2-1 **Environmental degradation:** these massive stone figures on Easter Island—some of them taller than the average five-story building—are the remains of the technology created by an ancient civilization of Polynesians. Their civilization collapsed because the people used up the trees (especially large palm trees) that were the basis of their livelihood. At least 300 of these huge stone statues, which can weigh as much as 89 metric tons (98 tons), once lined the coast of Easter Island.

Science is an adventure of the human spirit. It is essentially an artistic enterprise, stimulated largely by curiosity, served largely by disciplined imagination, and based largely on faith in the reasonableness, order, and beauty of the universe.

WARREN WEAVER

To help us develop more sustainable societies, we need to know about the nature of science and the matter and energy that make up the earth's living and nonliving resources—the subjects of this chapter. It discusses these questions:

- What is science, and what do scientists do?

- What are major components and behaviors of complex systems?

- What are the basic forms of matter, and what makes matter useful as a resource?

- What types of changes can matter undergo and what scientific law governs changes in matter?

- What are the major forms of energy, and what makes energy useful as a resource?

- What are two scientific laws governing changes of energy from one form to another?

- How are the scientific laws governing changes of matter and energy from one form to another related to resource use, environmental degradation, and sustainability?

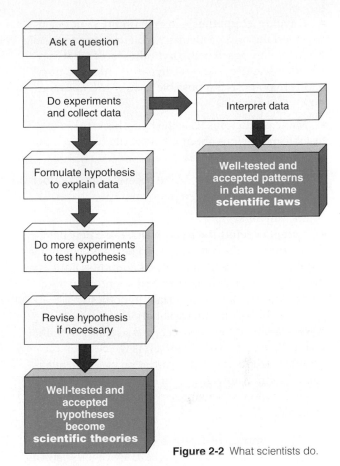

Figure 2-2 What scientists do.

THE NATURE OF SCIENCE

What Do Scientists Do?

Scientists collect data, form hypotheses, and develop theories, models, and laws about how nature works.

In essence, scientists ask and attempt to find answers about how the natural world works. **Science** is an attempt to discover order in the natural world and to use that knowledge to make predictions about what is likely to happen in nature. It is based on the assumption that events in the natural world follow orderly cause and effect patterns that can be understood through careful observation, experimentation, and modeling. Figure 2-2 summarizes the scientific process.

There is nothing mysterious about this process. You use it all the time in making decisions. Here is an example of applying the scientific process to an everyday situation:

Observation: You switch on your trusty flashlight and nothing happens.

Question: Why did the light not come on?

Hypothesis: Maybe the batteries are bad.

Test the hypothesis: Put in new batteries and switch on the flashlight.

Result: Flashlight still does not work.

New hypothesis: Maybe the bulb is burned out.

Experiment: Replace bulb with a new bulb.

Result: Flashlight works when switched on.

Conclusion: Second hypothesis is verified.

Here is a more formal outline of steps scientists often take in trying to understand nature, although not always in the order listed:

- **Ask a question or identify a problem to be investigated.** For example an environmental scientist might ask: "How are human actions affecting the natural resources and services (Figure 1-4, p. 9) provided by the earth?" An environmental economist might ask: "What is the estimated economic value of each of these goods and services?" And an environmental political scientist might ask: "How can we get people and governments to help sustain these resources and services?"

- **Collect data related to the question or problem by making observations and measurements.** Scientists often conduct **experiments** to study some phenomenon under known conditions. Scientists work hard to reduce the errors and uncertainty in their measurements, as discussed in Supplement 1 on p. S2.

- **Develop a hypothesis to explain the data.** Scientists working on a particular problem suggest possible explanations, or **scientific hypotheses,** of what they (or other scientists) observe in nature. In effect, a scientific hypothesis is a possible answer to a question posed by scientists.

- **Make predictions.** Use the hypothesis to make testable predictions about what should happen if the hypothesis is valid.

- **Test the predictions.** Make observations, conduct experiments, or develop a mathematical or other *model* to test the predictions.

- **Accept or reject the hypothesis.** If the new data do not support the hypothesis, come up with another testable explanation. This process continues until there is general agreement among scientists in the field being studied that a certain hypothesis provides the best useful explanation of the data. A well-tested and widely accepted scientific hypothesis is called a **scientific theory.** It is the best and most useful answer to a scientific question based on available scientific knowledge and a great deal of research and evaluation by scientists in the field or fields involved.

Three important features of the scientific process are *skepticism, peer review* of results by other scientists, and *reproducibility.* Scientists tend to be highly skeptical of new data and hypotheses until they can be verified. Peer review happens when scientists publish details of the methods they used, the results of their experiments and models, and the reasoning behind their hypotheses for other scientists working in the same field (their peers) to examine and criticize (Figure 2-3). Ideally, other scientists repeat and analyze the work to see if the data can be reproduced and whether the proposed hypotheses are reasonable and useful.

Science should not be confused with technology. Science is a search for understanding of how the natural world works. *Technology* is the development of devices, processes, and products to benefit human beings.

Scientific Theories and Laws: The Most Important Results of Science

A widely tested and accepted scientific hypothesis becomes a scientific theory and a scientific law describes what we find happening in nature over and over again.

If an overwhelming body of observations and measurements or tests using models supports a hypothesis, it becomes a scientific theory. *Scientific theories are not to be taken lightly.* They have been tested widely, are supported by extensive evidence, and are accepted by most scientists in a particular field or related fields of study.

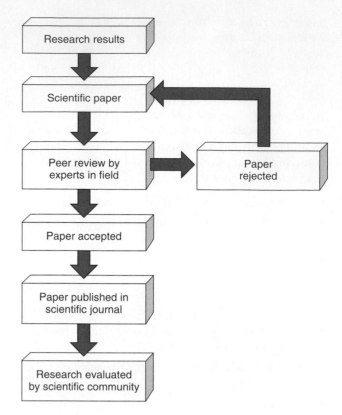

Figure 2-3 Scientists use a peer review process to help identify sound science.

Nonscientists often use the word *theory* incorrectly when they actually mean *scientific hypothesis,* a tentative explanation that needs further evaluation. The statement, "Oh, that's just a theory," made in everyday conversation, implies a lack of knowledge and careful testing—the opposite of the scientific meaning of the word.

? *THINKING ABOUT HYPOTHESES AND THEORIES* Try to find an example in the news media or everyday conversation in which the term "theory" is used when it should have been "hypothesis."

Another important result of science is a **scientific,** or **natural, law:** a description of what we find happening in nature over and over in the same way. For example, after making thousands of observations and measurements over many decades, scientists discovered the second law of thermodynamics. It says that heat always flows spontaneously from hot to cold—something you learned the first time you touched a hot object. Scientific laws describe repeated, consistent findings in nature, whereas scientific theories are widely accepted explanations of data, hypotheses, and laws.

A scientific law is no better than the accuracy of the observations or measurements upon which it is based (see Figure 1 on p. S3 in Supplement 1). But if the data are accurate, a scientific law cannot be broken. See Supplement 26 on pp. S73–S80 on How to Analyze a Scientific Article.

Testing Hypotheses

Scientists test hypotheses by using controlled experiments and by running mathematical models on high-speed computers.

Many *variables* or *factors* influence most processes or parts of nature that scientists seek to understand. Scientists conduct *controlled experiments* to try to isolate and study the effect of a one variable at a time.

To do such *single-variable analysis*, scientists set up two groups. One is an *experimental group* in which the chosen variable is changed in a known way. The other is a *control group* in which the chosen variable is not changed. If the experiment is designed properly, any difference between the two groups should result from the variable that was changed in the experimental group.

A basic problem is that many environmental phenomena involve a huge number of interacting variables. Sometimes this limitation is overcome by using *multivariable analysis*—that is, by running mathematical models on high-speed computers to analyze the interactions of many variables without having to carry out traditional controlled experiments.

Scientific Reasoning and Creativity

Scientists use inductive reasoning to convert observations and measurements to a general conclusion and deductive reasoning to convert a generalization to a specific conclusion.

Scientists arrive at certain conclusions with varying degrees of certainty by using inductive and deductive reasoning. **Inductive reasoning** involves using specific observations and measurements to arrive at a general conclusion or hypothesis. It is a form of *"bottom-up"* reasoning that involves going from the specific to the general. For example, suppose we observe that a variety of different objects fall to the ground when we drop them from various heights. We might then use inductive reasoning to conclude that *all objects fall to the earth's surface when dropped.*

Depending on the number of observations made, there may be a high degree of certainty in this conclusion. However, what we are really saying is that "All objects that we or other observers have dropped from various heights fall to the earth's surface." Although it is extremely unlikely, we cannot be *absolutely sure* someone will drop an object that does not fall to the earth's surface.

Deductive reasoning involves using logic to arrive at a specific conclusion based on a generalization or premise. It is a form of *"top-down"* reasoning that goes from the general to the specific. For example,

> *Generalization or premise:* All birds have feathers.
>
> *Example:* Eagles are birds.
>
> *Deductive conclusion:* All eagles have feathers.

The conclusion of this *syllogism* (a series of logically connected statements) is valid as long as the premise is correct and we do not use faulty logic to arrive at the conclusion.

Deductive and inductive reasoning and critical thinking skills (pp. 2–4) are important scientific tools. But scientists also use intuition, imagination, and creativity to explain some of their observations in nature. Often such ideas defy conventional logic and current scientific knowledge. According to physicist Albert Einstein, "There is no completely logical way to a new scientific idea." Intuition, imagination, and creativity are as important in science as they are in poetry, art, music, and other great adventures of the human spirit, as reflected by scientist Warren Weaver's quotation found at the opening of this chapter.

Paradigm Shifts: Major Changes in Scientific Theories

Occasionally new information or ideas can disprove and overthrow a well-accepted scientific theory in what is known as a paradigm shift.

Scientific theories may have a high probability of being valid. But they are not infallible. Occasionally new discoveries and new ideas can overthrow a well-accepted scientific theory. Thomas Kuhn in his 1962 book *The Structure of Scientific Revolutions* calls such revolutions in scientific thinking **paradigm shifts.** Such a shift occurs when the majority of scientists in a field or related fields agree that a new explanation or theory is better than the old one.

In the second century AD, the Greek astronomer and geographer Ptolemy proposed that the earth was the center of the universe. He and other scientists made extensive measurements and mathematical calculations that explained the movements of the planets based on his hypothesis. This became a widely accepted scientific theory that lasted for 1,400 years.

But this theory was disproved in 1543 by Nicolaus Copernicus. He proposed that the sun was the center of our solar system and showed that this new hypothesis explained the planetary movements better than the Ptolemaic theory. This sun-centered view has been a widely accepted theory for more than 460 years.

Paradigm shifts can also take place in our environmental worldviews. Some analysts detect an increasing shift from the planetary management worldview to the stewardship and environmental wisdom world-views (p. 23). It has been argued that if about 10% of the world's people make such a paradigm shift, it will speed up the transition to more sustainable economies and societies.

Frontier Science, Sound Science, and Junk Science

Scientific results fall into two categories: those that have not been confirmed (frontier science) and those that have been well tested and widely accepted (sound science).

News reports often focus on two things: so-called scientific breakthroughs, and disputes between scientists over the validity of preliminary and untested data, hypotheses, and models. These preliminary results, called **frontier science,** are often controversial because they have not been widely tested and accepted by peer review. Some of these results will be discredited. At the frontier stage, it is normal and healthy for reputable scientists to disagree about the meaning and accuracy of data and the validity of various hypotheses.

By contrast, **sound science,** or **consensus science,** consists of data, theories, and laws that are widely accepted by scientists who are considered experts in the field. The results of sound science are based on the self-correcting process of open peer review (Figure 2-3) and reproducibility. To find out what scientists generally agree on, you can seek out reports by scientific bodies such as the U.S. National Academy of Sciences and the British Royal Society, which attempt to summarize consensus among experts in key areas of science. New evidence and widely accepted theories may discredit sound science. But until that happens, sound science is the best explanation that we have.

Junk science consists of scientific results or hypotheses that are presented as sound science without having undergone the rigors of peer review, or that have been discarded as a result of peer review. Here are some critical thinking questions you can use to uncover junk science:

- What data support the proposed hypotheses? Have these data been verified? (*Are they reproducible?*)

- Do the conclusions and hypotheses follow logically from the data?

- Does the explanation account for all of the observations? Are there alternative explanations?

- Are the investigators unbiased in their interpretations of the results? Are they free of a hidden agenda? Is their funding from an unbiased source?

- Have the conclusions been verified by impartial peer review?

- Are the conclusions of the research widely accepted by other experts in this field?

If "yes" is answer to each of these questions, then the results can be classified as sound science. Otherwise, the results may represent frontier science that needs further testing and evaluation, or they can be classified as junk science.

Reporters sometimes mislead their audiences by presenting sound or consensus science along with a quote from a scientist in the field who disagrees with the consensus view or from someone who is not an expert in the field. This can cause the public to distrust well-established sound science and sometimes believe in ideas not widely accepted by the scientific community. See the Guest Essay on environmental reporting by Andrew C. Revkin on the website for this chapter.

Limitations of Environmental Science

Inadequate data and scientific understanding limit environmental science and make some of its results controversial.

Before we begin our study of environmental science, we need to recognize some of its limitations, as well as those of science in general. For example, scientists can disprove things but cannot prove anything absolutely because there is always some degree of uncertainty in scientific measurements (see Figure 1 on p. S3 in Supplement 1), observations, and models. Instead scientists try to establish that a particular model, theory, or law has a very high *probability* (90–99%) of being true. Most scientists rarely say something like, "Cigarettes cause lung cancer." Rather, they might say, "Overwhelming evidence from thousands of studies indicates that there is a significant relationship between cigarette smoking and lung cancer."

? *THINKING ABOUT SCIENTIFIC PROOF* Does the fact that science can never prove anything absolutely mean that its results are not valid or useful? Explain.

Second, scientists are human and cannot be expected to be totally free of bias about their results and hypotheses. However, bias can be minimized and often uncovered by the high standards of evidence required through peer review.

A third limitation especially important to environmental science involves validity of data. There is no way to measure accurately how many metric tons of soil are eroded worldwide, for example. Instead, scientists use statistical sampling and methods to estimate such numbers. However, such environmental data should not be dismissed as "only estimates" because they can indicate important trends.

Another limitation is that most environmental problems are difficult to understand completely because they involve many variables and highly complex interactions. Much progress has been made, but we still know too little about how the earth works, its current state of environmental health, and the environmental impacts of our activities. Filling in these gaps are important and urgent *research frontiers*.

MODELS AND BEHAVIOR OF SYSTEMS

Usefulness of Models

Scientists predict the behavior of a complex system by developing a model of its inputs, throughputs (flows), and outputs of matter, energy, and information.

A **system** is a set of components that function and interact in some regular and theoretically understandable manner. For example, a system might be the human body, a population of tigers, a river, an economy, or the entire earth. Most *systems* have the following key components: **inputs** from the environment, **flows** or **throughputs** within the system at certain rates, and **outputs** to the environment as shown in Figure 1-10, p. 17.

Scientists use *models* or approximate representations or simulations to find out how systems work and to evaluate ideas or hypotheses. Some of our most powerful and useful technologies are mathematical models, which are used to supplement our mental models.

Making a mathematical model usually requires going through three steps many times. *First*, make a guess and write down some equations. *Second*, compute the likely behavior of the system implied by the equations. *Third*, compare the system's projected behavior with observations and behavior projected by mental models, existing experimental data, and scientific hypotheses, laws, and theories. Mathematical models are particularly useful when there are many interacting variables, when the time frame is long, and when controlled experiments are impossible, too slow, or too expensive to conduct.

After building and testing a mathematical model, scientists use it to predict what is *likely* to happen under a variety of conditions. In effect, they use mathematical models to answer *if–then* questions: "*If* we do such and such, *then* what is likely to happen now and in the future?" This process can give us a variety of projections or scenarios of possible futures or outcomes based on different assumptions. Mathematical models (like all other models) are no better than the assumptions on which they are built and the data fed into them.

Feedback Loops: How Systems Respond to Change

Outputs of matter, energy, or information fed back into a system can cause the system to do more of what it was doing (positive feedback) or less (negative feedback).

When people ask you for feedback, they are seeking information that they can feed back into their mental processes to help them make a decision or carry out some action. All systems undergo change as a result of feedback loops. A **feedback loop** occurs when an output of matter, energy, or information is fed back into the system as an input and leads to changes in that system.

A **positive feedback loop** causes a system to change further in the same direction. One example involves depositing money in a bank at compound interest and leaving it there. The interest increases the balance, which through a positive feedback loop leads to more interest and an even higher balance.

A **negative,** or **corrective, feedback loop** causes a system to change in the opposite direction. An example is recycling aluminum cans. This involves melting aluminum and feeding it back into an economic system to make new aluminum products. This negative feedback loop of matter reduces the need to find, extract, and process virgin aluminum ore. It also reduces the flow of waste matter (discarded aluminum cans) into the environment.

The tragedy on Easter Island discussed at the beginning of the chapter involved the coupling of positive and negative feedback loops. As the abundance of trees turned to a shortage of trees, a positive feedback loop (more births than deaths) became weaker as death rates rose. Eventually a negative feedback loop (more deaths than births) dominated and caused a dieback of the island's human population.

 ? *THINKING ABOUT EASTER ISLAND AND OUR CURRENT ENVIRONMENTAL SITUATION* Give an example of two harmful positive feedback loops in today's world and two negative feedback loops that could be used to prevent us from repeating the Easter Island environmental tragedy.

Time Delays: Instant Response Is Rare

Sometimes corrective feedback takes so long to work that a system can cross a threshold and change its normal behavior.

Complex systems often show **time delays** between the input of a stimulus and the response to it. A long time delay can mean that corrective action comes too late. For example, a smoker exposed to cancer-causing chemicals in cigarette smoke may not get lung cancer for 20 years or more. And Easter Islanders (Core Case Study, p. 28) cut the island's trees for hundreds of years before depleting this essential resource.

Time delays can allow a problem to build up slowly until it reaches a *threshold level*, or *tipping point*, and causes a fundamental shift in the behavior of a system. Prolonged delays dampen the negative feedback mechanisms that might slow, prevent, or halt environmental problems. Examples are population growth, leaks from toxic waste dumps, and degradation of forests from prolonged exposure to air pollutants.

? THINKING ABOUT TIME DELAYS Give an example of a time delay effect not discussed in this book. Try to come up with one related to your life.

Synergy: Amplifying Responses

Sometimes processes and feedbacks in a system can interact to amplify the results.

A **synergistic interaction,** or **synergy,** occurs when two or more processes interact so that the combined effect is greater than the sum of their separate effects. Scientific studies reveal a synergistic interaction between smoking and inhalation of tiny particles of asbestos in causing lung cancer. Lifetime smokers have ten times the risk that nonsmokers have of getting lung cancer. And individuals exposed to particles of asbestos for long periods increase their risk of getting lung cancer fivefold. But people who smoke and are exposed to asbestos increase their risk of getting lung cancer fiftyfold.

Synergy can result when two people work together to accomplish a task. For example, suppose you and I need to move a 140-kilogram (300-pound) tree that has fallen across the road. By ourselves, each of us can lift only, say, 45 kilograms (100 pounds). But if we work together and use our muscles properly, we can move the tree out of the way. That is using synergy to solve a problem.

? THINKING ABOUT SYNERGY AND EASTER ISLAND What are two examples of beneficial synergy that could help us avoid the tragedy of Easter Island (Core Case Study, p. 28)?

Unintended Harmful Results of Human Activities

Human actions in a complex system often lead to unintended harmful results and environmental surprises.

One of the four guidelines for living more sustainably, based on the principles of sustainability (Figure 1-16, p. 24) is *we can never do just one thing.* Any action in a complex system has multiple, unintended, and often unpredictable effects. As a result, most of the environmental problems we face today are unintended results of activities designed to increase the quality of human life.

For example, clearing land of trees to plant crops increases food production and can improve nutrition. But it can also lead to deforestation, soil erosion, and a loss of biodiversity, as Easter Islanders (Core Case Study, p. 28) and other civilizations (see Supplement 6 on p. S25) learned the hard way.

One factor that can lead to an environmental surprise is a *discontinuity* or abrupt change in a previously stable system when some *environmental threshold* is crossed. For example, you may be able to lean back in a chair and balance yourself on two of its legs for a long time with only minor adjustments. But if you pass a certain threshold of movement, your balanced system suffers a discontinuity, or sudden shift, and you may find yourself on the floor.

Scientific evidence indicates that we are crossing an increasing number of environmental thresholds or tipping points. For example, when we exceed the sustainable yield of a fishery for a number of years the available fish decline to the point where it is not profitable to harvest them. A similar collapse can happen when many trees in a forest start dying after being weakened because of depleted soil nutrients and decades of exposure to a cocktail of air pollutants. Other examples are coral reefs dying, glaciers melting, and seas rising because of global warming.

F RESEARCH FRONTIER Tipping points for various environmental systems such as fisheries, forests, climate, and coral reefs

TYPES AND STRUCTURE OF MATTER

Elements and Compounds

Matter exists in chemical forms as elements and compounds.

Matter is anything that has mass (the amount of material in an object) and takes up space. There are two two chemical forms of matter. One is **elements:** the distinctive building blocks of matter that make up every material substance. The other consists of **compounds:** two or more different elements held together in fixed proportions by attractive forces called *chemical bonds.* (See Supplement 7 on p. S27 for an expanded discussion of basic chemistry.)

To simplify things, chemists represent each element by a one- or two-letter symbol. Examples used in this book are hydrogen (H), carbon (C), oxygen (O), nitrogen (N), phosphorus (P), sulfur (S), chlorine (Cl), fluorine (F), bromine (Br), sodium (Na), calcium (Ca), lead (Pb), mercury (Hg), arsenic (As), and uranium (U). These elements are the only ones you need to know to understand the material in this book.

Chemists refer to compounds of two or more elements by combining their symbols. For example, one carbon (C) and two oxygen (O_2) atoms join in nature to form CO_2, called carbon dioxide. Chemists have developed a way to classify the elements according to their chemical behavior, in what is called the *periodic table of elements* (see Figure 1 on p. S27 in Supplement 7).

Four elements—oxygen, carbon, hydrogen, and nitrogen—make up about 96.3% of your body weight. From a chemical standpoint, how much are you worth? Not much. If we add up the market price per kilogram for each element in someone weighing 70 kilograms (154 pounds), the total value comes to about $120. Not very uplifting, is it?

Of course, you are worth much more because your body is not just a bunch of chemicals enclosed in a bag of skin. From a scientific standpoint, you are an incredibly complex system in which air, water, soil nutrients, energy-storing chemicals, and food chemicals interact in millions of ways to keep you alive and healthy.

Atoms

Atoms are basic building blocks of matter.

One basic building block of matter is an **atom:** the smallest unit of matter that exhibits the characteristics of an element. If you had a supermicroscope, you would find that each different type of atom contains a certain number of *subatomic particles*. There are three types of these atomic building blocks: positively charged **protons (p)**, uncharged **neutrons (n)**, and negatively charged **electrons (e).**

Each atom consists of an extremely small center, or **nucleus**, and one or more electrons in rapid motion somewhere outside the nucleus. The nucleus contains one or more protons, and in most cases neutrons. Atoms are incredibly small. In fact, more than 3 million hydrogen atoms could sit side by side on the period at the end of this sentence.

Each atom has equal numbers of positively charged protons inside its nucleus and negatively charged electrons whirling around outside its nucleus. Because these electrical charges cancel one another, *the atom as a whole has no net electrical charge.*

Each element has a unique **atomic number,** equal to the number of protons in the nucleus of each of its atoms. The simplest element, hydrogen (H), has only 1 proton in its nucleus, so its atomic number is 1. Carbon (C), with 6 protons, has an atomic number of 6, whereas uranium (U), a much larger atom, has 92 protons and an atomic number of 92.

Because electrons have so little mass compared with the masses of protons or neutrons, *most of an atom's mass is concentrated in its nucleus.* The mass of an atom is described in terms of its **mass number:** the total number of neutrons and protons in its nucleus. For example, a hydrogen atom with 1 proton and no neutrons in its nucleus has a mass number of 1, and a uranium atom with 92 protons and 143 neutrons in its nucleus has a mass number of 235 (92 + 143 = 235).

All atoms of an element have the same number of protons in their nuclei. But they may have different numbers of uncharged neutrons in their nuclei, and

thus different mass numbers. Various forms of an element having the same atomic number but different mass numbers are called **isotopes** of that element. Scientists identify isotopes by attaching their mass numbers to the name or symbol of the element. For example, hydrogen has three isotopes: hydrogen-1 (H-1, with one proton and no neutrons in its nucleus), hydrogen-2 (H-2, common name *deuterium*, with one proton and one neutron in its nucleus), and hydrogen-3 (H-3, common name *tritium*, with one proton and two neutrons). How many protons and neutrons are there in the nucleus of a uranium-238 atom? Figure 2-4 shows a simplified model of a carbon-12 atom.

Ions and pH

Ions are another building block of matter and scientists use pH as a measure of the hydrogen ions in water solutions.

An *ion* is an atom or groups of atoms with one or more net positive (+) or negative (−) electrical charges. (For details on how ions form see p. S27 in Supplement 7.)

The number of positive or negative charges on an ion is shown as a superscript after the symbol for an atom or a group of atoms. Examples encountered in this book include *positive* hydrogen ions (H^+), sodium ions (Na^+), calcium ions (Ca^{2+}), aluminum ions (Al^{3+}), and ammonium ions (NH_4^+), and *negative* chloride ions (Cl^-), hydroxide ions (OH^-), nitrate ions (NO_2^-), sulfate ions (SO_4^{2-}), and phosphate ions (PO_4^{3-}).

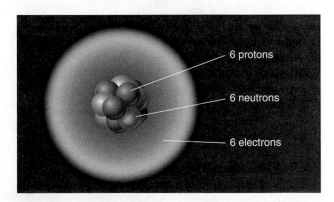

6 protons

6 neutrons

6 electrons

Figure 2-4 Greatly simplified model of a carbon-12 atom. It consists of a nucleus containing 6 positively charged protons and 6 neutral neutrons. There are 6 negatively charged electrons found outside its nucleus. We cannot determine the exact location of the electrons. Instead, we can estimate the *probability* that they will be found at various locations outside the nucleus—sometimes called an *electron probability cloud*. This is somewhat like saying that there are six airplanes flying around inside a cloud. We don't know their exact location, but the cloud represents an area where we can probably find them. **QUESTION:** *How would a model of an atom of carbon-14 differ from that of carbon-12?*

Scientists use **pH** as a measure of the acidity of a solution based on its concentration of hydrogen ions (H^+). Pure water (not tap water or rainwater) has an equal number of H^+ and OH^- ions. It is called a **neutral solution** and has a pH of 7. An **acidic solution** has more hydrogen ions than hydroxide ions and has a pH less than 7. A **basic solution** has more hydroxide ions than hydrogen ions and has a pH greater than 7. Figure 2-5 shows the approximate pH and hydrogen ion concentration per liter of solution (right side) for various common substances.

Compounds and Chemical Formulas: Chemical Shorthand

Chemical formulas are shorthand ways to show the atoms and ions in a chemical compound.

A third building block of matter is a **molecule:** a combination of two or more atoms of the same or different elements held together by chemical bonds. (For more details on chemical bonds see p. S28 in Supplement 7.) Molecules are the building blocks of *compounds*.

Chemists use a **chemical formula** to show the number of atoms or ions of each type in a compound.

This shorthand contains the symbol for each element present and uses subscripts to represent the number of atoms or ions of each element in the compound's basic structural unit. Examples of compounds and their formulas encountered in this book are sodium chloride (NaCl), water (H_2O, read as "H-two-O"), oxygen (O_2), ozone (O_3), nitrogen (N_2), nitrous oxide (N_2O), nitric oxide (NO), hydrogen sulfide (H_2S), carbon monoxide (CO), carbon dioxide (CO_2), nitrogen dioxide (NO_2), sulfur dioxide (SO_2), ammonia (NH_3), sulfuric acid (H_2SO_4), nitric acid (HNO_3), methane (CH_4), and glucose ($C_6H_{12}O_6$). (See Figures 2, 4, and 5 on pp. S28–S29 in Supplement 7 for more information on chemical compounds and the bonds that hold them together.)

 Examine atoms—their parts, how they work, and how they bond together to form molecules— at ThomsonNOW.

Organic Compounds: Carbon Rules

Organic compounds contain carbon atoms combined with one another and with various other atoms such as hydrogen, nitrogen, or chlorine.

Table sugar, vitamins, plastics, aspirin, penicillin, and most of the chemicals in your body are **organic compounds** that contain at least two carbon atoms combined with each other and with atoms of one or more other elements, such as hydrogen, oxygen, nitrogen, sulfur, phosphorus, chlorine, and fluorine. One exception, methane (CH_4), has only one carbon atom. All other compounds are called **inorganic compounds.**

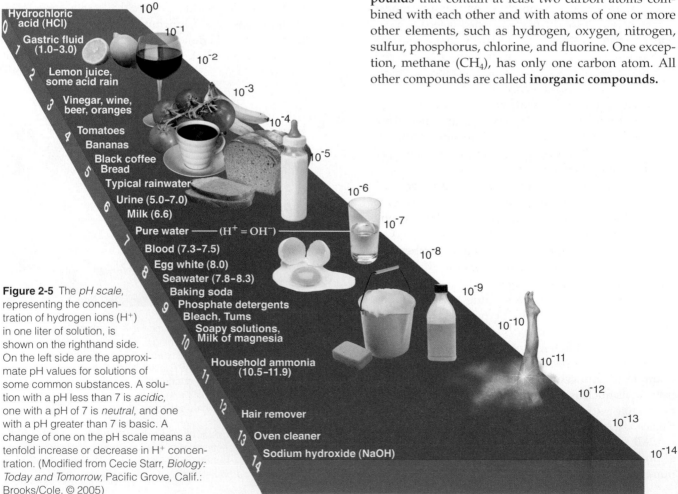

Figure 2-5 The *pH scale*, representing the concentration of hydrogen ions (H^+) in one liter of solution, is shown on the righthand side. On the left side are the approximate pH values for solutions of some common substances. A solution with a pH less than 7 is *acidic*, one with a pH of 7 is *neutral*, and one with a pH greater than 7 is basic. A change of one on the pH scale means a tenfold increase or decrease in H^+ concentration. (Modified from Cecie Starr, *Biology: Today and Tomorrow*, Pacific Grove, Calif.: Brooks/Cole, © 2005)

Labels on figure:

- 0 Hydrochloric acid (HCl) — 10^0
- 1 Gastric fluid (1.0–3.0) — 10^{-1}
- 2 Lemon juice, some acid rain — 10^{-2}
- 3 Vinegar, wine, beer, oranges — 10^{-3}
- 4 Tomatoes, Bananas — 10^{-4}
- 5 Black coffee, Bread, Typical rainwater — 10^{-5}
- 6 Urine (5.0–7.0), Milk (6.6) — 10^{-6}
- 7 Pure water ($H^+ = OH^-$) — 10^{-7}
- 8 Blood (7.3–7.5), Egg white (8.0), Seawater (7.8–8.3) — 10^{-8}
- 9 Baking soda, Phosphate detergents, Bleach, Tums — 10^{-9}
- 10 Soapy solutions, Milk of magnesia — 10^{-10}
- 11 Household ammonia (10.5–11.9) — 10^{-11}
- 12 — 10^{-12}
- 13 Hair remover — 10^{-13}
- 13 Oven cleaner
- 14 Sodium hydroxide (NaOH) — 10^{-14}

The millions of known organic (carbon-based) compounds include the following:

- *Hydrocarbons:* compounds of carbon and hydrogen atoms. An example is methane (CH_4), the main component of natural gas, and the simplest organic compound.

- *Chlorinated hydrocarbons:* compounds of carbon, hydrogen, and chlorine atoms. An example is the insecticide DDT ($C_{14}H_9Cl_5$).

- *Simple carbohydrates* (simple sugars): certain types of compounds of carbon, hydrogen, and oxygen atoms. An example is glucose ($C_6H_{12}O_6$), which most plants and animals break down in their cells to obtain energy. (See Figure 6 on p. S30 in Supplement 7 for a closer look at the structure of a glucose molecule and how these molecules can link together to form complex carbohydrates.)

Cells: The Fundamental Units of Life

Cells are the basic structural and functional units of all forms of life

All living things are composed of **cells:** minute compartments containing chemicals necessary for life and within which most of the processes of life take place. Cells are the structural and functional units of all life. Organisms may consist of a single cell (bacteria, for instance) or plants and animals that contain huge numbers of cells.

On the basis of their cell structure, organisms can be classified as either *eukaryotic* or *prokaryotic.* Each cell of a **eukaryotic** organism is surrounded by a membrane and has a distinct *nucleus* (a membrane-bounded structure containing genetic material in the form of DNA), and several other internal parts called *organelles* (Figure 2-6b). Most organisms consist of eukaryotic cells.

A membrane surrounds the cell of a **prokaryotic** organism, but the cell contains no distinct nucleus or organelles enclosed by membranes (Figure 2-6a). A prokaryotic cell is much simpler and usually much smaller than a eukaryotic cell. All bacteria are single-celled prokaryotic organisms.

Macromolecules, DNA, Genes, and Chromosomes: Life's Building Blocks

Complex organic molecules are the basic building blocks of life found in genes and chromosomes.

Larger and more complex organic compounds, called **macromolecules,** make up the basic molecular units found in living organisms. Three of these molecules are *polymers,* formed when a number of simple organic molecules (*monomers*) are linked together by chemical bonds, somewhat like rail cars linked in a freight train. The three major types of organic polymers are **(1)** *com-*

(a) Prokaryotic Cell

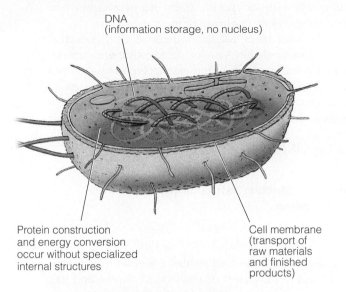

DNA (information storage, no nucleus)

Protein construction and energy conversion occur without specialized internal structures

Cell membrane (transport of raw materials and finished products)

(b) Eukaryotic Cell

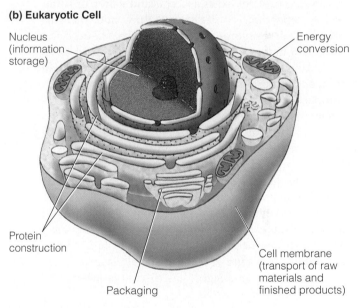

Nucleus (information storage)

Energy conversion

Protein construction

Packaging

Cell membrane (transport of raw materials and finished products)

Figure 2-6 Natural capital: (a) generalized structure of a *prokaryotic cell.* Note that a prokaryotic cell lacks the distinct nucleus and generalized structure of (b) a *eukaryotic cell.* The parts and internal structure of cells in various types of organisms such as plants and animals differ somewhat from this generalized model.

plex carbohydrates such as cellulose and starch that consist of two or more monomers of simple sugars such as glucose linked together, **(2)** *proteins* formed by linking together monomers of amino acids, and **(3)** *nucleic acids* (such as DNA and RNA) formed by linking monomers called nucleotides. *Lipids* are a fourth type of macromolecule found in living organisms. Figures 6, 7, 8, 9, and 10 on pp. S30–S31 in Supplement 7 give more details on the structures of these four types of macromolecules.

Genes consist of specific sequences of nucleotides found within a DNA molecule (see Figures 8 and 9 on pp. S30–S31 in Supplement 7) that contain information to make specific proteins. These coded units of genetic information about specific traits are passed on from parents to offspring during reproduction.

Chromosomes are combinations of genes that make up a single DNA molecule, together with a number of proteins. Each chromosome typically contains thousands of genes. Genetic information coded in your chromosomal DNA is what makes you different from an oak leaf, an alligator, or a flea, and from your parents. The relationships of genetic material to cells are depicted in Figure 2-7.

States of Matter: Solids, Liquids, Gases, and Plasma

Matter exists in solid, liquid, and gaseous physical states and a fourth state known as plasma.

The atoms, ions, and molecules that make up matter are found in three *physical states:* solid, liquid, and gas. For example, water exists as ice, liquid water, or water vapor depending on its temperature and the surrounding air pressure. The three physical states of any sample of matter differ in the spacing and orderliness of its atoms, ions, or molecules. A solid has the most compact and orderly arrangement and a gas the least compact and orderly arrangement. Liquids are somewhere in between.

A fourth state of matter, called *plasma,* is a high-energy mixture of positively charged ions and negatively charged electrons. The sun and all stars consist mostly of plasma, which makes it the most abundant form of matter in the universe.

There is little natural plasma on the earth, with most of it found in lightning bolts and flames. But scientists have learned how to run a high-voltage electric current through a gas to make artificial plasmas in fluorescent lights, neon signs, gas discharge lasers, and TV and computer screens.

Matter Quality

Matter can be classified as having high or low quality depending on how useful it is to us as a resource.

Matter quality is a measure of how useful a form of matter is to humans as a resource, based on its availability and concentration, as shown in Figure 2-8. **High-quality matter** is concentrated, is typically found near the earth's surface, and has great potential for use as a matter resource. **Low-quality matter** is dilute, is often located deep underground or dispersed in the ocean or the atmosphere, and usually has little potential for use as a material resource.

An aluminum can is a more concentrated, higher-quality form of aluminum than aluminum ore containing the same amount of aluminum. It takes less energy, water, and money to recycle an aluminum can than to make a new can from aluminum ore.

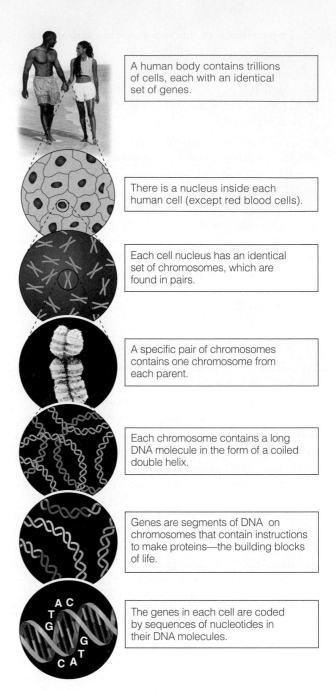

A human body contains trillions of cells, each with an identical set of genes.

There is a nucleus inside each human cell (except red blood cells).

Each cell nucleus has an identical set of chromosomes, which are found in pairs.

A specific pair of chromosomes contains one chromosome from each parent.

Each chromosome contains a long DNA molecule in the form of a coiled double helix.

Genes are segments of DNA on chromosomes that contain instructions to make proteins—the building blocks of life.

The genes in each cell are coded by sequences of nucleotides in their DNA molecules.

Figure 2-7 Natural capital: relationships among cells, nuclei, chromosomes, DNA, and genes.

Material efficiency, or **resource productivity,** is the total amount of material needed to produce each unit of goods or services. For example, the material efficiency of an aluminum drink can is higher than it used to be because it contains less aluminum. Business expert Paul Hawken and physicist Amory Lovins contend that the resource productivity of most materials used in developed countries could be improved by 75–90% within two decades using existing technologies.

F **RESEARCH FRONTIER** Improving the resource productivity of commonly used materials

Solid   Gas

Salt   Solution of salt in water

Coal   Coal-fired power plant emissions

Gasoline   Automobile emissions

Aluminum can   Aluminum ore

Figure 2-8 Examples of differences in matter quality. *High-quality matter* (left column) is fairly easy to extract and is concentrated; *low-quality matter* (right column) is more difficult to extract and is more widely dispersed than high-quality matter.

CHANGES IN MATTER

Physical and Chemical Changes

Matter can change from one physical form to another or change its chemical composition.

When a sample of matter undergoes a **physical change,** its chemical composition does not change. A piece of aluminum foil cut into small pieces is still aluminum foil. When solid water (ice) melts or liquid water boils, none of the H_2O molecules involved changes; instead, the molecules are organized in different spatial (physical) patterns.

In a **chemical change,** or **chemical reaction,** there is a change in the chemical compositions of the elements or compounds involved. Chemists use shorthand chemical equations to represent what happens in a chemical reaction. For example, when coal burns completely, the solid carbon (C) in the coal combines with oxygen gas (O_2) from the atmosphere to form the gaseous compound carbon dioxide (CO_2).

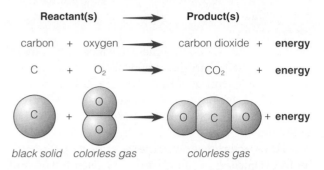

Reactant(s)	$\longrightarrow$	Product(s)
carbon + oxygen	$\longrightarrow$	carbon dioxide + **energy**
C + O_2	$\longrightarrow$	CO_2 + **energy**

black solid   *colorless gas*   *colorless gas*   + **energy**

Energy is given off in this reaction, which explains why coal is a useful fuel. The reaction also shows how the complete burning of coal (or any of the carbon-containing compounds in wood, natural gas, oil, and gasoline) produces carbon dioxide, which helps warm the lower atmosphere.

The Law of Conservation of Matter: There Is No "Away"

When a physical or chemical change occurs, no atoms are created or destroyed.

We may change elements and compounds from one physical or chemical form to another, but we can never create or destroy any of the atoms involved in any physical or chemical change. All we can do is re-arrange the elements and compounds into different spatial patterns (physical changes) or combinations (chemical changes). This statement, based on many thousands of measurements, is known as the **law of conservation of matter.** In describing chemical reactions, chemists use a shorthand system to account for all of the atoms, which they then use to balance chemical equations, as described on pp. S31–S32 in Supplement 7.

The law of conservation of matter means there is no "away" as in "to throw away." *Everything we think we have thrown away remains here with us in some form.* We can collect dust and soot from the smokestacks of industrial plants, but these solid wastes must then be put somewhere. We can remove substances from polluted water at a sewage treatment plant, but then we must burn them (producing some air pollution), bury them (possibly contaminating underground water supplies), or clean them up and apply the gooey sludge to the land as fertilizer (dangerous if the sludge contains nondegradable toxic metals such as lead and mercury).

↶ ⑦ *THINKING ABOUT THE LAW OF CONSERVATION OF MATTER AND EASTER ISLAND* Derive two rules from our understanding of the law of conservation of matter that could help keep us from repeating the tragedy of Easter Island (Core Case Study, p. 28).

Types of Pollutants

We will always generate some pollutants, but we can prevent much pollution and clean up some of what we do produce.

We can make the environment cleaner and convert some potentially harmful chemicals into less harmful physical or chemical forms. But the law of conservation of matter means we will always face the problem of what to do with some quantity of wastes and pollutants.

Three factors determine the severity of a pollutant's harmful effects: its *chemical nature,* its *concentration,* and its *persistence.*

The amount of a substance in a unit volume of air, water, or other medium is called its *concentration.* Concentration, is sometimes expressed in terms of *parts per million (ppm);* 1 ppm corresponds to 1 part pollutant per million parts of the gas, liquid, or solid mixture in which the pollutant is found. Smaller concentration units are parts per billion (ppb) and parts per trillion (ppt).

We can reduce the concentration of a pollutant by dumping it into the air or into a large volume of water, but there are limits to the effectiveness of this dilution approach. For example, the water flowing in a river can dilute or disperse some of the wastes dumped into it. If we dump in too much waste, however, this natural cleansing process does not work.

Persistence is a measure of how long the pollutant stays in the air, water, soil, or body. Pollutants can be classified into four categories based on their persistence:

- **Degradable pollutants** are broken down completely or reduced to acceptable levels by natural physical, chemical, and biological processes.

- **Biodegradable pollutants** are complex chemical pollutants that living organisms (usually specialized bacteria) break down into simpler chemicals. Human sewage in a river, for example, is biodegraded fairly quickly by bacteria if the sewage is not added faster than it can be broken down.

- **Slowly degradable pollutants** take decades or longer to degrade. Examples include the insecticide DDT and most plastics.

- **Nondegradable pollutants** are chemicals that natural processes cannot break down. Examples include the toxic elements lead, mercury, and arsenic. Ideally, we should try not to use these chemicals. If we do, we should figure out ways to keep them from getting into the environment. *Green Career:* Environmental chemist

Nuclear Changes: Radioactive Decay, Fission, and Fusion

Nuclei of some atoms can spontaneously lose particles or give off high-energy radiation, split apart, or fuse together.

In addition to physical and chemical changes, matter can undergo **nuclear changes.** This occurs when nuclei of certain isotopes spontaneously change or are made to change into nuclei of different isotopes. There are three types of nuclear change: natural radioactive decay, nuclear fission, and nuclear fusion.

Natural radioactive decay is a nuclear change in which unstable isotopes spontaneously emit fast-moving chunks of matter (alpha particles or beta particles), high-energy radiation (gamma rays), or both at a fixed rate. The unstable isotopes are called **radioactive isotopes** or **radioisotopes.**

Each type of radioisotope spontaneously decays at a characteristic rate into a different isotope. This rate of decay can be expressed in terms of **half-life:** the time needed for *one-half* of the nuclei in a given quantity of a radioisotope to decay and emit their radiation to form a different isotope. The decay continues, often producing a series of different radioisotopes, until a nonradioactive stable isotope is formed.

Half-lives range from a few millionths of a second to several billion years. An isotope's half-life cannot be changed by temperature, pressure, chemical reactions, or any other known factor.

Half-life can be used to estimate how long a radioisotope sample must be stored safely before it decays to a safe level. A rule of thumb is that such decay takes about 10 half-lives. Thus people must be protected from radioactive waste containing iodine-131 (which concentrates in the thyroid gland and has a half-life of 8 days) for 80 days (10 × 8 days). Plutonium-239, which is produced in nuclear reactors and used as the explosive in some nuclear weapons, can cause lung cancer when its particles are inhaled in even minute amounts. Its half-life is 24,000 years. Thus it must be stored safely for 240,000 years (10 × 24,000 years)—much longer than our species (*Homo sapiens sapiens*) has existed.

Exposure to alpha particles, beta particles, or gamma rays can alter DNA molecules in cells and in some cases can lead to genetic defects in one or more generations of offspring. Such exposure can also damage body tissues and cause burns, miscarriages, eye cataracts, and certain cancers.

 Learn more about half-lives and how doctors use radioactive particles to help us at ThomsonNOW.

Nuclear fission is a nuclear change in which the nuclei of certain isotopes with large mass numbers (such as uranium-235) are split apart into lighter nuclei when struck by neutrons; each fission releases two or three more neutrons plus energy (Figure 2-9). Each of these neutrons, in turn, can trigger an additional fission reaction. For multiple fissions to take place, enough fissionable nuclei must be present to provide the **critical mass** needed for efficient capture of these neutrons.

Multiple fissions within a critical mass produce a **chain reaction,** which releases an enormous amount of energy (Figure 2-9). This is somewhat like a room in which the floor is covered with spring-loaded mousetraps, each topped by a Ping-Pong ball. Open the door, throw in a single Ping-Pong ball, and watch the action in this simulated chain reaction of snapping mousetraps and balls flying around in every direction.

In an atomic bomb, an enormous amount of energy is released in a fraction of a second in an uncontrolled nuclear fission chain reaction. In the reactor of a nuclear power plant, the rate at which the nuclear fission chain reaction takes place is controlled. The heat released produces high-pressure steam to spin turbines, thereby generating electricity.

Nuclear fusion is a nuclear change in which two isotopes of light elements, such as hydrogen, are forced together at extremely high temperatures until they fuse to form a heavier nucleus. A tremendous amount of energy is released in this process. Fusion of hydrogen nuclei to form helium nuclei is the source of energy in the sun and other stars.

After World War II, the principle of *uncontrolled nuclear fusion* was used to develop extremely powerful hydrogen, or thermonuclear, weapons. These weapons use the D–T fusion reaction, in which a hydrogen-2, or deuterium (D), nucleus and a hydrogen-3 (tritium, T) nucleus are fused to form a helium-4 nucleus, a neutron, and energy, as shown in Figure 2-10 (p. 42).

Scientists have also tried to develop *controlled nuclear fusion,* in which the D–T reaction is used to produce heat that can be converted into electricity. After more than 50 years of research, this process is still in the laboratory stage. Even if it becomes technologically and economically feasible, many energy experts do not expect it to be a practical source of energy until 2030, if then.

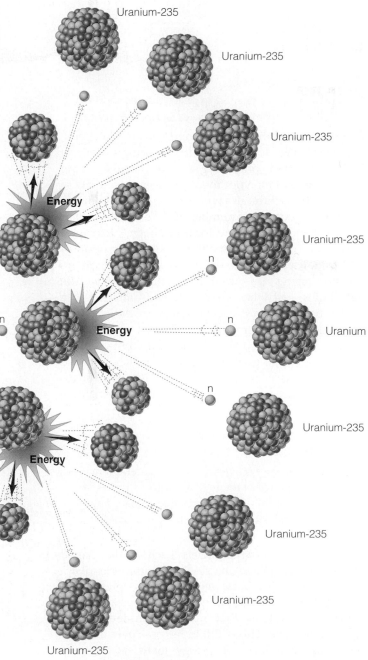

Figure 2-9 Fission of a uranium-235 nucleus by a neutron (*n*) releases more neutrons that can cause multiple fissions in a *nuclear chain reaction.*

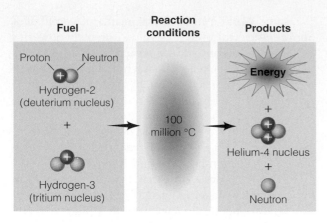

Fuel **Reaction conditions** **Products**

Figure 2-10 The deuterium–tritium (D–T) *nuclear fusion* reaction takes place at extremely high temperatures.

ENERGY

What Is Energy?

Energy is the work needed to move matter and the heat that flows from hot to cooler samples of matter.

Energy is the ability to do work and transfer heat. Using energy to do work means moving or lifting something such as this book, propelling a car or plane, warming your room, cooking your food, and using electricity to move electrons through a wire and light your room.

Energy exists in a number of different forms that can be changed from one form to another. Examples are *electrical energy* from the flow of electrons, *mechanical energy* used to move or lift matter, *light* or *electromagnetic* energy produced by sunlight and electric light bulbs, *heat* when energy flows from a hot to a colder body, *chemical energy* stored in the chemical bonds holding matter such as coal or oil together, and *nuclear* energy stored in the nuclei of atoms. During our brief time on earth we have learned how to tap into various sources of energy that have given us increasing control over nature (Case Study, below).

Case Study: Human Energy Use—A Brief History

Early humans were scavengers and hunter–gatherers whose main source of energy was muscle power. A human living at this basic survival level needs about 2,000 kilocalories of energy per day, most of it in the form of food.

In a modern industrial society such as the United States the average person uses 2,000 kilocalories of energy per day for basic energy needs, plus about 600,000 kilocalories of energy per day used by machines and systems that maintain an individual's complex lifestyle. This 300-fold increase over the minimum survival level of energy gives us immense power to alter and control nature for our own benefit.

Our first step along this energy path began with the discovery of fire that hunter–gatherers used to cook food and to light and heat their dwellings. Later they learned to use fire to burn grasslands to stampede animals they hunted over cliffs.

After settling down as farmers about 12,000 years ago we learned how to domesticate wild animals and use their muscle power to plow fields, carry loads for us, and transport us from place to place.

Later we learned to tap into energy from the wind to pump up underground water and to transport people and goods in sailing ships. We also used the power of flowing water to move goods and people on ships, to power mills for grinding grain, and eventually to produce electricity.

About 275 years ago we began inventing machines such as the steam engine to power ships, tractors, locomotives, and factory machinery. Renewable firewood provided about 91% of the energy used for heating and for running steam engines. But in 1850, this began changing as many forests were depleted.

We survived this early energy crisis by learning how to burn coal for heating and for running factories and trains. By 1900, wood provided only about 18% of our energy and coal 73%. In 1859 we learned how to pump oil out of the ground and later invented ways to convert it to fuels such as gasoline and heating oil.

In 1885, Carl Benz invented the internal combustion engine to power cars and other vehicles that could run on gasoline. By 1900, we got 40% of our energy from oil, 38% from coal, and 18% from natural gas—all nonrenewable resources.

In the 1950s, we learned how to get an enormous amount of energy by splitting the nuclei of certain types of uranium atoms (Figure 2-9) and to use this energy to produce electricity. Today we continue to live in a *fossil fuel era* with 82% of our energy coming from nonrenewable resources: oil (33%), coal (22%), natural gas (21%), and nuclear power (6%). The remaining 18% of our energy comes from several renewable resources—about 11% of it from wood and other forms of biomass. The rest comes from hydropower (electricity from flowing water), geothermal energy from the earth's interior, and energy from the sun and wind.

There is a wide gap in energy use per person in developed and developing countries. Today an average American uses 300 times more energy than the average Ethiopian uses.

Kinetic and Potential Energy

Energy can be moving (kinetic energy) or stored (potential energy) for possible use.

There are two major types of energy. One is **kinetic energy,** possessed by matter because of its mass and its speed or velocity. Examples of this energy in motion are wind (a moving mass of air), flowing streams, and electricity (flowing electrons).

Another type of moving energy is **heat:** the total kinetic energy of all moving atoms, ions, or molecules within a given substance, excluding the overall motion of the whole object. When two objects at different temperatures contact one another, kinetic energy in the form of heat flows from the hotter object to the cooler object until both objects reach the same temperature.

In **electromagnetic radiation,** another type of moving or kinetic energy, energy travels in the form of a *wave* as a result of the changes in electric and magnetic fields. Many different forms of electromagnetic radiation exist, each having a different *wavelength* (distance between successive peaks or troughs in the wave) and *energy content,* as shown in Figure 2-11. Such radiation travels through space at the speed of light—about 300,000 kilometers per second (186,000 miles per second). Visible light makes up most of the spectrum of electromagnetic radiation emitted by the sun (Figure 2-12). Organisms vary in their abilities to sense various parts of the electromagnetic spectrum.

 Find out how color, wavelengths, and energy intensities of visible light are related at ThomsonNOW.

The second type of energy is **potential energy,** which is stored and potentially available for use. Examples of potential energy include a rock held in your hand, an unlit match, still water behind a dam, the chemical energy stored in gasoline molecules, and the nuclear energy stored in the nuclei of atoms.

Potential energy can be changed to kinetic energy. Drop this book on your foot, and the book's potential

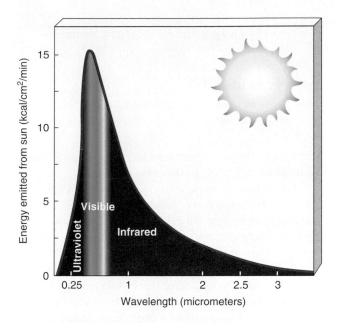

ThomsonNOW **Active Figure 2-12** **Solar capital:** the spectrum of electromagnetic radiation released by the sun consists mostly of visible light. *See an animation based on this figure and take a short quiz on the concept.*

energy when you held it changes into kinetic energy. When a car engine burns gasoline, the potential energy stored in the chemical bonds of gasoline molecules changes into heat, light, and mechanical (kinetic) energy that propels the car. Potential energy stored in a flashlight's batteries becomes kinetic energy in the form of light when the flashlight is turned on. Potential energy stored in various molecules such as carbohydrates becomes kinetic energy when your body uses it to do work.

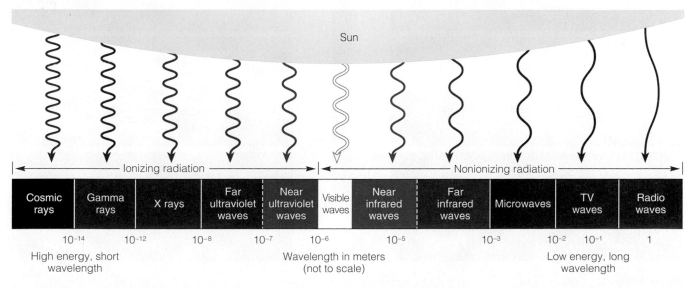

ThomsonNOW **Active Active Figure 2-11** The *electromagnetic spectrum:* the range of electromagnetic waves, which differ in wavelength (distance between successive peaks or troughs) and energy content. *See an animation based on this figure and take a short quiz on the concept.*

Witness how kinetic and potential energy might be used by a Martian at ThomsonNOW.

Energy Quality

Energy can be classified as having high or low quality depending on how useful it is to us as a resource.

Energy quality is a measure of an energy source's ability to do useful work, as described in Figure 2-13.

High-quality energy is concentrated and can perform much useful work. Examples include electricity, the chemical energy stored in coal and gasoline, concentrated sunlight, and nuclei of uranium-235 used as fuel in nuclear power plants.

By contrast, **low-quality energy** is dispersed and has little ability to do useful work. An example is heat dispersed in the moving molecules of a large amount of matter (such as the atmosphere or an ocean) so that its temperature is low.

For example, the total amount of heat stored in the Atlantic Ocean is greater than the amount of high-quality chemical energy stored in all of Saudi Arabia's vast oil deposits. Yet because the ocean's heat is so widely dispersed, it cannot be used to move things or to heat things to high temperatures.

ENERGY LAWS: TWO RULES WE CANNOT BREAK

The First Law of Thermodynamics: We Cannot Create or Destroy Energy

In a physical or chemical change, we can change energy from one form to another, but we can never create or destroy any of the energy involved.

Scientists have observed energy being changed from one form to another in millions of physical and chemical changes. But they have never been able to detect the creation or destruction of any energy (except in nu-

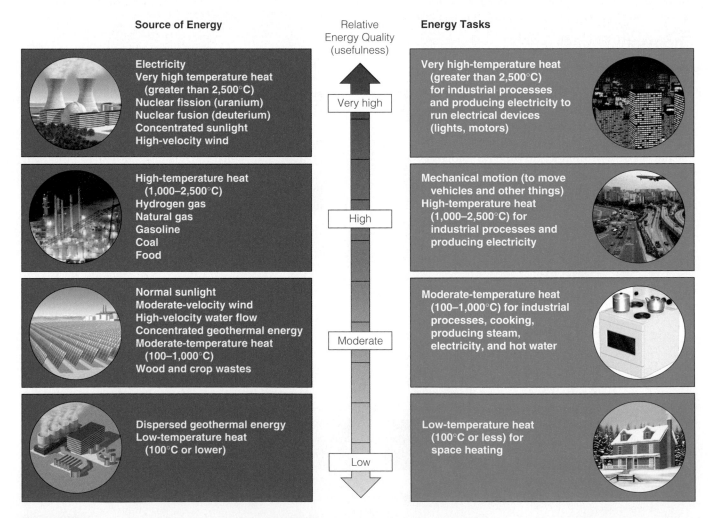

Figure 2-13 Natural capital: categories of the qualities of different sources of energy. *High-quality energy* is concentrated and has great ability to perform useful work. *Low-quality energy* is dispersed and has little ability to do useful work. To avoid unnecessary energy waste, you should match the quality of an energy source with the quality of energy needed to perform a task.

clear changes). The results of their experiments have been summarized in the **law of conservation of energy,** also known as the **first law of thermodynamics:** *In all physical and chemical changes, energy is neither created nor destroyed, although it may be converted from one form to another.*

This scientific law tells us that when one form of energy is converted to another form in any physical or chemical change, *energy input always equals energy output.* No matter how hard we try or how clever we are, we cannot get more energy out of a system than we put in; in other words, *we cannot get something for nothing in terms of energy quantity.* This is one of Mother Nature's basic rules that we have to live with.

The Second Law of Thermodynamics: Energy Quality Always Decreases.

Whenever energy changes from one form to another, we always end up with less usable energy than we started with.

Because the first law of thermodynamics states that energy can be neither created nor destroyed, you may be tempted to think there will always be enough energy. Yet if you fill a car's tank with gasoline and drive around or use a flashlight battery until it is dead, something has been lost. But what is it? The answer is *energy quality,* the amount of energy available that can perform useful work (Figure 2-13).

Countless experiments have shown that when energy changes from one form to another, a decrease in energy quality or ability to do useful work always occurs. The results of these experiments have been sum-

marized in the **second law of thermodynamics:** *When energy changes from one form to another, some of the useful energy is always degraded to lower-quality, more dispersed, less useful energy.* This degraded energy usually takes the form of heat given off at a low temperature to the surroundings (environment). There it is dispersed by the random motion of air or water molecules and becomes even less useful as a resource.

In other words, *we cannot break even in terms of energy quality because energy always goes from a more useful to a less useful form when it changes from one form to another.* No one has ever found a violation of this fundamental scientific law. It is another one of Mother Nature's basic rules.

Consider three examples of the second law of thermodynamics in action. *First,* when you drive a car, only about 6% of the high-quality chemical energy available in its gasoline fuel actually moves the car, according to physicist and energy expert Amory Lovins. (See his Guest Essay on the website for this chapter.) The remaining 94% is degraded to low-quality heat that is released into the environment and eventually lost into space. Thus, 94% of the money you spend for gasoline is not used to transport you anywhere.

Second, when electrical energy in the form of moving electrons flows through filament wires in an incandescent light bulb, it changes into about 5% useful light and 95% low-quality heat that flows into the environment. In other words, the *light bulb* is really a *heat bulb. Good news.* Scientists have developed compact fluorescent bulbs that are four times more efficient than incandescent bulbs, and even more efficient versions are on the way.

Third, in living systems, solar energy is converted into chemical energy (food molecules) and then into mechanical energy (moving, thinking, and living). During each conversion, high-quality energy is degraded and flows into the environment as low-quality heat. Trace the flows and energy conversions in Figure 2-14 to see how.

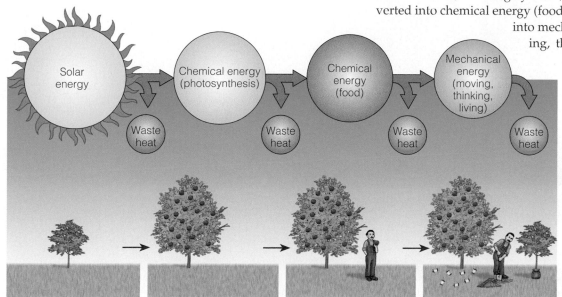

ThomsonNOW™ Active Figure 2-14 The second law of thermodynamics in action in living systems. Each time energy changes from one form to another, some of the initial input of high-quality energy is degraded, usually to low-quality heat that is dispersed into the environment. *See an animation based on this figure and take a short quiz on the concept.*

The second law of thermodynamics also means that *we can never recycle or reuse high-quality energy to perform useful work.* Once the concentrated energy in a serving of food, a liter of gasoline, a lump of coal, or a chunk of uranium is released, it is degraded to low-quality heat that is dispersed into the environment.

Energy efficiency, or **energy productivity,** is a measure of how much useful work is accomplished by a particular input of energy into a system. *Good news.* There is plenty of room for improving energy efficiency. Scientists estimate that only 16% of the energy used in the United States ends up performing useful work. The remaining 84% is either unavoidably wasted because of the second law of thermodynamics (41%) or unnecessarily wasted (43%).

Thermodynamics teaches us an important lesson: The cheapest and quickest way to get more energy is to stop wasting almost half of the energy we use. We can do so by driving gas-efficient motor vehicles and by living in well-insulated houses that have energy-efficient lights, heating and cooling systems, and appliances. Ideally, our houses and other buildings should get much of their energy for heating, cooling, lighting, and running appliances from the sun and from electricity produced by renewable flowing water (hydropower), wind, the earth's internal heat (geothermal energy), and biofuels. This involves using the first principle of sustainability (Figure 1-16, p. 24).

 See examples of how the first and second laws of thermodynamics apply in our world at ThomsonNOW.

SUSTAINABILITY AND MATTER AND ENERGY LAWS

Unsustainable High-Throughput Economies: Working In Straight Lines

Most nations increase their economic growth by converting the world's resources to goods and services in ways that add large amounts of waste, pollution, and low-quality heat to the environment.

As a result of the law of conservation of matter and the second law of thermodynamics, individual resource use automatically adds some waste heat and waste matter to the environment. Most of today's advanced industrialized countries have **high-throughput (high-waste) economies** that attempt to boost economic growth by increasing the one-way flow of matter and energy resources through their economic systems (Figure 2-15). These resources flow through their economies into planetary *sinks* (air, water, soil, organisms), where pollutants and wastes can accumulate to harmful levels.

What happens if more people continue to use and waste more energy and matter resources at an increasing rate? In other words, what happens if most of the world's people become infected with the affluenza virus?

The law of conservation of matter and the two laws of thermodynamics discussed in this chapter tell us that eventually this consumption will exceed the capacity of the environment to provide sufficient renewable resources and dilute and degrade waste matter and absorb waste heat. This could lead to environmen-

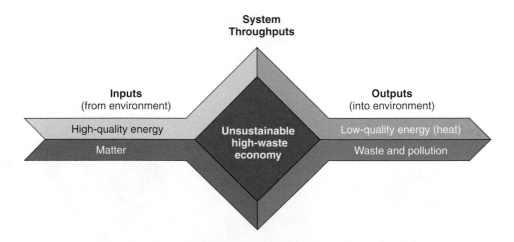

ThomsonNOW **Active Figure 2-15** The *high-throughput economies* of most developed countries rely on continually increasing the rates of energy and matter flow. This practice produces valuable goods and services but also converts high-quality matter and energy resources into waste, pollution, and low-quality heat. *See an animation based on this figure and take a short quiz on the concept.*

tal and economic unsustainability. However, these scientific laws do not tell us how close we are to reaching such limits.

Matter-Recycling-and-Reuse Economies: Working in Circles

Recycling and reusing more of the earth's matter resources slow down our depletion of nonrenewable matter resources and reduce our environmental impact.

A temporary solution to this problem is to convert a linear high-throughput economy into a circular **matter-recycling-and-reuse economy,** which mimics nature by recycling and reusing most of our matter outputs instead of dumping them into the environment. This involves applying another of the four scientific principles of sustainability (Figure 1-16, p. 24).

Although changing to a matter-recycling-and-reuse economy will buy some time, it does not allow ever more people to use ever more resources indefinitely, even if all resources were somehow perfectly recycled and reused. The reason is that the two laws of thermodynamics tell us that recycling and reusing mat-

ter resources always requires using high-quality energy (which cannot be recycled) and adds waste heat to the environment.

Sustainable Low-Throughput Economies: Learning from Nature

We can live more sustainably by reducing the throughput of matter and energy in our economies, not wasting matter and energy resources, recycling and reusing most of the matter resources we use, and stabilizing the size of our population.

The three scientific laws governing matter and energy changes and the four scientific principles of sustainability (Figure 1-16, p. 24) suggest that the best long-term solution to our environmental and resource problems is to shift from an economy based on maximizing matter and energy flow (throughput) to a more sustainable **low-throughput (low-waste) economy,** as summarized in Figure 2-16.

 Compare how energy is used in high- and low-throughput economies at ThomsonNOW.

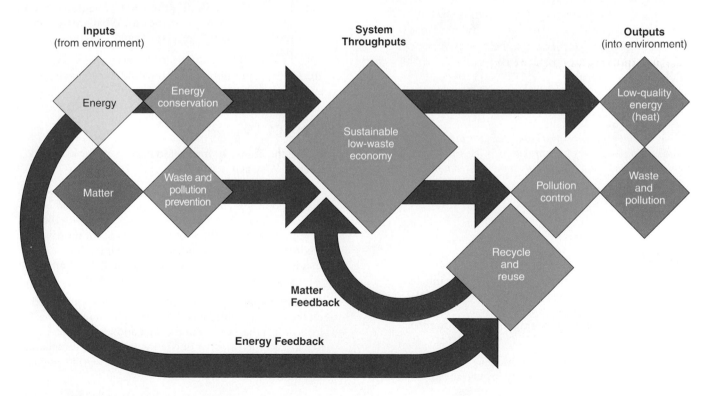

ThomsonNOW‾ Active Figure 2-16 Solutions: lessons from nature. A *low-throughput economy,* based on energy flow and matter recycling, works with nature to reduce the throughput of matter and energy resources (items shown in green). This is done by **(1)** reusing and recycling most nonrenewable matter resources, **(2)** using renewable resources no faster than they are replenished, **(3)** using matter and energy resources efficiently, **(4)** reducing unnecessary consumption, **(5)** emphasizing pollution prevention and waste reduction, and **(6)** controlling population growth. *See an animation based on this figure and take a short quiz on the concept.*

Revisiting Easter Island and Sustainability

Was Easter Island (Core Case Study, p. 28) a high-throughput economy? In fact, it was a low-throughput economy that proved to be unsustainable. This should make us think even harder about how long we can sustain the world's current high-throughput, high-waste societies.

The earth, an island in space, is many times larger than Easter Island. But its resources, like those of smaller islands, are finite and can be degraded and exhausted.

What Earth islanders have that Easter Islanders lacked is the benefits of scientific inquiry and learning. The Easter Islanders were not aware of the positive feedback loop that resulted in a growing population using up the trees that supported their lifestyle.

Perhaps the benefits of science and an understanding and use of the four scientific principles of sustainability (Figure 1-16, p. 24) will help Earth islanders to avoid the same fate.

Since living systems are composed of matter and energy the laws governing changes in matter and energy also apply to them. The next five chapters apply the three basic scientific laws of matter and thermodynamics to living systems and look at some *biological principles* that can also teach us how to live more sustainably by working with nature.

The second law of thermodynamics holds, I think, the supreme position among laws of nature. . . . If your theory is found to be against the second law of thermodynamics, I can give you no hope.

ARTHUR S. EDDINGTON

CRITICAL THINKING

1. Explain how the the use of resources by the inhabitants of Easter Island is related to the concepts of **(a)** exponential growth of population and resource use (Core Case Study, p. 28) and **(b)** the tragedy of the commons (p. 12).

2. Respond to the following statements:
   a. Scientists have not absolutely proven that anyone has ever died from smoking cigarettes.
   b. The greenhouse theory—that certain gases (such as water vapor and carbon dioxide) warm the atmosphere—is not a reliable idea because it is just a scientific theory.

3. Find an advertisement or an article describing some aspect of science in which **(a)** the concept of scientific proof is misused, **(b)** a consensus or sound scientific finding is dismissed or downplayed because it is "only a theory," and **(c)** an example of sound science is labeled as junk science for political purposes.

4. A tree grows and increases its mass. Explain why this phenomenon is not a violation of the law of conservation of matter.

5. If there is no "away," why is the world not filled with waste matter?

6. Suppose you have 100 grams of radioactive plutonium-239 with a half-life of 24,000 years. How many grams of plutonium-239 will remain after **(a)** 12,000 years, **(b)** 24,000 years, and **(c)** 96,000 years?

7. Someone wants you to invest money in an automobile engine that will produce more energy than the energy in the fuel (such as gasoline or electricity) used to run the motor. What is your response? Explain.

8. Use the second law of thermodynamics to explain why a barrel of oil can be used only once as a fuel.

9. **a.** Imagine you have the power to revoke the law of conservation of matter for one day. What are the three most important things you would do with this power?
   b. Imagine you have the power to violate the first law of thermodynamics for one day. What are the three most important things you would do with this power?

10. List two questions that you would like to have answered as a result of reading this chapter.

PROJECTS

1. Use the library or Internet to find an example of junk science. Why is it junk science?

2. **(a)** List two examples of negative feedback loops not discussed in this chapter, one that is beneficial and one that is detrimental. Compare your examples with those of your classmates. **(b)** Give two examples of positive feedback loops not discussed in this chapter. Include one that is beneficial and one that is detrimental. Compare your examples with those of your classmates.

3. Many papers and scientific ideas are now published on the Internet without peer review. This makes it hard to establish the validity of the data and ideas presented. Try to find such a paper or comment on the Internet and use the principles of critical thinking (pp. 2–4) to evaluate its validity.

4. If you have the use of a sensitive chemical balance (check with the chemistry department), try to demonstrate the law of conservation of mass in a physical change. Weigh a container with a lid (a glass jar will do), add an ice cube and weigh it again, and then allow the ice to melt and weigh it again. Explain how your results obey the law of conservation of matter.

5. Use the library or Internet to find examples of various perpetual motion machines and inventions that allegedly violate the two laws of thermodynamics by producing more high-quality energy than the high-quality energy needed to make them run. What has happened to these schemes and machines?

6. Make a concept map of this chapter's major ideas using the section heads, subheads, and key terms (in bold-face). Look on the website for this book for information about making concept maps.

LEARNING ONLINE

The website for this book contains study aids and many ideas for further reading and research. They include a chapter summary, review questions for the entire chapter, flash cards for key terms and concepts, a multiple-choice practice quiz, interesting Internet sites, references, information about green careers, and a guide for accessing thousands of InfoTrac® College Edition articles. Log into

www.thomsonedu.com/biology/miller

Then choose Chapter 2, and select a learning resource. For access to animations, additional quizzes, chapter outlines and summaries, register and log into

at **www.thomsonedu.com** using the access code card in the front of your book.

CORE CASE STUDY

Have You Thanked the Insects Today?

Insects have a bad reputation. We classify many as *pests* because they compete with us for food, spread human diseases such as malaria, and invade our lawns, gardens, and houses. Some people have "bugitis": they fear all insects and think the only good bug is a dead bug. They fail to recognize the vital roles insects play in helping sustain life on earth.

Many of the earth's plant species depend on insects to pollinate their flowers (Figure 3-1, left). Without the natural service of pollination, plants cannot reproduce sexually, and no plant species would be around for long. Without pollinating insects, we would have very few fruits and vegetables to enjoy.

Insects that eat other insects—such as the praying mantis (Figure 3-1, right)—help control the populations of at least half the species of insects we call pests. This free pest control service is an important part of the earth's natural capital that helps sustain us.

Insects have been around for at least 400 million years and are phenomenally successful forms of life. Some insects can reproduce at an astounding rate. For example, a single housefly and her offspring can theoretically produce about 5.6 trillion flies in only one year.

Insects can rapidly develop new genetic traits, such as resistance to pesticides. They also have an exceptional ability to evolve into new species when faced with new environmental conditions, and they are very resistant to extinction. This is fortunate because according to ant specialist and biodiversity expert E. O. Wilson, if all insects disappeared, humanity probably could not last more than a few months.

The environmental lesson: although insects can thrive without newcomers such as us, we and most other land organisms would perish without them.

Learning about insects' roles in nature requires us to understand how insects and other organisms living in a *biological community*, such as a forest or pond, interact with one another and with the nonliving environment. *Ecology* is the science that studies such relationships and interactions in nature, as discussed in this and the following six chapters.

Figure 3-1 Natural capital: the monarch butterfly, feeding on pollen in a flower (left), and other insects pollinate flowering plants that serve as food for many plant eaters. The praying mantis, eating a house cricket (right), and many other insect species help control the populations of at least half of the insect species we classify as pests.

The world is truly a complex system, and we are part of it, still dependent on its renewable productivity, which we ourselves are beginning to stifle.

NILES ELDRIDGE

This chapter describes the major components of ecosystems and the processes that sustain them. It discusses these questions:

- What is ecology?
- What basic processes keep us and other organisms alive?
- What are the major components of an ecosystem?

- What happens to energy in an ecosystem?
- What are soils and how are they formed?
- What happens to matter in an ecosystem?
- How do scientists study ecosystems?

THE NATURE OF ECOLOGY

What Is Ecology?

Ecology is a study of connections in nature.

Ecology (from the Greek words *oikos*, "house" or "place to live," and *logos*, "study of") is the study of how organisms interact with one another and with their nonliving environment. In effect, it is a study of *connections in nature*—the house for the earth's life.

To enhance their understanding of nature, scientists classify matter into levels of organization from atoms to cells to the biosphere. Ecologists focus on trying to understand the interactions among organisms, populations, communities, ecosystems, and the biosphere (Figure 3-2).

Organisms and Species

Organisms, the different forms of life on earth, can be classified into different species based on certain characteristics.

An **organism** is any form of life. It is the most fundamental unit of ecology. The cell is the basic unit of life in organisms (Figure 2-6, p. 37). Some organisms such as bacteria consist of a single cell but most consist of many cells.

Organisms can be classified into **species,** groups of organisms that resemble one another in appearance, behavior, chemistry, and genetic makeup. Scientists use a specific system to classify and name each species, as discussed in Supplement 8 on p. S34.

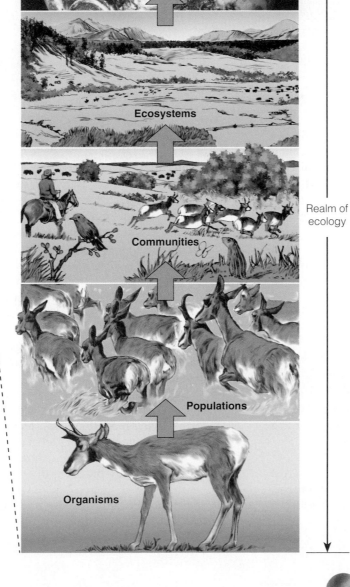

Realm of ecology

ThomsonNOW Active Figure 3-2 Natural capital: levels of organization of matter in nature. Ecology focuses on five of these levels. *See an animation based on this figure and take a short quiz on the concept.*

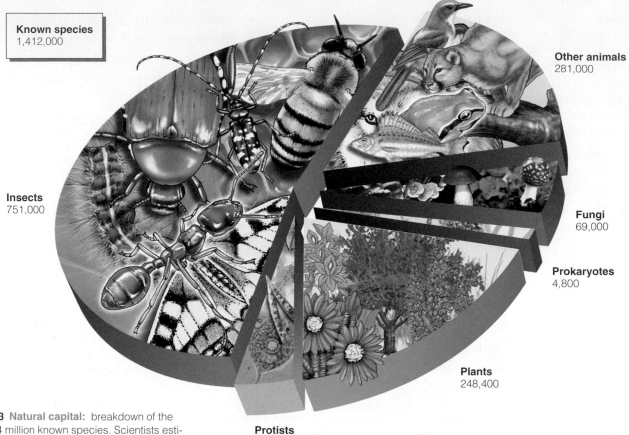

Other animals
281,000

Insects
751,000

Fungi
69,000

Prokaryotes
4,800

Plants
248,400

Protists
57,700

Figure 3-3 **Natural capital:** breakdown of the earth's 1.4 million known species. Scientists estimate that there are 4 million to 100 million species.

How many species are on the earth? We do not know. Estimates range from 4 million to 100 million species—most of them microorganisms too small to be seen with the naked eye. A best guess is that we share the planet with 10–15 million other species. So far biologists have identified and named about 1.4 million species, most of them insects (Figure 3-3 and Core Case Study, p. 50). If you went into a tropical forest with a net, within a few hours you could probably catch an unidentified insect species that could be named after you.

Case Study: Which Species Run the World?

Multitudes of tiny microbes such as bacteria, protozoa, fungi, and yeast help keep us alive.

They are everywhere and there are trillions of them. Trillions are found inside your body, on your body, in a handful of soil, and in a cup of ocean water.

These mostly invisible rulers of the earth and our bodies are *microbes* (or *microorganisms*), catchall terms for many thousands of species of bacteria, protozoa, fungi, and yeasts—most too small to be seen with the naked eye.

Microbes do not get the respect they deserve. Most of us think of them as threats to our health in the form of infectious bacteria or "germs," fungi that cause athlete's foot and other skin diseases, and protozoa that cause diseases such as malaria. But these harmful microbes are in the minority.

You are alive because of multitudes of microbes toiling away mostly out of sight. Soil bacteria convert nitrogen gas in the atmosphere into forms that plants can take up from the soil as nutrients. They also help produce foods such as bread, cheese, yogurt, vinegar, tofu, soy sauce, beer, and wine. Bacteria and fungi in the soil decompose organic wastes into nutrients that can be taken up by plants that we and most other animals eat. Without these wee creatures, we would be up to our eyeballs in waste matter.

Microbes, particularly those in the ocean, dominate the earth's biodiversity, account for 90% of the earth's living mass (biomass), provide the planet with oxygen, and help combat global warming.

Microbes, especially bacteria, help purify the water you drink by breaking down wastes. Bacteria in your intestinal tract break down the food you eat. Some microbes in your nose prevent harmful bacteria from reaching your lungs.

Other bacteria are the sources of disease-fighting antibiotics, including penicillin, erythromycin, and streptomycin. Scientists are working on using microbes to develop new medicines and fuels, and genetic engineers are developing microbes that can extract metals from ores, and help clean up polluted water and soils.

Some microbes help control diseases that affect plants and populations of insects that attack our food crops. Relying more on these microbes for natural pest control can reduce the use of potentially harmful chemical pesticides.

We spend much more money on learning about the moon and Mars than on understanding the microbes that sustain us and other forms of life.

Populations, Communities, and Ecosystems

Members of a species interact in groups called populations; populations of different species living and interacting in an area form a community; and a community interacting with its physical environment of matter and energy is an ecosystem.

A **population** is a group of interacting individuals of the same species occupying a specific area (Figure 3-4).

Figure 3-4 Natural capital: population of monarch butterflies. The geographic distribution of this butterfly coincides with that of the milkweed plant, on which monarch larvae and caterpillars feed.

Figure 3-5 Natural capital: the *genetic diversity* among individuals of one species of Caribbean snail is reflected in the variations in shell color and banding patterns.

Examples include sunfish in a pond, white oak trees in a forest, and people in a country. In most natural populations, individuals vary slightly in their genetic makeup, which is why they do not all look or act alike. This is called a population's **genetic diversity** (Figure 3-5).

The place where a population (or an individual organism) normally lives is its **habitat.** It may be as large as an ocean or as small as the intestine of a termite.

The area over which we can find a species is called its **distribution** or **range.** Many species, such as some tropical plants, have a small range and may be found on only a single hillside. Other species such the grizzly bear have large ranges.

A **community,** or **biological community,** consists of all the populations of different species that live and interact in a particular area.

An **ecosystem** is a community where populations of different species interact with one another and with their nonliving environment of matter and energy. Ecosystems can range in size from a puddle of water to a stream, or from a patch of woods to an entire forest. Ecosystems can be natural or artificial (human created). Examples of artificial ecosystems include crop fields, tree farms (see photo 1 in the Detailed Contents), farm ponds, and reservoirs.

All of the earth's ecosystems together make up the **biosphere,** the global ecosystem where all life is interconnected. The key ecological lesson from studying the biosphere: *everything is linked to everything else.*

 Learn more about how the earth's life is organized on five levels in the study of ecology at ThomsonNOW.

THE EARTH'S LIFE-SUPPORT SYSTEMS

The Earth's Life-Support Systems: Four Spheres

The earth is made up of interconnected spherical layers that contain air, water, soil, minerals, and life.

We can think of the earth's life-support system, or *biosphere*, as consisting of several spherical layers (Figure 3-6). The **atmosphere** is a thin envelope or membrane of air around the planet. Its inner layer, the **troposphere,** extends only about 17 kilometers (11 miles) above sea level. It contains the majority of the planet's air, mostly nitrogen (78%) and oxygen (21%) by volume.

The next layer, stretching 17–48 kilometers (11–30 miles) above the earth's surface, is the **stratosphere.** Its lower portion contains enough ozone (O_3) to filter out most of the sun's harmful ultraviolet radiation. This allows life to exist on land and in the surface layers of bodies of water.

The **hydrosphere** consists of the earth's water. It is found as *liquid water* (on the planet's surface and underground), *ice* (polar ice, icebergs, and ice in frozen soil layers called *permafrost*), and *water vapor* in the atmosphere.

The earth consists of an intensely hot *core,* a thick *mantle* composed mostly of rock, and a thin outer *crust.* The **lithosphere** is the earth's crust and upper mantle.

It contains nonrenewable fossil fuels and minerals we use as well as renewable soil chemicals (nutrients) needed for plant life.

If the earth were an apple, the biosphere would be no thicker than the apple's skin. *The goal of ecology is to understand the interactions in this thin, life-supporting global skin of air, water, soil, and organisms.*

What Sustains Life on Earth?

Solar energy, the cycling of matter, and gravity sustain the earth's life.

Life on the earth depends on three interconnected factors, shown in Figure 3-7:

- The *flow of high-quality energy* from the sun through materials and living things in their feeding interactions, into the environment as low-quality energy (mostly heat dispersed into air or water molecules at a low temperature), and eventually back into space as heat. No round-trips are allowed because energy cannot be recycled.

- The *cycling of matter or nutrients* (the atoms, ions, or compounds needed for survival by living organisms) through parts of the biosphere. Because the earth is closed to significant inputs of matter from space, its essentially fixed supply of nutrients must be continually recycled to support life. Nutrient trips in ecosystems are round-trips.

- *Gravity,* which allows the planet to hold on to its atmosphere and helps enable the movement of chemicals between the air, water, soil, and organisms in the matter cycles.

What Happens to Solar Energy Reaching the Earth?

Solar energy flowing through the biosphere warms the atmosphere, evaporates and recycles water, generates winds, and supports plant growth.

Energy from the sun, a gigantic nuclear fusion reactor, supports most life on the earth by lighting and warming the planet. It also supports *photosynthesis*, the process in which green plants, algae, and some bacteria absorb light and use it to make compounds such as carbohydrates that keep them alive and feed most other organisms. The sun also powers the cycling of matter and drives the climate and weather systems that distribute heat and freshwater over the earth's surface.

About one-billionth of the sun's output of energy reaches the earth—a tiny sphere in the vastness of space—in the form of electro-

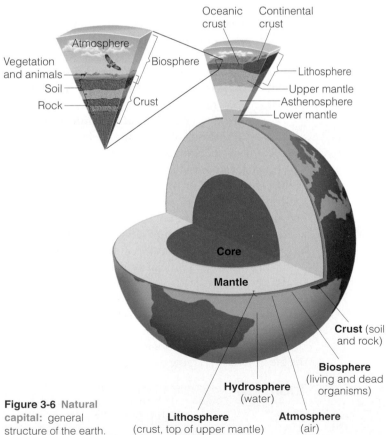

Figure 3-6 Natural capital: general structure of the earth.

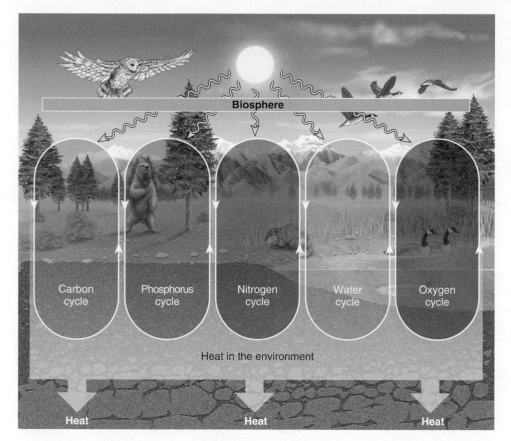

Active Figure 3-7
Natural capital: life on the earth depends on the *flow of energy* (wavy arrows) from the sun through the biosphere and back into space, the *cycling of crucial elements* (solid arrows around ovals), and *gravity*, which keeps atmospheric gases from escaping into space and helps recycle nutrients through air, water, soil, and organisms. This simplified model depicts only a few of the many cycling elements. *See an animation based on this figure and take a short quiz on the concept.*

magnetic waves, mostly as visible light (Figure 2-12, p. 43). Much of this energy is either reflected away or absorbed by chemicals, dust, and clouds in the planet's atmosphere (Figure 3-8). The amount of energy reaching the earth from the sun equals the amount of heat energy the earth reflects or radiates back into space. Otherwise, the earth would be too hot for life as we know it.

About 80% of the energy that gets through warms the troposphere and evaporates and cycles water through the biosphere. Approximately 1% of this incoming energy generates winds, and green plants, algae, and bacteria use less than 0.1% to produce their food through photosynthesis.

Most solar radiation making it through the atmosphere is degraded into longer-wavelength infrared radiation. This infrared radiation encounters the so-called *greenhouse gases* (such as water vapor, carbon dioxide, methane, nitrous oxide, and ozone) in the troposphere. The radiation causes these gaseous molecules to vibrate and release infrared radiation with even longer wavelengths into the troposphere. As this radiation interacts with molecules in the air, it increases their kinetic energy, helping warm the troposphere and the earth's surface.

Active Figure 3-8 Solar capital: flow of energy to and from the earth. *See an animation based on this figure and take a short quiz on the concept.*

Without this **natural greenhouse effect,** the earth would be too cold for survival of the forms of life we find on the earth today.

 Learn more about the flow of energy— from sun to earth and within the earth's systems—at ThomsonNOW.

ECOSYSTEM COMPONENTS

Biomes and Aquatic Life Zones: Where Organisms Live

Life exists on land systems called biomes and in freshwater and ocean aquatic life zones.

Viewed from outer space, the earth resembles an enormous jigsaw puzzle consisting of large masses of land and vast expanses of ocean (Figure 1-1, p. 6).

Biologists have classified the terrestrial (land) portion of the biosphere into **biomes** ("BY-ohms"). They are large regions such as forests, deserts, and grasslands with distinct climates and specific species (espe-

cially vegetation) adapted to them. (See Figure 1 on pp. S8–S9 in Supplement 4). Figure 3-9 shows different major biomes along the 39th parallel spanning the United States.

Scientists divide the watery parts of the biosphere into **aquatic life zones,** each containing numerous ecosystems. Examples include *freshwater life zones* (such as lakes and streams) and *ocean* or *marine life zones* (such as coral reefs, coastal estuaries, and the deep ocean).

Nonliving and Living Components of Ecosystems

Ecosystems consist of nonliving (abiotic) and living (biotic) components.

Two types of components make up the biosphere and its ecosystems: One type, called **abiotic,** consists of nonliving components such as water, air, nutrients, and solar energy. The other type, called **biotic,** consists of biological components such as *producers* (mostly plants and floating algae phytoplankton that produce their own food), *consumers* (animals that get their food

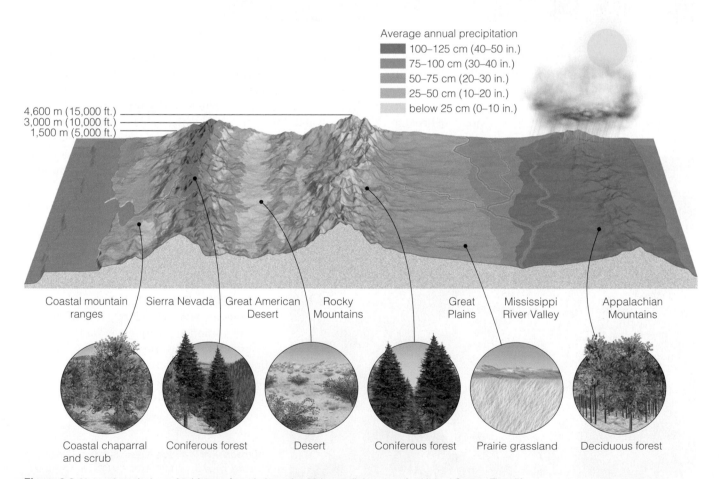

Figure 3-9 Natural capital: major biomes found along the 39th parallel across the United States. The differences reflect changes in climate, mainly differences in average annual precipitation and temperature.

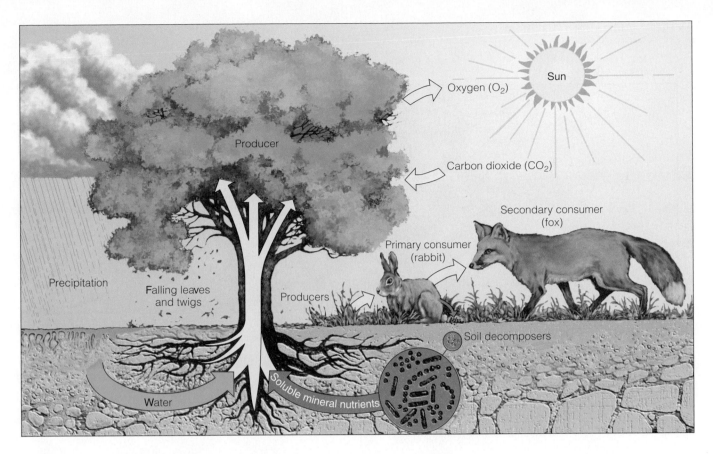

ThomsonNOW **Active Figure 3-10** **Natural capital:** major components of an ecosystem in a field. *See an animation based on this figure and take a short quiz on the concept.*

by eating plants or other animals), and *decomposers* (mostly bacteria that break down the dead remains of plants and animals and recycle them into the soil or water for reuse by producers). Figure 3-10 is a greatly simplified diagram of some of the biotic and abiotic components in a terrestrial ecosystem.

Different species and their populations thrive under different physical and chemical conditions. Some need bright sunlight; others flourish in shade. Some need a hot environment; others prefer a cool or cold one. Some do best under wet conditions; others thrive under dry conditions.

Each population in an ecosystem has a **range of tolerance** to variations in its physical and chemical environment, as shown in Figure 3-11 (p. 58). Individuals within a population may also have slightly different tolerance ranges for temperature or other factors because of small differences in genetic makeup, health, and age. For example, a trout population may do best within a narrow band of temperatures (*optimum level or range*), but a few individuals can survive above and below that band. Of course, if the water becomes much too hot or too cold, none of the trout can survive.

A species may have a wide range of tolerance to some factors and a narrow range of tolerance to others. Most organisms are least tolerant during juvenile or reproductive stages of their life cycles. Highly tolerant species can live in a variety of habitats with widely different conditions. Figure 3-12 (p. 58) shows how environmental physical conditions can limit the distribution of a particular species.

Factors That Limit Population Growth

Availability of matter and energy resources can limit the number of organisms in a population.

A variety of factors can affect the number of organisms in a population. Sometimes one factor, known as a **limiting factor,** is more important in regulating population growth than other factors. This ecological principle is called the **limiting factor principle:** *Too much or too little of any abiotic factor can limit or prevent growth of a population, even if all other factors are at or near the optimum range of tolerance.* This principle describes the major way in which population control—a scientific principle of sustainability (Figure 1-16, p. 24)—is implemented.

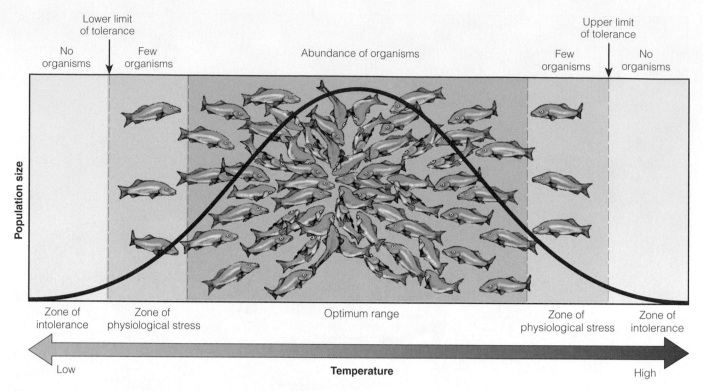

Lower limit of tolerance

Few organisms

Abundance of organisms

Upper limit of tolerance

Few organisms

No organisms

Population size

Zone of intolerance

Zone of physiological stress

Optimum range

Zone of physiological stress

Zone of intolerance

Low

Temperature

High

Figure 3-11 Natural capital: range of tolerance for a population of organisms, such as fish, to an abiotic environmental factor—in this case, temperature. These restrictions keep particular species from taking over an ecosystem by keeping their population size in check.

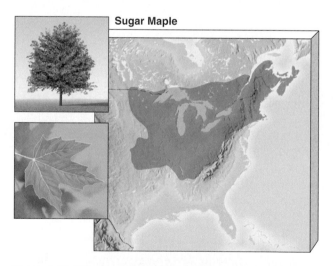

Sugar Maple

Figure 3-12 The physical conditions of the environment can limit the distribution of a species. The green area shows the current range of sugar maple trees in eastern North America. (Data from U.S. Department of Agriculture)

On land, precipitation often is the limiting factor. Lack of water in a desert limits plant growth. Soil nutrients also can act as a limiting factor on land. Suppose a farmer plants corn in phosphorus-poor soil. Even if water, nitrogen, potassium, and other nutrients are at optimum levels, the corn will stop growing when it uses up the available phosphorus.

Too much of an abiotic factor can also be limiting. For example, too much water or fertilizer can kill plants—both common mistakes made by many beginning gardeners.

Important limiting factors for aquatic life zones include temperature, sunlight, and nutrient availability, and the low solubility of oxygen gas in water (*dissolved oxygen content*). Another limiting factor in aquatic life zones is *salinity*—the amounts of various inorganic minerals or salts dissolved in a given volume of water.

Producers: Basic Source of All Food

Some organisms in ecosystems can produce the food (carbohydrates) they need from chemicals in their environment. Most need sunlight to produce but some do not.

The earth's organisms either produce or consume food. **Producers,** sometimes called **autotrophs** (self-feeders), make their own food from compounds and energy obtained from their environment.

On land, most producers are green plants. In freshwater and marine ecosystems, algae and plants are the major producers near shorelines. In open water, the dominant producers are *phytoplankton*—mostly microscopic organisms that float or drift in the water.

Most producers capture sunlight to produce carbohydrates (such as glucose, $C_6H_{12}O_6$) by **photosynthesis.**

Although hundreds of chemical changes take place during photosynthesis, the overall chemical reaction can be summarized as follows:

carbon dioxide + water + **solar energy** $\longrightarrow$ glucose + oxygen

$$6\,CO_2 + 6\,H_2O + \text{solar energy} \longrightarrow C_6H_{12}O_6 + 6\,O_2$$

See pp. S31–S32 in Supplement 7 for information on how to balance chemical equations such as this one. The Spotlight below gives more details on photosynthesis.

A few producers, mostly specialized bacteria, can convert simple inorganic compounds from their

Photosynthesis: A Closer Look

In photosynthesis, sunlight powers a complex series of chemical reactions that combine water taken up by plant roots and carbon dioxide from the air to produce sugars such as glucose. This process converts low-quality solar energy into high-quality chemical energy in sugars for use by plant cells. Figure 3-A is a greatly simplified summary of the photosynthesis process.

Photosynthesis takes place within tiny organelles called *chloroplasts* found within plant cells. Chlorophyll, a special compound in chloroplasts, absorbs incoming visible light mostly in the violet and red wavelengths. The green light that is not absorbed is reflected back, which is why photosynthetic plants look green. The absorbed wavelengths of solar energy initiate a sequence of chemical reactions with other molecules in what are called *light-dependent reactions*.

This series of reactions splits water into hydrogen ions (H^+) and oxygen (O_2) which is released into the atmosphere. It also produces small ADP molecules that absorb the energy released and store it as chemical energy in ATP molecules (see Figure 12 on p. S32 in Supplement 7). The chemical energy released by the ATP molecules drives a series of *light-independent reactions* that can take place in the darkness of plant cells. In this second sequence of reactions, carbon atoms stripped from carbon

dioxide combine with hydrogen and oxygen to produce sugars such as glucose ($C_6H_{12}O_6$) that plant cells can use as a source of energy.

Critical Thinking

What main types of life would exist on the earth today if organisms that produce biomass by photosynthesis had not developed?

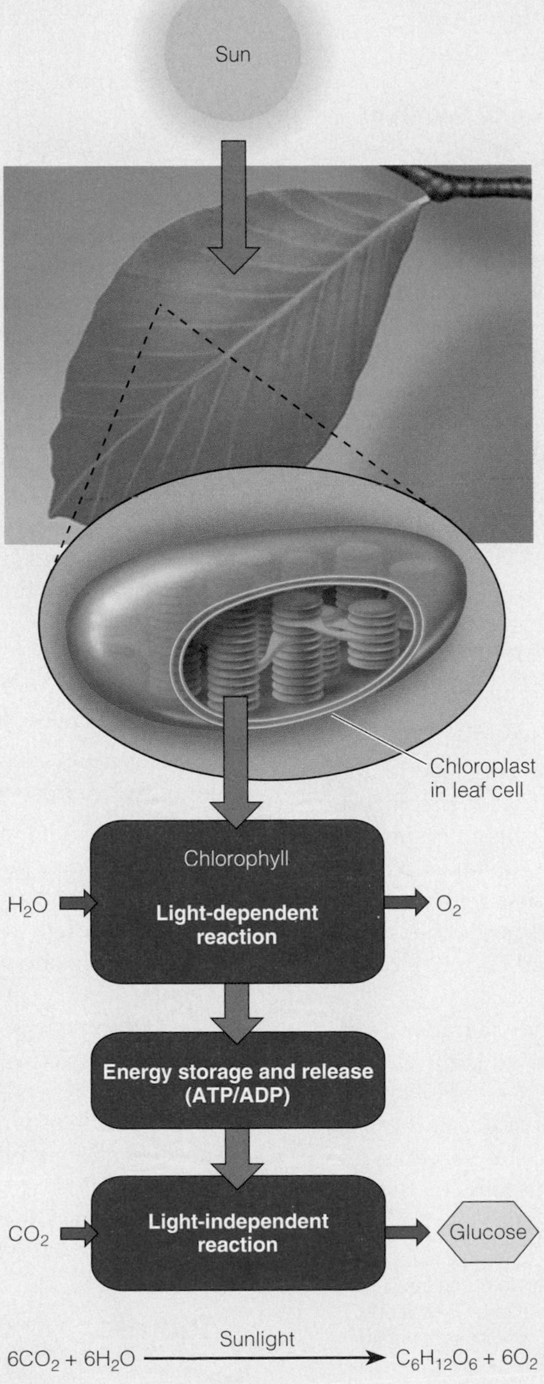

Sun

Chloroplast in leaf cell

Chlorophyll

H_2O → **Light-dependent reaction** → O_2

Energy storage and release (ATP/ADP)

CO_2 → **Light-independent reaction** → Glucose

$$6CO_2 + 6H_2O \xrightarrow{\text{Sunlight}} C_6H_{12}O_6 + 6O_2$$

Figure 3-A Simplified overview of *photosynthesis.* In this process, chlorophyll molecules in the chloroplasts of plant cells absorb solar energy. This initiates a complex series of chemical reactions in which carbon dioxide and water are converted to sugars, such as glucose, and oxygen.

environment into more complex nutrient compounds without sunlight, through a process called **chemosynthesis.** In 1977, scientists discovered a community of bacteria living in the extremely hot water around *hydrothermal vents,* which survived by chemosynthesis. These bacteria serve as producers for these ecosystems without the use of sunlight. They draw energy and produce carbohydrates from hydrogen sulfide (H_2S) gas escaping through fissures in the ocean floor. Most of the earth's organisms get their energy indirectly from the sun. But chemosynthetic organisms in these dark and deep sea habitats survive indirectly on *geothermal energy* in the earth's interior and represent an exception to the first scientific principle of sustainability.

Consumers: Eating and Recycling to Survive

Consumers get their food by eating or breaking down all or parts of other organisms or their remains.

All other organisms in an ecosystem are **consumers,** or **heterotrophs** ("other-feeders") that get the energy and nutrients they need by feeding on other organisms or their remains. **Primary consumers** or **herbivores,** such as rabbits and zooplankton, eat producers. **Secondary consumers** or **carnivores,** such as foxes and fish, feed on herbivores. **Third and higher level consumers** are carnivores that feed on other carnivores. These relationships are shown in Figure 3-10 (p. 57).

Omnivores play dual roles by feeding on both plants and animals. Examples are pigs, rats, foxes, bears, cockroaches, and humans.

? *THINKING ABOUT WHAT YOU EAT* When you had lunch today were you an herbivore, a carnivore, or an omnivore?

Decomposers (mostly certain types of bacteria and fungi) are specialized organisms that recycle nutrients in ecosystems. They secrete enzymes that digest or biodegrade living or dead organisms into simpler inorganic compounds that producers can take up from the soil and water and use as nutrients. Other consumers, called **detritivores,** are insects and other scavengers that feed on the wastes or dead bodies of other organisms.

Hordes of these scavengers and degraders can transform a fallen tree trunk into a powder and finally into simple inorganic molecules that plants can absorb as nutrients (Figure 3-13). *In natural ecosystems, there is little or no waste.* One organism's wastes serve as resources for other organisms, as the nutrients that make life possible are recycled again and again.

 ? *THINKING ABOUT SCAVENGER INSECTS* Note that the scavenger organisms in Figure 3-13 are insects (Core Case Study, p. 50). How would your life be changed if these scavenger insects disappeared? Why do timber companies want to eliminate many of these scavenger insects?

Aerobic and Anaerobic Respiration: Getting Energy for Survival

Organisms break down carbohydrates and other organic compounds in their cells to obtain the energy they need.

Producers, consumers, and decomposers use the chemical energy stored in glucose and other organic compounds to fuel their life processes. In most cells this energy is released by **aerobic respiration,** which uses oxygen to convert organic nutrients back into carbon dioxide and water. The net effect of the hundreds of steps in this complex process is represented by the following chemical reaction:

$$\text{glucose} + \text{oxygen} \longrightarrow \text{carbon dioxide} + \text{water} + \text{energy}$$

$$C_6H_{12}O_6 + 6\,O_2 \longrightarrow 6\,CO_2 + 6\,H_2O + \text{energy}$$

Although the detailed steps differ, the net chemical change for aerobic respiration is the opposite of that for photosynthesis.

Some decomposers get the energy they need by breaking down glucose (or other organic compounds) in the absence of oxygen. This form of cellular respiration is called **anaerobic respiration,** or **fermentation.** Instead of carbon dioxide and water, the end products of this process are compounds such as methane gas (CH_4, the main component of natural gas), ethyl alcohol (C_2H_6O), acetic acid ($C_2H_4O_2$, the key component of vinegar), and hydrogen sulfide (H_2S, when sulfur compounds are broken down).

Two Secrets of Survival: Energy Flow and Matter Recycling

An ecosystem survives by a combination of energy flow and matter recycling.

The survival of any individual organism depends on the *one-way flow of matter and energy* through its body. However, an ecosystem as a whole survives primarily through a combination of *matter recycling* (rather than one-way flow) and *one-way energy flow* (Figure 3-14).

Decomposers complete the cycle of matter by breaking down organic matter into inorganic nutrients that can be reused by producers. These nutrient recyclers provide us with this crucial ecological service and never send us a bill. Without decomposers, the entire world would be knee-deep in plant litter, dead animal bodies, animal wastes, and garbage, and most life as we know it would no longer exist.

 Explore the components of ecosystems, how they interact, the roles of bugs and plants, and what a fox will eat at ThomsonNOW.

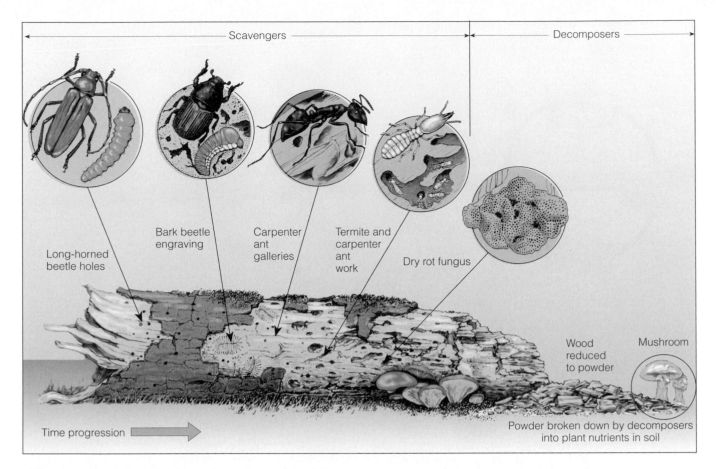

Scavengers | Decomposers

Long-horned beetle holes

Bark beetle engraving

Carpenter ant galleries

Termite and carpenter ant work

Dry rot fungus

Wood reduced to powder

Mushroom

Time progression

Powder broken down by decomposers into plant nutrients in soil

Figure 3-13 Natural capital: various scavengers (detritivores) and decomposers (mostly fungi and bacteria) can "feed on" or digest parts of a log and eventually convert its complex organic chemicals into simpler inorganic nutrients that can be taken up by producers.

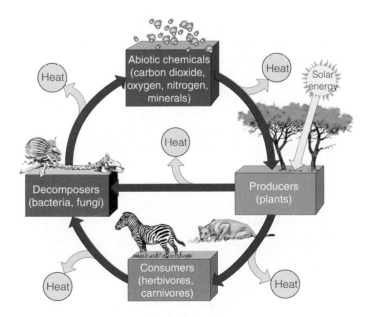

ThomsonNOW™ Active Figure 3-14 Natural capital: the main structural components of an ecosystem (energy, chemicals, and organisms). Matter recycling and the flow of energy—first from the sun, then through organisms, and finally into the environment as low-quality heat—links these components. *See an animation based on this figure and take a short quiz on the concept.*

BIODIVERSITY

The Diversity of Life: A Crucial Resource

A vital renewable resource is the biodiversity found in the earth's variety of genes, species, ecosystems, and ecosystem processes.

Biological diversity, or **biodiversity,** is one of the earth's most important renewable resources. It includes four components, as shown in Figure 3-15 (p. 62). Photos 5, 6, 7, and 8 in the Detailed Contents show species found in tropical forests that are part of the earth's species diversity.

? *THINKING ABOUT BIODIVERSITY* Get together a group of people including a biologist and identify the biodiversity in your backyard or in an area of your school.

Biodiversity Loss and Species Extinction: Remember HIPPO

Human activities are destroying and degrading the habitats for many wild species and driving some of them to premature extinction.

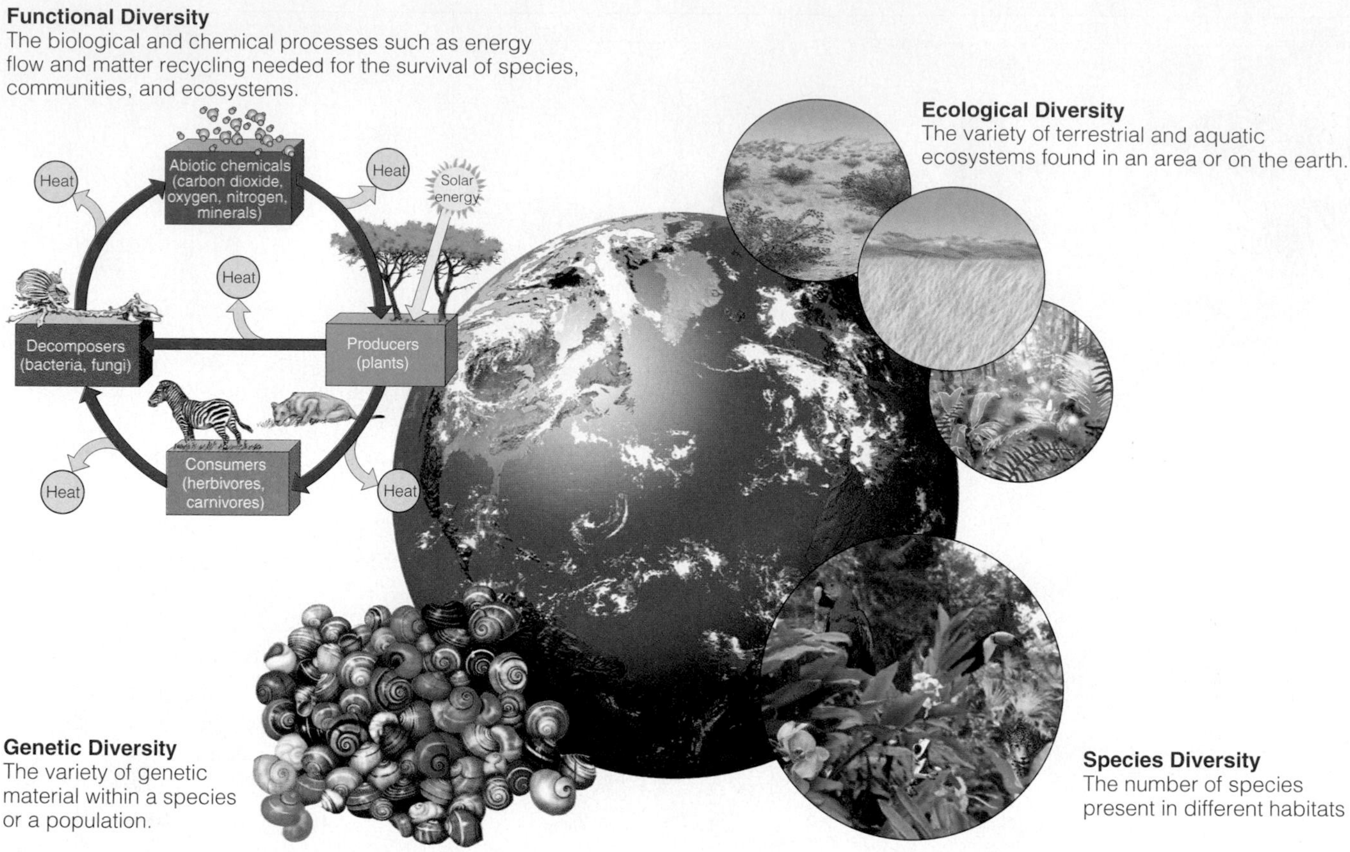

Functional Diversity
The biological and chemical processes such as energy flow and matter recycling needed for the survival of species, communities, and ecosystems.

Ecological Diversity
The variety of terrestrial and aquatic ecosystems found in an area or on the earth.

Genetic Diversity
The variety of genetic material within a species or a population.

Species Diversity
The number of species present in different habitats

Figure 3-15 Natural capital: the major components of the earth's *biodiversity*—one of the earth's most important renewable resources. Some people also include *human cultural diversity* as part of the earth's biodiversity. Each human culture has developed various ways to deal with changing environmental conditions.

Biodiversity at all levels is being eliminated or degraded by human activities. Sooner or later all species become extinct because they cannot respond successfully to changing environmental conditions. But studies such as the 2005 Millennium Ecosystem Assessment indicate that current extinction rates are 100 to 10,000 times higher than the world's natural rate of extinction, because of human activities.

Scientists use the acronym **HIPPO** to help us remember five major causes of species decline and premature extinction:

- **H** for *habitat destruction and degradation*—the leading cause.

- **I** for *invasive species* that we deliberately or accidentally introduce into ecosystems—the second most important cause.

- **P** for *pollution,* including human induced changes in global and regional climates.

- **P** for *human population growth* and the accompanying resource consumption that are crowding out wild species and degrading the places where they live.

- **O** for *overexploitation.* This includes *overhunting* of species with valuable parts, such as the ivory tusks of elephants and the skins of tigers and *overconsumption* of resources that wild species need for their survival.

You will learn more about these threats to many of the earth's ecosystems and species in Chapters 10, 11, and 12.

Why Should We Care about Biodiversity?

Biodiversity provides us with natural resources (such as food, wood, energy, and medicines), and natural services (such as air and water purification, soil fertility, waste disposal, and pest control), and it gives us great pleasure.

The earth's biodiversity is the biological wealth or capital that helps keep us alive and supports our economies (Figure 1-4, p. 9). It supplies us with food, wood, fibers, energy, raw materials, industrial chemicals, and medicines—all of which pour hundreds of billions of dollars into the world economy each year. It also helps preserve the quality of the air and water, maintain the fertility of soils, dispose of wastes, and control populations of pests that attack crops and forests.

 THINKING ABOUT INSECTS AND BIODIVERSITY Summarize the importance of insects (Core Case Study, p. 50) in the earth's biodiversity.

Figure 3-16 outlines the goals, strategies, and tactics involved in working to sustain the earth's biodiversity. These efforts are based on Aldo Leopold's (Individuals Matter, p. 23) ethical principle that something is right when it tends to maintain or sustain the earth's life support systems for us and other species and wrong when it does not.

RESEARCH FRONTIER Studying the successes and failures of the tactics listed in Figure 3-16 and using this information to improve these tactics

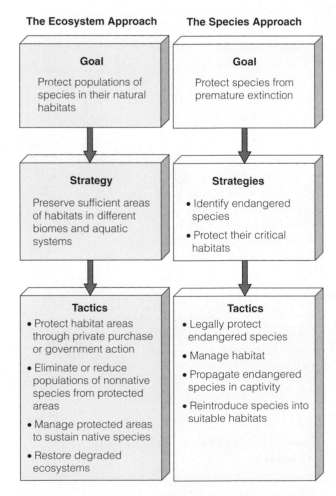

Figure 3-16 Solutions: goals, strategies, and tactics for protecting biodiversity.

ENERGY FLOW IN ECOSYSTEMS

Food Chains and Food Webs: Who Eats and Decomposes Whom

Food chains and webs show how eaters, the eaten, and the decomposed are connected to one another in an ecosystem.

All organisms, whether dead or alive, are potential sources of food for other organisms. A caterpillar eats a leaf, a robin eats the caterpillar, and a hawk eats the robin. Decomposers consume the leaf, caterpillar, robin, and hawk after they die. As a result, *there is little matter wasted in natural ecosystems.*

A sequence of organisms, each of which is a source of food for the next, is called a **food chain.** It determines how energy and nutrients move from one organism to another through an ecosystem, as shown in Figure 3-17, p. 64).

Ecologists assign each organism in an ecosystem to a *feeding level,* or **trophic level** (from the Greek word *trophos,* meaning "nourishment"), depending on whether it is a producer or a consumer and on what it eats or decomposes. Producers belong to the first trophic level, primary consumers to the second trophic level, secondary consumers to the third, and so on. Detritivores and decomposers process detritus from all trophic levels.

In real ecosystems most consumers feed on more than one type of organism, and most organisms are eaten by more than one type of consumer. Because most species participate in several different food chains, the organisms in most ecosystems form a complex network of interconnected food chains called a **food web,** as shown in Figure 3-18 (p. 65). Trophic levels can be assigned in food webs just as in food chains. A food web shows how eaters, the eaten, and the decomposed are connected to one another. This summary of feeding relations, communities, and ecosystems is a map of life's interdependence.

Energy Flow in an Ecosystem: Losing Energy in Food Chains and Webs

There is a decrease in the amount of energy available to each succeeding organism in a food chain or web.

Each trophic level in a food chain or web contains a certain amount of **biomass,** the dry weight of all organic matter contained in its organisms. In a food chain or web, the chemical energy stored in biomass is transferred from one trophic level to another.

Energy transfer through food chains and food webs is not very efficient because with each transfer some usable energy is degraded and lost to the environment as low-quality heat, in accordance with the

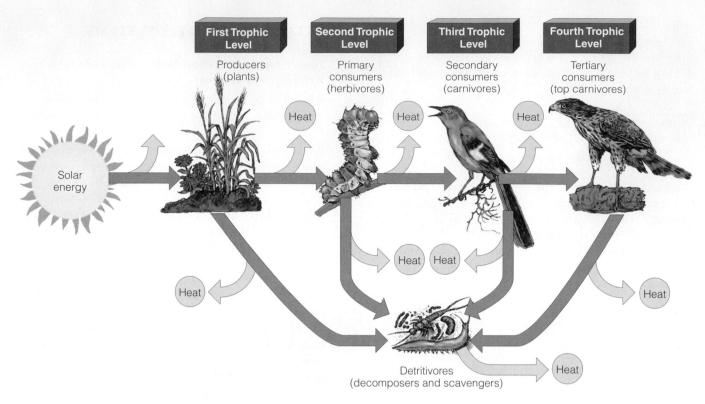

First Trophic Level
Producers (plants)

Second Trophic Level
Primary consumers (herbivores)

Third Trophic Level
Secondary consumers (carnivores)

Fourth Trophic Level
Tertiary consumers (top carnivores)

Solar energy

Heat

Detritivores (decomposers and scavengers)

ThomsonNOW™ **Active Figure 3-17 Natural capital:** a *food chain*. The arrows show how chemical energy in food flows through various *trophic levels* in energy transfers; most of the energy is degraded to heat, in accordance with the second law of thermodynamics. *See an animation based on this figure and take a short quiz on the concept.*

second law of thermodynamics. Thus only a small portion of what is eaten and digested is actually converted into an organism's bodily material or biomass, and the amount of usable energy available to each successive trophic level declines.

The percentage of usable energy transferred as biomass from one trophic level to the next is called **ecological efficiency.** It ranges from 2% to 40% (that is, a loss of 60–98%) depending on the types of species and the ecosystem involved, but 10% is typical.

Assuming 10% ecological efficiency (90% loss) at each trophic transfer, if green plants in an area manage to capture 10,000 units of energy from the sun, then only about 1,000 units of energy will be available to support herbivores and only about 100 units to support carnivores.

The more trophic levels in a food chain or web, the greater the cumulative loss of usable energy as energy flows through the various trophic levels. The **pyramid of energy flow** in Figure 3-19 (p. 66) illustrates this energy loss for a simple food chain, assuming a 90% energy loss with each transfer.

Energy flow pyramids explain why the earth can support more people if they eat at lower trophic levels by consuming grains, vegetables, and fruits directly

rather than passing such crops through another trophic level and eating grain eaters such as cattle.

The large loss in energy between successive trophic levels also explains why food chains and webs rarely have more than four or five trophic levels. In most cases, too little energy is left after four or five transfers to support organisms feeding at these high trophic levels.

Thomson NOW! Examine how energy flows among organisms at different trophic levels and through food webs in rain forests, prairies, and other ecosystems at ThomsonNOW.

❓ *THINKING ABOUT TIGERS AND INSECTS* Use Figure 3-16 to help explain **(a)** why there are not many tigers in the world and why they are vulnerable to premature extinction because of human activities, and **(b)** why there are so many insects (Core Case Study, p. 50) in the world.

Productivity of Producers: The Rate Is Crucial

Different ecosystems use solar energy to produce and use biomass at different rates.

The *rate* at which an ecosystem's producers convert solar energy into chemical energy as biomass is the eco-

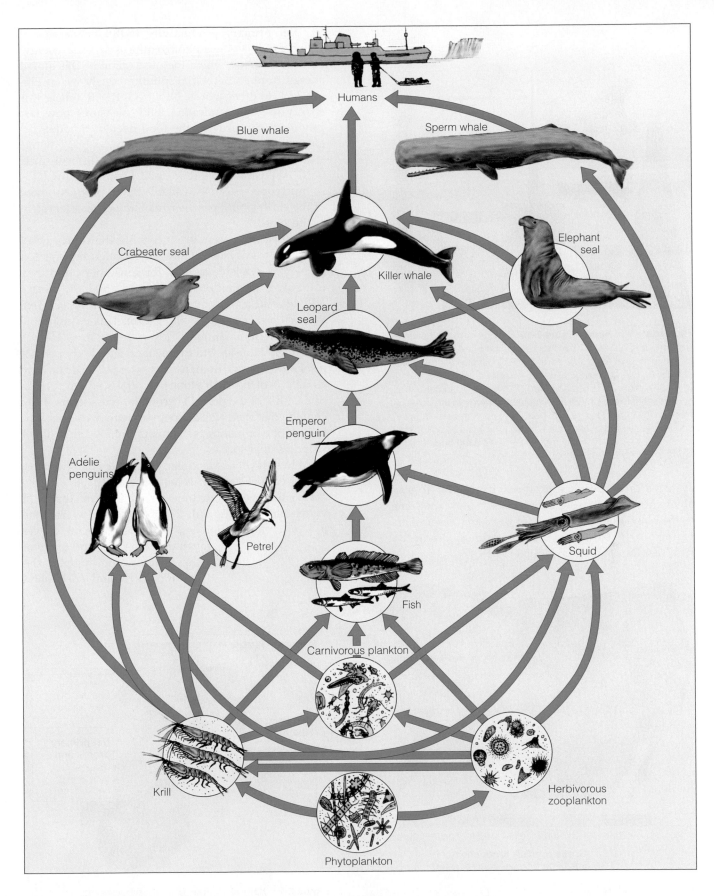

Humans

Blue whale

Sperm whale

Crabeater seal

Killer whale

Elephant seal

Leopard seal

Emperor penguin

Adélie penguins

Petrel

Squid

Fish

Carnivorous plankton

Krill

Herbivorous zooplankton

Phytoplankton

ThomsonNOW™ **Active Figure 3-18 Natural capital:** a greatly simplified *food web* in the Antarctic. Many more participants in the web, including an array of decomposer organisms, are not depicted here. *See an animation based on this figure and take a short quiz on the concept.*

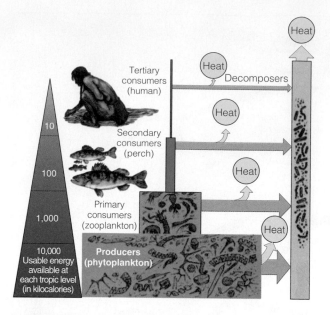

Active Figure 3-19 Natural capital: generalized *pyramid of energy flow* showing the decrease in usable energy available at each succeeding trophic level in a food chain or web. In nature, ecological efficiency varies from 2% to 40%, with 10% efficiency being common. This model assumes a 10% ecological efficiency (90% loss in usable energy to the environment, in the form of low-quality heat) with each transfer from one trophic level to another. QUESTION: *Why is it a scientific error to call this a pyramid of energy? See an animation based on this figure and take a short quiz on the concept.*

system's **gross primary productivity (GPP).** To stay alive, grow, and reproduce, producers must use some of the biomass they produce for their own respiration. Figure 3-20 (p. 66) shows the gross primary productivity across the continental United States.

Net primary productivity (NPP) is the *rate* at which producers use photosynthesis to store energy *minus* the *rate* at which they use some of this stored energy through aerobic respiration as shown in Figure 3-21. In other words, NPP = GPP − R, where R is energy used in respiration. NPP measures how fast producers can provide the food needed by other organisms (consumers) in an ecosystem.

Various ecosystems and life zones differ in their NPP, as illustrated in Figure 3-22. Despite its low NPP, so much open ocean is available that it produces more of the earth's NPP per year than any other ecosystem or life zone.

As we have seen, producers are the source of all food in an ecosystem. Only the biomass represented by NPP is available as food for consumers, and they use only a portion of this amount. Thus, *the planet's NPP ultimately limits the number of consumers (including humans) that can survive on the earth.* This is an important lesson from nature.

Peter Vitousek, Stuart Rojstaczer, and other ecologists estimate that humans now use, waste, or destroy about 27% of the earth's total potential NPP and 10–55% of the NPP of the planet's terrestrial ecosystems. They contend that this is the main reason why we are crowding out or eliminating the habitats and food supplies of so many other species.

Physicist Paul MacCready estimates that humans, their livestock, and pets now make up 98% of the earth's total vertebrate biomass. This means that wild vertebrates make up only 2% of the planet's vertebrate biomass, as humans have taken over much of the planet. And many of these remaining tigers, elephants, birds, and small mam-

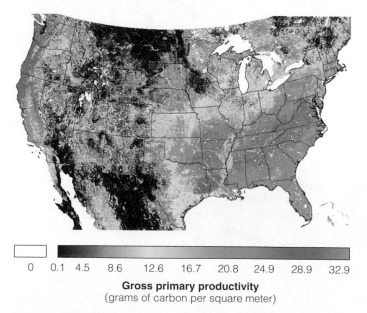

0   0.1   4.5   8.6   12.6   16.7   20.8   24.9   28.9   32.9

Gross primary productivity
(grams of carbon per square meter)

Figure 3-20 Natural capital: gross primary productivity across the continental United States based on remote satellite data. The differences roughly correlate with variations in moisture and soil types. (NASA's Earth Observatory)

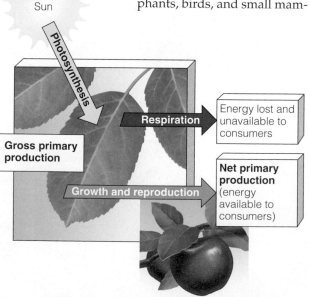

Figure 3-21 Natural capital: distinction between gross primary productivity and net primary productivity. A plant uses some of its gross primary productivity to survive through respiration. The remaining energy is available to consumers.

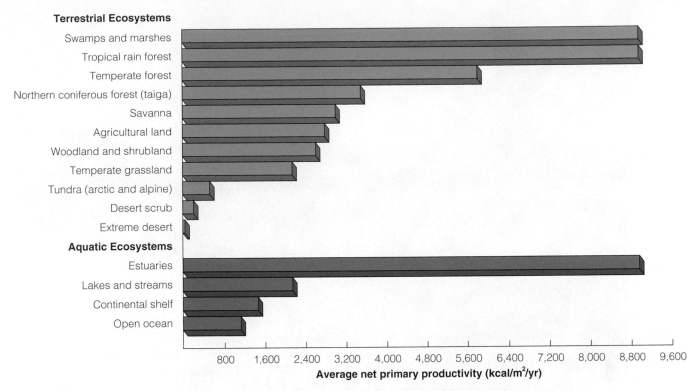

Terrestrial Ecosystems
Swamps and marshes
Tropical rain forest
Temperate forest
Northern coniferous forest (taiga)
Savanna
Agricultural land
Woodland and shrubland
Temperate grassland
Tundra (arctic and alpine)
Desert scrub
Extreme desert
Aquatic Ecosystems
Estuaries
Lakes and streams
Continental shelf
Open ocean

800   1,600   2,400   3,200   4,000   4,800   5,600   6,400   7,200   8,000   8,800   9,600

Average net primary productivity (kcal/m²/yr)

Figure 3-22 Natural capital: estimated annual average *net primary productivity* per unit of area in major life zones and ecosystems, expressed as kilocalories of energy produced per square meter per year (kcal/m²/yr). **QUESTION:** *What are nature's three most productive and three least productive systems?* (Data from *Communities and Ecosystems,* 2nd ed., by R. H. Whittaker, 1975. New York: Macmillan)

mals face extinction as the human ecological footprint expands and becomes heavier.

? *THINKING ABOUT RESOURCE CONSUMPTION* What might happen to us and to other consumer species as the human population grows over the next 40–50 years and per capita consumption of resources such as food, timber, and grassland rises sharply? What are three ways to prevent this from happening?

SOIL: A RENEWABLE RESOURCE

What Is Soil and Why Is It Important?

Soil is a slowly renewed resource that provides most of the nutrients needed for plant growth and also helps purify water.

Soil is a thin covering over most land that is a complex mixture of eroded rock, mineral nutrients, decaying organic matter, water, air, and billions of living organisms, most of them microscopic decomposers. Soil formation begins when bedrock is broken down into rock fragments and particles by physical, chemical, and biological processes called **weathering.** Organisms such as lichen that live on the rock fragments add nutrients,

and when they die their decaying bodies add organic matter to the soil. Over hundreds to thousands of years various types of life build up layers of inorganic and organic matter on a soil's original bedrock.

Figure 3-23 (p. 68) shows a profile of different-aged soils. Although soil is a renewable resource, it is renewed very slowly. Depending mostly on climate, the formation of just 1 centimeter (0.4 inch) of soil can take from 15 years to hundreds of years.

Soil is the base of life on land. Producers that supply food for us and other consumers get the nutrients they need from soil and water. Indeed, you are mostly composed of soil nutrients imported into your body by the food you eat. And soil helps cleanse water percolating downward through it. Soil also helps decompose and recycle biodegradable wastes and is a major component of the earth's water recycling and water storage processes. In addition, it helps control the earth's climate by removing carbon dioxide from the atmosphere and storing it as carbon compounds.

Since the beginning of agriculture, human activities have accelerated natural soil erosion, which can convert this renewable resource into a nonrenewable resource. Entire civilizations have collapsed because they mismanaged the topsoil that supported their populations (Core Case Study, p. 28 and Supplement 6, p. S25) Studies indicate that one-third to one-half of

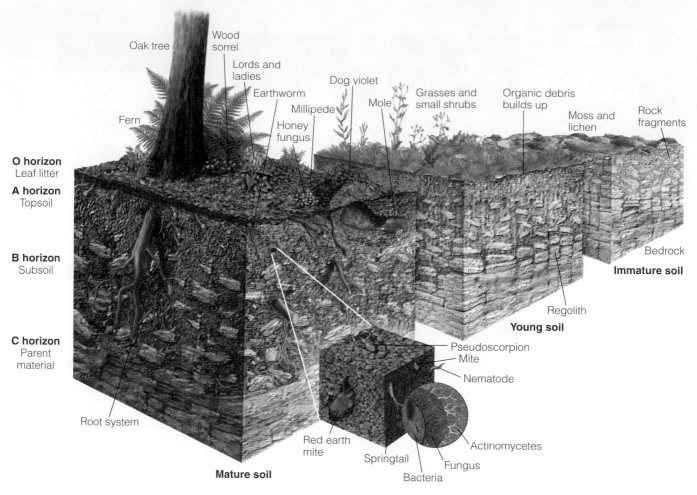

O horizon
Leaf litter

A horizon
Topsoil

B horizon
Subsoil

C horizon
Parent
material

Oak tree
Wood
sorrel
Lords and
ladies
Earthworm
Dog violet
Fern
Millipede
Mole
Grasses and
small shrubs
Organic debris
builds up
Honey
fungus
Moss and
lichen
Rock
fragments

Root system

Red earth
mite
Springtail
Bacteria

Pseudoscorpion
Mite
Nematode
Actinomycetes
Fungus

Bedrock

Immature soil

Regolith
Young soil

Mature soil

ThomsonNOW **Active Figure 3-23 Natural capital:** soil formation and generalized soil profile.
Horizons, or layers, vary in number, composition, and thickness, depending on the type of soil. *See
an animation based on this figure and take a short quiz on the concept.* (Used by permission of
Macmillan Publishing Company from Derek Elsom, *Earth*, New York: Macmillan, 1992. Copyright ©
1992 by Marshall Editions Developments Limited)

the world's croplands are losing topsoil faster than it is
being renewed by natural processes.

Layers in Mature Soils

Most soils developed over a long time consist of
several layers containing different materials.

Mature soils, or soils that have developed over a long
time, are arranged in a series of horizontal layers
called **soil horizons**, each with a distinct texture and
composition that varies with different types of soils. A
cross-sectional view of the horizons in a soil is called a
soil profile. Most mature soils have at least three of the
possible horizons (Figure 3-23). Think of them as floors
in the geological building of life underneath your feet.

The top layer is the *surface litter layer,* or *O horizon.*
It consists mostly of freshly fallen undecomposed or
partially decomposed leaves, twigs, crop wastes, animal waste, fungi, and other organic materials. Normally, it is brown or black.

The *topsoil layer,* or *A horizon,* is a porous mixture
of the partially decomposed bodies of dead plants and
animals, called **humus,** and inorganic materials such
as clay, silt, and sand. A fertile soil that produces high
crop yields has a thick topsoil layer with lots of humus. This helps topsoil hold water and nutrients taken
up by plant roots.

The roots of most plants and the majority of a soil's
organic matter are concentrated in a soil's two upper
layers. As long as vegetation anchors these layers, a
soil stores water and releases it in a nourishing trickle.

The two top layers of most well-developed soils
teem with bacteria, fungi, earthworms, and small insects (Core Case Study, p. 50) that interact in complex
food webs. Bacteria and other decomposer microorganisms found by the billions in every handful of topsoil break down some of its complex organic compounds into simpler inorganic compounds soluble in
water. Soil moisture carrying these dissolved nutrients
is drawn up by the roots of plants and transported

through stems and into leaves as part of the earth's chemical cycling processes.

The color of its topsoil suggests how useful a soil is for growing crops. Dark brown or black topsoil is rich in both nitrogen and organic matter. Gray, bright yellow, and red topsoils are low in organic matter and need nitrogen enrichment to support most crops.

The *B horizon (subsoil)* and the *C horizon (parent material)* contain most of a soil's inorganic matter, mostly broken-down rock consisting of varying mixtures of sand, silt, clay, and gravel, much of it transported by water from the A horizon. The C horizon lies on a base of unweathered parent material, which is often *bedrock.*

The spaces, or pores, between the solid organic and inorganic particles in the upper and lower soil layers contain varying amounts of air (mostly nitrogen and oxygen gas) and water. Plant roots need the oxygen for cellular respiration.

Some precipitation that reaches the soil percolates through the soil layers and occupies many of the soil's open spaces or pores. This downward movement of water through soil is called **infiltration.** As the water seeps down, it dissolves various minerals and organic matter in upper layers and carries them to lower layers in a process called **leaching.**

Most of the world's crops are grown on soils exposed when grasslands and deciduous (leaf-shedding) forests are cleared. Worldwide there are many thousands of different soil types—at least 15,000 in the United States alone. Figure 3-24 profiles five important types of soil.

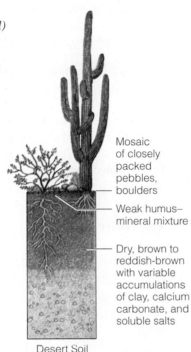

Mosaic of closely packed pebbles, boulders

Weak humus–mineral mixture

Dry, brown to reddish-brown with variable accumulations of clay, calcium carbonate, and soluble salts

Desert Soil
(hot, dry climate)

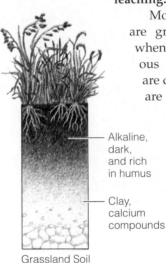

Alkaline, dark, and rich in humus

Clay, calcium compounds

Grassland Soil
(semiarid climate)

Acidic light-colored humus

Iron and aluminum compounds mixed with clay

Tropical Rain Forest Soil
(humid, tropical climate)

Forest litter leaf mold

Humus–mineral mixture

Light, grayish-brown, silt loam

Dark brown firm clay

Deciduous Forest Soil
(humid, mild climate)

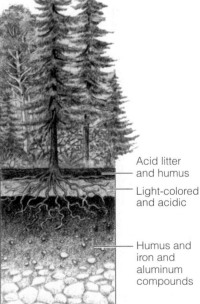

Acid litter and humus

Light-colored and acidic

Humus and iron and aluminum compounds

Coniferous Forest Soil
(humid, cold climate)

ThomsonNOW™ Active Figure 3-24 Natural capital: soil profiles of the principal soil types typically found in five types of terrestrial ecosystems. *See an animation based on this figure and take a short quiz on the concept.*

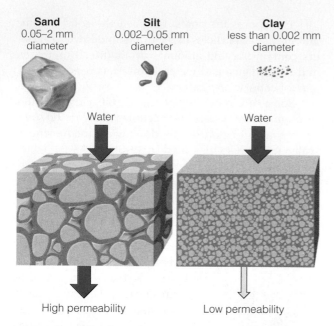

| **Sand** 0.05–2 mm diameter | **Silt** 0.002–0.05 mm diameter | **Clay** less than 0.002 mm diameter |

Water   Water

High permeability   Low permeability

Figure 3-25 Natural capital: the size, shape, and degree of clumping of soil particles determine the number and volume of spaces for air and water within a soil. Soils with more pore spaces (left) contain more air and are more permeable to water than soils with fewer pores (right).

Some Soil Properties

Soils vary in the size of the particles they contain, the amount of space between these particles, and how rapidly water flows through them.

Soils are mixtures of particles of three different sizes: very small *clay* particles, medium size *silt* particles, and larger *sand* particles (Figure 3-25, top). The relative amounts of the different sizes and types of these mineral particles determine **soil texture.**

To get an idea of a soil's texture, take a small amount of topsoil, moisten it, and rub it between your fingers and thumb. A gritty feel means it contains a lot of sand. A sticky feel means a high clay content, and you should be able to roll it into a clump. Silt-laden soil feels smooth, like flour. A loam topsoil is best suited for plant growth. It has a texture between these extremes—a crumbly, spongy feeling—with many of its particles clumped loosely together.

 Compare soil profiles from grassland, desert, and three types of forests at ThomsonNOW.

MATTER CYCLING IN ECOSYSTEMS

Nutrient Cycles: Global Recycling

Global cycles recycle nutrients through the earth's air, land, water, and living organisms and, in the process, connect past, present, and future forms of life.

Nutrients are the elements and compounds that organisms need to live, grow, and reproduce. These substances move through air, water, soil, rock, and living organisms in cycles called **biogeochemical cycles** (literally, life–earth–chemical cycles) or **nutrient cycles.** These cycles, prime examples of one of the four scientific principles of sustainability (Figure 1-16, p. 24), are driven directly or indirectly by incoming solar energy and gravity. They include the carbon, oxygen, nitrogen, phosphorus, and hydrologic (water) cycles (Figure 3-7, p. 55).

The earth's chemical cycles connect past, present, and future forms of life. Some of the carbon atoms in your skin may once have been part of a leaf, a dinosaur's skin, or a layer of limestone rock. Your grandmother, Ludwig Beethoven, Attila the Hun, or a hunter–gatherer who lived 25,000 years ago may have inhaled some of the oxygen molecules you just inhaled.

The Water Cycle

A vast global cycle collects, purifies, distributes, and recycles the earth's fixed supply of water.

Water is an amazing substance (Spotlight, at right). The **hydrologic cycle,** or **water cycle,** collects, purifies, distributes, and recycles the earth's fixed supply of water, as shown in Figure 3-26, p. 72. Trace the flows and paths in this diagram.

The water cycle is powered by energy from the sun, which evaporates water into the atmosphere, and by gravity, which draws the water back to the earth's surface as precipitation. About 84% of water vapor in the atmosphere comes from the oceans; the rest comes from land.

Some of the freshwater returning to the earth's surface as precipitation in this cycle is converted into ice stored in *glaciers.* And some infiltrates and percolates through soil and permeable rock formations to groundwater storage areas called *aquifers.* But most precipitation falling on terrestrial ecosystems becomes *surface runoff.* This water flows into streams and lakes, which eventually carry water back to the oceans, from which it can evaporate to repeat the cycle.

Besides replenishing streams and lakes, surface runoff causes natural soil erosion, which moves soil and rock fragments from one place to another. Water is the primary sculptor of the earth's landscape. Because water dissolves many nutrient compounds, it is a major medium for transporting nutrients within and between ecosystems.

Throughout the hydrologic cycle, many natural processes purify water. Evaporation and subsequent precipitation act as a natural distillation process that removes impurities dissolved in water. Water flowing above ground through streams and lakes and below ground in aquifers is naturally filtered and partially purified by chemical and biological processes, mostly by the actions of decomposer bacteria. *Thus the hydrologic cycle can be viewed as a cycle of natural renewal of water quality.*

Only about 0.024% of the earth's vast water supply is available to us as liquid freshwater in accessible groundwater deposits and in lakes, rivers, and streams. The rest is too salty for us to use, is tied up as ice, or is too deep underground to extract at affordable prices using current technology.

Effects of Human Activities on the Water Cycle

We alter the water cycle by withdrawing large amounts of freshwater, clearing vegetation and eroding soils, polluting surface and underground water, and contributing to climate change.

During the past 100 years, we have been intervening in the earth's current water cycle in four major ways. *First*, we withdraw large quantities of freshwater from streams, lakes, and underground sources, sometimes at rates faster than nature replaces it.

Second, we clear vegetation from land for agriculture, mining, road and building construction, and other activities and sometimes cover the land with buildings, concrete, or asphalt. This increases runoff, reduces infiltration that recharges groundwater supplies, increases the risk of flooding, and accelerates soil erosion and landslides. We also increase flooding by destroying wetlands, which act like sponges to absorb and slowly release overflows of water.

Third, we add nutrients (such as phosphates and nitrates found in fertilizers) and other pollutants to water. This overload of plant nutrients can change or impair natural ecological processes that purify water.

Fourth, according to a 2003 study by Ruth Curry and her colleagues, the earth's water cycle is speeding up as a result of a warmer climate caused partially by human inputs of carbon dioxide and other greenhouse gases into the atmosphere. This could change global precipitation patterns that affect the severity and frequency of droughts, floods, and storms. It can also intensify global warming by speeding up the input of water vapor—a powerful greenhouse gas—into the troposphere.

Water's Unique Properties

SPOTLIGHT

Water is a remarkable substance with a unique combination of properties:

- *There are strong forces of attraction* (called *hydrogen bonds*, see Figure 5 on p. S29 in Supplement 7) *between molecules of water.* These attractive forces are the major factor determining water's distinctive properties.

- *Water exists as a liquid over a wide temperature range because of the strong forces of attraction between its molecules.* Without water's high boiling point the oceans would have evaporated a long time ago.

- *Liquid water changes temperature slowly because it can store a large amount of heat without a large change in temperature.* This high heat capacity helps protect living organisms from temperature fluctuations. It also moderates the earth's climate and makes water an excellent coolant for car engines and power plants.

- *It takes a large amount of energy for water to evaporate because of the* strong forces of attraction between its molecules. Water absorbs large amounts of heat as it changes into water vapor and releases this heat as the vapor condenses back to liquid water. This helps distribute heat throughout the world and determine the climates of various areas. This property also makes evaporation a cooling process—explaining why you feel cooler when perspiration evaporates from your skin.

- *Liquid water can dissolve a variety of compounds.* (See Figure 3 on p. S28 in Supplement 7). This enables it to carry dissolved nutrients into the tissues of living organisms, flush waste products out of those tissues, serve as an all-purpose cleanser, and help remove and dilute the water-soluble wastes of civilization. This property also means that water-soluble wastes can easily pollute water.

- *Water filters out wavelengths of the sun's ultraviolet (UV) radiation that would harm some aquatic organisms.*

- *Attractive forces between the molecules of liquid water cause its surface to* contract and to adhere to a solid surface. These strong cohesive forces allow narrow columns of water to rise through a plant from its roots to its leaves (capillary action).

- *Unlike most liquids, water expands when it freezes.* This means that ice floats on water because it has a lower density (mass per unit of volume) than liquid water. Otherwise lakes and streams in cold climates would freeze solid and lose most of their current forms of aquatic life. Because water expands upon freezing, it can break pipes, crack a car's engine block (which is why we use antifreeze), break up street pavements, and fracture rocks.

Critical Thinking

Water is a bent molecule (see Figure 4 on p. S29 in Supplement 7). and this allows it to form strong hydrogen bonds between its molecules. Think of three ways in which your life would differ if water were a linear or straight molecule.

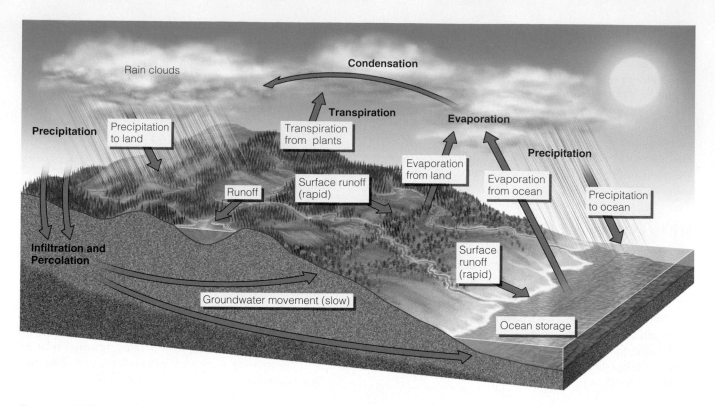

ThomsonNOW™ **Active Figure 3-26** **Natural capital:** simplified model of the *hydrologic cycle*. *See an animation based on this figure and take a short quiz on the concept.*

ThomsonNOW™ **Active Figure 3-27** **Natural capital:** simplified model of the global *carbon cycle*. Carbon moves through both marine ecosystems (left side) and terrestrial ecosystems (right side). Carbon reservoirs are shown as boxes; processes that change one form of carbon to another are shown in unboxed print. QUESTION: *What are three ways in which your lifestyle directly or indirectly affects the carbon cycle? See an animation based on this figure and take a short quiz on the concept.* (From Cecie Starr, *Biology: Concepts and Applications,* 4th ed., Pacific Grove, Calif.: Brooks/Cole, © 2000)

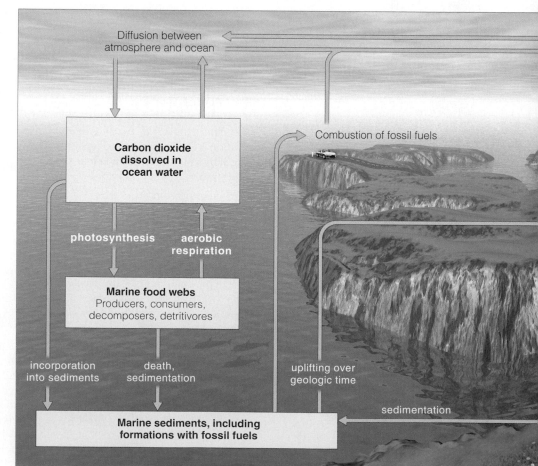

The Carbon Cycle: Part of Nature's Thermostat

Carbon cycles through the earth's air, water, soil, and living organisms and depends on photosynthesis and respiration.

Carbon, the basic building block of the carbohydrates, fats, proteins, DNA, and other organic compounds necessary for life, circulates through the biosphere in the **carbon cycle** shown in Figure 3-27.

The carbon cycle is based on carbon dioxide (CO_2) gas, which makes up 0.038% of the volume of the troposphere and is also dissolved in water. Carbon dioxide is a key component of nature's thermostat. If the carbon cycle removes too much CO_2 from the atmosphere, the atmosphere will cool; if it generates too much CO_2, the atmosphere will get warmer. Thus, even slight changes in this cycle can affect climate and ultimately help determine the types of life that can exist on various parts of the earth.

Terrestrial producers remove CO_2 from the atmosphere, and aquatic producers remove it from the water.

They then use photosynthesis to convert CO_2 into complex carbohydrates such as glucose ($C_6H_{12}O_6$).

The cells in oxygen-consuming producers, consumers, and decomposers then carry out aerobic respiration. This process breaks down glucose and other complex organic compounds and converts the carbon back to CO_2 in the atmosphere or water for reuse by producers. This linkage between *photosynthesis* in producers and *aerobic respiration* in producers, consumers, and decomposers circulates carbon in the biosphere. Oxygen and hydrogen—the other elements in carbohydrates—cycle almost in step with carbon.

Some carbon atoms take a long time to recycle. Over millions of years, buried deposits of dead plant matter and bacteria are compressed between layers of sediment, where they form carbon-containing *fossil fuels* such as coal and oil (Figure 3-27). This carbon is not released to the atmosphere as CO_2 for recycling until these fuels are extracted and burned, or until long-term geological processes expose these deposits to air. In only a few hundred years, and especially in the last 50 years, we have extracted and burned fossil fuels that took millions of years to form. This is why, on a human time scale, fossil fuels are nonrenewable resources.

Oceans play important roles in the carbon cycle. Some of the atmosphere's carbon dioxide dissolves in

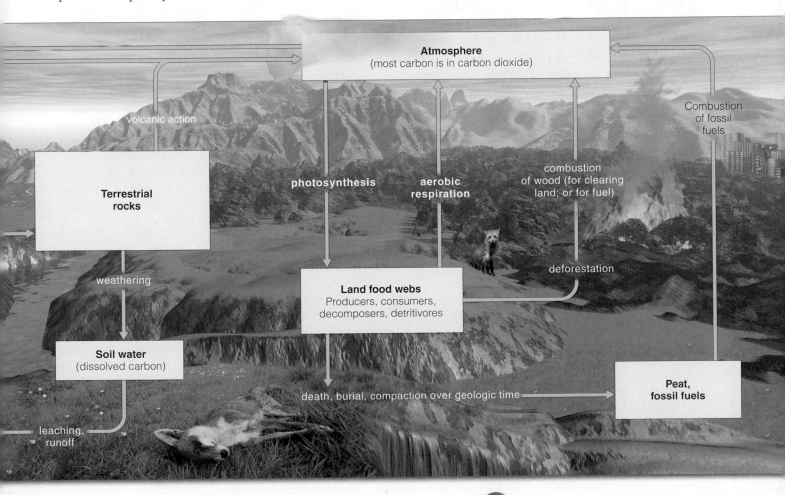

Atmosphere
(most carbon is in carbon dioxide)

volcanic action

Combustion of fossil fuels

Terrestrial rocks

photosynthesis

aerobic respiration

combustion of wood (for clearing land; or for fuel)

weathering

deforestation

Land food webs
Producers, consumers, decomposers, detritivores

Soil water
(dissolved carbon)

death, burial, compaction over geologic time

Peat, fossil fuels

leaching, runoff

ocean water, and the ocean's photosynthesizing producers remove some. On the other hand, as ocean water warms, some of its dissolved CO_2 returns to the atmosphere, just as carbon dioxide fizzes out of a carbonated beverage when it warms. The balance between these two processes plays a role in the earth's average temperature.

Some ocean organisms build their shells and skeletons by using dissolved CO_2 molecules in seawater to form carbonate compounds such as calcium carbonate ($CaCO_3$). When these organisms die, tiny particles of their shells and bone drift slowly to the ocean depths. There they are buried for eons (as long as 400 million years) in deep bottom sediments (Figure 3-27, left), where under immense pressure they are converted into limestone rock. Geological processes may eventually expose the limestone to the atmosphere and acidic precipitation and make its carbon available to living organisms once again.

Effects of Human Activities on the Carbon Cycle

Burning fossil fuels and clearing photosynthesizing vegetation faster than it is replaced can increase the earth's average temperature by adding excess carbon dioxide to the atmosphere.

Since 1800, and especially since 1950, we have been intervening in the earth's carbon cycle in two ways that add carbon dioxide to the atmosphere: *First,* in some areas we clear trees and other plants that absorb CO_2 through photosynthesis faster than they can grow back. *Second,* we add large amounts of CO_2 by burning fossil fuels (Figure 3-28) and wood.

Computer models of the earth's climate systems suggest that increased concentrations of atmospheric CO_2 and other gases could enhance the planet's *natural greenhouse effect* that helps warm the lower atmosphere (troposphere) and the earth's surface (Figure 3-8). The resulting *global warming* could disrupt global food production and wildlife habitats, alter temperature and precipitation patterns, and raise the average sea level in various parts of the world.

The Nitrogen Cycle: Bacteria in Action

Different types of bacteria help recycle nitrogen through the earth's air, water, soil, and living organisms.

Nitrogen is the atmosphere's most abundant element, with chemically unreactive nitrogen gas (N_2) making up about 78% of the volume of the troposphere. Nitrogen is a crucial component of proteins, many vitamins, and nucleic acids such as DNA. However, N_2 cannot be absorbed and used directly as a nutrient by multicellular plants or animals.

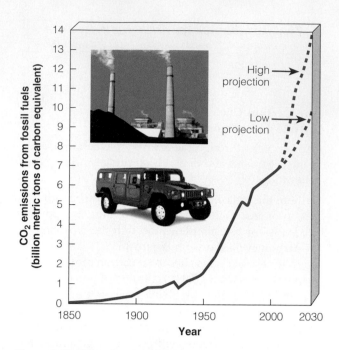

Figure 3-28 Natural capital degradation: human interference in the global carbon cycle from carbon dioxide emissions when fossil fuels are burned and forests are cleared, 1850 to 2006 and projections to 2030 (dashed lines). (Data from UN Environment Programme, British Petroleum, International Energy Agency, and U.S. Department of Energy)

Fortunately, two natural processes convert or *fix* N_2 into compounds useful as nutrients for plants and animals. One is atmospheric electrical discharges, or lightning. In the other process, certain types of bacteria—called *nitrogen-fixing bacteria*—in aquatic systems, in the soil, and in the roots of some plants complete this conversion as part of the **nitrogen cycle,** depicted in Figure 3-29.

The nitrogen cycle consists of several major steps. In *nitrogen fixation,* specialized bacteria in soil and aquatic environments convert (or fix) gaseous nitrogen (N_2) to ammonia (NH_3), which is converted to ammonium ions (NH_4^+) that can be used by plants.

Ammonia not taken up by plants may undergo *nitrification.* In this two-step process, specialized soil bacteria convert most of the NH_3 and NH_4^+ in soil first to *nitrite ions* (NO_2^-), which are toxic to plants, and then to *nitrate ions* (NO_3^-), which are easily taken up by the roots of plants. Animals, in turn, get their nitrogen by eating plants or plant-eating animals.

Plants and animals return nitrogen-rich organic compounds to the environment as wastes, cast-off particles, and through their bodies when they die. In *ammonification,* vast armies of specialized decomposer bacteria convert this organic material into simpler nitrogen-containing inorganic compounds such as ammonia (NH_3) and water-soluble salts containing ammonium ions (NH_4^+).

In *denitrification*, nitrogen leaves the soil as specialized bacteria in waterlogged soil and in the bottom sediments of lakes, oceans, swamps, and bogs convert NH_3 and NH_4^+ back into nitrite and nitrate ions, and then into nitrogen gas (N_2) and nitrous oxide gas (N_2O). These gases are released to the atmosphere to begin the nitrogen cycle again.

Effects of Human Activities on the Nitrogen Cycle

We add large amounts of nitrogen-containing compounds to the earth's air and water and remove nitrogen from the soil.

We intervene in the nitrogen cycle in several ways. *First,* we add large amounts of nitric oxide (NO) into the atmosphere when N_2 and O_2 combine as we burn any fuel at high temperatures. In the atmosphere, this gas can be converted to nitrogen dioxide gas (NO_2) and nitric acid (HNO_3), which can return to the earth's surface as damaging *acid deposition,* commonly called *acid rain.*

Second, we add nitrous oxide (N_2O) to the atmosphere through the action of anaerobic bacteria on livestock wastes and commercial inorganic fertilizers applied to the soil. This gas can warm the atmosphere and deplete ozone in the stratosphere.

Third, nitrate (NO_3^-) ions in inorganic fertilizers can leach through the soil and contaminate groundwater, which is harmful to drink, especially for infants and small children.

Fourth, we release large quantities of nitrogen stored in soils and plants as gaseous compounds into

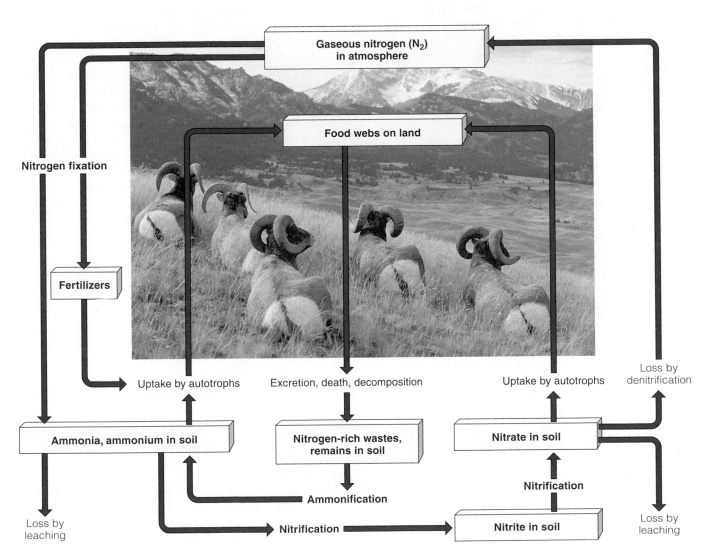

ThomsonNOW **Active Figure 3-29** **Natural capital:** simplified model of the *nitrogen cycle* in a terrestrial ecosystem. Nitrogen reservoirs are shown as boxes; processes changing one form of nitrogen to another are shown in unboxed print. **QUESTION:** *What are three ways in which your lifestyle directly or indirectly affects the nitrogen cycle?* See an animation based on this figure and take a short quiz on the concept. (Adapted from Cecie Starr, *Biology: Today and Tomorrow,* Brooks/Cole © 2005)

the troposphere through destruction of forests, grasslands, and wetlands.

Fifth, we upset aquatic ecosystems by adding excess nitrates to bodies of water through agricultural runoff and discharges from municipal sewage systems.

Sixth, we remove nitrogen from topsoil when we harvest nitrogen-rich crops, irrigate crops (washing it out of the soil), and burn or clear grasslands and forests before planting crops.

According to the 2005 Millennium Ecosystem Assessment, since 1950 human activities have more than doubled the annual release of nitrogen from the terrestrial portion of the earth into the rest of the environment (Figure 3-30). This excessive input of nitrogen into the air and water represents a serious local, regional, and global environmental problem that has attracted relatively little attention compared to problems such as global warming and depletion of ozone in the stratosphere. Princeton University physicist Robert Socolow calls for countries around the world to work out some type of nitrogen management agreement to help prevent this problem from reaching crisis levels.

The Phosphorus Cycle

Phosphorus cycles fairly slowly through the earth's water, soil, and living organisms.

Phosphorous is a key component of DNA and energy storage molecules such as ATP in cells. It circulates

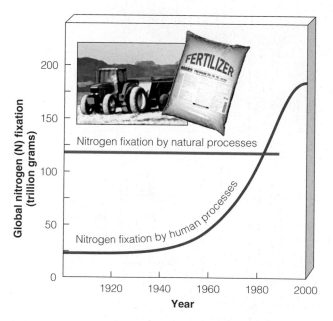

Figure 3-30 Natural capital degradation: human interference in the global nitrogen cycle. Human activities such as production of fertilizers now fix more nitrogen than all natural sources combined. (Data from UN Environment Programme, UN Food and Agriculture Organization, and U.S. Department of Agriculture)

through water, the earth's crust, and living organisms in the phosphorus cycle, depicted in Figure 3-31. Very little phosphorus circulates in the atmosphere because soil conditions do not allow bacteria to convert chemical forms of phosphorus to gaseous forms of phosphates. The phosphorus cycle is slow, and on a short human time scale much phosphorus flows one way from the land to the oceans.

Phosphorus typically is found as phosphate salts containing phosphate ions (PO_4^{3-}) in terrestrial rock formations and ocean bottom sediments. As water runs over phosphorus-containing rocks, it slowly erodes away inorganic compounds that contain phosphate ions.

Phosphate can be lost from the cycle for long periods when it washes from the land into streams and rivers and is carried to the ocean. There it can be deposited as sediment and remain trapped for millions of years. Some day the geological processes of uplift may expose these seafloor deposits, from which phosphate can be eroded to start the cycle again.

Plants obtain phosphorus as phosphate ions (PO_4^{3-}) directly from soil or water and incorporate it in various organic compounds. Animals get their phosphorus from plants and eliminate excess phosphorus in their urine.

Because most soils contain little phosphate, it is often the *limiting factor* for plant growth on land unless phosphorus (as phosphate salts mined from the earth) is applied to the soil as a fertilizer. Phosphorus also limits the growth of producer populations in many freshwater streams and lakes because phosphate salts are only slightly soluble in water.

Effects of Human Activities on the Phosphorus Cycle

We remove large amounts of phosphate from the earth to make fertilizer, reduce phosphorus in tropical soils by clearing forests, and add excess phosphates to aquatic systems.

We intervene in the earth's phosphorus cycle in three ways. *First,* we mine large quantities of phosphate rock to make commercial inorganic fertilizers and some detergents. *Second,* we reduce the available phosphate in tropical soils when we cut down areas of tropical forests. *Third,* we disrupt aquatic systems with phosphates from runoff of animal wastes and fertilizers and discharges from sewage treatment systems.

Since 1900 human activities have increased the natural rate of phosphorus release into the environment about 3.7-fold.

? THINKING ABOUT THE PHOSPHORUS CYCLE List three possible effects on your lifestyle if we continue to add excess phosphorus to the environment.

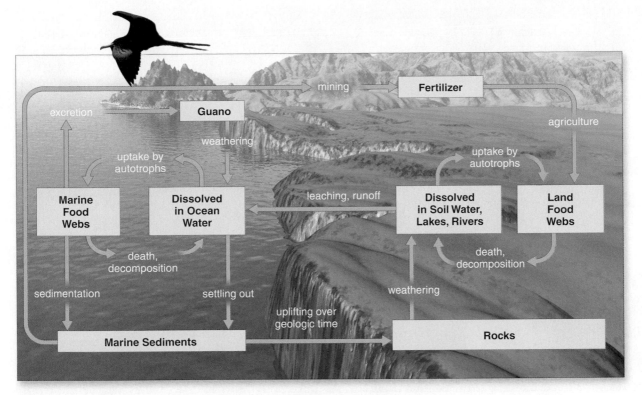

Figure 3-31 Natural capital: simplified model of the *phosphorus cycle*. Phosphorus reservoirs are shown as boxes; processes that change one form of phosphorus to another are shown in unboxed print. QUESTION: *What are three ways in which your lifestyle directly or indirectly affects the phosphorus cycle?* (From Cecie Starr and Ralph Taggart, *Biology: The Unity and Diversity of Life*, 9th ed., Belmont, Calif.: Wadsworth © 2001)

The Sulfur Cycle

Sulfur cycles through the earth's air, water, soil, and living organisms.

Sulfur circulates through the biosphere in the **sulfur cycle,** shown in Figure 3-32 (p. 78). Much of the earth's sulfur is stored underground in rocks and minerals, including sulfate (SO_4^{2-}) salts buried deep under ocean sediments.

Sulfur also enters the atmosphere from several natural sources. Hydrogen sulfide (H_2S)—a colorless, highly poisonous gas with a rotten-egg smell—is released from active volcanoes and from organic matter in flooded swamps, bogs, and tidal flats broken down by anaerobic decomposers. Sulfur dioxide (SO_2), a colorless and suffocating gas, also comes from volcanoes.

Particles of sulfate (SO_4^{2-}) salts, such as ammonium sulfate, enter the atmosphere from sea spray, dust storms, and forest fires. Plant roots absorb sulfate ions and incorporate the sulfur as an essential component of many proteins.

Certain marine algae produce large amounts of volatile dimethyl sulfide, or DMS (CH_3SCH_3). Tiny droplets of DMS serve as nuclei for the condensation of water into droplets found in clouds. In this way, changes in DMS emissions can affect cloud cover and climate.

In the atmosphere, DMS is converted to sulfur dioxide, some of which in turn is converted to sulfur trioxide gas (SO_3) and to tiny droplets of sulfuric acid (H_2SO_4). DMS also reacts with other atmospheric chemicals such as ammonia to produce tiny particles of sulfate salts. These droplets and particles fall to the earth as components of *acid deposition*, which along with other air pollutants can harm trees and aquatic life.

In the oxygen-deficient environments of flooded soils, freshwater wetlands, and tidal flats, specialized bacteria convert sulfate ions to sulfide ions (S^{2-}). The sulfide ions can then react with metal ions to form insoluble metallic sulfides, which are deposited as rock, and the cycle continues.

Effects of Human Activities on the Sulfur Cycle

Burning coal and oil, refining oil, and producing some metals from ores add sulfur dioxide to the atmosphere.

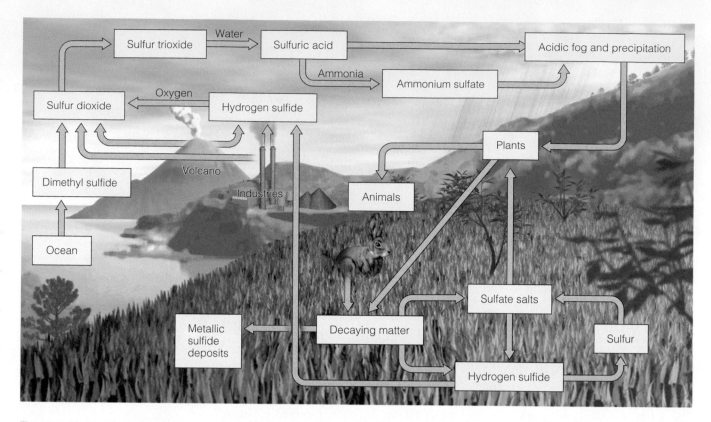

ThomsonNOW **Active Figure 3-32 Natural capital:** simplified model of the *sulfur cycle*. The movement of sulfur compounds in living organisms is shown in green, blue in aquatic systems, and orange in the atmosphere. **QUESTION:** *What are three ways in which your lifestyle directly or indirectly affects the sulfur cycle?* See an animation based on this figure and take a short quiz on the concept.

We add sulfur dioxide to the atmosphere in three ways. *First,* we burn sulfur-containing coal and oil to produce electric power. *Second,* we refine sulfur-containing petroleum to make gasoline, heating oil, and other useful products. *Third,* we convert sulfur-containing metallic mineral ores into free metals such as copper, lead, and zinc—an activity that releases large amounts of sulfur dioxide into the environment.

E *RESEARCH FRONTIER* How human activities affect the major nutrient cycles and how we can reduce these effects

 Learn more about the water, carbon, nitrogen, phosphorus, and sulfur cycles using interactive animations at ThomsonNOW.

The Gaia Hypothesis: Is the Earth Alive?

Some have proposed that the earth's various forms of life control or at least influence its chemical cycles and other earth-sustaining processes.

The cycling of matter and the flow of energy through the biosphere and its ecosystems connects the earth's past and current organisms. In this sense, the earth is an incredibly complex system that sustains itself and adapts to changing environmental conditions through an intricate network of positive and negative feedback loops (p. 33).

Some people believe the earth behaves like a single self-regulating, system in which living things affect the environment in ways that make it possible for life to persist and flourish on the earth. This idea is known as the *Gaia* (pronounced GUY–uh) *hypothesis,* named for the Greek goddess of the earth (Mother Earth). It was first proposed in 1979 by English inventor and atmospheric chemist James Lovelock.

The original Gaia hypothesis that life *controls* the earth's life-sustaining processes is known as the *strong Gaia hypothesis.* Few scientists support this hypothesis. The idea that life *influences* the earth's life-sustaining processes is called the *weak Gaia hypothesis.* Many scientists support this hypothesis but contend that we need to continue testing its validity. In 2006, Lovelock published a book, *The Revenge of Gaia,* in which he applies the Gaia hypotheses to global climate change.

HOW DO ECOLOGISTS LEARN ABOUT ECOSYSTEMS?

Field Research, Remote Sensing, and Geographic Information Systems

Ecologists go into ecosystems and learn what organisms live there and how they interact, use sensors on aircraft and satellites to collect data, and store and analyze geographic data in large databases.

Field research, sometimes called muddy-boots biology, involves going into nature and observing and measuring the structure of ecosystems and what happens in them. Most of what we know about the structure and functioning of ecosystems has come from such research. *Green Career:* Ecologist

Ecologists trek through forests, deserts, and grasslands and wade or boat through wetlands, lakes, and streams collecting and observing species. Sometimes they carry out controlled experiments by isolating and changing a variable in part of an area and comparing the results with nearby unchanged areas.

Tropical ecologists use tall construction cranes that stretch into the canopies of tropical forests to identify and observe the rich diversity of species living or feeding in these treetop habitats.

Increasingly, ecologists are using new technologies to collect field data. In *remote sensing* from aircraft, satellites, and space shuttles and *geographic information systems* (GISs), data gathered from broad geographic regions are stored in spatial databases (Figure 3-33). Computers and GIS software can analyze and manipulate the data and combine them with ground and other data. The result: computerized maps of forest cover, water resources, air pollution emissions, coastal changes, relationships between cancers and sources of pollution, gross primary productivity, and changes in global sea temperatures.

Data and models from remote sensing, supercomputers, and a global network of computer communication are increasingly being used by international networks of scientists working on global ecological problems. Such multidisciplinary research is used in weather forecasting, urban planning, agriculture, formulating international environmental policy, and judging the effectiveness of international environmental treaties. In 2005, scientists launched the Global Earth Observation System of Systems GEOSS)— a 10-year program to integrate the data from sensors, gauges, buoys, and satellites that monitor the earth's

surface, atmosphere, and oceans into a unified global whole. This is the first generation of scientists with the tools to study ecology and environmental problems on a global scale. *Green Careers:* Geographic Information Systems analyst and remote sensing analyst

F **RESEARCH FRONTIER** Applying GIS, remote sensing, supercomputers, and other high-tech tools to the environmental sciences

Studying Ecosystems in the Laboratory

Ecologists use aquarium tanks, greenhouses, and controlled indoor and outdoor chambers to study ecosystems.

During the past 50 years, ecologists have increasingly supplemented field research by using *laboratory research* to set up, observe, and make measurements of model ecosystems and populations under laboratory conditions. Such simplified systems have been created in containers such as culture tubes, bottles, aquarium

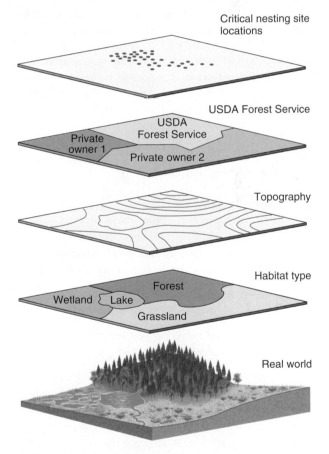

Figure 3-33 *Geographic information systems* (GISs) provide the computer technology for organizing, storing, and analyzing complex data collected over broad geographic areas. They enable scientists to overlay many layers of data (such as soils, topography, distribution of endangered populations, and land protection status).

tanks, greenhouses, and in indoor and outdoor chambers where temperature, light, CO_2, humidity, and other variables can be controlled carefully.

Such systems make it easier for scientists to carry out controlled experiments. In addition, such laboratory experiments often are quicker and cheaper than similar experiments in the field.

But there is a catch. We must consider whether scientific observations and measurements in a simplified, controlled system under laboratory conditions reflect what takes place under the more complex and dynamic conditions found in nature. Thus the results of laboratory research must be coupled with and supported by field research.

Systems Analysis

Ecologists develop mathematical and other models to simulate the behavior of ecosystems.

Since the late 1960s, ecologists have explored the use of *systems analysis* to develop mathematical and other models that simulate ecosystems. Computer simulations can help us understand large and very complex systems (such as rivers, oceans, forests, grasslands, cities, and climate) that cannot be adequately studied and modeled in field and laboratory research. Figure 3-34 outlines the major stages of systems analysis.

Researchers can change values of the variables in their computer models to project possible changes in environmental conditions, help anticipate environmental surprises, and analyze the effectiveness of various alternative solutions to environmental problems. *Green Career:* Systems analyst

---

F **RESEARCH FRONTIER** Improved computer modeling for understanding complex environmental systems

---

Of course, simulations and projections made using ecosystem models are no better than the data and assumptions used to develop the models. Clearly, careful field and laboratory ecological research must be used to provide the baseline data and determine the causal relationships between key variables needed to develop and test ecosystem models.

Importance of Baseline Ecological Data

We need baseline data on the world's ecosystems so we can see how they are changing and develop effective strategies for preventing or slowing their degradation.

Before we can understand what is happening in nature and how best to prevent harmful environmental changes, we need to know about current conditions. In other words, we need *baseline data* about the condition of the earth's ecosystems.

By analogy, your doctor would like to have baseline data on your blood pressure, weight, and func-

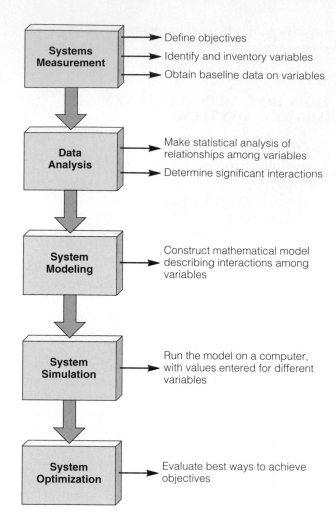

Figure 3-34 Major stages of systems analysis. (Modified data from Charles Southwick)

tioning of your organs and other systems as revealed by basic tests. If something happens to your health, the doctor can run new tests and compare the results with the baseline data to identify changes and come up with an effective treatment.

Bad news. According to a 2002 ecological study published by the Heinz Foundation and the 2005 Millennium Ecosystem Assessment, scientists have less than half of the basic ecological data they need to evaluate the status of ecosystems in the United States. Even fewer data are available for most other parts of the world.

Ecologists call for a massive program to develop baseline data for the world's ecosystems. Scientists are wiring up parts of the natural world with networks of tiny—often wireless—sensors, robots, cameras, and computers to collect basic environmental data. Using batteries and solar cells to power such devices means that scientists can use them in remote places and move them around. Scientists want to deploy millions of these environmental sensor networks (ESNs) throughout the United States and the world.

Revisiting Insects and Sustainability

This chapter applied two of the scientific principles of sustainability (Figure 1-16, p. 24) by which natural ecosystems have achieved *long-term* sustainability. *First,* almost all of them use *renewable solar energy* as their energy source. *Second,* they *recycle the chemical nutrients* their organisms need for survival, growth, and reproduction in complex networks of interdependency.

These two sustainability principles arise from the structure and function of natural ecosystems (Figures 3-7 and 3-14), the law of conservation of matter (pp. 39–40), and the two laws of thermodynamics (pp. 44–46).

This chapter started with a discussion of the importance of insects (Core Case Study, p. 50). Insects play a vital role in implementing these two scientific principles of sustainability. They rely on solar energy by consuming a vast amount of what producers produce, provide large quantities of food for consumers higher on the food chain, and take part in and depend on recycling of nutrients in the biosphere.

All things come from earth, and to earth they all return.
MENANDER (342–290 BC)

CRITICAL THINKING

1. How would you explain the importance of insects to someone who has "bugitis" and hates all insects and to a farmer whose crops are devoured by insect pests?

2. (a) A bumper sticker asks, "Have you thanked a green plant today?" Give two reasons for appreciating a green plant. **(b)** Trace the sources of the materials that make up the bumper sticker, and decide whether the sticker itself is a sound application of the slogan.

3. Explain why microbes are the real rulers of the earth. List two beneficial and two harmful effects of microbes on your health and lifestyle.

4. Make a list of the food you ate for lunch or dinner today. Trace each type of food back to a particular producer species.

5. Use the second law of thermodynamics (pp. 45–46) to explain why many poor people in developing countries live on a mostly vegetarian diet.

6. Why do farmers not need to apply carbon to grow their crops but often need to add fertilizer containing nitrogen and phosphorus?

7. What changes might take place in the hydrologic cycle if the earth's climate becomes **(a)** hotter or **(b)** cooler? In each case explain how these changes might affect your lifestyle.

8. What would happen to an ecosystem if **(a)** all its decomposers and scavengers were eliminated, **(b)** all its producers were eliminated, or **(c)** all of its insects (Core Case Study, p. 50) were eliminated? Could a balanced ecosystem exist with only producers and decomposers and no consumers such as humans and other animals? Explain.

9. List two questions that you would like to have answered as a result of reading this chapter.

PROJECTS

1. Visit a nearby aquatic life zone or terrestrial ecosystem and try to identify its major producers, consumers, detritivores, and decomposers.

2. Write a brief scenario describing the sequence of consequences for us to other forms of life if each of the following nutrient cycles stopped functioning: **(a)** carbon, **(b)** nitrogen, **(c)** phosphorus, and **(d)** water.

3. Make a concept map of this chapter's major ideas using the section heads, subheads, and key terms (in boldface). Look on the website for this book for information about making concept maps.

LEARNING ONLINE

The website for this book contains study aids and many ideas for further reading and research. They include a chapter summary, review questions for the entire chapter, flash cards for key terms and concepts, a multiple-choice practice quiz, interesting Internet sites, references, information about green careers, and a guide for accessing thousands of InfoTrac® College Edition articles. Log into

www.thomsonedu.com/biology/miller

Then choose Chapter 3, and select a learning resource. For access to animations, additional quizzes, chapter outlines and summaries, register and log into

at **www.thomsonedu.com** using the access code card in the front of your book.

Active Graphing

Log into ThomsonNow at www.thomsonedu.com to explore the graphing exercise for this chapter.

Evolution and Biodiversity

CORE CASE STUDY

Earth: The Just-Right, Adaptable Planet

Life on the earth (Figure 4-1) as we know it needs a certain temperature range, which depends on the liquid water that dominates the earth's surface. Temperature is crucial because most life on the earth needs average temperatures between the freezing and boiling points of water.

The earth's orbit is the right distance from the sun to provide these conditions. If the earth were much closer to the sun, it would be too hot—like Venus—for water vapor to condense to form rain. If it were much farther away, the earth's surface would be so cold—like Mars—that its water would exist only as ice. The earth also spins; if it did not, the side facing the sun would be too hot and the other side too cold for water-based life to exist.

The size of the earth is also just right for life: It has enough gravitational mass to keep its iron and nickel core molten and to keep the light gaseous molecules (such as N_2, O_2, CO_2, and H_2O) in its atmosphere from flying off into space.

On a time scale of millions of years, the earth is enormously resilient and adaptive. During the 3.7 billion years since life arose, the average surface temperature of the earth has remained within the narrow range of 10–20°C (50–68°F), even with a 30–40% increase in the sun's energy output.

For several hundred million years oxygen has made up about 21% of the volume of earth's atmosphere. If the atmosphere's oxygen content dropped to about 15%, this would be lethal for most forms of life. If it increased to about 25%, oxygen in the atmosphere would probably ignite into a giant fireball.

And thanks to the development of photosynthesizing bacteria more than 2 billion years ago, an ozone sunscreen protects us and many other forms of life from an overdose of ultraviolet radiation. In short, this remarkable planet we live on is uniquely suited for life as we know it. And perhaps the two most astounding features of the planet are its incredible diversity of life (biodiversity) and its inherent ability to sustain life (sustainability).

Understanding how organisms adapt to changing environmental conditions is important for understanding how nature works, how our activities can affect the earth's life, and how we can prevent unnecessary loss of the planet's biodiversity. This chapter shows how each species here today represents a long chain of genetic changes in populations in response to changing environmental conditions and plays a unique ecological role in the earth's communities and ecosystems.

Figure 4-1 Natural capital: the earth, a blue and white planet in the black void of space. Currently, it has the right physical and chemical conditions to allow the development of life as we know it.

NASA

There is grandeur to this view of life . . . that, whilst this planet has gone cycling on . . . endless forms most beautiful and most wonderful have been, and are being, evolved.

CHARLES DARWIN

This chapter describes how most scientists believe life on earth arose and developed into the diversity of species we find today. It discusses these questions:

- How do scientists account for the development of life on the earth?

- What is biological evolution by natural selection, and how can it account for the current diversity of organisms on the earth?

- How can geologic processes, climate change, and catastrophes affect biological evolution?

- What is an ecological niche, and how does it help a population adapt to changing environmental conditions?

- How do extinction of species and formation of new species affect biodiversity?

- What is the future of evolution, and what role should humans play in this future?

- How did we become such a powerful species in a short time?

ORIGINS OF LIFE

Development of Life on the Primitive Earth: The Big Picture

Scientific evidence indicates that the earth's life is the result of about 1 billion years of chemical change to form the first cells, followed by about 3.7 billion years of biological change to produce the variety of species we find on the earth today.

How did life on the earth evolve to its present incredible diversity of between 4 and 100 million species? The scientific answer involves *biological evolution:* the description of how the earth's life changes over time.

Before modern science, the primary explanation was that all life was created by God (or gods). And many people still see this as the only possible explanation.

The idea that organisms change over time and are descended from a single common ancestor has been around in one form or another since the early Greek philosophers. But no one had come up with a credible explanation of how this could happen.

This changed in 1858 when naturalists Charles Darwin (1809–1882) and Alfred Russel Wallace (1823–

1913) independently proposed the concept of *natural selection* as a mechanism for *biological evolution* of the earth's huge variety of life-forms. Although Wallace also proposed the idea of natural selection, it was Darwin who meticulously gathered evidence for this idea and published it in 1859 in his book, *On the Origin of Species by Means of Natural Selection.*

Darwin and Wallace observed that organisms must constantly struggle to obtain enough food and other resources to survive and reproduce. They also observed that individuals in a population with some edge over other individuals are more likely to survive, reproduce, and have offspring with similar survival skills. They concluded that these survival traits would become more prevalent in future populations of the species through a process called **natural selection.**

A huge body of field and laboratory evidence has supported this idea. As a result, biological evolution through natural selection has become a *scientific theory* accepted by an overwhelming majority of biologists. However, it is not a complete theory, its details are continually being debated and tweaked, and there are other ways that new species can form.

There are still many unanswered questions and scientific debates over the details of evolution by natural selection. But considerable evidence suggests that life on the earth developed in two phases over the past 4.7 billion years (Figure 4-2, p. 84).

The first phase involved *chemical evolution* of the organic molecules, biopolymers, and systems of chemical reactions needed to form the first cells. It took about 1 billion years.

Evidence for this phase comes from chemical analysis and measurements of radioactive elements in primitive rocks and fossils. Chemists have also conducted laboratory experiments showing how simple inorganic compounds believed to be in the earth's early atmosphere might have reacted to produce amino acids, simple sugars, and other organic molecules used as building blocks for the proteins, complex carbohydrates, RNA, and DNA needed for life.

 Learn more about one of the most famous experiments exploring how the molecules necessary for early life might have formed at ThomsonNOW.

Fossil and other evidence indicates that chemical evolution was followed by *biological evolution* by natural selection from single-celled bacteria to multicellular protists, plants, fungi, and animals. This second phase has been going on for about 3.7 billion years, as summarized in Figure 4-3 (p. 84).

 Get a detailed look at early biological evolution by natural selection —the roots of the tree of life—at ThomsonNOW.

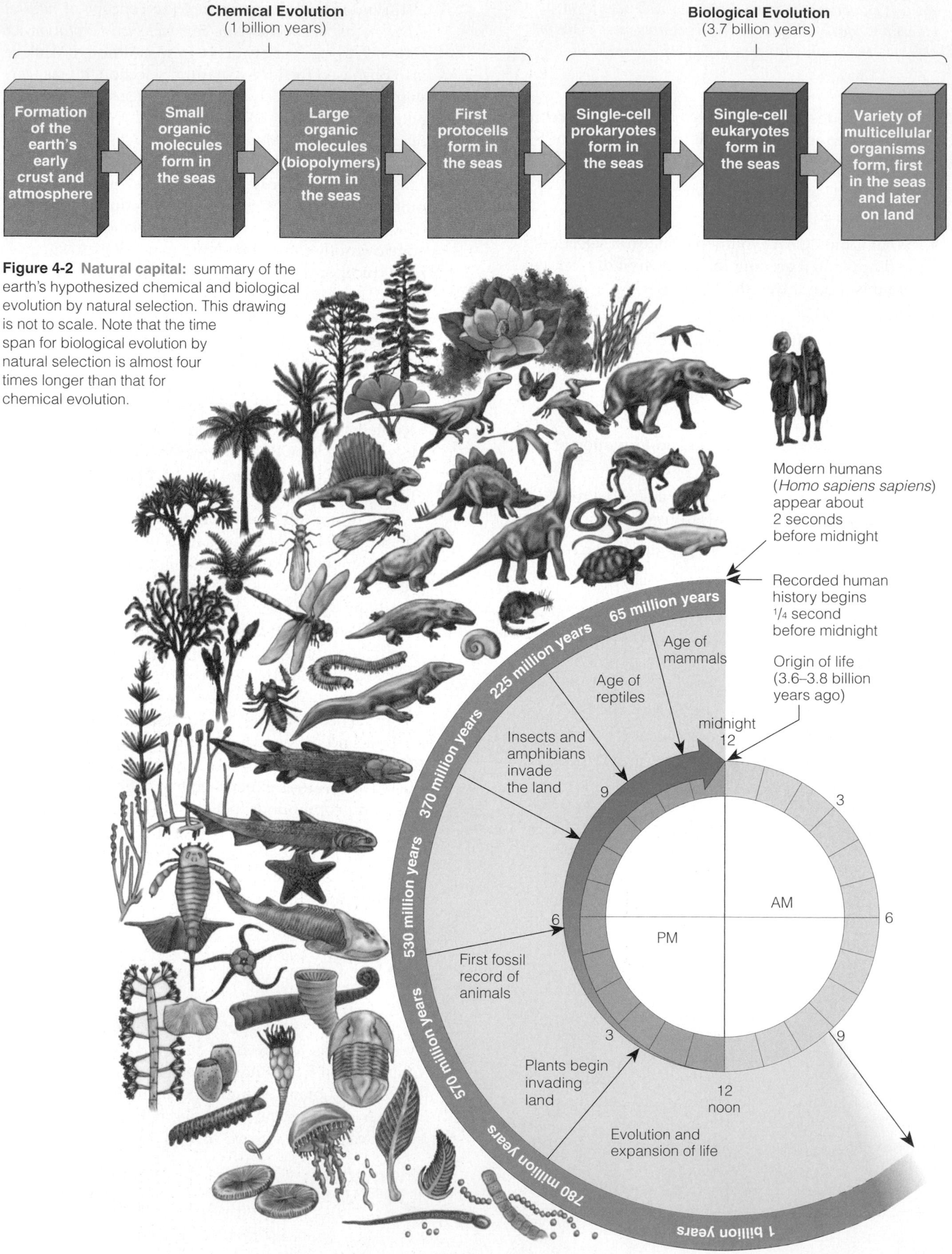

Figure 4-2 Natural capital: summary of the earth's hypothesized chemical and biological evolution by natural selection. This drawing is not to scale. Note that the time span for biological evolution by natural selection is almost four times longer than that for chemical evolution.

How Do We Know Which Organisms Lived in the Past?

Our knowledge about past life comes from fossils, chemical analysis, cores drilled out of buried ice, and DNA analysis.

Most of what we know of the earth's life history comes from **fossils:** mineralized or petrified replicas of skeletons, bones, teeth, shells, leaves, and seeds, or impressions of such items found in rocks. Fossils provide physical evidence of ancient organisms and reveal what their internal structures looked like (Figure 4-4). Scientists also drill cores from glacial ice and examine the kinds of life found at different layers. In addition, they compare the DNA of past and current organisms.

The world's cumulative body of fossils found is called the *fossil record*. This record is uneven and incomplete. Some forms of life left no fossils, and some fossils have decomposed. The fossils found so far probably represent only 1% of all species that have ever lived. Trying to reconstruct the development of life with so little evidence is a challenging scientific detective game for paleontologists. *Green Career:* Paleontologist

EVOLUTION, NATURAL SELECTION, AND ADAPTATION

Genetic Mutations: Changes in a Population's Gene Pool

The most widely accepted idea is that biological evolution by natural selection results from changes in a population's genetic makeup over time.

According to scientific evidence, **biological evolution** by natural selection involves the change in a population's genetic makeup through successive generations. Note that *populations—not individuals—evolve by becoming genetically different*. Religious and other groups may offer other explanations, but this is the most widely accepted *scientific explanation* of how life on earth has developed as a result of natural selection.

Figure 4-3 (facing page) **Natural capital:** greatly simplified overview of the biological evolution by natural selection of life on the earth, which was preceded by about 1 billion years of chemical evolution. Microorganisms (mostly bacteria) that lived in water dominated the early span of biological evolution on the earth, between about 3.7 billion and 1 billion years ago. Plants and animals evolved first in the seas. Fossil and recent DNA evidence suggests that plants began invading the land some 780 million years ago, and animals began living on land some 370 million years ago. Humans arrived on the scene only a very short time ago—equivalent to less than an eye blink of the earth's roughly 3.7-billion-year history of biological evolution.

Kevin Schafer/Peter Arnold, Inc.

Figure 4-4 Fossilized skeleton of an herbivore that lived during the Cenozoic era from 26–66 million years ago.

The first step in this process is the development of *genetic variability* in a population. Such genetic variety occurs through **mutations:** random changes in the structure or number of DNA molecules (see Figure 9 on p. S31 in Supplement 7) in a cell that can be inherited by offspring. Mutations can occur in two ways. One is by exposure of DNA to external agents such as radioactivity, X rays, and natural and human-made chemicals (called *mutagens*). The other results from random mistakes that sometimes occur in coded genetic instructions when DNA molecules are copied each time a cell divides and when an organism reproduces.

Mutations can occur in any cells, but only those in reproductive cells are passed on to offspring. Some mutations are harmless, but most are lethal. *Every so often, a mutation is beneficial.* The result is new genetic traits that give an individual and its offspring better chances for survival and reproduction under existing environmental conditions or when such conditions change.

Natural Selection and Adaptation: Leaving More Offspring With Beneficial Genetic Traits

Some members of a population may have genetic traits that enhance their ability to survive and produce offspring with these traits.

The next step in conventional biological evolution is **natural selection.** It occurs when some individuals of a population have genetically based traits that increase their chances of survival and their ability to produce offspring with the same traits. For example, in the face of snow and cold, individuals with thicker skin than others of its species might live longer.

As those individuals mate, genes for thicker skin spread throughout the population and individuals with those genes increase in number and pass this

with those genes increase in number and pass this helpful trait on to their offspring. Thus, natural selection explains how populations adapt to changes in environmental conditions.

Three conditions are necessary for the biological evolution of a population by natural selection. *First,* there must be enough *genetic variability* for a trait to exist in a population. *Second,* the trait must be *heritable,* meaning that it can be passed from one generation to another. *Third,* the trait must lead to **differential reproduction.** This means it must enable individuals with the trait to leave more offspring than other members of the population. Note that natural selection acts on individuals, but evolution occurs in populations.

 Learn more about two special types of natural selection, one stabilizing and the other disruptive, at ThomsonNOW.

An **adaptation,** or **adaptive trait,** is any heritable trait that enables an organism to survive through natural selection and reproduce better under prevailing environmental conditions. Natural selection tends to preserve beneficial adaptations in populations and discard harmful ones.

When faced with a change in environmental conditions, a population of a species has three possibilities: *adapt* to the new conditions through natural selection, *migrate* (if possible) to an area with more favorable conditions, or *become extinct.*

The process of biological evolution by natural selection can be summarized simply: *Genes mutate, individuals are selected, and populations evolve that are better adapted to survive and reproduce under existing environmental conditions.* Figure 1 on p. S34 in Supplement 8 gives an overview of how life evolved into six different kingdoms of species as a result of natural selection.

 ? *THINKING ABOUT NATURAL SELECTION AND THE EARTH'S ADAPTABILITY* Explain how natural selection can contribute to the earth's ability to adapt to environmental changes (Core Case Study, p. 82).

 How many moths can you eat? Find out and learn more about adaptation at ThomsonNOW.

Coevolution: A Biological Arms Race

Interacting species can engage in a back-and-forth genetic contest in which each gains a temporary genetic advantage over the other.

Some biologists have proposed that when populations of two different species interact over a long time, changes in the gene pool of one species can lead to changes in the gene pool of the other. This process is called **coevolution.** In this give-and-take evolutionary game, each species is in a genetic race to produce the largest number of surviving offspring.

Consider the interactions between bats and moths. Bats like to eat moths, and they hunt at night and use echolocation to navigate and locate their prey. To do so, they emit extremely high frequency, high-intensity pulses of sound. They analyze the returning echoes to create a sonic "image" of their prey. (We have copied this natural technology by using sonar to detect submarines, whales, and schools of fish.)

As a countermeasure, some moth species have evolved ears that are especially sensitive to the sound frequencies that bats use to find them. When the moths hear the bat frequencies, they try to escape by falling to the ground or flying evasively.

Some bat species evolved ways to counter this defense by switching the frequency of their sound pulses. In turn, some moths evolved their own high-frequency clicks to jam the bats' echolocation system (we have also learned to jam radar). Some bat species then adapted by turning off their echolocation system and using the moths' clicks to locate their prey.

Coevolution is like an arms race between interacting populations of different species. Sometimes the predators surge ahead; at other times the prey get the upper hand.

Hybridization and Gene Swapping: Other Ways to Exchange Genes

Sometimes different species can crossbreed to form hybrids that can survive and reproduce, and some species can exchange genes without sexual reproduction.

According to conventional evolutionary theory by natural selection, different sexually reproducing species cannot crossbreed to produce live fertile offspring. However, recent evidence indicates that some new species can arise through *hybridization.* It occurs when individuals of two distinct species crossbreed to produce an individual or *hybrid* that in some cases has a better ability to survive than conventional offspring of the two parent species.

Biologists are also finding that some species (mostly microorganisms) can exchange genes without sexual reproduction. This gene swapping process, known as *horizontal gene transfer,* can occur when one species feeds upon, infects, or comes into close contact with another species (such as a bacterium or virus) and transfers bits of genetic information from one species to the other. Hybridization and gene transfers and the resulting adaptations can occur rapidly compared to the thousands to millions of years required for the conventional Darwinian evolution of sexually reproducing species through natural selection.

F *RESEARCH FRONTIER* Better understanding of the roles of hybridization and gene transfer in biological evolution

Limits on Adaptation through Natural Selection

A population's ability to adapt to new environmental conditions through natural selection is limited by its gene pool and how fast it can reproduce.

In the not too distant future, will adaptations through natural selection to new environmental conditions allow our skin to become more resistant to the harmful effects of ultraviolet radiation, our lungs to cope with air pollutants, and our liver to better detoxify pollutants?

The answer is *no* because of two limits to adaptations in nature through conventional natural selection. *First*, a change in environmental conditions can lead to adaptation through conventional natural selection only for genetic traits already present in a population's gene pool. You must have genetic dice to play the genetic natural selection dice game.

Second, even if a beneficial heritable trait is present in a population, the population's ability to adapt may be limited by its reproductive capacity. Populations of genetically diverse species that reproduce quickly—such as weeds, mosquitoes, rats, bacteria, or cockroaches—often adapt to a change in environmental conditions in a short time. In contrast, species that cannot produce large numbers of offspring rapidly—such as elephants, tigers, sharks, and humans—take a long time (typically thousands or even millions of years) to adapt through natural selection. It helps to be able to throw the genetic dice fast.

Here is some *bad news* for most members of a population. Even when a favorable genetic trait is present in a population, most of the population would have to die or become sterile so individuals with the trait could predominate and pass the trait on through natural selection. As a result, most players get kicked out of the genetic dice game before they have a chance to win. This means that most members of the human population would have to die prematurely for hundreds of thousands of generations for a new genetic trait to predominate through conventional natural selection. This is hardly a desirable solution to the environmental problems we face.

However, these limitations do not apply to development of new species through hybridization and the exchange of genes between different species without sexual reproduction.

Common Myths about Evolution through Natural Selection

Evolution through natural selection is about leaving the most descendants; organisms do not develop certain traits because they need them or want them; and there is no master plan leading to genetic perfection.

There are three common misconceptions about biological evolution through conventional natural selection. One is that "survival of the fittest" means "survival of the strongest." To biologists, *fitness* is a measure of reproductive success, not strength. Thus, the fittest individuals are those that leave the most descendants.

Another misconception is that organisms develop certain traits because they need or want them. A giraffe does not have a very long neck because it needs or wants it to feed on vegetation high in trees. Rather, some ancestor had a gene for long necks that gave it an advantage over other members of its population in getting food, and that giraffe produced more offspring with long necks.

A third misconception is that evolution by natural selection involves some grand plan of nature in which species become more perfectly adapted. From a scientific standpoint, no plan or goal of genetic perfection has been identified in the evolutionary process.

? *THINKING ABOUT EVOLUTION* Do you accept or reject the scientific theory of biological evolution by natural selection? If you reject it, what *specific* mechanism do you believe can account for the variety of life on the earth and what is the accepted and peer-reviewed scientific evidence for such a mechanism?

GEOLOGIC PROCESSES, CLIMATE CHANGE, CATASTROPHES, AND EVOLUTION

Geologic Processes and Evolution by Natural Selection

The very slow movement of huge solid plates making up the earth's surface, volcanic eruptions, and earthquakes can wipe out existing species and help form new ones.

The earth's surface has changed dramatically over its long history. Scientists have discovered that huge flows of molten rock within the earth's interior break the earth's surface into a series of gigantic solid plates, called *tectonic plates*. These plates have very slowly drifted back and forth across the planet's surface (Figure 4-5, p. 88) over thousands to millions of years— much like gigantic pieces of extremely slow moving ice floating on the surface of the ocean.

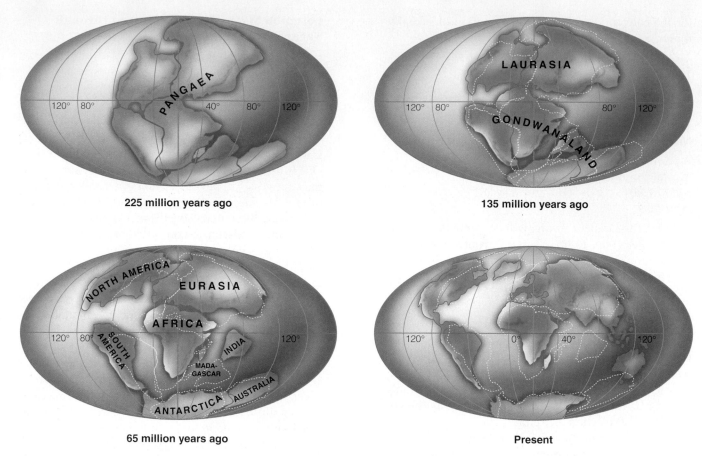

225 million years ago

135 million years ago

65 million years ago

Present

Figure 4-5 Geological processes and biological evolution. Over millions of years the earth's continents have moved very slowly on several gigantic tectonic plates. This process plays a role in the extinction of species as land areas split apart and promote the rise of new species when once isolated land areas combine. Rock and fossil evidence indicates that 200–250 million years ago all of the earth's present-day continents were locked together in a supercontinent called Pangaea (top left). About 180 million years ago, Pangaea began splitting apart as the earth's huge plates separated and eventually resulted in today's locations of the continents (bottom right).

This process has had two important effects on the evolution and location of life on the earth. *First,* the locations of continents and oceanic basins greatly influence the earth's climate and thus help determine where plants and animals can live.

Second, the movement of continents has allowed species to move, adapt to new environments, and form new species through natural selection. When continents join together populations can disperse to new areas and adapt to new environmental conditions. And when continents separate, populations must evolve under isolated conditions or become extinct.

Volcanic eruptions (see Figure 4 on p. S45 in Supplement 11) can also affect biological evolution by destroying habitats and reducing or wiping out populations of species. On the other hand, deposits of lava can yield a soil that can provide habitats for some species.

Earthquakes (see Figure 1 on p. S43 in Supplement 11) can separate and isolate populations of spe-

cies. Over long periods of time, this can lead to the formation of new species as each isolated population changes genetically in response to new environmental conditions.

Climate Change and Natural Selection

Changes in climate throughout the earth's history have shifted where plants and animals can live.

Throughout its long history the earth's climate has changed drastically. Sometimes it has cooled and covered much of the earth with ice. At other times it has warmed, melted ice, and drastically raised sea levels. Such alternating periods of cooling and heating have led to the retreat and advance of ice sheets at high latitudes over much of the northern hemisphere, as recently as 18,000 years ago (Figure 4-6).

These long-term climate changes have a major effect on biological evolution by determining where dif-

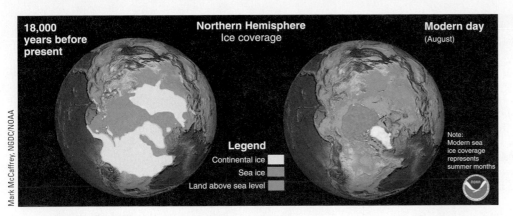

Mark McCaffrey, NGDC/NOAA

Figure 4-6 Changes in ice coverage in the northern hemisphere during the past 18,000 years. (Data from the National Oceanic and Atmospheric Administration)

ferent types of plants and animals can live and thrive and by changing the locations of different types of ecosystems such as deserts, grasslands, and forests. Some existing species became extinct because the climate changed too rapidly for them to survive and new species evolved.

Catastrophes and Natural Selection

Asteroids and meteorites hitting the earth and large upheavals of the earth's crust from geological processes have wiped out large numbers of species and created opportunities for the evolution by natural selection of new species.

Large asteroids have collided with the earth many times over the past 4.7 billion years. Sometimes such impacts have caused widespread destruction of ecosystems and wiped out large numbers of species.

Each major catastrophe resulted in long periods of extreme environmental stress. This changed the course of evolution by wiping out large numbers of existing species. But such mass extinctions opened up opportunities for the evolution by natural selection of new species and shifts in the locations of different types of ecosystems. On a long-term basis, the earth's four scientific principles of sustainability, especially its biodiversity (Figure 3-15, p. 62), have enabled the earth to adapt to drastic changes in environmental conditions.

ECOLOGICAL NICHES AND ADAPTATION

Ecological Niches: How Species Live and Coexist

Each species in an ecosystem has a specific role or way of life.

If asked what role a certain species, such as an alligator, plays in an ecosystem, an ecologist would describe its **ecological niche,** or simply **niche** (pronounced "nitch"). It is a species' way of life or role in a community or ecosystem and includes everything that affects its survival and reproduction.

An important principle of ecology is that *each species has a distinct niche or role to play in the ecosystems where it is found.* A species' **fundamental niche** consists of the full potential range of physical, chemical, and biological conditions and resources it could theoretically use if it could avoid direct competition from other species. Of course, in a particular ecosystem, different species often compete with one another for the same resources. In short, the niches of competing species overlap.

To survive and avoid competition for the same resources, a species usually occupies only part of its fundamental niche in a particular community or ecosystem—what ecologists call its **realized niche.** By analogy, you may be capable of being president of a particular company (your *fundamental professional niche*), but competition from others may mean you become only a vice president (your *realized professional niche*).

Generalist and Specialist Species: Broad and Narrow Niches

Some species have broad ecological roles and others have narrower or more specialized roles.

Scientists use the niches of species to classify them broadly as *generalists* or *specialists.* **Generalist species** have broad niches (Figure 4-7, p. 91, right curve). They can live in many different places, eat a variety of foods, and tolerate a wide range of environmental conditions. Flies, cockroaches (Spotlight, p. 90), rats, white-tailed deer, raccoons, coyotes, copperheads, starlings, humans, and many weeds are generalist species.

Specialist species occupy narrow niches (Figure 4-7, left curve). They may be able to live in only one type of habitat, use one or a few types of food, or tolerate a narrow range of climatic and other environmental conditions. This makes specialists more prone to extinction when environmental conditions change.

Cockroaches: Nature's Ultimate Survivors

Cockroaches (Figure 4-A), the bugs many people love to hate, have been around for 350 million years—longer than the dinosaurs lasted. One of evolution's great success stories, they have thrived because they are *generalists*.

The earth's 3,500 cockroach species can eat almost anything, including algae, dead insects, fingernail clippings, salts in tennis shoes, electrical cords, glue, paper, and soap. They can also live and breed almost anywhere except in polar regions.

Some cockroach species can go for a month without food, survive for a month on a drop of water from a dishrag, and withstand massive doses of radiation. One species can survive being frozen for 48 hours.

Cockroaches usually can evade their predators and a human foot in hot pursuit because most species have antennae that can detect minute movements of air, vibration sensors in their knee joints, and rapid response times (faster than

you can blink). Some even have wings. They also have compound eyes that allow them to see in almost all directions at once. Each eye has about 2,000 lenses, compared to one in each of your eyes.

Clemson University–USDA Cooperative Extension Slide Series. www.forestryimages.org

Figure 4-A As generalists, cockroaches are among the earth's most adaptable and prolific species. This is a photo of an American cockroach.

They also have high reproductive rates. In only a year, a single Asian cockroach (especially prevalent in Florida) and its offspring can add about 10 million new cockroaches to the world. Their high reproductive rate also helps them quickly de-

velop genetic resistance to almost any poison we throw at them.

Most cockroaches also sample food before it enters their mouths and learn to shun foul-tasting poisons. They also clean up after themselves by eating their own dead and, if food is scarce enough, their living.

Only about 25 species of cockroach live in homes. Such species can carry viruses and bacteria that cause diseases such as hepatitis, polio, typhoid fever, plague, and salmonella. They can also cause people to have allergic reactions ranging from watery eyes to severe wheezing. About 60% of Americans suffering from asthma are allergic to live or dead cockroaches.

On the other hand cockroaches play a role in nature's food webs. They are a tasty meal for birds and lizards.

Critical Thinking

If you could, would you exterminate all cockroach species? What might be some ecological consequences of this action?

For example, *tiger salamanders* breed only in fishless ponds where their larvae will not be eaten. China's *giant panda* is a highly endangered species because of a combination of habitat loss, low birth rate, and its highly specialized diet consisting mostly of various types of bamboo. Some shorebirds also occupy specialized

niches, feeding on crustaceans, insects, and other organisms on sandy beaches and their adjoining coastal wetlands (Figure 4-8).

Is it better to be a generalist or a specialist? It depends. When environmental conditions are fairly constant, as in a tropical rain forest, specialists have an

Black skimmer seizes small fish at water surface

Brown pelican dives for fish, which it locates from the air

Scaup and other diving ducks feed on mollusks, crustaceans, and aquatic vegetation

Avocet sweeps bill through mud and surface water in search of small crustaceans, insects, and seeds

Flamingo feeds on minute organisms in mud

Louisiana heron wades into water to seize small fish

Oystercatcher feeds on clams, mussels, and other shellfish into which it pries its narrow beak

advantage because they have fewer competitors. But under rapidly changing environmental conditions, the generalist usually is better off than the specialist.

Natural selection can lead to an increase in specialized species when several species must compete intensely for scarce resources. Over time one species may evolve into a variety of species with different adaptations that reduce competition and allow them to share limited resources.

Birds called honeycreepers that live on the island of Hawaii illustrate this *evolutionary divergence.* Starting from a single ancestor species, numerous honeycreeper

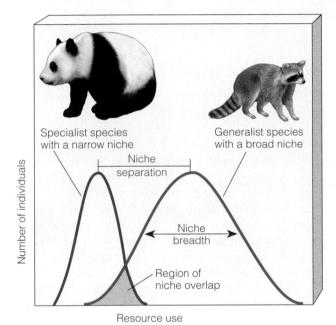

Figure 4-7 Overlap of the niches of two different species: a specialist and a generalist. In the overlap area, the two species compete for one or more of the same resources. As a result, each species can occupy only a part of its *fundamental niche;* the part it occupies is its *realized niche.* Generalist species such as a raccoon have a broad niche (right), and specialist species such as the giant panda have a narrow niche (left).

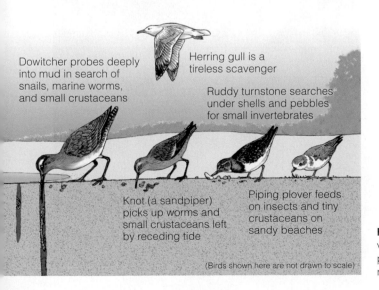

Dowitcher probes deeply into mud in search of snails, marine worms, and small crustaceans

Herring gull is a tireless scavenger

Ruddy turnstone searches under shells and pebbles for small invertebrates

Knot (a sandpiper) picks up worms and small crustaceans left by receding tide

Piping plover feeds on insects and tiny crustaceans on sandy beaches

(Birds shown here are not drawn to scale)

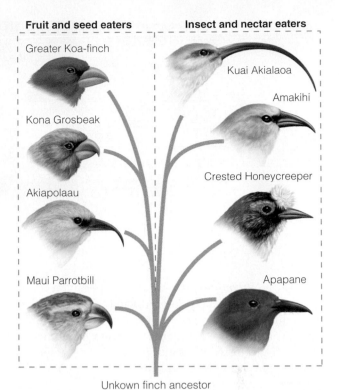

Fruit and seed eaters

Insect and nectar eaters

Greater Koa-finch

Kuai Akialaoa

Amakihi

Kona Grosbeak

Akiapolaau

Crested Honeycreeper

Maui Parrotbill

Apapane

Unkown finch ancestor

Figure 4-9 **Natural capital:** evolutionary divergence of honeycreepers into specialized ecological niches. Each species has a beak specialized to take advantage of certain types of food resources.

species evolved with different types of beaks specialized to feed on food sources such as specific types of insects, nectar from particular types of flowers, and certain types of seeds and fruit (Figure 4-9).

SPECIATION, EXTINCTION, AND BIODIVERSITY

How Do New Species Evolve?

A new species can arise when members of a population are isolated from other members for so long that changes in their genetic makeup prevent them from producing fertile offspring if they get together again.

Under certain circumstances, natural selection can lead to an entirely new species. In this process, called **speciation,** two species arise from one. For sexually reproducing species, a new species is formed when some members of a population can no longer breed with other members to produce fertile offspring.

Figure 4-8 **Natural capital:** specialized feeding niches of various bird species in a coastal wetland. Such resource partitioning reduces competition and allows sharing of limited resources.

Figure 4-10 *Geographic isolation* can lead to reproductive isolation, divergence of gene pools, and speciation.

Early fox population → Spreads northward and southward and separates

Northern population

Southern population

Arctic Fox Adapted to cold through heavier fur, short ears, short legs, short nose. White fur matches snow for camouflage.

Different environmental conditions lead to different selective pressures and evolution into two different species.

Gray Fox Adapted to heat through lightweight fur and long ears, legs, and nose, which give off more heat.

The most common mechanism of speciation (especially among sexually reproducing animals) takes place in two phases: geographic isolation and reproductive isolation. **Geographic isolation** occurs when different groups of the same population of a species become physically isolated from one another for long periods. For example, part of a population may migrate in search of food and then begin living in another area with different environmental conditions. Populations can become separated by a physical barrier (such as a mountain range, stream, lake, or road), by a change such as a volcanic eruption or earthquake, or by a few individuals being carried to a new area by wind or flowing water.

In **reproductive isolation,** mutation and change by natural selection operate independently in the gene pools of geographically isolated populations. If this process continues long enough, members of the geographically and reproductively isolated populations of sexually reproducing species may become so different in genetic makeup that they cannot produce live, fertile offspring if they get together again. Then one species has become two, and speciation has occurred (Figure 4-10).

For some rapidly reproducing organisms, this type of speciation may occur within hundreds of years. For most species, such speciation takes from tens of thousands to millions of years—making it difficult to observe and document the appearance of a new species. However, some species can speed up the process of evolution by forming new species through hybridization and gene swapping (p. 86).

Learn more about different types of speciation and ways in which they occur at ThomsonNOW.

 THINKING ABOUT SPECIATION AND THE EARTH'S ADAPTABILITY Explain how speciation can contribute to the earth's ability to adapt to environmental changes (Core Case Study, p. 82).

Extinction: Lights Out

A species becomes extinct when its populations cannot adapt to changing environmental conditions.

Another process affecting the number and types of species on the earth is **extinction,** in which an entire species ceases to exist. Species that are found in only one area are called **endemic species** and are especially vulnerable to extinction. They exist on islands and in other unique small areas, especially in tropical rainforests where most species are highly specialized. An example is the brilliantly colored golden toad (Figure 4-11) found only in a small area of lush cloud forests in Costa Rica's mountainous region. By 1989, it had apparently become extinct because of atmospheric warming that reduced the moisture in its forest habitat.

F *RESEARCH FRONTIER* Learning more about causes of extinction

Michael P. Fogden/Bruce Coleman USA

Figure 4-11 Depleted natural capital: male golden toad in Costa Rica's high-altitude Monteverde cloud forest. This species has recently become extinct because changes in climate dried up its habitat.

Background Extinction, Mass Extinction, and Mass Depletion

All species eventually become extinct, but drastic changes in environmental conditions can eliminate large groups of species.

Extinction is the ultimate fate of all species, just as death is for all individual organisms. Biologists estimate that 99.9% of all the species that ever existed are now extinct.

As local environmental conditions change, a certain number of species disappear at a low rate, called **background extinction.** Based on the fossil record and analysis of ice cores, biologists estimate that the average annual background extinction rate is one to five species for each million species on the earth.

In contrast, **mass extinction** is a significant rise in extinction rates above the background level. In such a catastrophic, widespread (often global) event, large groups of existing species (perhaps 25–70%) are wiped out in a geological period lasting up to 5 million years. Fossil and geological evidence indicates that the earth's species have experienced five mass extinctions (20–60 million years apart) during the past 500 million years (Figure 4-12).

Scientists have also identified periods of **mass depletion** in which extinction rates are higher than

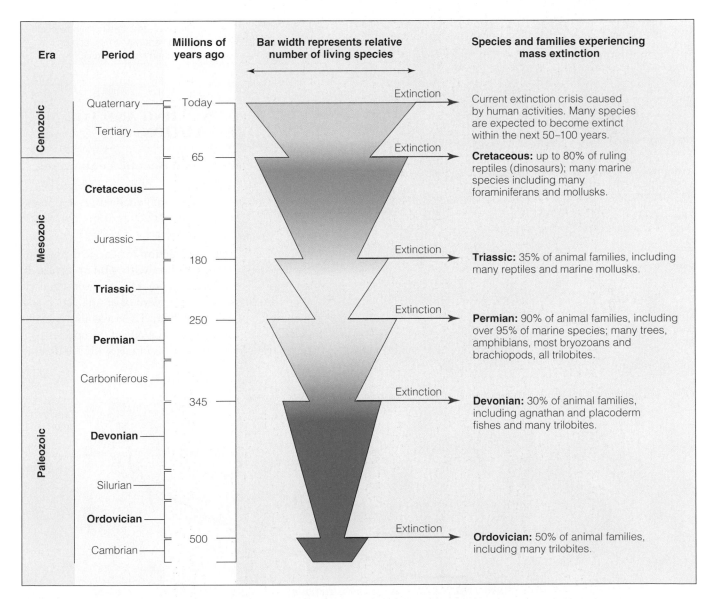

Figure 4-12 Fossils and radioactive dating indicate that five major *mass extinctions* (indicated by arrows) have taken place over the past 500 million years. Mass extinctions leave many organism roles (niches) unoccupied and create new niches. Each mass extinction has been followed by periods of recovery (represented by the wedge shapes) called *adaptive radiations*. During these periods, which last 10 million years or longer, new species evolve to fill new or vacated niches. Many scientists say that we are now in the midst of a *sixth mass extinction*, caused primarily by human activities.

normal but not high enough to classify as a mass extinction. In both types of events, large numbers of species have become extinct.

A mass extinction or mass depletion crisis for some species is an opportunity for other species that can fill unoccupied niches or newly created ones. The existence of millions of species today means that speciation, on average, has kept ahead of extinction, especially during the last 250 million years (Figure 4-13).

THINKING ABOUT NATURAL SELECTION AND THE EARTH'S ADAPTABILITY Explain how extinction can contribute to the earth's ability to adapt to environmental changes (Core Case Study, p. 82).

Effects of Human Activities on the Earth's Biodiversity: Are We a Wise Species?

The scientific consensus is that human activities are decreasing the earth's biodiversity.

Speciation minus extinction equals *biodiversity*, the planet's genetic raw material for future evolution in response to changing environmental conditions. Extinction is a natural process. But much evidence indicates that humans have become a major force in the premature extinction of a growing number of species.

During the twentieth century, according to biologists Stuart Primm and Edward O. Wilson and the 2005 Millennium Ecosystem Assessment, extinction rates increased by 100–1,000 times the natural background extinction rate. As human population and resource consumption increase over the next 50–100 years, we are expected to take over a larger share of the earth's surface and net primary productivity (NPP) (Figure 3-21, p. 66) and degrade or destroy more of the planet's wildlife habitats.

According to Wilson and Primm, this may cause the premature extinction of at least one-fifth of the earth's current species by 2030 and up to half of those species by the end of this century. This could constitute a new mass depletion and possibly a new mass extinction. Wilson says that if we make an "all-out effort to save the biologically richest parts of the world, the amount of loss can be cut at least by half."

On our short time scale, such major losses cannot be recouped by formation of new species; it took millions of years after each of the earth's past mass extinctions and depletions for life to recover to the previous level of biodiversity. We are also destroying or degrading ecosystems such as tropical forests, coral reefs, and wetlands that are centers for future speciation. See the Guest Essay on this topic by Norman Myers on the website for this chapter.

RESEARCH FRONTIER Better understanding of how evolutionary processes produce biodiversity and how ecological processes help sustain the resulting biodiversity

GENETIC ENGINEERING AND THE FUTURE OF EVOLUTION

Artificial Selection and Genetic Engineering

We selectively breed members of populations to produce offspring with certain genetic traits and use genetic engineering to transfer genes from one species to another.

We have used **artificial selection** to change the genetic characteristics of populations with similar genes. In this process, we select one or more desirable genetic traits in the population of a plant or animal, such as a type of wheat, fruit, or dog. Then we use *selective breeding* to end up with populations of the species containing large numbers of individuals with the desired traits.

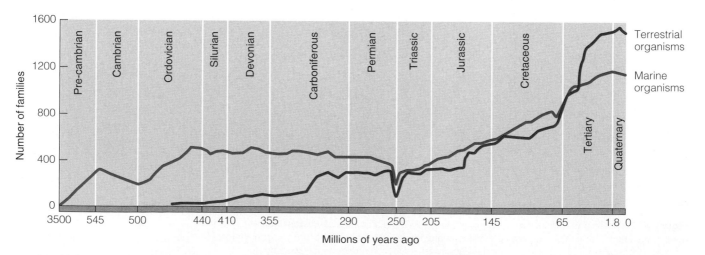

Figure 4-13 Natural capital: changes in the earth's biodiversity over geological time. The biological diversity of life on land and in the oceans has increased dramatically over the last 3.5 billion years, especially during the past 250 million years. During the last 1.8 million years this increase has leveled off.

Artificial selection has yielded food crops with higher yields, cows that give more milk, trees that grow faster, and many different types of dogs and cats. But traditional crossbreeding is a slow process. And it can combine traits only from species that are close to one another genetically.

Today scientists are using genetic engineering to speed up our ability to manipulate genes. **Genetic engineering,** or **gene splicing,** is the alteration of an organism's genetic material through adding, deleting, or changing segments of its DNA (see Figure 9 on p. S31 in Supplement 7), to produce desirable traits or eliminate negative ones. This enables scientists to transfer genes between different species that would not interbreed in nature. For example, genes from a fish species can be put into a tomato or strawberry.

A key tool used in genetic engineering is **recombinant DNA,** which is DNA that has been altered to contain genes or portions of genes from organisms of different species. Organisms that have been genetically engineered by use of recombinant DNA technology are called **genetically modified organisms (GMOs),** or **transgenic organisms.** Figure 4-14 outlines the steps involved in developing a genetically modified plant.

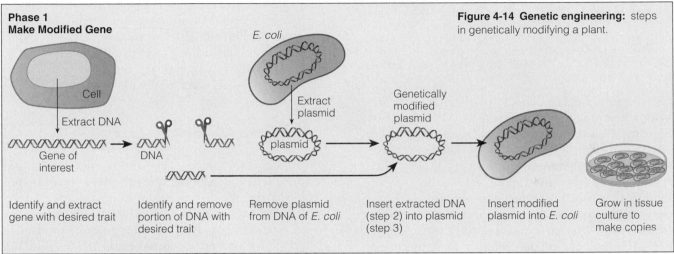

Phase 1
Make Modified Gene

E. coli

Cell

Extract DNA

Gene of interest

DNA

Extract plasmid

plasmid

Genetically modified plasmid

Figure 4-14 Genetic engineering: steps in genetically modifying a plant.

Identify and extract gene with desired trait

Identify and remove portion of DNA with desired trait

Remove plasmid from DNA of *E. coli*

Insert extracted DNA (step 2) into plasmid (step 3)

Insert modified plasmid into *E. coli*

Grow in tissue culture to make copies

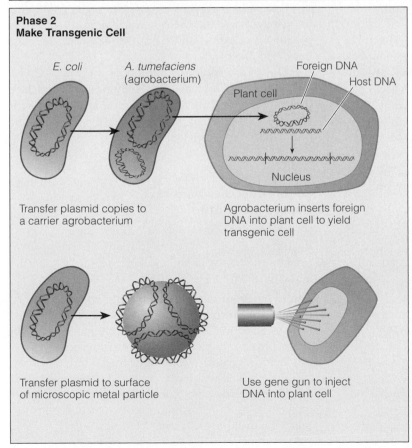

Phase 2
Make Transgenic Cell

E. coli

A. tumefaciens (agrobacterium)

Foreign DNA

Host DNA

Plant cell

Nucleus

Transfer plasmid copies to a carrier agrobacterium

Agrobacterium inserts foreign DNA into plant cell to yield transgenic cell

Transfer plasmid to surface of microscopic metal particle

Use gene gun to inject DNA into plant cell

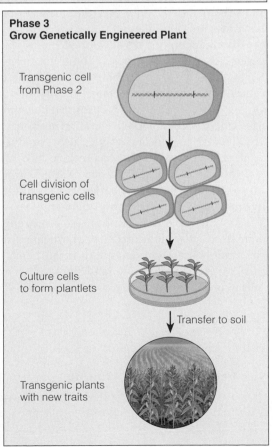

Phase 3
Grow Genetically Engineered Plant

Transgenic cell from Phase 2

Cell division of transgenic cells

Culture cells to form plantlets

Transfer to soil

Transgenic plants with new traits

Figure 4-15 An example of genetic engineering. The six-month-old mouse on the left is normal; the same-age mouse on the right has a human growth hormone gene inserted in its cells. Mice with the human growth hormone gene grow two to three times faster and reach a size twice that of mice without the gene.

Gene splicing takes about half as much time to develop a new crop or animal variety as traditional cross-breeding does and costs less. And genetic engineering allows us to transfer traits between different types of organisms without breeding them.

Scientists have used gene splicing to develop modified crop plants, genetically engineered drugs, pest-resistant plants, and animals that grow rapidly (Figure 4-15). They have also created genetically engineered bacteria to extract minerals such as copper from their underground ores and clean up oil spills and other toxic pollutants.

Bioengineers have developed many beneficial GMOs: chickens that lay low-cholesterol eggs, tomatoes with genes that can help prevent some types of cancer, and bananas and potatoes that contain oral vaccines to treat certain viral diseases in developing countries where needles and refrigeration are not available.

Genetic engineers have also produced the *Schwarzenegger mouse* that has muscle building genes and the *marathon mouse* that never seems to tire. And they are in hot pursuit of a *Methuselah mouse* that can live much longer than a conventional mouse.

Researchers envision using genetically engineered animals to act as biofactories for producing drugs, vaccines, antibodies, hormones, industrial chemicals such as plastics and detergents, and human body organs. This new field is called *biopharming*.

F **RESEARCH FRONTIER** Biopharming and other applications of genetic engineering

Synthetic Biology and Cloning

Biologists are learning to rebuild organisms from their cell components and to make identical copies or clones of organisms.

Some biologists are trying to get around the problems and uncertainty associated with genetic engineering. The goal of this new field, called *synthetic biology,* is to go beyond conventional genetic engineering, separate cells into their fundamental components, and use them to rebuild new organisms.

Some synthetic biologists have created a polio virus by stitching together various individual genes and others are working on getting *E. coli* bacteria to produce artenisin, a drug used to treat malaria that is now extracted from the wormwood tree. Israeli scientists have created the world's smallest computer by reengineering DNA to carry out mathematical functions. Synthetic biologists also plan to string various genes together to create from scratch unique organisms that can produce alternative fuels such as ethanol and hydrogen. Some biologists envision that biological engineers will one day sit at computers writing genetic instructions for making specific cells, like today's software developers. *Green Career:* Biological engineer

F **RESEARCH FRONTIER** Synthetic biology

Genetic engineers have also learned how to produce a *clone*—a genetically identical version of an individual in a population. Scientists have made clones of domestic animals such as sheep, horses, cows, and dogs and may someday be able to clone humans—a possibility that excites some people and horrifies others. Researchers also hope to use cloning of DNA samples from cells found in the bones, hair, teeth, or frozen specimens of extinct animals to bring vanished species back from the dead.

However, cloning experiments on animals have led to several problems for the clones. They include high miscarriage rates, rapid aging, a shortened life span, and defects of the kidneys, liver, heart, and brain. Researchers are working on ways to reduce such problems.

Some Concerns about the Genetic Revolution

Genetic engineering has great promise for improving the human condition, but it is an unpredictable process and raises a number of privacy, ethical, legal, and environmental issues.

The hype about genetic engineering suggests that its results are controllable and predictable. In reality genetic engineering is messy and unpredictable. Genetic engineers can insert a gene into the nucleus of a cell, but with current technology they do not know whether the cell will incorporate the new gene into its DNA. They also do not know where the new gene will be located in the DNA molecule's structure and how this will affect the organism.

Thus, conventional *genetic engineering is a trial-and-error process* with many failures and unexpected results. Indeed, the average success rate of current genetic engineering experiments is only about 1%. However, new techniques and advances in synthetic biology could overcome some of these problems.

Application of our increasing genetic knowledge is filled with great promise, but it raises some serious ethical and privacy issues. For example, some people have genes that make them more likely to develop certain genetic diseases or disorders. We now have the power to detect these genetic deficiencies, even before birth.

If gene therapy is developed for correcting these deficiencies, who will get it? Will it be reserved mostly for the rich? Will it lead to more abortions of genetically defective fetuses? Will health insurers refuse to insure people with certain genetic defects that could lead to health problems? Will employers refuse to hire them?

Soon we may enter the age of *designer babies* where people will be able to walk into a fertility clinic and choose the traits they want in their offspring from a genetic shopping list. Will generals and athletic coaches want to clone superior soldiers and athletes? Will some want to clone geniuses, people who are superior musicians, or people with great beauty? Will one gender be chosen more often and how will this affect population growth, marriage opportunities, and other social interactions? How will this affect the ratios of minorities in societies? Will the world become polarized into the few who can afford such procedures and the many who cannot? What will be the beneficial and harmful environmental effects of such a change in the reproductive process?

Some people dream of a day when our genetic engineering prowess could eliminate death and aging altogether. As one's cells, organs, or other parts wear out or are damaged, they would be replaced with new ones. These replacement parts might be grown in genetic engineering laboratories or in biopharms. Or people might choose to have a clone available for spare parts.

Is it moral to do this? Who decides this? Who regulates this? Will genetically designed humans and clones have the same legal rights as people?

X *HOW WOULD YOU VOTE?* Should we legalize the production of human clones if a reasonably safe technology for doing so becomes available? Cast your vote online at www.thomsonedu.com/biology/miller.

What might be the environmental impacts of such genetic developments on population, resource use, pollution, environmental degradation, and the earth's long-term ability to adapt to environmental changes (Core Case Study, p. 82)? If everyone could live with

good health as long as they wanted for a price, sellers of body makeovers would encourage customers to line up. Each of these affluent, long-lived people could have an enormous ecological footprint for perhaps centuries.

Controversy over Genetic Engineering

There are arguments over how much we should regulate genetic engineering research and development.

In the 1990s, a backlash developed against the increasing use of genetically modified food plants and animals. Some protesters argue against this new technology for mostly ethical reasons. Others advocate slowing down the technological rush and taking a closer look at the short- and long-term advantages and disadvantages of genetic technologies.

At the very least, they say, all genetically modified crops and animal products and foods containing such components should be clearly labeled as such. Such labels would give consumers a more informed choice, much like the food labels that list ingredients and nutritional information. Makers of genetically modified products strongly oppose such labeling because they say that such products are not harmful and fear that it would hurt sales.

? *THINKING ABOUT GENETICALLY MODIFIED FOODS* Should all genetically modified crops and animal products and foods containing such components be clearly labeled as such? Explain.

Supporters of genetic engineering and synthetic biology wonder why there is so much concern. After all, we have been genetically modifying plants and animals for centuries. Now we have a faster, better, and perhaps cheaper way to do it, so why not use it?

Proponents of more careful control of genetic engineering and synthetic biology counter that most new technologies have had unintended harmful consequences. For example, pesticides have helped protect crops from insect pests and disease. Wonderful. At the same time, their overuse has accelerated genetic evolution in many species, which have become resistant to many of the most widely used pesticides. The pesticides have also unintentionally wiped out many natural predator insects that helped keep pest populations under control.

The ecological lesson: whenever we intervene in nature, we must pause and ask, "What happens next?" That explains why many analysts urge caution before rushing into genetic engineering and other forms of biotechnology without more careful evaluation of their possible unintended consequences and more stringent regulation of these new technologies.

THINKING ABOUT GENETIC ENGINEERING AND THE EARTH'S RESILIENCY Do you think that widespread use of genetic engineering and synthetic biology will enhance or hinder the earth's long-term ability to adapt to environmental changes (Core Case Study, p. 82)? Explain.

F **RESEARCH FRONTIER** Learning more about the beneficial and harmful environmental impacts of genetic engineering and synthetic biology

Case Study: How Did We Become Such a Powerful Species So Quickly?

We have thrived as a species mostly because of our strong opposable thumbs, ability to walk upright, and complex brains.

Like many other species, humans have survived and thrived because we have certain traits that allow us to adapt to and modify the environment to increase our survival chances. What are these adaptive traits?

First, consider the traits we do *not* have. We lack exceptional strength, speed, and agility. We do not have weapons such as claws or fangs, and we lack a protective shell or body armor.

Our senses are unremarkable. We see only visible light—a tiny fraction of the spectrum of electromagnetic radiation that bathes the earth. We cannot see infrared radiation, as a rattlesnake can, or the ultraviolet light that guides some insects to their favorite flowers.

We cannot see as far as an eagle or see well in the night like some owls and other nocturnal creatures. We cannot hear the high-pitched sounds that help bats maneuver in the dark. Our ears cannot pick up low-pitched sounds that are the songs of whales as they glide through the world's oceans. We cannot smell as keenly as a dog or a wolf. We cannot respond to potential danger nearly as quickly as a cockroach. By such measures, our physical and sensory powers are pitiful.

Yet we have survived and flourished within less than a twitch of the 3.7 billion years that life has existed on the earth. Analysts attribute our success to three adaptations: *strong opposable thumbs* that allow us to grip and use tools better than the few other animals that have thumbs, an ability to *walk upright*, and a *complex brain*. These adaptations have helped us develop weapons, protective devices, and technologies that extend our limited senses and help make up for some of our deficiencies as a species.

In a short time, we have developed many powerful technologies to take over much of the earth's life-support systems and net primary productivity to meet our basic needs and rapidly growing wants. We named ourselves *Homo sapiens sapiens*—the doubly wise species. If we keep degrading the life-support system for us and other species, some say we should be called *Homo ignoramus*. During this century we will probably learn which of these names is more appropriate.

The *good news* is that we can change our ways. We can learn to work in concert with nature by understanding and copying the ways nature has sustained itself for several billion years despite major changes in environmental conditions (Figure 1-16, p. 24). This means heeding Aldo Leopold's call for us to become earth citizens, not earth rulers.

 Revisiting the Adaptable Earth and Sustainability

In this chapter, we have learned that through changes in their genes every species on the earth is related to every other species in a pattern of evolutionary interconnectedness. And these past and ongoing connections make the earth a habitable planet for life as we know it and allow it to adapt to changing environmental conditions, as described in the Core Case Study that opened this chapter.

The four scientific principles of sustainability underlie the earth's amazing ability to adapt to sometimes drastic changes in environmental conditions. Without the sun and chemical cycling, life as we know it would not exist. And life could not adapt to environmental changes without the diverse and changing array of genes, species, ecosystems, and ecosystem processes that make up the earth's biodiversity and the population control provided by multiple interactions among species.

We are fortunate to live on such an amazing and adaptable planet and should dedicate ourselves to working with—not against—its life-sustaining processes.

All we have yet discovered is but a trifle in comparison with what lies hid in the great treasury of nature.

ANTONI VAN LEEUWENHOEK

CRITICAL THINKING

1. Explain how the movements of tectonic plates, volcanic eruptions, earthquakes, and climate change can contribute to the earth's ability to adapt to changes in environmental conditions (Core Case Study, p. 82).

2. How would you respond to someone who tells you that he or she does not believe in biological evolution by natural selection because it is "just a theory"?

3. How would you respond to a statement that we should not worry about air pollution because natural selection will enable humans to develop lungs that can detoxify pollutants?

4. How would you respond to someone who says that because extinction is a natural process, we should not worry about the loss of biodiversity?

5. What role does each of the following processes play in helping implement the four scientific principles of sustainability: **(a)** natural selection, **(b)** speciation, and **(c)** extinction?

6. Describe the major differences between the ecological niches of humans and cockroaches. Are these two species in competition? If so, how do they manage to coexist?

7. Explain why you are for or against using genetic engineering and synthetic biology to **(a)** develop "superior" human beings, and **(b)** eliminate aging and death. How might doing these things enhance or hinder the ability of the earth to maintain conditions favorable to life as we know it (Core Case Study, p. 82)

8. Congratulations! You are in charge of the future evolution of life on the earth. What three things would you put on the top of your list to do?

9. List two questions that you would like to have answered as a result of reading this chapter.

PROJECTS

1. Use the library or the Internet to find out what, if any, controls on genetic engineering exist in the country where you live, and how well such controls are enforced.

2. Develop three guidelines based on the four scientific principles of sustainability (Figure 1-16, p. 24) for our use of genetic engineering and synthetic biology to modify species and ecosystems.

3. An important adaptation of humans is a strong opposable thumb, which allows us to grip and manipulate things with our hands. Fold each of your thumbs into the palm of its hand and then tape them securely in that position for an entire day. After the demonstration, make a list of the things you could not do without the use of your thumbs.

4. Make a concept map of this chapter's major ideas, using the section heads, subheads, and key terms (in boldface). Look on the website for this book for information about making concept maps.

LEARNING ONLINE

The website for this book includes review questions for the entire chapter, flash cards for key terms and concepts, a multiple-choice practice quiz, interesting Internet sites, references, information about green careers, and a guide for accessing thousands of InfoTrac® College Edition articles. Log into

www.thomsonedu.com/biology/miller

Then choose Chapter 4, and select a learning resource. For access to animations, additional quizzes, chapter outlines and summaries, register and log into

at **www.thomsonedu.com** using the access code card in the front of your book.

Active Graphing
Log into ThomsonNow at www.thomsonedu.com to explore the graphing exercise for this chapter.

CORE CASE STUDY
Blowing in the Wind: A Story of Connections

Wind, a vital part of the planet's circulatory system, connects most life on the earth. Without wind, the tropics would be unbearably hot and most of the rest of the planet would freeze.

Winds also transport nutrients from one place to another. Dust rich in phosphates and iron blows across the Atlantic from the Sahara Desert in Africa (Figure 5-1). This movement helps build up agricultural soils in the Bahamas and supplies nutrients for plants in the rain forest's upper canopy in Brazil. Dust blowing from China's Gobi Desert deposits iron into the Pacific Ocean between Hawaii and Alaska. The iron stimulates the growth of phytoplankton, the minute producers that support ocean food webs. This is the *good news.*

Now for the *bad news:* Dust storms in the Sahara Desert have increased tenfold since 1950 mostly because of drought due to climate change and overgrazing. Another reason is the *SUV connection.* Increasing numbers of four-wheel vehicles speeding over the sand break the desert's surface crust. This allows wind storms to blow the underlying dusty material into the atmosphere.

Wind also transports harmful viruses, bacteria, fungi, and particles of long-lived pesticides and toxic metals. Particles of reddish-brown soil and pesticides banned in the United States are blown from Africa's deserts and eroding farmlands into the sky over the U.S. state of Florida. This makes it difficult for the state to meet federal air pollution standards during summer months.

More *bad news.* Some types of fungi in this dust may play a role in degrading or killing coral reefs in the Florida Keys in the U.S. and in the Caribbean. Scientists are also studying possible links between contaminated African dust and a sharp rise in rates of asthma in the Caribbean since 1973.

Particles of iron-rich dust from Africa that enhance the productivity of algae have also been linked to outbreaks of toxic algal blooms—referred to as *red tides*—in Florida's coastal waters. People who eat shellfish contaminated by a toxin produced in red tides can become paralyzed or even die.

Pollution and dust from rapidly industrializing China and central Asia blow across the Pacific Ocean and degrade air quality over parts of the western United States. Asian pollution contributes as much as 10% to smog on the U.S. West Coast, a threat expected to increase as China industrializes.

There is also *mixed news.* Particles from volcanic eruptions ride the winds, circle the globe, and change the earth's temperature for a while. Emissions from the 1991 eruption of Mount Pinatubo in the Philippines cooled the earth slightly for 3 years, temporarily masking signs of global warming. And volcanic ash, like the blowing desert dust, adds valuable trace minerals to the soil where it settles.

The familiar lesson: *There is no away* because *everything is connected.* Wind acts as part of the planet's circulatory system for heat, moisture, plant nutrients, and long-lived pollutants we put into the air. Movement of soil particles from one place to another by wind and water is a natural phenomenon. When we disturb the soil and leave it unprotected, we hasten and intensify this natural process.

Wind is also an important factor in climate through its influence on global air circulation patterns. Climate, in turn, is crucial for determining what kinds of plant and animal life are found in the major biomes of the biosphere, as discussed in this chapter.

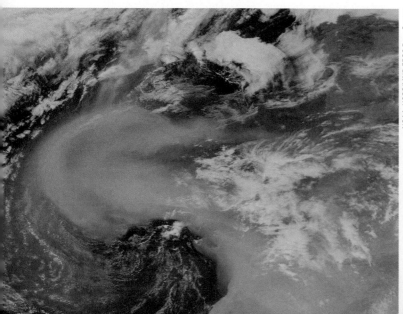

NOAA. USGS/MND EROS Data Center

Figure 5-1 Some of the dust shown here blowing from Africa's Sahara Desert can end up as soil nutrients in Amazonian rain forests and toxic air pollutants in the U.S. state of Florida and the Caribbean. This slowly depletes nutrients and reduces biological productivity in Africa.

To do science is to search for repeated patterns, not simply to accumulate facts, and to do the science of geographical ecology is to search for patterns of plant and animal life that can be put on a map.

ROBERT H. MACARTHUR

This chapter provides an introduction to the earth's climate and how it affects the types of life found in different parts of the earth. It addresses these questions:

- What factors influence the earth's climate?

- How does climate determine where the earth's major biomes are found?

- What are the major types of desert biomes?

- What are the major types of grassland biomes?

- What are the major types of forest and mountain biomes?

- How have human activities affected the world's desert, grassland, forest, and mountain biomes?

CLIMATE: A BRIEF INTRODUCTION

Weather and Climate

Weather is a local area's short-term physical conditions such as temperature and precipitation, and climate is a region's average weather conditions over a long time.

Weather is an area's temperature, precipitation, humidity, wind speed, cloud cover, and other physical conditions of the lower atmosphere over hours or days. Supplement 10 on p. S38 introduces you to weather basics.

Climate is a region's general pattern of atmospheric or weather conditions over a long time—years, decades, and centuries. As American writer and humorist Mark Twain once said, "Climate is what we expect, weather is what we get." *Average temperature* and *average precipitation* are the two main factors determining climate, along with the closely related factors of **latitude** (distance from the equator) and **elevation** (height above sea level). The earth's major climate zones are shown in Figure 5-2.

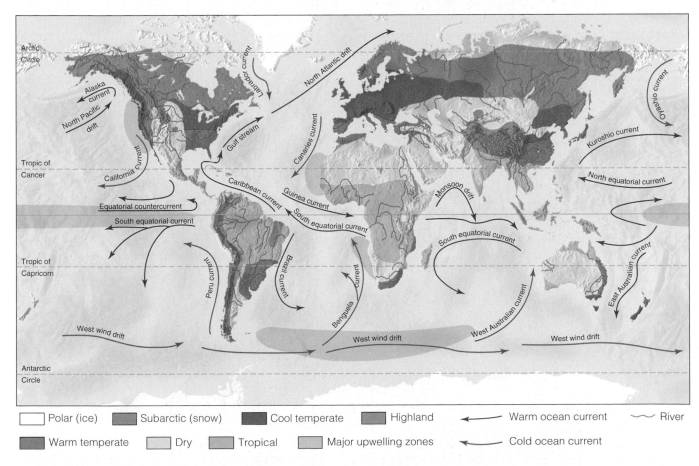

□ Polar (ice)	■ Subarctic (snow)	■ Cool temperate	■ Highland	← Warm ocean current	∿ River
■ Warm temperate	□ Dry	■ Tropical	■ Major upwelling zones	← Cold ocean current	

ThomsonNOW™ **Active Figure 5-2 Natural capital:** generalized map of the earth's current climate zones, showing the major contributing ocean currents and drifts and upwelling areas. *See an animation based on this figure and take a short quiz on the concept.*

Solar Energy and Global Air Circulation: Distributing Heat

Global air circulation is affected by the uneven heating of the earth's surface by solar energy, seasonal changes in temperature and precipitation, rotation of the earth on its axis, and the properties of air, water, and land.

Many factors contribute to a local climate, including the amount of solar radiation reaching the area, the earth's daily rotation and annual path around the sun, air circulation over the earth's surface, the global distribution of land masses and seas, the circulation of ocean currents, and the elevation of land masses.

Four major factors determine global air circulation patterns. The first factor is the *uneven heating of the earth's surface by the sun*. Air is heated much more at the earth's fattest part, the equator, where the sun's rays strike directly, than at the poles, where sunlight strikes at a slanted angle and spreads out over a much larger area. You can observe this effect by shining a flashlight in a darkened room on the middle of a spherical object such as a basketball and moving the light up and down.

These differences in the amount of incoming solar energy help explain why tropical regions near the equator are hot, polar regions are cold, and temperate regions in between generally have intermediate average temperatures. Temperature also becomes progressively colder as elevation above sea level increases.

A second factor is *seasonal changes in temperature and precipitation*. The earth's axis—an imaginary line connecting the north and south poles—is tilted with respect to the sun's rays. As a result, regions away from the equator are tipped toward or away from the sun most of the year, as the earth makes its annual revolution around the sun (Figure 5-3). This creates opposite seasons in the northern and southern hemispheres.

A third factor is *rotation of the earth on its axis*. As the earth rotates around its north–south axis, its equator spins faster than its polar regions. As a result, heated air masses rising above the equator and moving north and south to cooler areas are deflected to the west or east over different parts of the planet's surface—a phenomenon known as the *Coriolis effect* (Figure 5-4). The direction of air movement in the resulting huge atmospheric regions called *cells* sets up belts of *prevailing winds*—major surface winds that blow almost continuously and distribute air, moisture, and dust over the earth's surface (Core Case Study, p. 100).

The fourth factor affecting global air circulation is *properties of air, water, and land*. Heat from the sun evaporates ocean water and transfers heat from the oceans to the atmosphere, especially near the hot equator. This evaporation of water creates cyclical convection cells that circulate air, heat, and moisture both vertically and from place to place in the troposphere, as shown in Figure 5-5.

The earth's air circulation patterns, prevailing winds, and mixture of continents and oceans result in

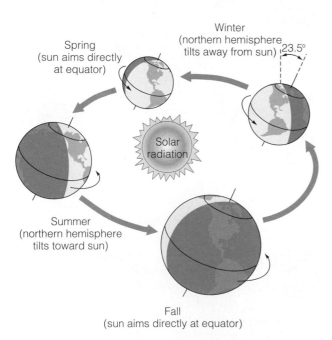

Figure 5-3 Natural capital: as the planet makes its annual revolution around the sun on an axis tilted about 23.5°, various regions are tipped toward or away from the sun. The resulting variations in the amount of solar energy reaching the earth create the seasons in the northern and southern hemispheres.

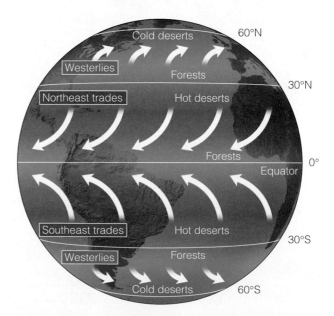

Figure 5-4 Natural capital: because of the Coriolis effect the earth's rotation deflects the movement of the air over different parts of the earth, creating global patterns of prevailing winds that help distribute heat and moisture in the troposphere.

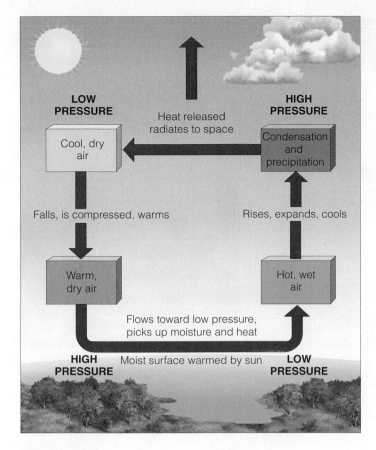

Figure 5-5 Natural capital: transfer of energy by convection in the troposphere. *Convection* occurs when hot and wet warm air rises, cools, and releases moisture as precipitation and heat (right side). Then the more dense cool and dry air sinks, gets warmer, and picks up moisture as it flows across the earth's surface to begin the cycle again.

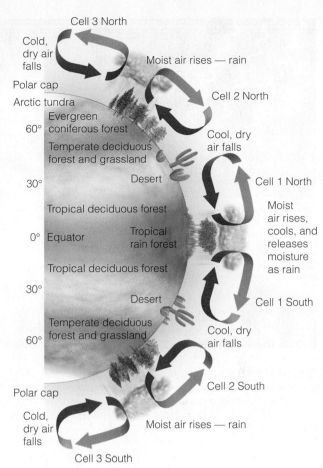

Figure 5-6 Natural capital: global air circulation and biomes. Heat and moisture are distributed over the earth's surface by vertical currents, which form six giant convection cells at different latitudes. The resulting uneven distribution of heat and moisture over the planet's surface leads to the forests, grasslands, and deserts that make up the earth's biomes.

six giant convection cells—three north of the equator and three south of the equator—in which warm, moist air rises and cools, and the cool, dry air sinks. This leads an irregular distribution of climates and patterns of vegetation, as shown in Figure 5-6

 Watch the formation of six giant convection cells and learn more about how they affect climates at ThomsonNOW.

? **THINKING ABOUT WINDS AND BIOMES** How might the distribution of the world's forests, grasslands, and deserts shown in Figure 5-6 differ if the prevailing winds shown in Figure 5-4 did not exist?

Ocean Currents: Distributing Heat and Nutrients

Ocean currents influence climate by distributing heat from place to place and mixing and distributing nutrients.

The major ocean currents (Figure 5-2) also affect the climates of regions. The oceans absorb heat from the air circulation patterns just described, with the bulk of this heat being absorbed near the warm tropical areas. This heat and differences in water density create warm and cold ocean currents. Irregularly shaped continents interrupt these currents and cause them to flow clockwise in roughly circular patterns between the continents in the northern hemisphere and counterclockwise in the southern hemisphere (Figure 5-2). Driven by winds (Figure 5-4) and the earth's rotation, these currents help redistribute heat received from the sun from one place to another, thereby influencing climate and vegetation, especially near coastal areas.

The warm Gulf Stream (Figure 5-2), for example, transports 25 times more water than all of the world's rivers combined. There has been a long-standing scientific myth that heat carried northward from the equator is the cause of northeastern Europe's mild climate.

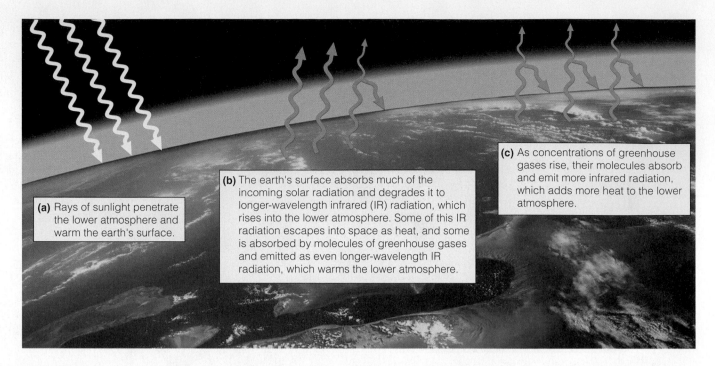

(a) Rays of sunlight penetrate the lower atmosphere and warm the earth's surface.

(b) The earth's surface absorbs much of the incoming solar radiation and degrades it to longer-wavelength infrared (IR) radiation, which rises into the lower atmosphere. Some of this IR radiation escapes into space as heat, and some is absorbed by molecules of greenhouse gases and emitted as even longer-wavelength IR radiation, which warms the lower atmosphere.

(c) As concentrations of greenhouse gases rise, their molecules absorb and emit more infrared radiation, which adds more heat to the lower atmosphere.

ThomsonNOW™ Active Figure 5-7 Natural capital: the *natural greenhouse effect*. When concentrations of greenhouse gases in the atmosphere rise, the average temperature of the troposphere rises. *See an animation based on this figure and take a short quiz on the concept.* (Modified by permission from Cecie Starr, *Biology: Concepts and Applications*, 4th ed., Pacific Grove, Calif.: Brooks/Cole, 2000)

In 2006, research scientist Richard Seager analyzed climate data and found the largest role in this mild climate is played by the troposphere, with the Gulf Stream playing a much lesser role. Ocean currents also help mix ocean waters and distribute the nutrients and dissolved oxygen needed by aquatic organisms.

 Learn more about how oceans affect air movements where you live and all over the world at ThomsonNOW.

Atmospheric Gases and Climate: The Natural Greenhouse Effect

Water vapor, carbon dioxide, and other gases influence climate by warming the lower troposphere and the earth's surface.

Small amounts of certain gases, including water vapor (H_2O), carbon dioxide (CO_2), methane (CH_4), and nitrous oxide (N_2O) also play a role in determining the earth's average temperatures and thus its climates. These **greenhouse gases** allow mostly visible light and some infrared radiation and ultraviolet (UV) radiation from the sun to pass through the troposphere. The earth's surface absorbs much of this solar energy and

transforms it to longer-wavelength infrared radiation (heat), which then rises into the troposphere.

Some of this heat escapes into space, but some is absorbed by molecules of greenhouse gases and emitted into the troposphere as even longer-wavelength infrared radiation. Some of this released energy radiates into space, and some warms the troposphere and the earth's surface. This natural warming effect of the troposphere is called the **greenhouse effect** (Figure 5-7). Without its current greenhouse gases (especially water vapor, which is found in the largest concentration), the earth would be a cold and mostly lifeless planet.

Human activities such as burning fossil fuels, clearing forests, and growing crops release carbon dioxide, methane, and nitrous oxide into the atmosphere. Considerable scientific evidence and climate models indicate that these large inputs of greenhouse gases into the troposphere can enhance the earth's natural greenhouse effect, lead to *global warming*, and change the climate in various areas of the earth during your lifetime. This could alter precipitation patterns, shift areas where we can grow crops, raise average sea levels, and change the areas where some types of plants and animals can live. *Green Career:* Climate change specialist

F *RESEARCH FRONTIER* Modeling and other research to learn more about how human activities affect climate

Witness the greenhouse effect and see how human activity has affected it at at ThomsonNOW.

Topography and Local Climate: Land Matters

Interactions between land and oceans and disruptions of airflows by mountains and cities affect local climates.

Heat is absorbed and released more slowly by water than by land. This difference creates land and sea breezes. As a result, the world's oceans and large lakes moderate the climate of nearby lands.

Various topographic features of the earth's surface create local and regional climatic conditions that differ from the general climate of a region. For example, mountains interrupt the flow of prevailing surface winds (Figure 5-4) and the movement of storms. When moist air blowing inland from an ocean reaches a mountain range, it cools and expands as it rises in altitude and then loses most of its moisture as rain and snow on the windward (wind-facing) slopes.

As the drier air mass flows down the leeward (away from the wind) slopes, it draws moisture out of the plants and soil below. The lower precipitation and the resulting semiarid or arid conditions on the leeward side of high mountains create the rain shadow effect (Figure 5-8), sometimes leading to the formation of deserts. Thus, winds (Core Case Study, p. 100) play a key role in forming some of the earth's deserts.

Continents lying north or south of warm oceans experience heavy rains called **monsoons** parts of the year. Intense heating of land near such oceans during summer creates vast low-pressure air masses that draw moisture from the ocean. This plus trade winds and equatorial heating lead to heavy rainfall and alternating wet seasons with flooding and dry seasons with drought conditions.

Cities also create distinct *microclimates*. Bricks, concrete, asphalt, and other building materials absorb and hold heat, and buildings block wind flow. Motor vehicles and the climate control systems of buildings release large quantities of heat and pollutants. As a result, cities tend to have more haze and smog, higher temperatures, and lower wind speeds than the surrounding countryside.

? *THINKING ABOUT WINDS AND BIOMES* List three changes in your lifestyle that would take place if there were no winds where you live.

BIOMES: CLIMATE AND LIFE ON LAND

Why Do Different Organisms Live in Different Places?

Different climates lead to different communities of organisms, especially vegetation.

Why is one area of the earth's land surface a desert, another a grassland, and another a forest? Why do different types of deserts, grasslands, and forests exist? The

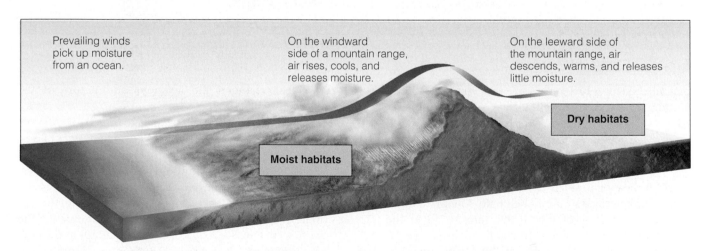

Prevailing winds pick up moisture from an ocean.

On the windward side of a mountain range, air rises, cools, and releases moisture.

On the leeward side of the mountain range, air descends, warms, and releases little moisture.

Dry habitats

Moist habitats

Figure 5-8 Natural capital: The *rain shadow effect* is a reduction of rainfall on the sides of mountains facing away from prevailing surface winds. Warm, moist air in prevailing onshore winds loses most of its moisture as rain and snow on the windward (wind-facing) slopes of a mountain range. This leads to semiarid and arid conditions on the leeward side of the mountain range and the land beyond. The Mojave Desert in the U.S. state of California and Asia's Gobi Desert were both created by this effect.

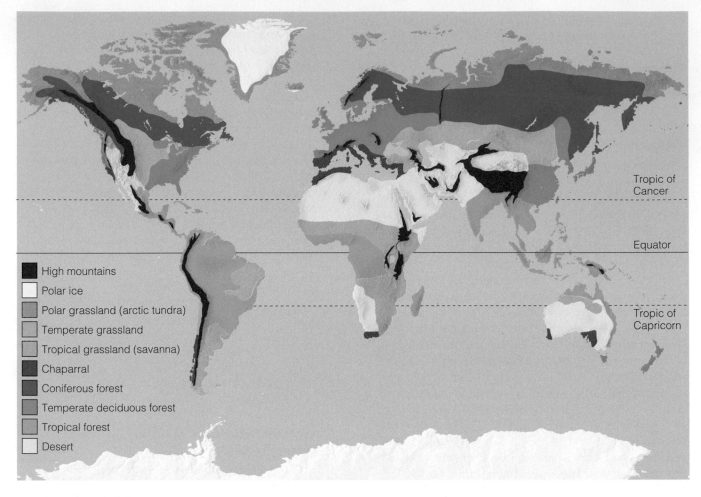

Legend:
- High mountains
- Polar ice
- Polar grassland (arctic tundra)
- Temperate grassland
- Tropical grassland (savanna)
- Chaparral
- Coniferous forest
- Temperate deciduous forest
- Tropical forest
- Desert

Tropic of Cancer

Equator

Tropic of Capricorn

ThomsonNOW **Active Figure 5-9 Natural capital:** the earth's major *biomes*—the main types of natural vegetation in various undisturbed land areas—result primarily from differences in climate. Each biome contains many ecosystems whose communities have adapted to differences in climate, soil, and other environmental factors. Human ecological footprints (Figures 3 and 4 on pp. S12–S15 in Supplement 4) have removed or altered much of the natural vegetation in some areas for farming, livestock grazing, lumber and fuelwood, mining, and construction. *See an animation based on this figure and take a short quiz on the concept.*

general answer is differences in *climate* (Figure 5-2), resulting mostly from differences in average temperature and precipitation caused by global air circulation (Figure 5-6). Different climates support different communities of organisms.

Figure 5-9 shows how scientists have divided the world into several major **biomes**—large terrestrial regions characterized by similar climate, soil, plants, and animals, regardless of where they are found in the world. What kind of biome do you live in?

Climate change is a part of the earth's environmental history. Over the past 4.7 billion years the planet's climate has changed drastically as a result of changes in solar output, emissions from volcanic eruptions, and continents moving (Figure 4-5, p. 88).

This has changed the locations and nature of the earth's biomes.

For example, 5,000 years ago much of Africa's Sahara Desert was fertile and covered with grasses. And 15,000 years ago much of the now arid Western United States was rainy and contained many lakes. What concerns environmental scientists today is the considerable evidence that we are changing the climate at such a fast rate (50–100 years) that much of the earth's current life may not be able to adapt to the new conditions or move to areas with more favorable climates as biomes change their vegetation and locations.

By comparing Figure 5-9 with Figure 5-2, you can see how the world's major biomes vary with climate. Figure 3-9 (p. 56) shows major biomes in the United

States that one would encounter moving through different climates along the 39th parallel.

[?] *THINKING ABOUT WINDS AND BIOMES* Use Figure 5-2 to determine the general type of climate where you live and Figure 5-9 to determine the general type of biome that should exist where you live. Then use Figures 3 and 4 on pp. S12–S15 in Supplement 4 to determine how human ecological footprints have effected the general type of biome where you live.

Average annual precipitation and temperature (as well as soil type, Figure 3-24, p. 69) are the most important factors in producing tropical (hot), temperate (moderate), or polar (cold) deserts, grasslands, and forests (Figure 5-10).

On maps such as the one in Figure 5-9, biomes are presented as having sharp boundaries and being covered with the same general type of vegetation. In reality, *biomes are not uniform.* They consist of a *mosaic of patches,* with somewhat different biological communities but with similarities unique to the biome. These patches occur mostly because the resources that plants and animals need are not uniformly distributed and because human activities remove and alter natural vegetation.

Figure 5-11 (p. 108) shows how climate and vegetation vary with *latitude* and *elevation.* If you climb a tall mountain from its base to its summit, you can observe changes in plant life similar to those you would encounter in traveling from the equator to the earth's poles.

 Find a map showing all the world's biomes and zoom in for details at ThomsonNOW.

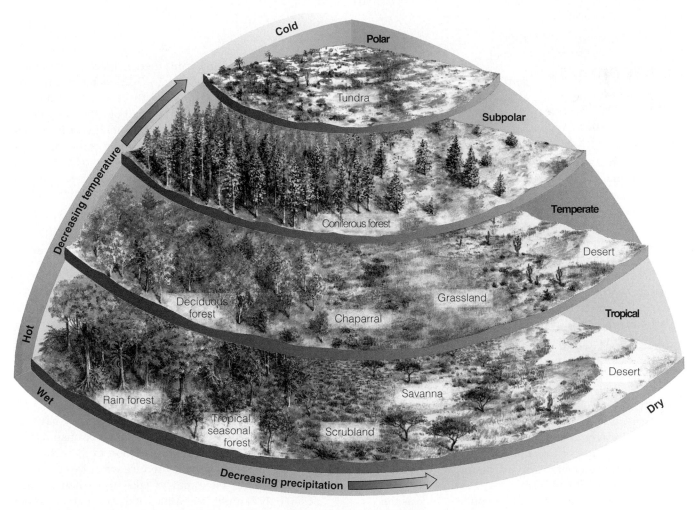

Figure 5-10 Natural capital: average precipitation and average temperature, acting together as limiting factors over a period of 30 or more years, determine the type of desert, grassland, or forest biome in a particular area. Although the actual situation is much more complex, this simplified diagram explains how climate determines the types and amounts of natural vegetation found in an area left undisturbed by human activities. (Used by permission of Macmillan Publishing Company, from Derek Elsom, *The Earth,* New York: Macmillan, 1992. Copyright © 1992 by Marshall Editions Developments Limited)

Figure 5-11 Natural capital: generalized effects of elevation (left) and latitude (right) on climate and biomes. Parallel changes in vegetation type occur when we travel from the equator to the poles or from lowlands to mountaintops.

DESERT BIOMES

Types of Deserts

Deserts have little precipitation and little vegetation and are found in tropical, temperate, and polar regions.

A **desert** is an area where evaporation exceeds precipitation. Annual precipitation is low in these driest of the earth's biomes and is often scattered unevenly throughout the year.

Deserts cover about 30% of the earth's land surface and are found mostly in tropical and subtropical regions (Figure 5-9). The largest deserts are found in the interiors of continents, far from moist sea air and moisture-bearing winds. Other, more local deserts form on the downwind sides of mountain ranges because of the rain shadow effect (Figure 5-8).

During the day, the baking sun warms the ground in the desert. At night, most of the heat stored in the ground radiates quickly into the atmosphere. Desert soils have little vegetation and moisture to help store the heat and the skies above deserts are usually clear. This explains why in a desert you may roast during the day but shiver at night.

A combination of low rainfall and different average temperatures creates tropical, temperate, and cold deserts (Figures 5-9 and 5-12).

Tropical deserts (Figure 5-12, top photo), such as the Sahara and Namib of Africa, are hot and dry most of the year (Figure 5-12, top graph). They have few plants and a hard, windblown surface strewn with rocks and some sand. They are the deserts we often see in movies.

In *temperate deserts* (Figure 5-12, center photo), such as the Mojave in the southern part of the U.S. state of California, daytime temperatures are high in sum-

mer and low in winter and there is more precipitation than in tropical deserts (Figure 5-12, center graph). The sparse vegetation consists mostly of widely dispersed, drought-resistant shrubs and cacti or other succulents adapted to the lack of water and temperature variations, as shown in Figure 5-12, center photo, and Figure 5-13 (p. 110). In *cold deserts,* such as the Gobi Desert in China, vegetation is sparse (Figure 5-12, bottom photo), winters are cold, summers are warm or hot, and precipitation is low (Figure 5-12, bottom graph).

How Do Desert Plants and Animals Survive?

Desert plants and animals have adaptations that help them stay cool and get enough water to survive.

Adaptations for survival in the desert have two themes: *beat the heat* and *every drop of water counts.*

Desert plants have evolved a number of strategies for doing this. During long hot and dry spells plants such as mesquite and creosote drop their leaves to survive in a dormant state. *Succulent* (fleshy) *plants,* such as the saguaro ("sah-WAH-ro") cactus (Figure 5-12, middle photo), have three adaptations: they have no leaves, which can lose water by evapotranspiration; they store water and synthesize food in their expandable, fleshy tissue; and they reduce water loss by opening their pores (stomata) to take up carbon dioxide (CO_2) only at night. The spines of these and many other desert plants guard them from being eaten by herbivores seeking the precious water they hold.

Some desert plants use deep roots to tap into groundwater. Others such as prickly pear (Figure 5-13, p. 110) and saguaro cacti use widely spread, shallow roots to collect water after brief showers and store it in their spongy tissue.

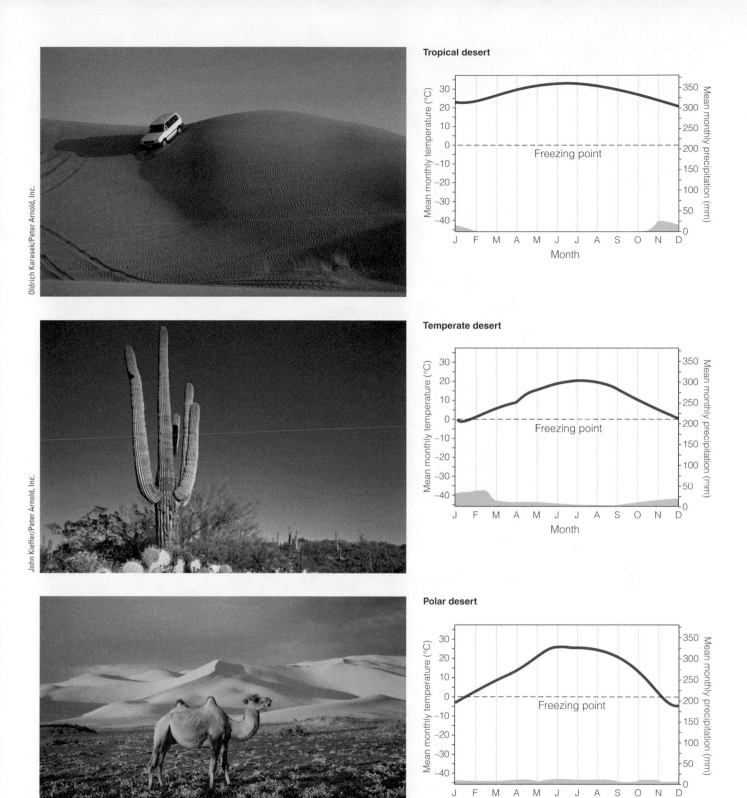

Figure 5-12 Natural capital: climate graphs showing typical variations in annual temperature (red) and precipitation (blue) in tropical, temperate, and cold deserts. Top photo shows a popular but destructive SUV rodeo in Saudi Arabia (tropical desert). Center photo shows saguaro cactus in the United States (temperate desert). Bottom photo shows a Bactrian camel in Mongolia's Gobi (cold) desert.

Figure 5-13 Natural capital:
some components and interactions in a *temperate desert ecosystem*. When these organisms die, decomposers break down their organic matter into minerals that plants use. Colored arrows indicate transfers of matter and energy between producers, primary consumers (herbivores), secondary or higher-level consumers (carnivores), and decomposers. Organisms are not drawn to scale

Red-tailed hawk

Gambel's quail

Yucca

Agave

Jack rabbit

Collared lizard

Prickly pear cactus

Roadrunner

Darkling beetle

Bacteria

Diamondback rattlesnake

Fungi

Kangaroo rat

Producer to primary consumer | Primary to secondary consumer | Secondary to higher-level consumer | All producers and consumers to decomposers

Evergreen plants conserve water by having wax-coated leaves that reduce water loss. Others, such as annual wildflowers and grasses, store much of their biomass in seeds that remain inactive, sometimes for years, until they receive enough water to germinate. Shortly after a rain these seeds germinate, grow, carpet some deserts with a dazzling array of colorful flowers, produce new seeds, and die, all in a few weeks.

Most desert animals are small. Some beat the heat by hiding in cool burrows or rocky crevices by day and coming out at night or in the early morning. Others become dormant during periods of extreme heat or drought.

Insects and reptiles have thick outer coverings to minimize water loss through evaporation, and their wastes are dry feces and a dried concentrate of urine.

Many spiders and insects get their water from dew or from the food they eat. Arabian oryxes survive by licking the dew that accumulates at night on rocks and on one another's hair.

Desert ecosystems are fragile. Their soils take a long time to recover from disturbances because of their slow plant growth, low species diversity, slow nutrient cycling (because of sparse bacterial activity in their soils), and lack of water. Tracks left by tanks practicing in California's Mojave in 1940 are still visible today. Desert vegetation destroyed by livestock overgrazing and off-road vehicles (Figure 5-12, top photo) may take decades to hundreds of years to grow back.

THINKING ABOUT WINDS AND DESERTS What roles do winds play in creating and sustaining deserts?

GRASSLANDS AND CHAPARRAL BIOMES

Types of Grasslands

Grasslands have enough precipitation to support grasses but not enough to support large stands of trees and are found in tropical, temperate, and polar regions.

Grasslands, or **prairies,** occur mostly in the interiors of continents in areas too moist for deserts and too dry for forests (Figure 5-9).

Grasslands persist because of a combination of seasonal drought, grazing by large herbivores, and occasional fires—all of which keep large numbers of shrubs and trees from growing. The three main types of grasslands—tropical, temperate, and polar (tundra)—result from combinations of low average precipitation and various average temperatures (Figures 5-10 and 5-14, p. 112).

Tropical Grasslands: Savannas

Savannas are tropical grasslands with scattered trees and enormous herds of hoofed animals.

Savanna is a type of tropical grassland dotted with widely scattered clumps of trees such as acacia, (Figure 5-14, top photo), which are covered with thorns to keep herbivores away. This biome usually has warm temperatures year-round and alternating dry and wet seasons (Figure 5-14, top graph). Drought during the dry season, occasional fires started by lightning during the beginning of the rainy season, and intense grazing inhibit the growth of trees and bushes. Tropical savanna is the birthplace of humankind. From this biome humans eventually migrated to every biome on the earth. Today, this biome stretches across parts of Africa, Australia, India, South America and other dry tropical regions (Figure 5-9).

Animals that are farsighted, swift, and stealthy have the best chance of surviving in this biome. Tropical savannas in East Africa have enormous herds of *grazing* (grass- and herb-eating) and *browsing* (twig- and leaf-nibbling) hoofed animals, including wildebeests (Figure 5-14, top photo), gazelles, zebras, giraffes, and antelopes and their predators such as lions, hyenas, and humans.

Large herds of these grazing and browsing animals migrate to find food and water in response to seasonal and year-to-year variations in rainfall (Figure 5-14, blue region in top graph) and food availability. Most grazing animals migrate to widely spaced water holes during the two rainy seasons.

As part of their niches, these and other large herbivores have evolved specialized eating habits that minimize competition between species for vegetation found on the savanna. For example, giraffes eat leaves and shoots from the tops of trees, elephants eat leaves and branches farther down, wildebeests prefer short grass, and zebras graze on longer grass and stems.

Savanna plants, like those in deserts, are adapted to survive drought and extreme heat or cold. Many have deep roots that can tap into groundwater.

Savannas in Africa, which has the most rapidly growing human population of any continent, are being rapidly converted to rangeland for domesticated grazing animals such as cattle. In some areas, overgrazing by these herds and the use of trees for firewood have converted savanna to desert.

Ecologists have shown that native African herbivores convert grass into meat more efficiently and with less soil destruction than do cattle. They call for providing meat by using domesticated herds of antelope and other large native grazers in place of cattle.

Temperate Grasslands: Fertile Soils

Temperate grasslands with cold winters and hot and dry summers have deep and fertile soils that make them widely used for growing crops and grazing cattle.

Temperate grasslands once covered vast expanses of plains and gently rolling hills in the interiors of North and South America, Europe, and Asia (Figure 5-9). They include the great North American prairies (Figure 5-14, center photo), the steppes of Eurasia, and the pampas of South America.

In these grasslands, winters are bitterly cold, summers are hot and dry, and annual precipitation is fairly sparse and falls unevenly through the year (Figure 5-14, center graph). Drought, occasional fires, and intense grazing inhibit the growth of trees and bushes, except along rivers and streams.

Because the aboveground parts of most of the grasses die and decompose each year, organic matter accumulates to produce a deep, fertile soil (Figure 3-24, top right, p. 69). This soil is held in place by a thick network of intertwined roots of drought-tolerant grasses unless the topsoil is plowed up and allowed to blow away by prolonged exposure to high winds found in these biomes. The natural grasses are also adapted to fires ignited by lightning or set deliberately. The fires burn the plants above the ground but do not harm the roots, from which new life can spring.

Temperate grasslands include the *short-grass prairies* (Figure 5-14, center photo) and the *tall-grass prairies* (Figure 5-15, p. 113) of the midwestern and western United States and Canada. Here winds blow almost continuously, and evaporation is rapid, often leading to fires in the summer and fall. This combination of winds (Core Case Study, p. 100) and fire helps maintain grasslands.

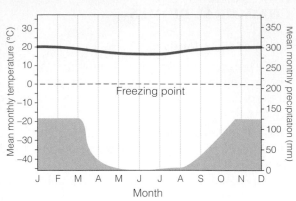

Tropical grassland (savanna)

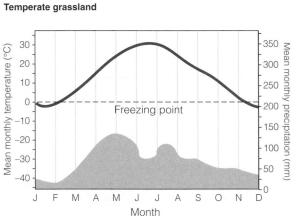

Temperate grassland

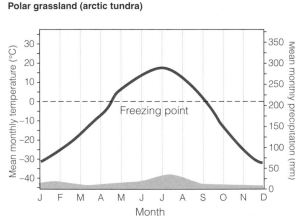

Polar grassland (arctic tundra)

Figure 5-14 Natural capital: climate graphs showing typical variations in annual temperature (red) and precipitation (blue) in tropical, temperate, and polar (arctic tundra) grasslands. Top photo shows wildebeests grazing on a savanna in Maasai Mara National Park in Kenya, Africa (tropical grassland). Center photo shows wildflowers in bloom on a prairie near East Glacier Park in the U.S. state of Montana (temperate grassland). Bottom photo shows arctic tundra with caribou in Alaska's Arctic National Wildlife Refuge (polar grassland).

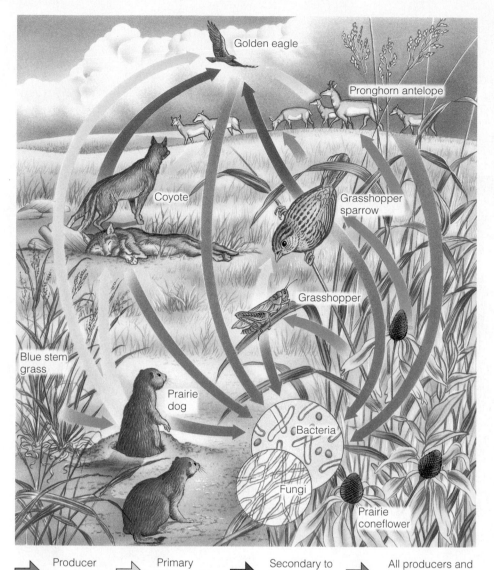

Golden eagle

Pronghorn antelope

Coyote

Grasshopper sparrow

Grasshopper

Blue stem grass

Prairie dog

Bacteria

Fungi

Prairie coneflower

ThomsonNOW Active Figure 5-15
Natural capital: some components and interactions in a *temperate tall-grass prairie ecosystem* in North America. When these organisms die, decomposers break down their organic matter into minerals that plants can use. Colored arrows indicate transfers of matter and energy between producers, primary consumers (herbivores), secondary or higher level consumers (carnivores), and decomposers. Organisms are not drawn to scale. *See an animation based on this figure and take a short quiz on the concept.*

➡ Producer to primary consumer ➡ Primary to secondary consumer ➡ Secondary to higher-level consumer ➡ All producers and consumers to decomposers

 Learn more about how plants and animals in a temperate prairie are connected in a food web at ThomsonNOW.

Temperate grasslands once covered large areas of the earth. However, because of their fertile soils much of this biome has been used to grow crops (Figure 5-16), raise cattle, and build towns and cities. Compare the idealized location of temperate grasslands in the midwestern United States (Figure 5-9) with how they have

Figure 5-16 Natural capital degradation: replacement of a biologically diverse temperate grassland with a monoculture crop in the U.S. state of California. When humans remove the tangled root network of natural grasses, the fertile topsoil becomes subject to severe wind erosion (see Case Study on pp. S23–S24 in Supplement 5) unless it is covered with some type of vegetation.

National Archives/EPA Documerica

Figure 5-17 Natural capital: some components and interactions in an *arctic tundra (polar grassland) ecosystem.* When these organisms die, decomposers break down their organic matter into minerals that plants use. Colored arrows indicate transfers of matter and energy between producers, primary consumers (herbivores), secondary or higher-level consumers (carnivores), and decomposers. Organisms are not drawn to scale.

Long-tailed jaeger

Grizzly bear

Caribou

Mosquito

Snowy owl

Horned lark

Arctic fox

Willow ptarmigan

Dwarf willow

Lemming

Mountain cranberry

Moss campion

| Producer to primary consumer | Primary to secondary consumer | Secondary to higher-level consumer | All producers and consumers to decomposers |

been disrupted by the human ecological footprint, as shown in Figure 4 on pp. S14–S15 in Supplement 4.

F **RESEARCH FRONTIER** How grazing and off-road vehicles affect grasslands and deserts

Polar Grasslands: Arctic Tundra

Polar grasslands are covered with ice and snow except during a brief summer.

Polar grasslands, or *arctic tundra* (Russian for "marshy plain"), lie south of the arctic polar ice cap (Figure 5-9).* During most of the year, these treeless plains are bit-

*Some ecologists classify tundra as a very cold (polar) desert because precipitation is low and most of the year water is frozen and unavailable to support life.

terly cold (Figure 5-14, bottom graph), swept by frigid winds, and covered with ice and snow. Winters are long and dark, and the scant precipitation falls mostly as snow.

Under the snow, this biome is carpeted with a thick, spongy mat of low-growing plants, primarily grasses, mosses, lichens, and dwarf shrubs (Figure 5-14, bottom photo). Trees or tall plants cannot survive in the cold and windy tundra because they would lose too much of their heat. Most of the annual growth of the tundra's plants occurs during the 6- to 8-week summer, when the sun shines almost around the clock. Figure 5-17 shows some of the components and interactions in this biome.

One outcome of the extreme cold is the formation of **permafrost,** underground soil in which captured water stays frozen for more than 2 consecutive years. During the long and cold winters the surface soil also freezes.

During the brief summer the permafrost layer keeps melted snow and ice from soaking into the ground. As a consequence, the waterlogged tundra forms a large number of shallow lakes, marshes, bogs, ponds, and other seasonal wetlands when the snow and frozen surface soil melt. Hordes of mosquitoes, black flies, and other insects thrive in these shallow surface pools. They serve as food for large colonies of migratory birds (especially waterfowl) that return from the south to nest and breed in the bogs and ponds. Animals in this biome survive the intense winter cold through adaptations such as thick coats of fur (arctic wolf, arctic fox, and musk oxen), feathers (snowy owl), and living underground (arctic lemming).

Global warming is causing some of the permafrost in parts of Alaska to melt. This disrupts these ecosystems and releases methane (CH_4) gas from the soil. Methane is a potent greenhouse gas that can accelerate global warming and cause more permafrost to melt—an example of harmful positive feedback in action.

Tundra is a fragile biome because its short growing season means its soil and vegetation recover very slowly from damage or disturbance. Human activities in Arctic tundra—mostly oil drilling sites, pipelines, mines, and military bases—leave scars that persist for centuries.

Another type of tundra, called *alpine tundra,* occurs above the limit of tree growth but below the permanent snow line on high mountains (Figure 5-11, left). The vegetation is similar to that found in arctic tundra, but it gets more sunlight than arctic vegetation and has no permafrost layer. During the brief summer alpine tundra can be covered with an array of beautiful wildflowers.

 THINKING ABOUT WINDS AND GRASSLANDS What roles do winds play in creating and sustaining grasslands?

Temperate Shrublands: Chaparral

Chaparral has a moderate climate but its dense thickets of spiny shrubs are subject to periodic fires.

In many coastal regions that border on deserts we find fairly small patches of a biome known as *temperate shrubland* or *chaparral* (Spanish for thicket). Closeness to the sea provides a slightly longer winter rainy season than nearby temperate deserts have, and fogs during the spring and fall reduce evaporation. These biomes are found along coastal areas of southern California in the United States, the Mediterranean Sea, central Chile, southern Australia, and southwestern South Africa (Figure 5-9).

Figure 5-18 Natural capital: chaparral vegetation.

Chaparral consists mostly of dense growths of low-growing evergreen shrubs and occasional small trees with leathery leaves that reduce evaporation (Figure 5-18). The soil is thin and not very fertile. Animal species of the chaparral include mule deer, chipmunks, jackrabbits, lizards, and a variety of birds.

During the long, warm, and dry summers, chaparral vegetation becomes very dry and highly flammable. In the late summer and fall, fires started by lightning or human activities spread with incredible swiftness. Research reveals that chaparral is adapted to and maintained by these periodic fires. Many of the shrubs store food reserves in their fire-resistant roots and have seeds that sprout only after a hot fire. With the first rain, annual grasses and wildflowers spring up and use nutrients released by the fire. New shrubs grow quickly and crowd out the grasses.

People like living in this biome because of its moderate, sunny climate with mild, wet winters and warm, dry summers. As a result, humans have moved in and modified this biome considerably. For example, compare the idealized chaparral biome found in southern Europe's Mediterranean area (Figure 5-9) with how this area has been modified by the human ecological footprint (see Figure 3 on pp. S12–S13 in Supplement 4). The downside of its favorable climate is that people living in chaparral assume the high risk of losing their homes and possibly their lives to frequent fires followed by mudslides during rainy seasons. Chaparral: nice climate, risky place to live.

Martin Harvey/Peter Arnold, Inc.

Tropical rain forest

Paul W. Johnson/Biological Photo Service

Temperate deciduous forest

Dave Powell, USDA Forest Service

Polar evergreen coniferous forest (boreal forest, taiga)

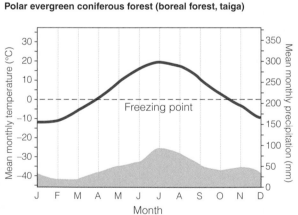

Figure 5-19 Natural capital: climate graphs showing typical variations in annual temperature (red) and precipitation (blue) in tropical, temperate, and polar (cold) forests. Top photo shows the closed canopy of a tropical rain forest in the western Congo Basin of Gabon, Africa. Middle photo shows a temperate deciduous forest in the U.S. state of Rhode Island during the fall. Photo 9 in the Detailed Contents shows this same area of forest during winter. Bottom photo shows a northern coniferous forest in the Malheur National Forest and Strawberry Mountain Wilderness in the U.S. state of Oregon.

FOREST BIOMES

Types of Forests

Forests have enough precipitation to support stands of trees and are found in tropical, temperate, and polar regions.

Undisturbed areas with moderate to high average annual precipitation tend to be covered with **forest,** which contains various species of trees and smaller forms of vegetation. The three main types of forest—tropical, temperate, and *boreal* (polar)—result from combinations of this precipitation level and various average temperatures (Figures 5-10 and 5-19, facing page).

Tropical Rain Forests: Threatened Centers of Biodiversity with Poor Soils

Tropical rain forests have heavy rainfall on most days and a rich diversity of species occupying a variety of specialized niches in distinct layers.

Tropical rain forests (Figure 5-19, top photo) are found near the equator (Figure 5-9), where hot, moisture-laden air rises and dumps its moisture. These forests have year-round uniformly warm temperatures, high humidity, and heavy rainfall almost daily (Figure 5-19, top graph).

Figure 5-20 shows some of the components and interactions in these extremely diverse ecosystems. These forests are dominated by a lush variety of *broadleaf evergreen plants*, which keep most of their leaves year-round. No single tree species dominates these forests. In fact, you could walk for several hundred meters and not come across two members of the same tree species.

This biome typically has huge trees with shallow roots and wide bases (called buttresses) that support their massive weight. The tops of the trees form a dense canopy (Figure 5-19, top photo) which blocks most light from reaching the forest floor, illuminating it with a dim greenish light.

The ground level of these forests have little vegetation, except near stream banks or where a fallen tree has opened up the canopy and let in sunlight. Many of the plants that do live at the ground level have enormous leaves to capture what little sunlight filters through to the dimly lit forest floor.

Individual trees may be draped with vines (called lianas) that reach tree tops to

Producer to primary consumer	Primary to secondary consumer	Secondary to higher-level consumer	All producers and consumers to decomposers

ThomsonNOW **Active Figure 5-20 Natural capital:** some components and interactions in a *tropical rain forest ecosystem.* When these organisms die, decomposers break down their organic matter into minerals that plants use. Colored arrows indicate transfers of matter and energy between producers, primary consumers (herbivores), secondary or higher-level consumers (carnivores), and decomposers. Organisms are not drawn to scale. *See an animation based on this figure and take a short quiz on the concept.*

Figure 5-21 Natural capital: stratification of specialized plant and animal niches in a *tropical rain forest*. Filling such specialized niches enables species to avoid or minimize competition for resources and results in the coexistence of a great variety of species.

gain access to sunlight. Once in the canopy, the vines grow from one tree to another, providing walkways for many of the species living in the canopy. When a large tree is cut down its vines can also pull down other nearby trees.

Tropical rain forests have a very high net primary productivity (Figure 3-22, p. 67), are teeming with life, and have incredible biological diversity (See photos 5, 6, 7, and 8 in the Detailed Contents).

The species in this biome occupy a variety of specialized niches in distinct layers—in the plants' case, based mostly on their need for sunlight, as shown in Figure 5-21. Species in this green tapestry of life are connected and supported by an intricate ecological symphony of feeding relationships. Much of the animal life, particularly insects, bats, and birds, lives in the sunny *canopy* layer, with its abundant shelter and supplies of leaves, flowers, and fruits. To study life in

the canopy, ecologists climb trees, use tall construction cranes, and build platforms and boardwalks in the upper canopy.

Stratification of specialized plant and animal niches in a tropical rain forest enables the coexistence of a great variety of species. Although tropical rain forests cover only 2% of the earth's land surface, biologists estimate that at least half of the earth's terrestrial species reside there. For example, a single tree in these forests may support several thousand insect species, many of them not identified by scientists. Biologists estimate that tropical rain forests may contain tens of millions of undiscovered insect species.

Because of the dense vegetation there is little wind in these forests to spread seeds and pollen. Thus, most rain forest plant species depend on bats, butterflies, birds, bees, and other species to pollinate their flowers and to spread seeds in their droppings. For example, the world's largest flower (Rafflesia,) is a leafless plant with a huge flower (See photo 5 in the Detailed Contents). It smells like rotting meat, presumably to attract flies and beetles that pollinate its huge flower. After blossoming once a year for a few weeks, the flower dissolves into a slimy black mass.

Dropped leaves, fallen trees, and dead animals decompose quickly because of the warm, moist conditions and hordes of decomposers. This rapid recycling of scarce soil nutrients explains the lack of plant litter on the ground.

Despite their incredible diversity of plant life, the soils in tropical rain forests tend to be acidic and low in nutrients (Figure 3-24, bottom left, p. 69). Instead of being stored in the soil, most mineral nutrients released by decomposition are taken up quickly by and stored by trees, vines, and other plants. As a result, tropical rain forest soils contain very few plant nutrients.

 Learn more about how plants and animals in a rain forest are connected in a food web at ThomsonNOW.

This explains why rain forests are not good places to clear and grow crops or graze cattle on a sustainable basis. Their nutrient-poor soils will support only a year or two of crops. Despite this ecological limitation, many of these forests are being cleared or degraded for logging, growing crops, grazing cattle, and mineral extraction. If the clearing is too extensive, the few nutrients in the soil are leached out by heavy rains and such areas are converted to shrubland.

So far at least 40% of these forests have been destroyed or disturbed by human activities and the pace of destruction and degradation of these centers of terrestrial biodiversity is increasing. Ecologists warn that without strong conservation measures, most of these forests will probably be gone within your lifetime. This will reduce the earth's biodiversity and help accelerate global warming by eliminating huge areas of trees that remove carbon dioxide from the troposphere.

[?] THINKING ABOUT TROPICAL RAIN FOREST DESTRUCTION

What harmful effects might the loss of most of the world's remaining tropical rain forests have on your life and lifestyle? What two things could you do to reduce this loss?

Tropical dry forests are found in tropical areas (Figure 5-9) with warm temperatures year round and wet and dry seasons. Tree heights are lower and tree canopies are less dense than in tropical rain forests.

Much of the world's tropical dry forests have been cleared for growing crops and grazing livestock because they are easier to clear than tropical rain forests

are. However, when large areas of these forests are cleared their soils are subject to severe erosion during the extremely heavy rainy season. Within a few years crop productivity and grazing vegetation fall sharply and much of the land is abandoned.

Temperate Deciduous Forests: Changing with the Seasons

Most of the trees in these forests survive winter by dropping their leaves, which decay and produce a nutrient-rich soil.

Temperate deciduous forests (Figure 5-19, center photo and photo 9 in the Detailed Contents) grow in areas with moderate average temperatures that change significantly with the season. These areas have long, warm summers, cold but not too severe winters, and abundant precipitation, often spread fairly evenly throughout the year (Figure 5-19, center graph).

Figure 5-22 (p. 120) shows some of the components and interactions in this ecosystem. This biome is dominated by a few species of *broadleaf deciduous trees* such as oak, hickory, maple, poplar, and beech. They survive cold winters by dropping their leaves in the fall and becoming dormant (see photo 9 in the Detailed Contents). Each spring they grow new leaves that change in the fall into an array of reds and golds before dropping (Figure 5-19, center photo).

Temperate deciduous forests have fewer tree species than tropical rain forests, but the penetration of more sunlight supports a richer diversity of plant life at ground level. Because of a slow rate of decomposition, these forests accumulate a thick layer of slowly decaying leaf litter that is a storehouse of nutrients (Figure 3-24, bottom middle, p. 69).

The temperate deciduous forests of the eastern United States were once home for such large predators as bears, wolves, foxes, wildcats, and mountain lions (pumas). Today most of the predators have been killed or displaced, and the dominant mammal species often is the white-tailed deer, along with smaller mammals such as squirrels, rabbits, opossums, raccoons, and mice.

Warblers, robins, and other bird species migrate to these forests during the summer to feed and breed. Many of these species are declining in numbers because of loss or fragmentation of their summer and winter habitats.

Temperate deciduous forests once covered the eastern half of the United States and western Europe (Figure 5-9). But as these areas were settled, industrialized, and urbanized most of the original forests were cleared. Today, on a worldwide basis, this biome has been disturbed by human activity more than any other terrestrial biome.

Figure 5-22 Natural capital: some components and interactions in a *temperate deciduous forest ecosystem*. When these organisms die, decomposers break down their organic matter into minerals that plants use. Colored arrows indicate transfers of matter and energy between producers, primary consumers (herbivores), secondary or higher-level consumers (carnivores), and decomposers. Organisms are not drawn to scale.

Broad-winged hawk

Hairy woodpecker

Gray squirrel

White oak

Metallic wood-boring beetle and larvae

White-footed mouse

White-tailed deer

Mountain winterberry

Shagbark hickory

May beetle

Racer

Fungi

Long-tailed weasel

Bacteria

Wood frog

Producer to primary consumer	Primary to secondary consumer	Secondary to higher-level consumer	All producers and consumers to decomposers

Evergreen Coniferous Forests: Cold Winters, Wet Summers, and Conifers

These forests consist mostly of cone-bearing evergreen trees that keep their needles year-round to help the trees survive long and cold winters.

Evergreen coniferous forests (Figure 5-19, bottom photo) are also called *boreal forests* and *taigas* ("TIE-guhs"). They are found just south of the arctic tundra in northern regions across North America, Asia, and Europe (Figure 5-9). In this subarctic climate, winters are long, dry, and extremely cold; in the northernmost taiga, sunlight is available only 6–8 hours a day. Summers are short, with cool to warm temperatures (Figure 5-19, bottom graph), and the sun shines up to 19 hours a day.

Figure 5-23 shows some of the components and interactions in this ecosystem. Most boreal forests are dominated by a few species of *coniferous* (cone-bearing) *evergreen trees* such as spruce, fir, cedar, hemlock, and pine that keep some of their narrow-pointed leaves (needles) all year long (Figure 5-19, bottom photo). The small, needle-shaped, waxy-coated leaves of these trees can withstand the intense cold and drought of winter when snow blankets the ground. Such trees are ready to take advantage of the brief summers in these areas without taking time to grow new needles. Plant diversity is low because few species can survive the winters when soil moisture is frozen.

Beneath the stands of trees is a deep layer of partially decomposed conifer needles and leaf litter. De-

Figure 5-23 Natural capital: some components and interactions in an *evergreen coniferous* (*boreal* or *taiga*) *forest ecosystem.* When these organisms die, decomposers break down their organic matter into minerals that plants use. Colored arrows indicate transfers of matter and energy between producers, primary consumers (herbivores), secondary or higher-level consumers (carnivores), and decomposers. Organisms are not drawn to scale.

Blue jay

Balsam fir

Great horned owl

Marten

Moose

White spruce

Wolf

Bebb willow

Pine sawyer beetle and larvae

Snowshoe hare

Fungi

Starflower

Bacteria

Bunchberry

➡ Producer to primary consumer   ➡ Primary to secondary consumer   ➡ Secondary to higher-level consumer   ➡ All producers and consumers to decomposers

composition is slow because of the low temperatures, waxy coating of conifer needles, and high soil acidity. The decomposing conifer needles make the thin, nutrient-poor soil acidic and prevent most other plants (except certain shrubs) from growing on the forest floor.

These biomes contain a variety of wildlife, as depicted in Figure 5-23. Year-round residents include bears, wolves, moose, lynx and many burrowing rodent species. During the brief summer the soil becomes waterlogged, forming acidic bogs, or *muskegs,* in low-lying areas of these forests. Warblers and other insect-eating birds that migrate from the tropics feed on hordes of flies, mosquitoes, and caterpillars.

Cone-bearing conifer forests are also found in regions with more moderate climates. An example is the economically important *southern pine forests* in the United States. Because these forests grow rapidly in the warm and moist southern climate, many of them have been cleared and converted to pine plantations with greatly reduced plant and animal diversity.

Temperate Rain Forests: Biodiversity Near Some Coastal Areas

Coastal areas support huge cone-bearing evergreen trees such as redwoods and Douglas fir in a cool and moist environment.

Coastal coniferous forests or *temperate rain forests* (Figure 5-24, p. 122)) are found in scattered coastal temperate areas with ample rainfall or moisture from dense

Figure 5-24 Natural capital: temperate rain forest in Olympic National Park in the U.S. state of Washington.

ocean fogs. The ocean moderates the temperature so winters are mild and summers are cool. Dense stands of large conifers such as Sitka spruce, Douglas fir, and redwoods once dominated undisturbed areas of these biomes along the coast of North America, from Canada to northern California in the United States.

Most of the trees are evergreen because the abundance of water means that they have no need to shed their leaves. Tree trunks and the ground are frequently covered with mosses and ferns in this cool and moist environment. As in tropical rain forests, little light reaches the forest floor.

[?] THINKING ABOUT WINDS AND FORESTS What roles do winds play in creating temperate and coniferous forests?

MOUNTAIN BIOMES

Mountains: Islands in the Sky

Mountains are high-elevation forested islands of biodiversity and often have snow-covered peaks that reflect solar radiation and gradually release water to lower-elevation streams and ecosystems.

Some of the world's most spectacular and important environments are on high mountains (Figure 5-25), which cover about one-fourth of the earth's land surface. Mountains are places where dramatic changes in altitude, climate, soil, and vegetation take place over a very short distance (Figure 5-11, left).

Because of the steep slopes, mountain soils are especially prone to erosion when the vegetation holding them in place is removed by natural disturbances (such as landslides and avalanches) or human activities (such as timber cutting and agriculture). Many freestanding mountains are *islands of biodiversity* surrounded by a sea of lower-elevation landscapes transformed by human activities.

Mountains play important ecological roles. They contain the majority of the world's forests, which are habitats for much of the world's terrestrial biodiversity. They often are habitats for endemic species found nowhere else on earth. And they serve as sanctuaries for animal species driven from lowland areas.

Mountains also help regulate the earth's climate. About 75% of the world's freshwater is stored in glacial ice, much of it in mountain areas. Mountaintops covered with ice and snow affect climate by reflecting solar radiation back into space. These mountains affect sea levels as a result of decreases or increases in glacial ice—most of which is locked up in Antarctica (the most mountainous of all continents).

Finally, mountains play a critical role in the hydrologic cycle by gradually releasing melting ice, snow, and water stored in the soils and vegetation of mountainsides to small streams.

Figure 5-25 Natural capital: Mountains such these in Mount Rainier National Park in the U.S. state of Washington play important ecological roles.

Despite their ecological, economic, and cultural importance, the fate of mountain ecosystems has not been a high priority of governments or of many environmental organizations. Mountain ecosystems are coming under increasing pressure from several human activities.

HUMAN IMPACTS ON TERRESTRIAL BIOMES

Increasing Human Disturbance

Human activities have damaged or disturbed to some extent more than half of the world's terrestrial ecosystems.

The human species dominates most of the planet. Ecologists estimate that we use, waste, or destroy about 10–55% of the net primary productivity of the earth's terrestrial ecosystems (Figure 3-22, p. 67). Recall that since plants produce the food for most animals, the planet's net primary productivity ultimately limits the number of consumers (including humans) that can survive on the earth—an important lesson from nature.

According to the the 2005 Millennium Ecosystem Assessment, 60% of the world's major terrestrial ecosystems are being degraded or used unsustainably (see Figures 3 and 4 on pp. S12–S15 in Supplement 4). And this destruction and degradation is increasing in many parts of the world.

How long can we keep eating away at the earth's natural capital without threatening our economies and the long-term survival of our own species and of other species we depend upon? No one knows. But there are increasing signs that we need to come to grips with this possibility.

A Survey of Our Harmful Ecological Impacts

Humans have had a number of specific harmful effects on the world's deserts, grassland, forests, and mountains.

In this chapter, we have discussed the nature and importance of the world's desert, grassland, forest, and mountain biomes. Figures 5-26, 5-27, 5-28 (p. 124), and 5-29 (p. 124) list some the specific harmful impacts humans are having on these biomes.

E *RESEARCH FRONTIER* Better understanding of the effects of human activities on terrestrial biomes

? *THINKING ABOUT SUSTAINABILITY* Develop four guidelines for preserving the earth's terrestrial biodiversity based on the four scientific principles of sustainability (Figure 1-16, p. 24).

Natural Capital Degradation
Deserts

Large desert cities

Soil destruction by off-road vehicles

Soil salinization from irrigation

Depletion of groundwater

Land disturbance and pollution from mineral extraction

Figure 5-26 Natural capital degradation: major human impacts on the world's deserts. QUESTION: *What are three direct and three indirect harmful effects of your lifestyle on deserts?*

Natural Capital Degradation
Grasslands

Conversion to cropland

Release of CO_2 to atmosphere from grassland burning

Overgrazing by livestock

Oil production and off-road vehicles in arctic tundra

Figure 5-27 Natural capital degradation: major human impacts on the world's grasslands. Some 70% of Brazil's tropical savanna—once the size of the Amazon—has been cleared and converted to the world's biggest grain growing area. QUESTION: *What are three direct and three indirect harmful effects of your lifestyle on grasslands?*

Natural Capital Degradation
Forests

Clearing for agriculture, livestock grazing, timber, and urban development

Conversion of diverse forests to tree plantations

Damage from off-road vehicles

Pollution of forest streams

Figure 5-28 Natural capital degradation: major human impacts on the world's forests. QUESTION: *What are three direct and three indirect effects of your lifestyle on forests?*

Natural Capital Degradation
Mountains

Agriculture

Timber extraction

Mineral extraction

Hydrolectric dams and reservoirs

Increasing tourism

Urban air pollution

Increased ultraviolet radiation from ozone depletion

Soil damage from off-road vehicles

Figure 5-29 Natural capital degradation: major human impacts on the world's mountains. QUESTION: *What are three direct and three indirect harmful effects of your lifestyle on mountains?*

↰ Revisiting Winds and Sustainability

This chapter's opening case study described how winds connect all parts of the planet to one another. Next time you feel or hear the wind blowing, think about these global connections. And as part of the global climate system, winds play important roles in creating and sustaining the world's deserts, grasslands, forests, and mountains.

Winds promote sustainability by helping distribute solar energy and the recycling of the earth's nutrients. In turn, this helps support the life-sustaining biodiversity of deserts, grassland, and forests and the diversity of species whose interactions in these ecosystems help control population sizes.

Gaining a better understanding of these climate and ecological connections is a vital step for learning how to help sustain these important components of the earth's natural capital.

When we try to pick out anything by itself, we find it hitched to everything else in the universe.

JOHN MUIR

CRITICAL THINKING

1. What would happen to the earth's species and your lifestyle if the winds stopped blowing?

2. List a limiting factor for each of the following ecosystems: **(a)** a desert, **(b)** arctic tundra, **(c)** alpine tundra, **(d)** the floor of a tropical rain forest, and **(e)** a temperate deciduous forest.

3. Why do deserts and arctic tundra support a much smaller biomass of animals than do tropical forests?

4. Some biologists have suggested restoring large herds of bison on public lands in the North American plains as a way of restoring remaining tracts of tall-grass prairie. Ranchers with permits to graze cattle and sheep on federally managed lands have strongly opposed this idea. Do you agree or disagree with the idea of restoring large numbers of bison to the plains of North America? Explain.

5. Why do most animals in a tropical rain forest live in its trees?

6. Why do most species living at high latitudes and high altitudes tend to be ecological generalists while those living the tropics tend to be ecological specialists?

7. What biomes are best suited for **(a)** raising crops and **(b)** grazing livestock? Use the four scientific principles of sustainability (p. 24) to come up with four guidelines for growing food and grazing livestock in these biomes on a more sustainable basis.

8. To some, the widespread destruction and degradation of prairie is justified because they believe that these parts of nature are being underutilized and should be put to use by humans. To others, that belief is a symptom of our lack of understanding of how nature works and an indication of our increasing separation and alienation from the rest of nature since the beginning of agriculture about 10,000 years ago. Which view do you support? Why?

9. Congratulations! You are in charge of the world. What are the three most important features of your plan to help sustain the earth's terrestrial biodiversity?

10. List two questions that you would like to have answered as a result of reading this chapter.

PROJECTS

1. How has the climate changed in the area where you live during the past 50 years? Investigate the beneficial and harmful effects of these changes. How have these changes benefited or harmed you personally?

2. What type of biome do you live in? How have human activities over the past 50 years affected the characteristic vegetation and animal life normally found where you live? How is your lifestyle affecting this biome?

3. Make a concept map of this chapter's major ideas, using the section heads, subheads, and key terms (in boldface). Look on the website for this book for information about making concept maps.

LEARNING ONLINE

The website for this book contains study aids and many ideas for further reading and research. They include a chapter summary, review questions for the entire chapter, flash cards for key terms and concepts, a multiple-choice practice quiz, interesting Internet sites, references, information about green careers, and a guide for accessing thousands of InfoTrac® College Edition articles. Log into

www.thomsonedu.com/biology/miller

Then choose Chapter 5, and select a learning resource. For access to animations, additional quizzes, chapter outlines and summaries, register and log into

at **www.thomsonedu.com** using the access code card in the front of your book.

Why Should We Care about Coral Reefs?

Coral reefs form in clear, warm coastal waters of the tropics and subtropics (Figure 6-1, left). These stunningly beautiful natural wonders are among the world's oldest, most diverse, and most productive ecosystems. In terms of biodiversity, they are the marine equivalents of tropical rain forests.

Coral reefs are formed by massive colonies of tiny animals called *polyps* (close relatives of jellyfish). They slowly build reefs by secreting a protective crust of limestone (calcium carbonate) around their soft bodies. When the polyps die, their empty crusts remain behind as a platform for more reef growth. The resulting elaborate network of crevices, ledges, and holes serves as calcium carbonate "condominiums" for a variety of marine animals.

Coral reefs are the result of a mutually beneficial relationship between the polyps and tiny single-celled algae called *zooxanthellae* ("zoh-ZAN-thel-ee") that live in the tissues of the polyps. The algae provide the polyps with color, food, and oxygen through photosynthesis, and help produce calcium carbonate, which forms the coral skeleton. The polyps, in turn, provide the algae with a well-protected home and some of their nutrients.

Although coral reefs occupy only about 0.1% of the world's ocean area, they provide numerous free ecological and economic services. They help moderate atmospheric temperatures by removing CO_2 from the atmosphere, act as natural barriers that help protect 15% of the world's coastlines from erosion by battering waves and storms, and provide habitats for a variety of marine organisms.

Economically, they produce about one-tenth of the global fish catch, one-fourth of the catch in developing countries, and provide jobs and building materials for some of the world's poorest countries. They also support fishing and tourism industries worth billions of dollars each year.

Finally, these biological treasures give us an underwater world to study and enjoy. Each year more than 1 million scuba divers and snorkelers visit coral reefs to experience these wonders of biodiversity.

Bad news. According to a 2005 report by the World Conservation Union, 20% of the world's coral reefs have been lost to coastal development, pollution, overfishing, warmer ocean temperatures, and other stresses. And if we don't take action now, another 30% of these aquatic oases of biodiversity will be seriously depleted within the next 20–40 years.

One problem is *coral bleaching* (Figure 6-1, right). It occurs when a coral becomes stressed and the algae on which it depends for food and color die out, leaving an underlying white or bleached skeleton of calcium carbonate. Two causes of bleaching are increased water temperature and runoff of silt from the land (usually from forest clearing) that covers the coral and prevents photosynthesis.

The decline and degradation of these colorful oceanic sentinels is a warning about aquatic health, the subject of this chapter.

Sergio Hanquet/Peter Arnold, Inc

Figure 6-1 Natural capital: a healthy coral reef in the Red Sea covered by colorful algae (left) and a bleached coral reef that has lost most of its algae (right) because of changes in the environment (such as cloudy water or too warm temperatures). With the algae gone, the white limestone of the coral skeleton becomes visible. If the environmental stress is not removed and no other alga species fill the abandoned niche, the corals die. These diverse and productive ecosystems are being damaged and destroyed at an alarming rate.

If there is magic on this planet, it is contained in water.

LOREN EISLEY

This chapter provides an introduction to aquatic life. It addresses these questions:

- What are the basic types of aquatic life zones, and what factors influence the kinds of life they contain?

- What are the major types of saltwater life zones, and how do human activities affect them?

- What are the major types of freshwater life zones, and how do human activities affect them?

AQUATIC ENVIRONMENTS

The Water Planet: Saltwater and Freshwater

Saltwater and freshwater aquatic life zones cover almost three-fourths of the earth's surface.

We live on the water planet, with a precious film of water—most of it saltwater—covering about 71% of the earth's surface (Figure 6-2). Thus, a more accurate name for Earth would be Ocean.

The major types of organisms found in aquatic environments are determined by the water's *salinity*—the amounts of various salts such as sodium chloride

Ocean hemisphere   Land–ocean hemisphere

Figure 6-2 Natural capital: the ocean planet. The salty oceans cover 71% of the earth's surface. About 97% of the earth's water is in the interconnected oceans, which cover 90% of the planet's mostly ocean hemisphere (left) and 50% of its land–ocean hemisphere (right). Freshwater systems cover less than 1% of the earth's surface.

(NaCl) dissolved in a given volume of water. As a result, aquatic life zones are classified into two major types: *saltwater* or *marine* (estuaries, coastlines, coral reefs [Core Case Study, p. 126], coastal marshes, mangrove swamps, and oceans) and *freshwater* (lakes, ponds, streams, rivers, and inland wetlands).

Figure 6-3 shows the global distribution of the world's major oceans, lakes, rivers, coral reefs, and mangroves. These aquatic systems play vital roles in

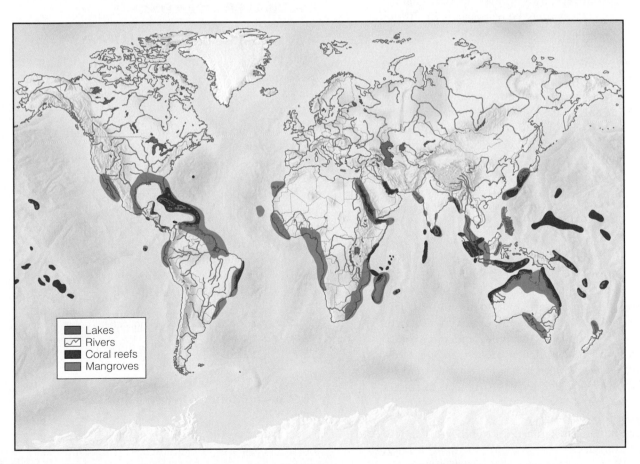

- Lakes
- Rivers
- Coral reefs
- Mangroves

Figure 6-3 Natural capital: distribution of the world's major saltwater oceans, coral reefs, mangroves, and freshwater lakes and rivers.

the earth's biological productivity, climate, biogeo-chemical cycles, and biodiversity and provide us with fish, shellfish, oil, natural gas, minerals, recreation, transportation routes, and many other economically important goods and services.

What Kinds of Organisms Live in Aquatic Life Zones?

Aquatic systems contain floating, drifting, swimming, bottom-dwelling, and decomposer organisms.

Saltwater and freshwater life zones contain several major types of organisms. One group consists of weakly swimming, free-floating **plankton.** One type is *phytoplankton* ("FIE-toe-plank-ton," Greek for "drifting plants"), or *plant plankton,* which includes many types of algae. They and various rooted plants near shore-lines are producers that support most aquatic food chains and food webs.

Another type is **zooplankton** ("ZOO-oh-plank-ton," Greek for "drifting animals"), or *animal plankton.* They consist of primary consumers (herbivores) that feed on phytoplankton and secondary consumers that feed on other zooplankton. They range from single-celled protozoa to large invertebrates such as jellyfish.

There are also huge populations of much smaller plankton called **ultraplankton.** These extremely small photosynthetic bacteria may be responsible for 70% of the primary productivity near the ocean surface.

A *second group* of organisms consists of **nekton,** strongly swimming consumers such as fish, turtles, and whales. A *third group,* called **benthos,** are bottom dwellers such as barnacles and oysters that anchor themselves to one spot, worms that burrow into the sand or mud, and lobsters and crabs that walk about on the bottom. A *fourth group* consists of **decomposers** (mostly bacteria) that break down the organic com-pounds in the dead bodies and wastes of aquatic organ-isms into simple nutrient compounds for use by aquatic producers.

Life in Layers

Life in most aquatic systems is found in surface, middle, and bottom layers.

Most aquatic life zones can be divided into three lay-ers: *surface, middle,* and *bottom.* Important environmen-tal factors that determine the types and numbers of organisms found in these layers are *temperature, access to sunlight for photosynthesis, dissolved oxygen content,* and *availability of nutrients* such as carbon (as dissolved CO_2 gas), nitrogen (as NO_3^-), and phosphorus (mostly as PO_4^{3-}) for producers.

In deep aquatic systems, photosynthesis is largely confined to the upper layer, or *euphotic zone,* through which sunlight can penetrate. The depth of the eu-photic zone in oceans and deep lakes can be reduced when excessive algal growth (algal blooms) clouds the water. Dissolved O_2 levels are higher near the surface because oxygen-producing photosynthesis takes place there. At lower depths O_2 levels fall because of aerobic respiration by aquatic animals and decomposers and because less oxygen gas dissolves in the deeper and colder water than in warmer surface water.

In shallow waters in streams, ponds, and oceans, ample supplies of nutrients for primary producers are usually available. By contrast, in the open ocean, ni-trates, phosphates, iron, and other nutrients often are in short supply and limit net primary productivity (NPP). However, NPP is much higher in parts of the open ocean where upwellings (Figure 5-2, p. 101 and Fig-ure 2 on p. S39 in Supplement 10) bring nutrients from the ocean bottom to the surface for use by producers.

Most creatures living on the bottom of the deep and dark ocean and deep lakes depend on animal and plant plankton that die and drift downward into deep waters. Because this food is limited, deep-dwelling fish species tend to reproduce slowly. This makes them especially vulnerable to depletion from overfishing.

SALTWATER LIFE ZONES

Why Should We Care about the Oceans?

The oceans that occupy most of the earth's surface provide many ecological and economic services.

The world's oceans provide many important ecologi-cal and economic services (Figure 6-4).

As land dwellers, we have a distorted and limited view of the blue aquatic wilderness that covers most of the earth's surface. We know more about the surface of the moon than about the earth's oceans. We also have far too little understanding of the planet's freshwater aquatic systems. According to aquatic scientists, the scientific investigation of poorly understood marine and freshwater aquatic systems could yield immense ecological and economic benefits.

F **RESEARCH FRONTIER** Discovering, cataloging, and study-ing the huge number of unknown aquatic species and their interactions

---

The Coastal Zone: Where Most of the Action Is

The coastal zone makes up less than 10% of the world's ocean area but contains 90% of all marine species.

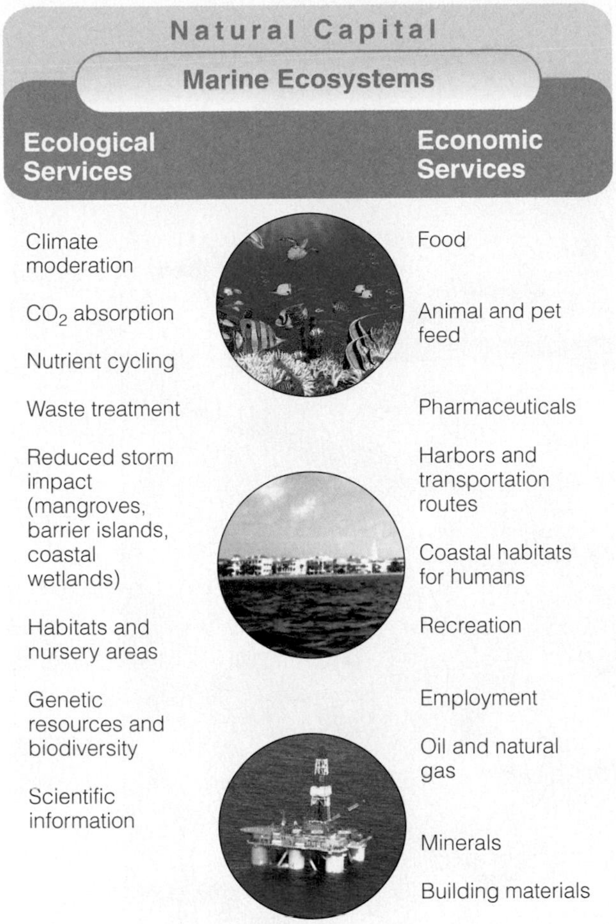

Natural Capital

Marine Ecosystems

Ecological Services	Economic Services
Climate moderation	Food
CO$_2$ absorption	Animal and pet feed
Nutrient cycling	Pharmaceuticals
Waste treatment	
Reduced storm impact (mangroves, barrier islands, coastal wetlands)	Harbors and transportation routes
	Coastal habitats for humans
Habitats and nursery areas	Recreation
Genetic resources and biodiversity	Employment
	Oil and natural gas
Scientific information	Minerals
	Building materials

Figure 6-4 Natural capital: major ecological and economic services provided by marine systems. Scientists estimate that marine systems provide $21 trillion in goods and services per year—70% more than terrestrial ecosystems. QUESTION: *Which two ecological services and which two economic services do you think are the most important?*

Oceans have two major life zones: the *coastal zone* and the *open sea* (Figure 6-5, p. 130). The **coastal zone** is the warm, nutrient-rich, shallow water that extends from the high-tide mark on land to the gently sloping, shallow edge of the *continental shelf* (the submerged part of the continents). This zone has numerous interactions with the land, so human activities easily affect it.

 Learn about ocean provinces where all ocean life exists at ThomsonNOW.

Although it makes up less than a 10% of the world's ocean area, the coastal zone contains 90% of all marine species and is the site of most large commercial marine fisheries. Most ecosystems found in the coastal zone have a high net primary productivity per unit of area, thanks to the zone's ample supplies of sunlight and plant nutrients that flow from land and are distributed by tidal flows and ocean currents.

Estuaries and Coastal Wetlands: Centers of Productivity

Estuaries and coastal wetlands are highly productive ecosystems.

One highly productive area in the coastal zone is an **estuary** where rivers meet the sea. In these partially enclosed bodies of water, seawater mixes with freshwater as well as with nutrients and pollutants from rivers, streams, and runoff from land (Figure 6-6, p. 130).

Estuaries and their associated **coastal wetlands**—land areas covered with water all or part of the year—include river mouths, inlets, bays, sounds, salt marshes (Figure 6-7, p. 131) in temperate zones, and mangrove forests in tropical zones. Life in these biologically productive coastal aquatic systems must adapt to significant daily and seasonal changes in tidal and river flows and land runoff of eroded soil sediment and other pollutants.

Estuaries and coastal marshes are some of the world's most productive ecosystems (Figure 3-22, p. 67) because of high nutrient inputs from rivers and nearby land, rapid circulation of nutrients by tidal flows, presence of many producer plants, and ample sunlight penetrating the shallow waters.

Mangrove forests are the tropical equivalent of salt marshes. They are found along about 70% of gently sloping sandy and silty coastlines in tropical and subtropical regions (Figure 6-3). The dominant organisms in these nutrient-rich coastal forests are mangroves. These trees that can grow in salt water have extensive roots that often extend above the water, where they can obtain oxygen and support the plant in changing water levels (Figure 6-8, p. 131).

These coastal aquatic systems provide important ecological and economic services. They filter toxic pollutants, excess plant nutrients, sediments, and other pollutants. They reduce storm damage by absorbing waves and storing excess water produced by storms and tsunamis (see Figures 2 and 3 on p. S44 in Supplement 11). And they provide food, habitats, and nursery sites for a variety of aquatic species.

According to a 2006 study by the UN Environment Programme, intact coastal mangroves are worth $200,000–$900,000 per square kilometer (0.4 square mile) per year, depending on location. This estimate is based mostly on their sustainable use for fishing and fuelwood and does not include the much larger value of their ecological services. This is 10–45 times more than the income of about $20,000 per hectare ($8,100 per acre) when mangroves are cleared for aquaculture. And protecting mangroves costs only about $1,000 per square kilometer ($2,600 per square mile) a year.

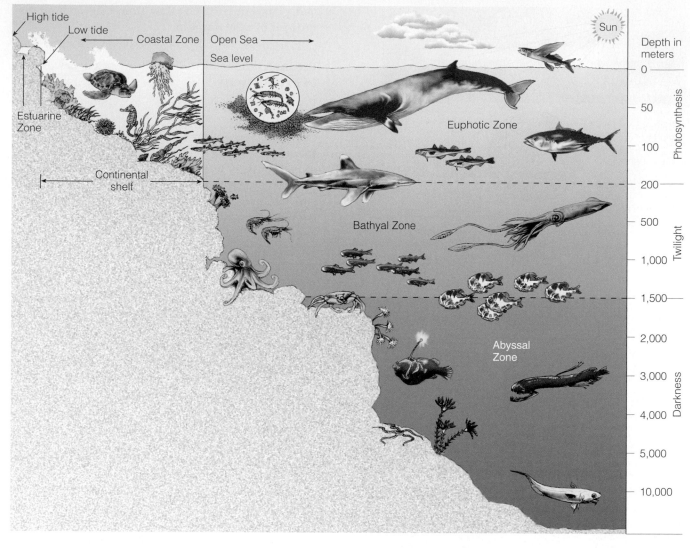

Figure 6-5 Natural capital: major life zones in an ocean (not drawn to scale). Actual depths of zones may vary.

Bad news. Researchers estimate that more than a third of the world's mangrove forests have been destroyed—mostly for shrimp farms, crops, and coastal development projects. Countries such as Bangladesh and the Phillipines have lost almost three-fourths of their mangrove forests.

[?] THINKING ABOUT MANGROVE FORESTS How can clearing mangrove forests increase the economic damage and loss of human life from tsunamis (see Figures 2, and 3 on p. S44 in Supplement 11) and tropical cyclones (Figure 6, on p. S41 in Supplement 10).

Figure 6-6 Natural capital degradation: view of an *estuary* taken from space. The photo shows the sediment plume at the mouth of Madagascar's Betsiboka River as it flows through the estuary and into the Mozambique Channel. Because of its topography, heavy rainfall, and the clearing of forests for agriculture, Madagascar is the world's most eroded country.

Herring gulls

Peregrine falcon

Snowy egret

Cordgrass

Short-billed dowitcher

Marsh periwinkle

Phytoplankton

Smelt

Zooplankton and small crustaceans

Soft-shelled clam

Clamworm

Bacteria

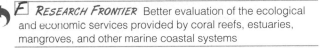

Producer to primary consumer → Primary to secondary consumer → Secondary to higher-level consumer → All consumers and producers to decomposers

© age fotostock/SuperStock

Figure 6-7 Natural capital: some components and interactions in a *salt marsh ecosystem* in a temperate area such as the United States. When these organisms die, decomposers break down their organic matter into minerals used by plants. Colored arrows indicate transfers of matter and energy between consumers (herbivores), secondary or higher-level consumers (carnivores), and decomposers. Organisms are not drawn to scale. The photo below shows a salt marsh in Peru.

Theo Allofs/Visuals Unlimited

Figure 6-8 Natural capital: mangrove forest in Daintree National Park in Queensland, Australia. The tangle of roots and dense vegetation in these coastal forests act like shock absorbers to reduce damage from storms and tsunamis (see pp. S44–S45 in Supplement 11).

F RESEARCH FRONTIER Better evaluation of the ecological and economic services provided by coral reefs, estuaries, mangroves, and other marine coastal systems

Rocky and Sandy Shores: Living with the Tides

Organisms experiencing daily low and high tides have evolved a number of ways to survive under harsh and changing conditions.

The gravitational pull of the moon and sun causes *tides* to rise and fall about every 6 hours in specific coastal areas. The area of shoreline between low and high tides is called the **intertidal zone.** Organisms living in this zone must avoid being swept away or crushed by waves, and deal with being immersed during high tides and left high and dry (and much hotter) at low tides. They must also survive changing levels of salinity when heavy rains dilute saltwater. To deal with

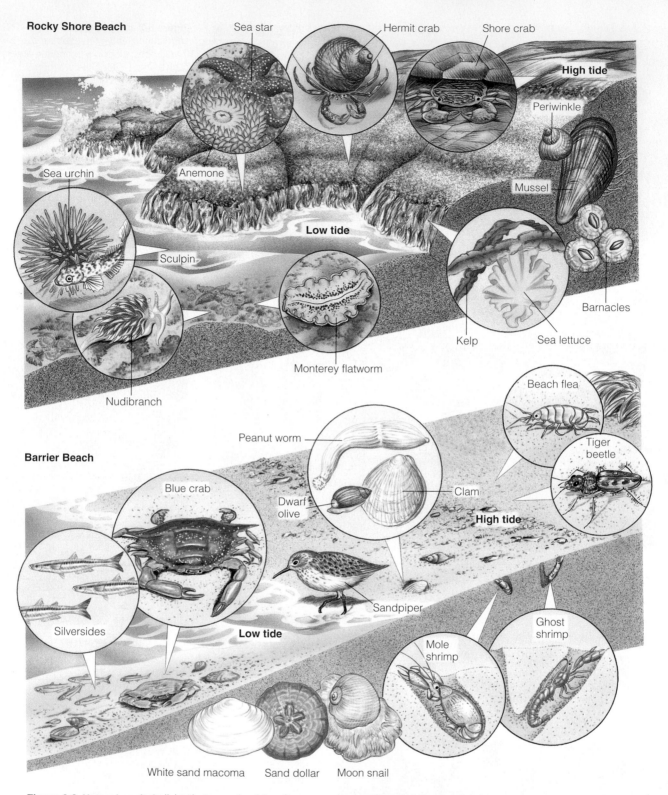

Rocky Shore Beach

Sea star · Hermit crab · Shore crab

High tide

Periwinkle

Sea urchin

Anemone

Mussel

Sculpin

Low tide

Barnacles

Nudibranch

Monterey flatworm

Kelp · Sea lettuce

Barrier Beach

Beach flea

Peanut worm

Tiger beetle

Blue crab

Dwarf olive · Clam

High tide

Silversides

Sandpiper

Low tide

Mole shrimp

Ghost shrimp

White sand macoma · Sand dollar · Moon snail

Figure 6-9 Natural capital: living between the tides. Some organisms with specialized niches found in various zones on rocky shore beaches (top) and barrier or sandy beaches (bottom). Organisms are not drawn to scale.

such stresses, most intertidal organisms hold on to something, dig in, or hide in protective shells.

On some coasts, steep *rocky shores* are pounded by waves. The numerous pools and other niches in their intertidal zones contain a remarkable variety of species that occupy different niches in response to daily and seasonal changes in environmental conditions such as temperature, water flows, and salinity (Figure 6-9, top).

Other coasts have gently sloping *barrier beaches,* or *sandy shores,* with niches for different marine organisms (Figure 6-9, bottom). Most of them are hidden from view and survive by burrowing, digging, and tunneling in the sand. These sandy beaches and their adjoining coastal wetlands are also home to a variety of shorebirds that feed in specialized niches on crustaceans, insects, and other organisms (Figure 4-8, p. 91).

Barrier islands are low, narrow, sandy islands that form offshore from a coastline. These beautiful but limited pieces of real estate are prime targets for development. Examples in the United States are Atlantic City, New Jersey, and Palm Beach, Florida. Living on these islands can be risky. Sooner or later many of the structures humans build on low-lying barrier islands are damaged or destroyed by flooding, severe beach erosion, or major storms (including tropical cyclones; see Figure 6 on p. S41, in Supplement 10). According to climate models, many of the world's barrier islands will be under water by the end of this century as a result of rising sea levels caused mostly by global warming.

Undisturbed barrier beaches have one or more rows of natural sand dunes in which the sand is held in place by the roots of grasses (Figure 6-10). These dunes are the first line of defense against the ravages of the sea. Such real estate is so scarce and valuable that coastal developers frequently remove the protective dunes or build behind the first set of dunes and cover them with buildings and roads. Large storms can then flood and even sweep away seaside buildings and severely erode the sandy beaches. Some people inaccurately call these human-influenced events "natural disasters."

? **THINKING ABOUT LIVING IN RISKY PLACES** Should governments help subsidize property insurance and rebuilding costs for dwellings and beach replenishment on coasts, near major rivers, in earthquake zones, or in other risky areas? Explain your position.

Threats to Coral Reefs: Increasing Stresses

Biologically diverse and productive coral reefs are being stressed by human activities.

Coral reefs (Core Case Study, p. 126, Figure 6-1, and photo 10 in the Detailed Contents) form in clear, warm coastal waters of the tropics and subtropics (Figure 6-3). These dazzling oases of aquatic biodiversity provide homes for one-fourth of all marine species (Figure 6-11, p. 134).

Coral reefs are vulnerable to damage because they grow slowly and are disrupted easily. They also thrive only in clear, warm, and fairly shallow water of constant high salinity. Corals can live only in water with a temperature of 18–30°C (64–86°F). Coral bleaching (Figure 6-1, right) can be triggered by an increase of just 1 C degree (1.8 F degree) above this maximum temperature.

The biodiversity of coral reefs can be reduced by natural disturbances such as severe storms, freshwater floods, and invasions of predatory fish. Today the biggest threats to the survival and biodiversity of many of the world's coral reefs come from the human activities listed in Figure 6-12 (p. 135).

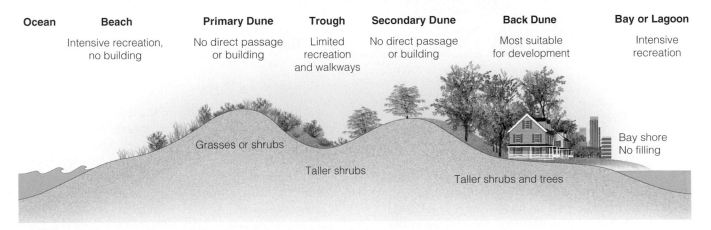

Ocean	Beach	Primary Dune	Trough	Secondary Dune	Back Dune	Bay or Lagoon
	Intensive recreation, no building	No direct passage or building	Limited recreation and walkways	No direct passage or building	Most suitable for development	Intensive recreation

Grasses or shrubs

Taller shrubs

Taller shrubs and trees

Bay shore No filling

Figure 6-10 Natural capital: primary and secondary dunes on gently sloping sandy barrier beaches help protect land from erosion by the sea. The roots of grasses that colonize the dunes help hold the sand in place. Ideally, construction is allowed only behind the second strip of dunes, and walkways to the beach are built over the dunes to keep them intact. This helps preserve barrier beaches and protect buildings from damage by wind, high tides, beach erosion, and flooding from storm surges. Such protection is rare because the short-term economic value of oceanfront land is incorrectly considered much higher than its long-term ecological value. Rising sea levels from global warming may put many barrier beaches under water by the end of this century.

Figure 6-11 Natural capital: some components and interactions in a *coral reef ecosystem.* When these organisms die, decomposers break down their organic matter into minerals used by plants. Colored arrows indicate transfers of matter and energy between producers, primary consumers (herbivores), secondary or higher-level consumers (carnivores), and decomposers. Organisms are not drawn to scale. See photos of a coral reef in Figure 6-1 and photo 10 in the Detailed Contents.

| Producer to primary consumer | Primary to secondary consumer | Secondary to higher-level consumer | All consumer and producers to decomposers |

? *THINKING ABOUT CORAL REEFS* What are two direct and two indirect harmful effects of your lifestyle on coral reefs?

In 2004, 240 experts from 96 countries estimated that 20% of the world's coral reefs are so damaged that they are unlikely to recover. They also projected that by 2050 another 30–50% of the world's coral reefs could be lost due to climate change, habitat loss, pollution (especially eroded sediment), and overfishing. Only about 300 of the world's 6,000 coral reefs are protected (at least on paper) as reserves or parks.

According to a 2006 study by the UN Environment Programme, coral reefs are worth $100,000–$600,000 per square kilometer ($260,000–$1.6 million per square mile) per year, depending on location. These estimates are based on using the reefs mainly for sustainable small-scale fishing, tourism, and selling aquarium fish and do not include their much larger ecological values. If these economic and ecological values are included, it is much cheaper to protect coral reefs than to damage them and use them unsustainably.

Some encouraging news: there is growing evidence that coral reefs can recover when protected by restricting fishing and reducing inputs of nutrients and other pollutants. Also, in 2004, scientists found that coral can form relationships with more heat-tolerant types of algae. This could allow coral in some areas to survive at higher temperatures.

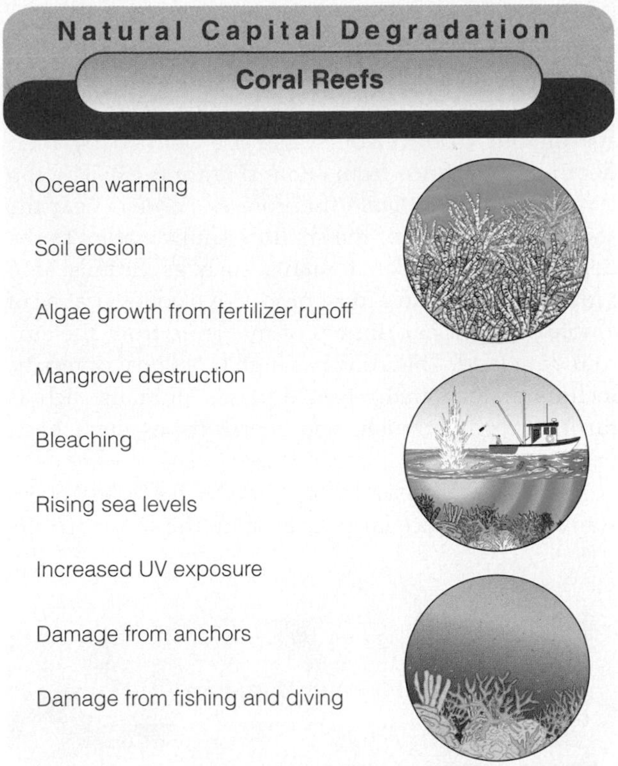

Ocean warming

Soil erosion

Algae growth from fertilizer runoff

Mangrove destruction

Bleaching

Rising sea levels

Increased UV exposure

Damage from anchors

Damage from fishing and diving

Figure 6-12 Natural capital degradation: major threats to coral reefs. QUESTION: *Which three of these threats do you think are the most serious?*

F *RESEARCH FRONTIER* Learning more about the harmful human impacts on coral reefs and how to reduce these impacts

Biological Zones in the Open Sea: Light Rules

The open ocean consists of a brightly lit surface layer, a dimly lit middle layer, and a dark bottom zone.

The sharp increase in water depth at the edge of the continental shelf separates the coastal zone from the vast volume of the ocean called the **open sea.** Primarily on the basis of the penetration of sunlight, it is divided into the three vertical zones (Figure 6-5).

The *euphotic zone* is the brightly lit upper zone where floating and drifting phytoplankton carry out photosynthesis. Nutrient levels are low (except around upwellings), and levels of dissolved oxygen are high. Large, fast-swimming predatory fish such as swordfish, sharks, and bluefin tuna populate this zone.

The *bathyal zone* is the dimly lit middle zone that does not contain photosynthesizing producers because of a lack of sunlight. Zooplankton and smaller fish, many of which migrate to feed on the surface at night, populate this zone.

The lowest zone, called the *abyssal zone,* is dark and very cold and has little dissolved oxygen. Nevertheless, the ocean floor contains enough nutrients to support a large number of species.

Parts of the ocean floor are as complex as the varying topography we find on land. Rock canyons as steep as the Grand Canyon in the U.S. state of Arizona are found just offshore of some continents. In some areas, there are ocean trenches deeper than the height of Mount Everest. The Mid-Atlantic Ridge that runs the entire length of the Atlantic Ocean is the planet's highest mountain range. And in some areas underwater volcanoes shoot out enough molten rock (magma) to build up volcanic islands such as those in Hawaii.

Most organisms of the deep waters and ocean floor get their food from showers of dead and decaying organisms—called *marine snow*—drifting down from upper lighted levels of the ocean. Some of these organisms, including many types of worms, are *deposit feeders,* which take mud into their guts and extract nutrients from it. Others such as oysters, clams, and sponges are *filter feeders,* which pass water through or over their bodies and extract nutrients from it.

Average primary productivity and NPP per unit of area are quite low in the open sea except at an occasional equatorial upwelling, where currents bring up nutrients from the ocean bottom (Figure 5-2, p. 101 and Figure 2, on p. S39 in Supplement 10). However, because the open sea covers so much of the earth's surface, it makes the largest contribution to the earth's overall NPP.

Mostly because of an abundance of producers, we generally find more ocean life as we move toward the poles (Figure 3-18, p. 65) than we find as we go toward the equator—the opposite of what we find with life on the land. Can you explain why?

Effects of Human Activities on Marine Systems: Red Alert

Human activities are destroying or degrading many ecological and economic services provided by the world's coastal areas.

In their desire to live near the coast, people are destroying or degrading the natural resources and services (Figure 6-4) that make coastal areas so enjoyable and economically and ecologically valuable. In 2006, about 45% of the world's population and more than half of the U.S. population lived along or near coasts. By 2010, as much as 80% of the world's people are projected to be living in or near the coastal zones. Figure 6-13 (p. 136) lists major human impacts on marine systems. *Green Career:* Marine scientist

Natural Capital Degradation

Marine Ecosystems

Half of coastal wetlands lost to agriculture and urban development

Over one-third of mangrove forests lost to agriculture, development, and aquaculture shrimp farms

Beaches eroding because of coastal development and rising sea level

Ocean bottom habitats degraded by dredging and trawler fishing

At least 20% of coral reefs severely damaged and 30–50% more threatened

Figure 6-13 Natural capital degradation: major human impacts on the world's marine systems. QUESTION: *Which two of these threats do you think are the most serious?*

FRESHWATER LIFE ZONES

Freshwater Systems

Freshwater life zones provide important ecological and economic services.

Freshwater life zones include *standing* (lentic) bodies of freshwater such as lakes, ponds, and inland wetlands and *flowing* (lotic) systems such as streams and rivers. Although freshwater systems cover less than 1% of the earth's surface, they provide a number of important ecological and economic services (Figure 6-14).

Lakes: Water-Filled Depressions

Life in lakes is found in several different layers.

Lakes are large natural bodies of standing freshwater formed when precipitation, runoff, and groundwater seepage fill depressions in the earth's surface. Causes of such depressions include glaciation (the Great Lakes of North America), crustal displacement (Lake Nyasa in East Africa), and volcanic activity (Crater Lake in Oregon). Lakes are supplied with water from rainfall, melting snow, and streams that drain the surrounding watersheds.

Freshwater lakes vary tremendously in size, depth, and nutrient content. Deep lakes normally consist of four distinct zones that are defined by their depth and distance from shore (Figure 6-15). The top layer, called the *littoral* ("LIT-tore-el") *zone* is near the shore and consists of the shallow sunlit waters to the depth at which rooted plants such as cattails stop growing. It is a lake's most productive zone because of ample sunlight and inputs of nutrients from the surrounding land. This zone has high biological diversity, including algae, many rooted plants, animals such as turtles, frogs, crayfish, and many fishes such bass, perch, and carp.

Next is the *limnetic zone* ("lim-NET-ic"): the open, sunlit water surface layer away from the shore that ex-

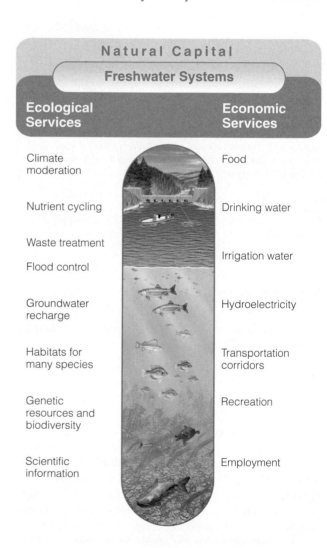

Figure 6-14 Natural capital: major ecological and economic services provided by freshwater systems. QUESTION: *Which two ecological services and which two economic services do you think are the most important?*

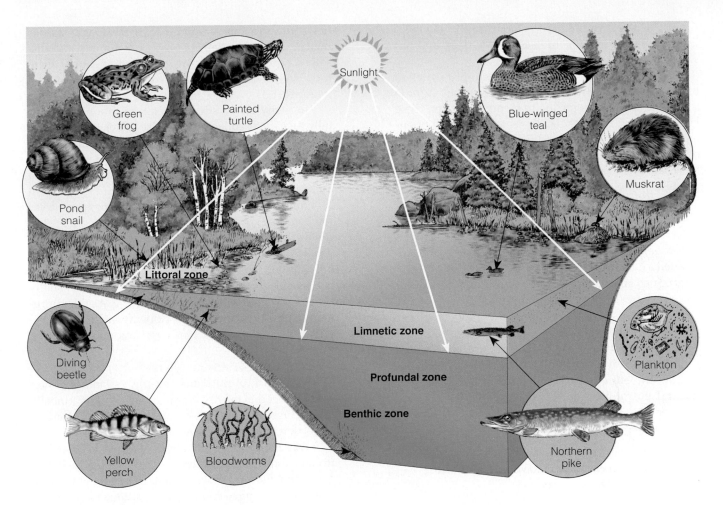

ThomsonNOW **Active Figure 6-15** **Natural capital:** distinct zones of life in a fairly deep temperate zone lake. *See an animation based on this figure and take a short quiz on the concept.*

tends to the depth penetrated by sunlight. The main photosynthetic body of the lake, this zone produces the food and oxygen that support most of the lake's consumers. Its primary organisms are microscopic phytoplankton and zooplankton. Fairly large fish spend most of their time in this zone, with occasional visits to the littoral zone to feed and reproduce.

Next comes the *profundal* ("pro-FUN-dahl") *zone:* the deep, open water where it is too dark for photosynthesis. Without sunlight and plants, oxygen levels are low here. Fish adapted to the lake's cooler and darker water are found in this zone.

The bottom of the lake contains the *benthic* ("BEN-thic") *zone.* Mostly decomposers, detritus feeders, and fish that swim from one zone to the other inhabit it. The benthic zone is nourished mainly by dead matter that falls from the littoral and limnetic zones and by sediment washing into the lake.

During the summer and winter, the water in deep temperate zone lakes becomes stratified into different temperature layers, which do not mix. In the fall and again in spring, the waters at all layers of these lakes mix in events called *overturns,* which equalize the temperatures at all depths. Overturns bring oxygen from the surface water to the lake bottom and nutrients from the lake bottom to the surface waters.

Thomson NOW! Learn more about the zones of a lake, how its water turns over between seasons, and how lakes differ below their surfaces at ThomsonNOW.

Effects of Plant Nutrients on Lakes: Too Much of a Good Thing

Plant nutrients from a lake's environment affect the types and numbers of organisms it can support.

Ecologists classify lakes according to their nutrient content and primary productivity. A newly formed lake generally has a small supply of plant nutrients and is called an **oligotrophic** (poorly nourished) **lake** (Figure 6-16, left). This type of lake is often deep and has steep banks.

Glaciers and mountain streams supply water to many such lakes. Because these sources contain little sediment or microscopic life to cloud the water, these lakes usually have crystal-clear water and small populations of phytoplankton and fish (such as smallmouth bass and trout). Because of their low levels of nutrients, these lakes have a low net primary productivity.

Over time, sediment, organic material, and inorganic nutrients wash into most oligotrophic lakes, and plants grow and decompose to form bottom sediments. A lake with a large or excessive supply of nutrients (mostly nitrates and phosphates) needed by producers is called a **eutrophic** (well-nourished) **lake** (Figure 6-16, right). Such lakes typically are shallow and have murky brown or green water with poor visibility. Because of their high levels of nutrients, these lakes have a high net primary productivity.

Human inputs of nutrients from the atmosphere and from nearby urban and agricultural areas can accelerate the eutrophication of lakes, a process called **cultural eutrophication.** Many lakes fall somewhere between the two extremes of nutrient enrichment. They are called **mesotrophic lakes.**

Freshwater Streams and Rivers: From the Mountains to the Oceans

Water flowing from mountains to the sea creates different aquatic conditions and habitats.

Precipitation that does not sink into the ground or evaporate is **surface water.** It becomes **runoff** when it flows into streams. A **watershed,** or **drainage basin** is the land area that delivers runoff, sediment, and dissolved substances to a stream. Small streams join to form rivers, and rivers flow downhill to the ocean (Figure 6-17).

In many areas, streams begin in mountainous or hilly areas that collect and release water falling to the earth's surface as rain or as snow that melts during warm seasons. The downward flow of surface water and groundwater from mountain highlands to the sea takes place in three different aquatic life zones with different environmental conditions: the *source zone,* the *transition zone,* and the *floodplain zone* (Figure 6-17). Rivers in various areas can differ somewhat from this generalized model.

In the first, narrow *source zone* (Figure 6-17, top), headwaters, or mountain highland streams are usually shallow, cold, clear, and swiftly flowing. As this turbulent water flows and tumbles downward over waterfalls and rapids, it dissolves large amounts of oxygen from the air.

Most of these streams are not very productive because of a lack of nutrients and phytoplankton. Their nutrients come mostly from organic matter (mostly leaves, branches, and the bodies of living and dead insects) that falls into the stream from nearby land.

This zone is populated by cold-water fish (such as trout in some areas), which need lots of dissolved oxygen. Many fish and other animals in fast-flowing headwater streams have compact and flattened bodies that allow them to live under stones. Others have streamlined and muscular bodies that allow them to swim in the rapid and strong currents. Most plants are algae and mosses attached to rocks.

Figure 6-16 Natural capital degradation: the effect of nutrient enrichment on a lake. Crater Lake in Oregon (left) is an example of an *oligotrophic lake* that is low in nutrients. Because of the low density of plankton, its water is quite clear. The lake on the right, found in western New York, is a *eutrophic lake.* Because of an excess of plant nutrients, its surface is covered with mats of algae and cyanobacteria.

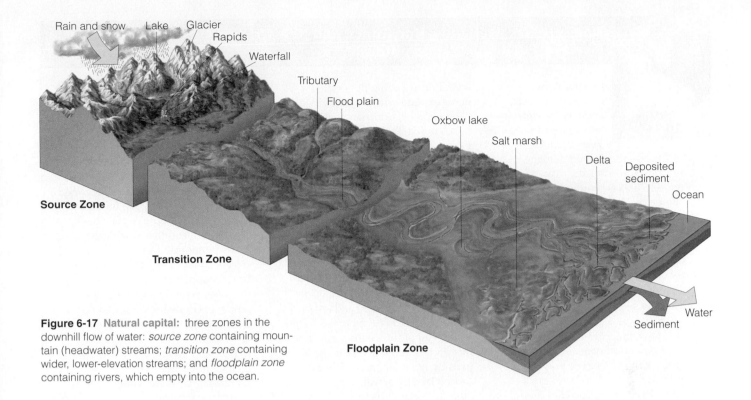

Rain and snow Lake Glacier
Rapids
Waterfall
Tributary
Flood plain
Oxbow lake
Salt marsh
Delta
Deposited sediment
Ocean

Source Zone

Transition Zone

Floodplain Zone

Water
Sediment

Figure 6-17 Natural capital: three zones in the downhill flow of water: *source zone* containing mountain (headwater) streams; *transition zone* containing wider, lower-elevation streams; and *floodplain zone* containing rivers, which empty into the ocean.

In the *transition zone* (Figure 6-17, middle), headwater streams merge to form wider, deeper, and warmer streams that flow down gentler slopes with fewer obstacles. They can be more cloudy (from suspended sediment), slower flowing, and have less dissolved oxygen than headwater streams. The warmer water and other conditions in this zone support more producers (phytoplankton) and cool-water and warm-water fish species (such as black bass) with slightly lower oxygen requirements.

In the *floodplain zone* (Figure 6-17, bottom), streams join into wider and deeper rivers that flow across broad, flat valleys. Water in this zone usually has higher temperatures and less dissolved oxygen than water in the two higher zones. These slow-moving rivers sometimes support fairly large populations of producers such as algae and cyanobacteria and rooted aquatic plants along the shores.

Because of increased erosion and runoff over a larger area, water in this zone often is muddy and contains high concentrations of suspended particulate matter (silt). The main channels of these slow-moving, wide, and murky rivers support distinctive varieties of fish (such as carp and catfish), whereas their backwaters support species similar to those present in lakes. At its mouth, a river may divide into many channels as it flows through deltas built up by deposited sediment and coastal wetlands and estuaries, where the river water mixes with ocean water (Figure 6-6).

Coastal deltas and wetlands and inland floodplains are important parts of the earth's natural capital. They absorb and slow the velocity of floodwaters from storms, tropical cyclones (Figure 6, on p. S41 in Supplement 10), and tsunamis (Figures 2 and 3 on p. S44 in Supplement 11). Protective coastal deltas are built up by deposits of sediments and nutrients at the mouths of rivers. Undeveloped barrier islands along some coastlines also provide natural protection from storms, hurricanes, typhoons, and tsunamis

Bad news: human activities have degraded and destroyed these natural protectors of coastal communities. As a result, hurricanes, typhoons, and tsunamis are partially *unnatural disasters* whose effects are worsened by human activities that reduce the naturally protective ecological services provided by these coastal systems (Case Study, p. 140).

As streams flow downhill, they shape the land through which they pass. Over millions of years the friction of moving water may level mountains and cut deep canyons, and the rock and soil removed by the water are deposited as sediment in low-lying areas.

Streams receive many of their nutrients from bordering land ecosystems. Such nutrient inputs come from falling leaves, animal feces, insects, and other forms of biomass washed into streams during heavy rainstorms or by melting snow. To protect a stream or river system from excessive inputs of nutrients and pollutants, we must focus on its watershed.

Dams, Deltas, Wetlands, Hurricanes, and New Orleans (Science, Economics, and Politics)

Coastal deltas, mangrove forests, and coastal wetlands provide considerable natural protection against flood damage from coastal storms, hurricanes, typhoons, and tsunamis.

Remove or degrade these speed bumps and sponges and the damage from a natural disaster such as a hurricane or typhoon is intensified. As a result, flooding in places like New Orleans, Louisiana (USA), the U.S. Gulf Coast, and Venice, Italy, are largely self-inflicted unnatural disasters.

We have built dams and levees to control water flows and provide electricity (hydroelectric power plants) along most of the world's rivers. This helps reduce flooding along rivers.

However, this reduces flood protection provided by the coastal deltas and wetlands found at the mouths of rivers because the naturally sinking deltas do not get inputs of sediment to build them back up.

As a result, most of the world's river deltas are sinking rather than rising and their protective coastal wetlands are being flooded. This helps explain why the U.S. city of New Orleans, Louisiana, which flooded in a 2005 hurricane, is 3 meters (10 feet) below sea level and in the not-too-distant future will probably be 6 meters (20 feet) below sea level. Add to this the destruction or reduction of the protective effects of coastal wetlands, mangrove forests, and barrier islands and you have a recipe for a major unnatural disaster.

To make matters worse, global sea levels have risen almost 0.3 meters (1 foot) since 1900 and are projected to rise 0.3–0.9 meter (1–3 feet) by the end of this century. Most of this projected rise is due to the expansion of water and melting ice caused by global warming—another unnatural disaster helped along mostly by our burning of fossil fuels and clearing of large areas of the world's tropical forests.

Governments can spend hundreds of billions of dollars building or rebuilding higher levees around cities such as New Orleans. But some scientists warn that sooner or later increasingly stronger hurricanes and typhoons will overwhelm these defenses and cause even greater damage and loss of life.

For example, much of New Orleans is a 3 meter- (10 foot-) deep bathtub or bowl. According to engineers, even if we build levees high enough to make it a 6-meter (20-foot-) deep bathtub a Category 5 hurricane and rising sea levels will eventually overwhelm such defenses and lead to a much more serious unnatural disaster.

The good news is that we now understand some of the connections between dams, deltas, wetlands, barrier islands, sea level rise, and hurricanes. The question is whether we will use such ecological and geological wisdom to change our ways or suffer the increasingly severe consequences of our own actions.

Critical Thinking

Do you think that a sinking city such as New Orleans, Louisiana, should be rebuilt and protected with higher levees or should the lower parts of the city be allowed to revert to wetlands that help protect nearby coastal areas? Explain.

Freshwater Inland Wetlands: Vital Sponges

Inland wetlands act like natural sponges that absorb and store excess water from storms and provide a variety of wildlife habitats.

Inland wetlands are lands covered with freshwater all or part of the time (excluding lakes, reservoirs, and streams) and located away from coastal areas. They include *marshes* (without trees), *swamps* (with trees, Figure 6-18), and *prairie potholes*—small shallow ponds in depressions carved out by ancient glaciers. Other examples are *floodplains,* which receive excess water during heavy rains and floods, and the wet *arctic tundra* in summer. Some wetlands are huge and some are small.

Some inland wetlands are covered with water year-round. Others, called *seasonal wetlands,* remain under water or soggy for only a short time each year. The latter include prairie potholes, floodplain wetlands, and bottomland hardwood swamps. Some stay dry for years before being covered with water again. In such cases, scientists must use the composition of the soil or the presence of certain plants (such as cattails, bulrushes, or red maples) to determine that a particular area is a wetland.

Wetland plants are highly productive because of an abundance of nutrients. Many of these wetlands are important habitats for game fish, muskrats, otters, beavers, migratory waterfowl, and many other bird species.

Inland wetlands provide a number of important and free ecological and economic services. They

- filter and degrade toxic wastes and pollutants. The Audubon Society conservatively estimates that inland wetlands in the United States provide water-quality protection worth at least $1.6 billion per year.

- reduce flooding and erosion by absorbing stormwater and releasing it slowly and by absorbing overflows from streams and lakes. In the United

Francois Suchel/Peter Arnold, Inc.

Figure 6-18 Natural capital: cypress swamp inland wetland in U.S. state of Tennessee.

States, scientists estimate that such natural floodwater storage is worth $3–4 billion a year.

- help replenish stream flows during dry periods.

- help recharge groundwater aquifers.

- help maintain biodiversity by providing habitats for a variety of species.

- supply valuable products such as fish and shellfish, blueberries, cranberries, wild rice, and timber.

- provide recreation for birdwatchers, nature photographers, boaters, anglers, and waterfowl hunters.

? *THINKING ABOUT INLAND WETLANDS* Which two ecological and which two economic services provided by inland wetlands do you believe are the most important? Why? List two ways that your lifestyle directly or indirectly degrades inland wetlands.

Impacts of Human Activities on Freshwater Systems

Dams, cities, farmlands, and filled-in wetlands alter and degrade freshwater habitats.

Human activities affect freshwater systems in four major ways. *First,* dams, diversions, or canals fragment about 40% of the world's 237 large rivers. They alter and destroy wildlife habitats along rivers and in coastal deltas and estuaries by reducing water flow and increasing damage from coastal storms (Case Study, p. 140).

Second, flood control levees and dikes built along rivers alter and destroy aquatic habitats. *Third,* cities and farmlands add pollutants and excess plant nutrients to nearby streams and rivers. *Fourth,* many inland wetlands have been drained or filled to grow crops or have been covered with concrete, asphalt, and buildings.

Case Study: Inland Wetland Losses in the United States (Science and Politics)

Since the 1600s, over half of the wetlands in the United States have been drained and converted to other uses—mostly for growing crops.

About 95% of the wetlands in the United States contain freshwater and are found inland. The remaining 5% are saltwater or coastal wetlands. Alaska has more of the nation's inland wetlands than the other 49 states put together.

More than half of the inland wetlands estimated to have existed in the continental United States during the 1600s no longer exist. About 80% of lost wetlands were destroyed to grow crops (Figure 6-19). The rest were lost to mining, forestry, oil and gas extraction, highways, and urban development. The heavily farmed U.S. state of Iowa has lost about 99% of its original inland wetlands.

This loss of natural capital has been an important factor in increased flood and drought damage in the United States—more examples of unnatural disasters. Many other countries have suffered similar losses. For example, 80% of all wetlands in Germany and France have been destroyed.

F *RESEARCH FRONTIER* Learning more about the harmful human impacts on freshwater aquatic biodiversity and how to reduce these impacts

U.S. Fish and Wildlife Service

Figure 6-19 Natural capital degradation: these prairie pothole wetlands have been ditched and drained for conversion to cropland.

Revisiting Coral Reefs and Sustainability

This chapter's opening case study pointed out the ecological and economic importance of the world's incredibly diverse coral reefs. They are living examples of the four scientific principles of sustainability in action. They survive on solar energy, participate in the cycling of carbon and other chemicals, are a prime example of aquatic biodiversity, and have a network of interactions between species that helps maintain sustainable population sizes.

In this chapter, we have seen that coral reefs and other aquatic systems are being severely stressed by a variety of human activities. Research shows when such harmful human activities are reduced, coral reefs and other stressed aquatic systems can recover fairly quickly.

In other words, we know what to do from a scientific standpoint. Whether we act is primarily a political and ethical problem. This requires educating leaders and citizens about the ecological and economic importance of the earth's aquatic ecosystems. It also involves individual citizens putting pressure on elected officials and business leaders to change the ways we treat these important parts of the earth's natural capital.

. . . the sea, once it casts its spell, holds one in its net of wonders forever.

JACQUES-YVES COUSTEAU

CRITICAL THINKING

1. Someone tries to sell you several brightly colored pieces of dry coral. Explain in biological terms why this transaction is probably fraudulent.

2. How would your lifestyle be affected if all of the world's coral reefs disappeared?

3. Why do aquatic plants such as phytoplankton tend to be very small, whereas most terrestrial plants such as trees tend to be larger and have more specialized structures such as stems and leaves for growth?

4. Why are some aquatic animals, especially marine mammals such as whales, extremely large compared with terrestrial animals?

5. How would you respond to someone who proposes that we use the deep portions of the world's oceans to deposit our radioactive and other hazardous wastes because the deep oceans are vast and are located far away from human habitats? Give reasons for your response.

6. List four factors in your lifestyle that contribute directly or indirectly to the destruction and degradation of coastal and inland wetlands.

7. You are a defense attorney arguing in court for sparing an undeveloped old-growth tropical rain forest and a coral reef from severe degradation or destruction by development. Give your three most important arguments for the defense of each of these ecosystems. If the judge decides you can save only one of the ecosystems, which one would you choose, and why?

8. Congratulations! You are in charge of the world. What are the three most important features of your plan to help sustain the earth's aquatic biodiversity?

9. List two questions that you would like to have answered as a result of reading this chapter.

PROJECTS

1. Develop three guidelines for preserving the earth's aquatic biodiversity based on the four scientific principles of sustainability (Figure 1-16, p. 24).

2. If possible, visit a nearby lake or reservoir. Would you classify it as oligotrophic, mesotrophic, or eutrophic? What are the primary factors contributing to its nutrient enrichment? Which of these factors are related to human activities?

3. Developers want to drain a large area of inland wetland in your community and build a large housing development. List **(a)** the main arguments the developers would use to support this project and **(b)** the main arguments ecologists would use in opposing it. If you were an elected city official, would you vote for or against this project? Can you come up with a compromise plan?

4. Make a concept map of this chapter's major ideas, using the section heads, subheads. and key terms (in boldface). Look on the website for this book for information about making concept maps.

LEARNING ONLINE

The website for this book contains study aids and many ideas for further reading and research. They include a chapter summary, review questions for the entire chapter, flash cards for key terms and concepts, a multiple-choice practice quiz, interesting Internet sites, references, information about green careers, and a guide for accessing thousands of InfoTrac® College Edition articles. Log into

www.thomsonedu.com/biology/miller

Then choose Chapter 6, and select a learning resource. For access to animations, additional quizzes, chapter outlines and summaries, register and log into

at **www.thomsonedu.com** using the access code card in the front of your book.

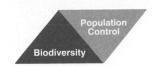

7 Community Ecology

CORE CASE STUDY
Why Should We Care about the American Alligator?

The American alligator (Figure 7-1), North America's largest reptile, has no natural predators except for humans and plays a number of important roles in the ecosystems where it is found. This species, which has survived for nearly 200 million years, has been able to adapt to numerous changes in the earth's environmental conditions.

This changed when hunters began killing large numbers of these animals for their exotic meat and their supple belly skin, used to make shoes, belts, and pocketbooks.

Other people hunted alligators for sport or out of hatred. Between 1950 and 1960, hunters wiped out 90% of the alligators in the U.S. state of Louisiana. By the 1960s, the alligator population in the Florida Everglades was also near extinction.

People who say "So what?" are overlooking the alligator's important ecological role—its *niche*—in subtropical wetland communities. Alligators dig deep depressions, or gator holes. These holes hold freshwater during dry spells, serve as refuges for aquatic life, and supply freshwater and food for many animals.

Large alligator nesting mounds provide nesting and feeding sites for species of herons and egrets.

Alligators eat large numbers of gar (a predatory fish). This helps maintain populations of game fish such as bass and bream.

As alligators move from gator holes to nesting mounds, they help keep areas of open water free of invading vegetation. Without these free ecosystem services, freshwater ponds and coastal wetlands found where alligators live would be filled in with shrubs and trees, and dozens of aquatic species and bird species would disappear from these ecosystems.

Some ecologists classify the American alligator as a *keystone species* because of its important ecological roles in helping to maintain the structure and function of the communities where it is found.

In 1967, the U.S. government placed the American alligator on the endangered species list. Protected from hunters, the population had made a strong comeback in many areas by 1975—too strong, according to those who find alligators in their backyards and swimming pools and to duck hunters whose retriever dogs are sometimes eaten by alligators.

In 1977, the U.S. Fish and Wildlife Service reclassified the American alligator as a *threatened* species in Florida, Louisiana, and Texas, where 90% of the animals live. In 1987, this reclassification was extended to seven more states.

The recent increase in demand for alligator meat and hides has created a booming business for alligator farms, especially in Florida. Such farms reduce the need for illegal hunting of wild alligators.

To biologists, the comeback of the American alligator is an important success story in wildlife conservation. In this chapter, we will look at how these and other species interact and how biological communities respond to changes in environmental conditions.

Figure 7-1 Natural capital: The American alligator plays important ecological roles in its marsh and swamp habitats in the southeastern United States. Since being classified as an endangered species in 1967, it has recovered enough to have its status changed from endangered to threatened species—an outstanding success story in wildlife conservation.

Animal and vegetable life is too complicated a problem for human intelligence to solve, and we can never know how wide a circle of disturbance we produce in the harmonies of nature when we throw the smallest pebble into the ocean of organic life.

GEORGE PERKINS MARSH

This chapter looks at the roles and interactions of species in a community, the ways in which communities respond to changes in environmental conditions, and the long-term sustainability of communities and ecosystems. It discusses these questions:

- What determines the number of species in a community?

- How can we classify species according to their roles in a community?

- How do species interact with one another?

- How do communities respond to changes in environmental conditions?

- Does high species biodiversity increase the stability and sustainability of a community?

COMMUNITY STRUCTURE AND SPECIES DIVERSITY

Community Structure: Appearance Matters

Biological communities differ in their structure and physical appearance.

One way that ecologists distinguish between biological communities is by describing their overall *physical appearance:* the relative sizes, stratification, and distribution of the populations and species in each community. Figure 7-2 shows these differences for various terrestrial communities. There are also differences in the physical structures and zones of communities in aquatic life zones such as oceans, rocky shores and sandy beaches, lakes, river systems, and inland wetlands.

The physical structure within a particular type of community or ecosystem can also vary. Most large terrestrial communities and ecosystems consist of a mosaic of different-sized vegetation patches that change in response to changing environmental conditions. Life is patchy because physical conditions, resources, and species vary greatly from place to place.

Likewise, community structure varies around its *edges* where one type of community makes a transition to a different type of community. For example, the edge area between a forest and an open field may be sunnier, warmer, and drier than the forest interior and support different species than do forest and field interiors.

Increasing the edge area through habitat fragmentation makes many species more vulnerable to stresses such as predators and fire. It also creates barriers that can prevent some species from colonizing new areas and finding food and mates.

Figure 7-2 Natural capital: generalized types, relative sizes, and stratification of plant species in various terrestrial communities.

Species Diversity and Niche Structure: Different Species Playing Different Roles

Biological communities differ in the types and numbers of species they contain and the ecological roles those species play.

Biological communities are shaped by the individual species that live in them, by interactions among these species, and by how the species interact with their physical environment.

An important characteristic of a community's structure is its **species diversity:** the number of different species it contains (**species richness**) combined with the abundance of individuals within each of those species (**species evenness**).

For example, two communities with a total of 20 different species and 200 individuals have the same species diversity. But these communities could differ in their species richness and species evenness. For example, community A might have 10 individuals in each of its 20 species. Community B might have 10 species, each with 2 individuals, and 10 other species, each with 18 individuals. Which community has the highest species richness?

Another community characteristic is its *niche structure:* how many potential ecological niches occur, how they resemble or differ from one another, and how the species occupying different niches interact.

A third factor is a community's *geographical location.* For most terrestrial plants and animals, species diversity is highest in the tropics and declines as we move from the equator toward the poles.

Major reasons for this are that most species in the tropics have a fairly constant daily climate and a more reliable supply of food sources. Thus, tropical species tend to be specialists with narrow niches (Figure 4-7, left, p. 91) and live in microhabitats. In contrast, many species living at high latitudes where weather is cold and variable tend to be generalist species with wide niches (Figure 4-7, right, p. 91). They have adaptations that enable them to thrive in a wide range of environments and occur over large expanses of territory.

❓ *THINKING ABOUT THE AMERICAN ALLIGATOR'S NICHE*
Does the American alligator (Core Case Study, p. 143) have a specialist or a generalist niche? Explain.

The most species-rich environments are tropical rain forests, coral reefs, the deep sea, and large tropical lakes. A community such as a tropical rain forest or a coral reef with a large number of different species (high species richness) generally has only a few members of each species (low species evenness). For example, biologist Terry Erwin found an estimated 1,700 different beetle species in a single tree in a tropical forest in Panama. Can you explain why? (Hint: think climate.) Scientists have also investigated species diversity on islands (Case Study, p. 146).

 Learn about how latitude affects species diversity and about the differences between big and small islands at ThomsonNOW.

TYPES OF SPECIES

Types of Species in Communities

Native, nonnative, indicator, keystone, and foundation species play different ecological roles in communities.

Ecologists often use labels such as *native, nonnative, indicator, keystone,* or *foundation* to describe the major niches filled by various species in communities. Any given species may play more than one of these five roles in a particular community.

Native species are those species that normally live and thrive in a particular community. Other species that migrate into or are deliberately or accidentally introduced into a community are called **nonnative species, invasive species,** or **alien species.**

Many people tend to think of nonnative species as villains. In fact, most introduced and domesticated species of crops and animals such as chickens, cattle, and fish from around the world are beneficial to us.

Sometimes, however, a nonnative species can reduce some or most of a community's native species and cause unintended and unexpected consequences. In 1957, for example, Brazil imported wild African bees to help increase honey production. Instead, the bees displaced domestic honeybees and reduced the honey supply.

Since then, these nonnative bee species—popularly known as "killer bees"—have moved northward into Central America and parts of the southwestern United States. They are still heading north but should be stopped eventually by the harsh winters in the central United States, unless they can adapt genetically to cold weather.

The wild African bees are not the fearsome killers portrayed in some horror movies, but they are aggressive and unpredictable. They have killed thousands of domesticated animals and an estimated 1,000 people in the western hemisphere. Most of their human victims died because they were allergic to bee stings or because they fell down or became trapped and could not flee.

Species Diversity on Islands

In the 1960s, Robert MacArthur and Edward O. Wilson began studying communities on islands to discover why large islands tend to have more species of a certain category such as insects, birds, or ferns than do small islands. To explain these differences in species richness with island size, MacArthur and Wilson proposed what is called the **species equilibrium model,** or the **theory of island biogeography.** According to this widely accepted model, the number of different species found on an island is determined by a balance between two factors: the rate at which new species immigrate to the island and the rate at which existing species become extinct on the island.

The model projects that at some point the rates of immigration and extinction should reach an equilibrium point that determines the island's average number of different species. (The ThomsonNOW online site for this book has a great interactive animation of this model. Check it out.)

According to the model, two features of an island affect its immigration and extinction rates and thus its species diversity. One is the island's *size,* with a small island tending to have fewer different species than a large. One reason is that a small island generally has a lower immigration rate because it is a smaller target for potential colonizers. In addition, a small island should have a higher extinction rate because it usually has fewer resources and less diverse habitats for its species.

A second factor is an island's *distance from the nearest mainland.*

Suppose we have two islands about equal in size and other factors. According to the model, the island closer to a mainland source of immigrant species should have the higher immigration rate and thus a higher species richness—assuming that extinction rates on both islands are about the same.

In recent years, scientists have used this widely accepted scientific theory to help protect wildlife in *habitat islands* such as national parks surrounded by a sea of developed and fragmented land.

Critical Thinking

Suppose we have two national parks surrounded by development. One is a large park and the other is much smaller. Which park is likely to have the highest species richness?

Indicator Species: Biological Smoke Alarms

Some species can alert us to harmful environmental changes taking place in biological communities.

Species that serve as early warnings of damage to a community or an ecosystem are called **indicator species.** For example, the presence or absence of trout species in water at temperatures within their range of tolerance (Figure 3-11, p. 58) is an indicator of water quality because trout need clean water with high levels of dissolved oxygen.

Birds are excellent biological indicators because they are found almost everywhere and are affected quickly by environmental change such as loss or fragmentation of their habitats and introduction of chemical pesticides. The populations of many bird species are declining. Butterflies are also good indicator species because their association with various plant species makes them vulnerable to habitat loss and fragmentation.

Using a living organism to monitor environmental quality is not new. Coal mining is a dangerous occupation, partly because of the underground presence of poisonous and explosive gases, many of which have no detectable odor. In the 1800s and early 1900s, coal miners took caged canaries into mines to act as early-warning sentinels. These birds sing loudly and often. If they quit singing for a long period and appeared to be dis-

tressed, miners took this as an indicator of the presence of poisonous or explosive gases and got out of the mine.

The latest idea is to use indicator species to warn of terrorist attacks involving harmful chemical and biological agents. Some scientists are genetically engineering species of weedy plants to change their color rapidly when exposed to a harmful biological or chemical agent. Genes from these plants could be inserted into evergreen trees, backyard shrubs, cheap houseplants, or even pond algae to turn them into early-warning systems against attacks involving biological or chemical weapons.

Case Study: Why Are Amphibians Vanishing?

Disappearing amphibian species may indicate a decline in environmental quality in many parts of the world.

Amphibians (frogs, toads, and salamanders) live part of their lives in water and part on land. Populations of some amphibians, also believed to be indicator species, are declining throughout the world.

Frogs, for example, are especially vulnerable to environmental disruption at various points in their life cycle, shown in Figure 7-3.

As tadpoles, frogs live in water and eat plants; as adults, they live mostly on land and eat insects that

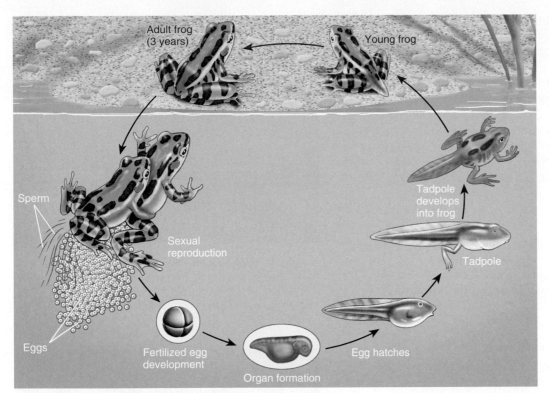

Figure 7-3 Typical *life cycle of a frog.* Populations of various frog species can decline because of the effects of harmful factors at different points in their life cycle. Such factors include habitat loss, drought, pollution, increased ultraviolet radiation, parasitism, disease, overhunting for food (frog legs), and nonnative predators and competitors.

Labels within figure: Adult frog (3 years); Young frog; Sperm; Sexual reproduction; Eggs; Fertilized egg development; Organ formation; Egg hatches; Tadpole; Tadpole develops into frog

can expose them to pesticides. The eggs of frogs have no protective shells to block ultraviolet (UV) radiation or pollution. As adults, they take in water and air through their thin, permeable skins, which can readily absorb pollutants from water, air, or soil.

Since 1980, populations of hundreds of the world's estimated 5,743 amphibian species have been vanishing or declining in almost every part of the world, even in protected wildlife reserves and parks (Figure 4-11, p. 92). According to the 2004 Global Amphibian Assessment, about 33% of all known amphibian species are threatened with extinction and populations of 43% of the species are declining. More than 80% of the Caribbean's amphibian species are threatened by severe habitat loss and disease.

No single cause has been identified to explain these amphibian declines. However, scientists have identified a number of factors that can affect frogs and other amphibians at various points in their life cycles:

- *Habitat loss and fragmentation* (especially from draining and filling of inland wetlands, deforestation, and development)

- *Prolonged drought* (dries up breeding pools so few tadpoles survive)

- *Pollution* (especially exposure to pesticides, which can make frogs more vulnerable to bacterial, viral, and fungal diseases and can cause sexual abnormalities)

- *Increases in ultraviolet radiation* caused by reductions in stratospheric ozone (can harm embryos of amphibians in shallow ponds)

- *Parasites*

- *Viral and fungal diseases* (especially the chytrid fungus that attacks the skin of frogs)

- *Climate change* (Figure 4-11, p. 92): A 2005 study found an apparent correlation between global warming and the extinction of about two-thirds of the 110 known species of harlequin frog in tropical forests in Central and South America. Global warming evaporates water and increases cloud cover in tropical forests. This lowers daytime temperatures and makes nights warmer and creates favorable conditions for the spread of the deadly chytrid fungus to frogs.

- *Overhunting* (especially in Asia and France, where frog legs are a delicacy)

- *Natural immigration or deliberate introduction of nonnative predators and competitors* (such as fish)

A combination of such factors probably is responsible for the decline or disappearance of most amphibian species.

F **RESEARCH FRONTIER** Learning more about why amphibians are disappearing and applying this knowledge to other threatened species

Why should we care if some amphibian species become extinct? Scientists give three reasons. *First,* this trend suggests that environmental health is deteriorating in parts of the world because amphibians are sensitive biological indicators of changes in environmental conditions such as habitat loss and degradation, air and water pollution, UV exposure, and climate change.

Second, adult amphibians play important ecological roles in biological communities. For example, amphibians eat more insects (including mosquitoes) than do birds. In some habitats, extinction of certain amphibian species could lead to extinction of other species, such as reptiles, birds, aquatic insects, fish, mammals, and other amphibians that feed on them or their larvae.

Third, amphibians are a genetic storehouse of pharmaceutical products waiting to be discovered. Compounds in secretions from amphibian skin have been isolated and used as painkillers and antibiotics and as treatment for burns and heart disease.

The plight of some amphibian indicator species is a warning signal. They may not need us, but we and other species need them.

Keystone Species: Major Players

Keystone species help determine the types and numbers of other species in a community.

A keystone is the wedge-shaped stone placed at the top of a stone archway. Remove this stone and the arch collapses. In some communities, ecologists hypothesize that **keystone species** play a similar role. Research indicates that they have a much larger effect on the types and abundances of other species in a community than their numbers would suggest. Eliminating a keystone species may dramatically alter the structure and function of a community.

Keystone species play critical ecological roles in helping sustain a community. One role is *pollination* of flowering plant species by bees, butterflies (Figure 3-1, left, p. 50), hummingbirds, bats, and other species. In addition, *top predator* keystone species feed on and help regulate the populations of other species. Examples are the wolf, leopard, lion, alligator (Core Case Study, p. 143), and some shark species (Case Study, p. 149).

Have you thanked a *dung beetle* (Figure 7-4) today? Perhaps you should: These keystone species rapidly remove, bury, and recycle dung as a food source for their newly hatched larvae. These beetles also churn and aerate the soil, making it more suitable for plant life. Without them, in many places we would be up to our eyeballs in animal wastes and many plants would be starved for nutrients.

Ecologist Robert Paine conducted a controlled experiment along the rocky Pacific coast of the U.S. state of Washington that demonstrated the keystone role of the top-predator sea star *Piaster orchaceus* (Figure 7-5) in an intertidal zone community (Figure 6-9, top, p. 132). Paine removed the mussel-eating *Piaster* sea stars from one rocky shoreline community but not

Figure 7-4 **Natural capital:** this dung beetle has rolled up a ball of fresh dung. They roll the balls into tunnels where they have laid eggs. When the eggs hatch, the larvae have an easily accessible food supply. These hardworking recyclers play keystone roles in many communities.

Figure 7-5 **Natural capital:** *Piaster orchaceus* sea stars. This species helps control mussel populations in intertidal zone communities in the Pacific northwest of the United States. Some scientists view them as keystone species.

from an adjacent community, which served as a control group. Mussels took over and crowded out most other species in the community without the *Piaster* sea stars.

The loss of a keystone species can lead to population crashes and extinctions of other species in a community that depends on it for certain services. This explains why identifying and protecting keystone species is one of the key goals of conservation biologists.

 Thinking about the American Alligator What species might disappear or suffer sharp population declines if the American alligator (Core Case Study, p, 143) became extinct in subtropical wetland ecosystems?

Why Should We Protect Sharks?

CASE STUDY

The world's 370 shark species vary widely in size. The smallest is the dwarf dog shark, about the size of a large goldfish. The largest, the whale shark, can grow to 15 meters (50 feet) long and weigh as much as two full-grown African elephants.

Shark species, feeding at the top of food webs, remove injured and sick animals from the ocean, and thus play an important ecological role. Without their services, the oceans would be teeming with dead and dying fish.

Many people—influenced by movies, popular novels, and widespread media coverage of a fairly small number of shark attacks per year—think of sharks as people-eating monsters. In reality, the three largest species—the whale shark, basking shark, and megamouth shark—are gentle giants. They swim through the water with their mouths open, filtering out and swallowing huge quantities of *plankton.*

Media coverage of shark attacks greatly distorts the danger from sharks. Every year, members of a few species—mostly great white, bull, tiger, gray reef, lemon, hammerhead, shortfin mako, and blue sharks—injure 60–100 people

worldwide. Since 1990, sharks have killed an average of seven people per year. Most attacks involve great white sharks, which feed on sea lions and other marine mammals and sometimes mistake divers and surfers for their usual prey.

For every shark that injures a person, we kill at least 1 million sharks, or a total of about 100 million sharks each year. Sharks are caught mostly for their valuable fins and then thrown back alive into the water, fins removed, to bleed to death or drown because they can no longer swim.

Shark fins are widely used in Asia as a soup ingredient and as a pharmaceutical cure-all. A top (dorsal) fin from a large whale shark can fetch up to $10,000. In high-end restaurants in China, a bowl of shark fin soup can cost $100 or more. Ironically, shark fins have been found to contain dangerously high levels of toxic mercury.

Sharks are also killed for their livers, meat, hides, and jaws, and because we fear them. Some sharks die when they are trapped in nets or lines deployed to catch swordfish, tuna, shrimp, and other species.

In addition to their important ecological roles, sharks save human lives. We may learn from sharks how to fight cancer, which they almost never get. Scientists are also studying their highly effective immune system, which allows

wounds to heal without becoming infected.

Sharks are especially vulnerable to overfishing because they grow slowly, mature late, and have only a few young each generation. Today, they are among the most vulnerable and least protected animals on earth.

In 2003, experts at the National Aquarium in Baltimore, Maryland, estimated that populations of some shark species have decreased by 90% since 1992. Eight of the world's shark species are considered critically endangered or endangered and 82 species are threatened with extinction.

In response to a public outcry over depletion of some species, the United States and several other countries have banned hunting sharks for their fins in their territorial waters. But such bans are difficult to enforce.

Sharks have been around for more than 400 million years. Sustaining this portion of the earth's biodiversity begins with the knowledge that sharks may not need us, but we and other species need them.

✗ *HOW WOULD YOU VOTE?*

Do we have an ethical obligation to protect shark species from premature extinction and to treat them humanely? Cast your vote online at www.thomsonedu.com/biology/miller.

Foundation Species: Other Major Players

Foundation species can create and enhance habitats that can benefit other species in a community.

Some ecologists think the keystone species category should be expanded to include **foundation species,** which play a major role in shaping communities by creating and enhancing their habitats in ways that benefit other species. For example, elephants push over, break, or uproot trees, creating forest openings in the savanna grasslands and woodlands of Africa. This promotes the growth of grasses and other forage plants that benefit smaller grazing species such as antelope. It also accelerates nutrient cycling rates.

Some bat and bird foundation species can regenerate deforested areas and spread fruit plants by depositing plant seeds in their droppings. And beavers acting as "ecological engineers" create wetlands used by other species by building dams in streams.

Proponents of the foundation species hypothesis say that Paine's study of the role of the sea star *Piaster orchaceus* (Figure 7-5) as a keystone species in an intertidal zone community did not take into account the role of mussels as *foundation species.*

According to one hypothesis, mussel beds are homes to hundreds of invertebrate species that do poorly in the presence of mussel competitors such as sea stars. Scientists measured the overall diversity of the species in a tide pool rather than just the 18 species observed by Paine. They found that the overall diversity of species was greater when the keystone sea star species was absent. Its absence allowed the number of mussel species and the species they interact with to expand. From this point of view, the mussels should be seen as a *foundation species* that expand species richness. This type of research and questioning of scientific hypotheses is a good example of science in action.

F *RESEARCH FRONTIER* Identifying and protecting keystone and foundation species

SPECIES INTERACTIONS: COMPETITION AND PREDATION

How Do Species Interact?

Species can interact through competition, predation, parasitism, mutualism, and commensalism.

When different species in a community have activities or resource needs in common, they may interact with one another. Members of these species may be harmed, helped, or unaffected by the interaction. Ecologists identify five basic types of interactions between species: *interspecific competition, predation, parasitism, mutualism,* and *commensalism.*

Such interactions play important roles in a biological community. Predation and competition help limit population size. Interactions between populations of different species also influence the abilities of both to survive and reproduce and thus serve as agents of natural selection. In other words, species interactions that help control population sizes illustrate one of the four scientific principles of sustainability (Figure 1-16, p. 24) in action.

The most common interaction between species is *competition* for shared or limited resources such as space and food. Ecologists call such competition between species **interspecific competition.** Instead of fighting for resources, most competition involves the ability of one species to become more efficient in acquiring food or other resources.

No two species can share the same vital and limited resource for long. When intense competition for limited resources such as food, sunlight, water, soil nutrients, and nesting sites occurs, one of the competing species must migrate (if possible) to another area, shift its feeding habits or behavior through natural selection, suffer a sharp population decline, or become extinct in that area.

Humans are in competition with many other species for space, food, and other resources. As our ecological footprints grow and spread and we convert more of the earth's land and aquatic resources and net primary productivity to our uses, we deprive many other species of resources they need to survive.

Reducing or Avoiding Competition: Sharing Resources

Some species evolve adaptations that allow them to reduce or avoid competition for resources with other species.

Over a time scale long enough for natural selection to occur, some species competing for the same resources evolve adaptations to reduce or avoid competition. One way this happens is through **resource partitioning.** It occurs when species competing for similar scarce resources evolve more specialized traits that allow them to use shared resources at different times, in different ways, or in different places.

Thus, through natural selection, the fairly broad niches of two competing species (Figure 7-6, top) can become more specialized (Figure 7-6, bottom) so that the species can share limited resources. When lions and leopards live in the same area, for example, lions take mostly larger animals as prey, and leopards take smaller ones. Hawks and owls feed on similar prey, but hawks hunt during the day and owls hunt at night.

Figure 7-7 shows resource partitioning by some insect-eating bird species. Figure 4-8 (p. 91) shows the specialized feeding niches of bird species in a coastal wetland.

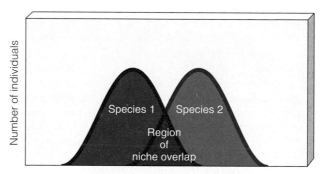

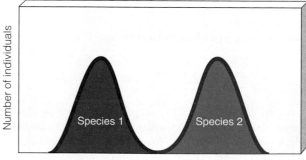

Figure 7-6 Natural capital: *resource partitioning* and *niche specialization* as a result of competition between two species. The top diagram shows the overlapping niches of two competing species. The bottom diagram shows that through natural selection the niches of the two species become separated and more specialized (narrower) so they avoid competing for the same resources.

Predators and Prey: Eating and Being Eaten

Species called predators feed on other species called prey.

In **predation,** members of one species (the *predator*) feed directly on all or part of a living organism of another species (the *prey*). Together, the two kinds of organisms, such as lions (the predator or hunter) and zebras (the prey or hunted), form a **predator–prey relationship.** Such relationships are depicted in Figures 3-10 (p. 57) and 3-18 (p. 65). Most predation occurs unseen at the microscopic level in soils and in the sediments of aquatic systems.

Sometimes predator–prey relationships can surprise us. During the summer months the mighty grizzly bears of the Greater Yellowstone ecosystem eat huge amounts of army cutworm moths, which huddle in masses high on remote mountain slopes. One grizzly bear can dig out and lap up as many as 40,000 of the moths a day. Consisting of 50–70% fat, the moths offer a nutrient that the bear can store in its fatty tissues and draw on during its winter hibernation.

At the individual level, members of the prey species are clearly harmed. At the population level, predation plays a role in evolution by natural selection. Predators, for example, tend to kill the sick, weak, aged, and least fit members of a population because they are the easiest to catch. This leaves behind individuals with better defenses against predation. Such individuals tend to survive longer and leave more offspring with adaptations that help them avoid predation.

Some people view predators with contempt. When a hawk tries to capture and feed on a rabbit, some root for the rabbit. Yet the hawk, like all predators, is merely trying to get enough food for itself and its young. In doing so, it plays an important ecological role in controlling rabbit populations.

Sensing the Environment to Find Food and Mates

Organisms use their senses to locate objects and prey and to attract pollinators and mates.

Most organisms sense various types of electromagnetic radiation (Figure 2-12, p. 43), especially visible light. Our eyes detect only visible light—a small part of the spectrum of electromagnetic radiation that flows around us. Many birds and insects such as bees can sense ultraviolet light that we cannot detect. Many animals such as owls have acute night vision to help them locate prey. Pit vipers, a group of reptiles that includes rattlesnakes, can locate potential prey by detecting infrared radiation (heat) given off by their bodies.

Animals with ears sense *sound*—pressure waves created in the air or water by objects that vibrate, move, or collide with other objects. Sound detection is enhanced by ear shapes and the ability of organisms

Figure 7-7 Sharing the wealth: *resource partitioning* of five species of insect-eating warblers in the spruce forests of the U.S. state of Maine. Each species minimizes competition with the others for food by spending at least half its feeding time in a distinct portion (shaded areas) of the spruce trees, and by consuming somewhat different insect species. (After R. H. MacArthur, "Population Ecology of Some Warblers in Northeastern Coniferous Forests," *Ecology* 36 (1958): 533–536)

such as lions and tigers to move their ears in different directions to pinpoint where sounds are coming from. Bats use a sonar system to find their way around and locate prey. They emit high-pitch pulses of sound (usually higher than we can hear) and use their large ears to sense the sounds that bounce back from objects in their environment.

Some predators locate prey by smelling volatile chemicals their prey give off. Plants, such as the flesh flower (photo 5 in the Detailed Contents), emits smells to attract pollinators. The females of many insects and some mammals give off sex attractant molecules to entice males for reproduction. Snakes navigate by flicking their tongues on the ground to detect chemicals.

How Do Predators Increase Their Chances of Getting a Meal?

Some predators are fast enough to catch their prey, some hide and lie in wait, and some inject chemicals to paralyze their prey.

In addition to using their senses, predators have a variety of methods that help them capture prey. *Herbivores* can simply walk, swim, or fly up to the plants they feed on.

Carnivores feeding on mobile prey have two main options: *pursuit* and *ambush*. Some, such as the cheetah, catch prey by running fast; others, such as the American bald eagle, can fly and have keen eyesight; still others, such as wolves and African lions, cooperate to capture their prey by hunting in packs.

Other predators use *camouflage* to hide in plain sight and ambush their prey. For example, praying mantises (Figure 3-1, right, p. 50) sit in flowers of a similar color and ambush visiting insects. White ermines (a type of weasel) and snowy owls hunt in snow-covered areas. People camouflage themselves to hunt wild game and use camouflaged traps to ambush wild game.

Some predators use chemical warfare to attack their prey. For example, spiders and poisonous snakes use venom to paralyze their prey and to deter their predators.

How Do Prey Defend against or Avoid Predators?

Some prey escape their predators or have protective shells or thorns, some camouflage themselves, and some use chemicals to repel or poison predators.

Prey species have evolved many ways to avoid predators, including the abilities to run, swim, or fly fast, and a highly developed sense of sight or smell that alerts them to the presence of predators. Other avoidance adaptations include protective shells (as on armadillos, which roll themselves up into an armorplated ball, and turtles), thick bark (giant sequoia), spines (porcupines), and thorns (cacti and rosebushes). Many lizards have brightly colored tails that break off when they are attacked, often giving them enough time to escape.

Other prey species use the camouflage of certain shapes or colors or the ability to change color (chameleons and cuttlefish). Some insect species have shapes that look like twigs (Figure 7-8a), bark, thorns, or even bird droppings on leaves. A leaf insect can be almost invisible against its background (Figure 7-8b), and the same is true for an arctic hare with its white winter fur against the snow.

Chemical warfare is another common strategy. Some prey species discourage predators with chemicals that are *poisonous* (oleander plants), *irritating* (stinging nettles and bombardier beetles, Figure 7-8c), *foul smelling* (skunks, skunk cabbages, and stinkbugs), or *bad tasting* (buttercups and monarch butterflies, Figure 7-8d). When attacked, some species of squid and octopus emit clouds of black ink to confuse predators and allow themselves to escape.

Scientists have identified more than 10,000 defensive chemicals made by plants. Some are *herbivore poisons* such as cocaine, caffeine, cyanide, opium, strychnine, peyote, nicotine, and rotenone, some of which we use as an insecticide. Others are *herbivore repellents* such as pepper, mustard, nutmeg, oregano, cinnamon, and mint, all of which we use to flavor or spice up our food.

Major pharmaceutical and pesticide companies view the plant world as a vast drugstore to study as a source for new medicines to treat human diseases and as a source of natural pesticides. Scientists going into nature to find promising natural chemicals are called *bioprospectors*. **Green Career:** Bioprospector

---

F̲ RESEARCH FRONTIER Identifying chemicals from plant and animal species for use as medicines and natural pesticides

---

Many bad-tasting, bad-smelling, toxic, or stinging prey species have evolved *warning coloration*, brightly colored advertising that enables experienced predators to recognize and avoid them. They flash a warning, "Eating me is risky." Examples are brilliantly colored poisonous frogs (Figure 7-8e), red-, yellow-, and black-striped coral snakes, and foul-tasting monarch butterflies (Figure 7-8d).

Based on coloration, biologist Edward O. Wilson gives us two rules for evaluating possible danger from an unknown animal species we encounter in nature. *First,* if it is small and strikingly beautiful, it is probably poisonous. *Second,* if it is strikingly beautiful and easy to catch, it is probably deadly.

Some butterfly species, such as the non-poisonous viceroy (Figure 7-8f), gain some protection by looking and acting like the monarch, a protective device known as *mimicry*. Other prey species use *behavioral strategies* to avoid predation. Some attempt to scare off predators by puffing up (blowfish), spreading their wings (peacocks), or mimicking a predator (Figure 7-8h). Some moths have wings that look like the eyes of much larger animals (Figure 7-8g). Other prey species gain some protection by living in large groups such as schools of fish, herds of antelope, and flocks of birds.

Animals such as ants live in complex social societies containing millions to trillions of individuals. Under such crowded conditions they survive onslaughts by harmful infectious bacteria and fungi by having glands that secrete antibiotics and antifungals. Because of these secretions, the outer surface of an ant is almost free of bacteria and fungi and is much cleaner than most human skin.

Some biologists have begun exploring the use of antibiotics produced by ants to treat human infectious diseases. So far two antibiotic patents have been filed based on studying ants, and there are more to come.

? *THINKING ABOUT THE AMERICAN ALLIGATOR*
What traits does the American alligator (Core Case Study, p. 143) have that help it **(a)** catch prey and **(b)** avoid being preyed upon?

SPECIES INTERACTIONS: PARASITISM, MUTUALISM, AND COMMENSALISM

Parasites: Sponging Off Others

Although parasites can harm their host organisms, they can also promote community biodiversity.

Parasitism occurs when one species (the *parasite*) feeds on part of another organism (the *host*), usually by living on or in the host. In this relationship, the parasite benefits and the host is harmed.

Parasitism can be viewed as a special form of predation. But unlike a conventional predator, a parasite usually is much smaller than its host (prey) and rarely kills its host. Also, most parasites remain closely associated with, draw nourishment from, and may gradually weaken their hosts over time.

(a) Span worm

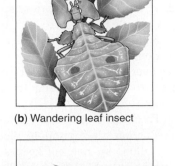
(b) Wandering leaf insect

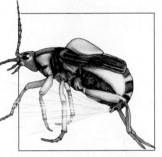
(c) Bombardier beetle

(d) Foul-tasting monarch butterfly

(e) Poison dart frog

(f) Viceroy butterfly mimics monarch butterfly

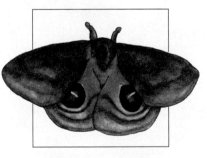
(g) Hind wings of Io moth resemble eyes of a much larger animal.

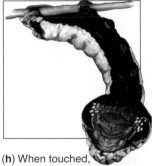
(h) When touched, snake caterpillar changes shape to look like head of snake.

Figure 7-8 Natural capital: some ways in which prey species avoid their predators: **(a, b)** *camouflage,* **(c–e)** *chemical warfare,* **(d, e)** *warning coloration,* **(f)** *mimicry,* **(g)** *deceptive looks,* and **(h)** *deceptive behavior.*

Worms such as tapeworms, disease-causing microorganisms (pathogens), and other parasites live *inside* their hosts. Other parasites attach themselves to the *outside* of their hosts. Examples include ticks, fleas, mosquitoes, mistletoe plants, and sea lampreys that use their sucker-like mouths to attach themselves to their fish hosts and feed on their blood. Some parasites move from one host to another, as fleas and ticks do; others, such as tapeworms, spend their adult lives with a single host.

Some parasites have little contact with their host. For example, North American cowbirds parasitize or take over the nests of other birds by laying their eggs in them and then letting the host birds raise their young.

From the host's point of view, parasites are harmful. But parasites promote biodiversity and control populations by helping keep some species from becoming so plentiful that they eliminate other species.

Mutualism: Win–Win Relationships

Species can interact in ways that benefit both of them.

In **mutualism,** two species or a network of species interact in a way that benefits both. Such benefits include having pollen and seeds dispersed for reproduction, being supplied with food, or receiving protection.

For example, honeybees, caterpillars, butterflies (Figure 3-1, left, p. 50), and other insects may feed on a male flower's nectar, picking up pollen in the process,

and then pollinate female flowers when they feed on them. And coral reefs survive by a mutualistic relationship between reef-building coral animals and bacteria that live in their tissues.

Figure 7-9 shows three examples of mutualistic relationships that combine *nutrition* and *protection*. One involves birds that ride on the backs of large animals like African buffalo, elephants, and rhinoceroses (Figure 7-9a). The birds remove and eat parasites and pests (such as ticks and flies) from the animal's body and often make noises warning the animal when predators approach.

A second example involves the clownfish species, which live within sea anemones, whose tentacles sting and paralyze most fish that touch them (Figure 7-9b). The clownfish, which are not harmed by the tentacles, gain protection from predators and feed on the detritus left from the meals of the anemones. The sea anemones benefit because the clownfish protect them from some of their predators.

A third example is the highly specialized fungi that combine with plant roots to form mycorrhizae (from the Greek words for fungus and roots). The fungi get nutrition from the plant's roots. In turn, the fungi benefit the plant by using their myriad networks

Figure 7-9 Natural capital: examples of *mutualism.* **(a)** Oxpeckers (or tickbirds) feed on parasitic ticks that infest large, thick-skinned animals such as the endangered black rhinoceros. **(b)** A clownfish gains protection and food by living among deadly stinging sea anemones and helps protect the anemones from some of their predators. **(c)** Beneficial effects of mycorrhizal fungi attached to roots of juniper seedlings on plant growth compared to **(d)** growth of such seedlings in sterilized soil without mycorrhizal fungi.

(a) Oxpeckers and black rhinoceros

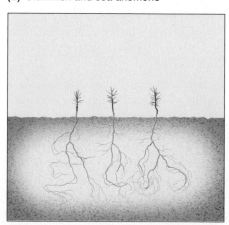

(b) Clownfish and sea anemone

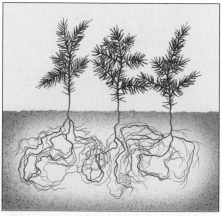

(c) Mycorrhizal fungi on juniper seedlings in normal soil

(d) Lack of mycorrhizal fungi on juniper seedlings in sterilized soil

of hairlike extensions to improve the plant's ability to extract nutrients and water from the soil (Figure 7-9c).

In *gut inhabitant mutualism,* vast armies of bacteria in the digestive systems of animals break down (digest) their food. The bacteria receive a sheltered habitat and food from their host. In turn, they help break down (digest) their host's food. Hundreds of millions of bacteria in your gut help digest the food you eat.

Another example involves termites that eat cellulose. After the termites chew up the cellulose, an array of bacteria and protozoans finding shelter and food in their hindgut convert it to chemical energy needed by the termites. Without abundant termites there would be no decay of cellulose in these tropical communities and a thick carpet of dead plants would quickly cover the landscape.

It is tempting to think of mutualism as an example of cooperation between species. In reality, each species benefits by exploiting the other.

Commensalism: Using without Harming

Some species interact in a way that helps one species but has little or no effect on the other.

Commensalism is an interaction that benefits one species but has little, if any, effect on the other species. One example is some kinds of silverfish insects that move along with columns of army ants to share the food left over during their raids. The army ants receive no apparent harm or benefit from the silverfish.

Birds can benefit from trees by making their nests in them. But generally this does not affect the trees in any way. Another example is plants called *epiphytes* (such as some types of orchids and bromeliads), which attach themselves to the trunks or branches of large trees in tropical and subtropical forests (Figure 7-10). These *air plants* benefit by having a solid base on which to grow. They also live in an elevated spot that gives them better access to sunlight, water from the humid air and rain, and nutrients falling from the tree's upper leaves and limbs. Their presence apparently does not harm the tree.

 Review the way species can interact and see the results of an experiment on species interaction at ThomsonNOW.

ECOLOGICAL SUCCESSION: COMMUNITIES IN TRANSITION

Ecological Succession: How Communities Change Over Time

New environmental conditions allow one group of species in a community to replace other groups.

Figure 7-10 Natural capital: *commensalism.* This bromeliad—an epiphyte or air plant in Brazil's Atlantic tropical rain forest—roots on the trunk of a tree rather than the soil without penetrating or harming the tree. In this interaction, the epiphyte gains access to water, other nutrient debris, and sunlight; the tree apparently remains unharmed.

All communities change their structure and composition in response to changing environmental conditions. The gradual change in species composition of a given area is called **ecological succession.** During succession *colonizing* or *pioneer species* arrive first. As environmental conditions change, they are replaced by others, and later these species may be replaced by another set of species.

Ecologists recognize two types of ecological succession, depending on the conditions present at the beginning of the process. **Primary succession** involves the gradual establishment of various biotic communities in lifeless areas where there is no soil in a terrestrial community or no bottom sediment in an aquatic community. The other more common type of ecological succession is called **secondary succession** in which a series of communities with different species develop in places containing soil or bottom sediment.

Ecologists are still learning about how one community gradually replaces another. In some cases, species in one community may modify the environment (such as by producing soil), which makes it easier for other species to move in. This process is called *facilitation.* In other cases, some species in earlier and later communities can coexist because they are not in direct competition for resources.

Primary Succession: Starting from Scratch

Over long periods, a series of communities with different species can develop in lifeless areas where there is no soil or bottom sediment.

Primary succession begins with an essentially lifeless area where there is no soil in a terrestrial ecosystem (Figure 7-11) or no bottom sediment in an aquatic ecosystem. Examples include bare rock exposed by a retreating glacier or severe soil erosion, newly cooled lava, an abandoned highway or parking lot, and a newly created shallow pond or reservoir.

Primary succession usually takes a long time because there is no fertile soil to provide the nutrients needed to establish a plant community. The slow process of soil formation begins when *pioneer* or *early successional species* arrive and attach themselves to inhospitable patches of bare rock. Examples are lichens and mosses whose seeds or spores are distributed by the wind or on the coats of animals.

These tough species start the soil formation process by trapping wind-blown soil particles and tiny pieces of detritus, producing tiny bits of organic matter, secreting mild acids that slowly fragment and break down the rock, and adding their own wastes and dead bodies. The chemical breakdown or weathering is hastened by physical weathering such as the fragmentation of rock that occurs when water freezes in cracks and expands.

After hundreds to thousands of years, the soil may be deep and fertile enough to store the moisture and nutrients needed to support the growth of *midsuccessional plant species* such as herbs, grasses, and low shrubs. Trees that need lots of sunlight and are adapted to the area's climate and soil usually replace these species.

As these tree species grow and create shade, they are replaced by *late successional plant species* (mostly trees) that can tolerate shade. Unless fire, flooding, severe erosion, tree cutting, climate change, or other natural or human processes disturb the area, what was once bare rock becomes a complex forest community (Figure 7-11).

Primary succession can also take place in newly created small ponds as a result of an influx of sediments and nutrients in runoff from the surrounding land. This sediment can support seeds or spores of plants reaching the pond by winds, birds, or other animals. Over time this process can transform the pond first into a marsh and eventually to dry land.

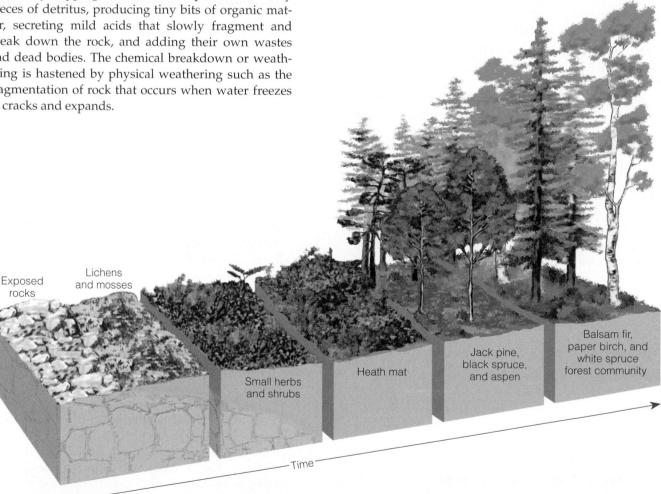

Figure 7-11 Natural capital: *primary ecological succession* over several hundred years of plant communities on bare rock exposed by a retreating glacier on Isle Royale, Michigan (USA) in northern Lake Superior. The details vary from one site to another.

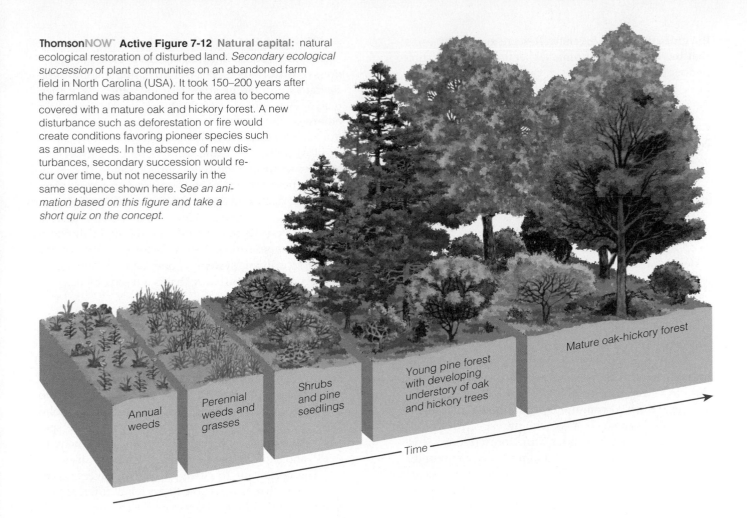

ThomsonNOW Active Figure 7-12 **Natural capital:** natural ecological restoration of disturbed land. *Secondary ecological succession* of plant communities on an abandoned farm field in North Carolina (USA). It took 150–200 years after the farmland was abandoned for the area to become covered with a mature oak and hickory forest. A new disturbance such as deforestation or fire would create conditions favoring pioneer species such as annual weeds. In the absence of new disturbances, secondary succession would recur over time, but not necessarily in the same sequence shown here. *See an animation based on this figure and take a short quiz on the concept.*

Annual weeds

Perennial weeds and grasses

Shrubs and pine seedlings

Young pine forest with developing understory of oak and hickory trees

Mature oak-hickory forest

Time

Secondary Succession: Starting Over with Some Help

A series of communities with different species can develop in places containing some soil or bottom sediment.

Secondary succession begins in an area where the natural community of organisms has been disturbed, removed, or destroyed, but some soil or bottom sediment remains. Candidates for secondary succession include abandoned farmland, burned or cut forests, heavily polluted streams, and land that has been flooded. Because some soil or sediment is present, new vegetation usually can germinate within a few weeks from seeds in the soil and those brought in from nearby plants by wind or by birds and other animals.

In the central or Piedmont region of North Carolina (USA), Europeans settling in North America cleared many of the mature native oak and hickory forests and planted the land with crops. Later they abandoned some of this farmland because of erosion and loss of soil nutrients. Figure 7-12 shows one way that such abandoned farmland has undergone secondary succession over 150–200 years.

Thomson NOW! Explore the difference between primary and secondary succession at ThomsonNOW.

Descriptions of ecological succession usually focus on changes in vegetation. But these changes in turn affect food and shelter for various types of animals. Thus as succession proceeds, the numbers and types of animals and decomposers also change.

During primary or secondary succession, disturbances such as natural or human-caused fires, deforestation, and overgrazing by livestock can convert a particular stage of succession to an earlier stage. Such disturbances create new conditions that encourage some species and discourage or eliminate others.

But many ecologists contend that in the long run some types of disturbances such as fires can be beneficial for the species diversity of some communities and ecosystems. Such disturbances create new conditions that can discourage or eliminate some species but encourage others by releasing nutrients and creating unfilled niches.

According to the *intermediate disturbance hypothesis,* fairly frequent but moderate disturbances lead to the greatest species diversity. Researchers hypothesize that there is sufficient time between moderate disturbances for colonization by a variety of species. Some field experiments support this hypothesis.

Can We Predict the Path of Succession, and Is Nature in Balance?

Scientists cannot predict the course of succession or view it as preordained progress toward a stable climax community that is in balance with its environment.

According to the traditional view, succession proceeds in an orderly sequence along an expected path until a certain stable type of *climax community* occupies an area. Such a community is dominated by a few long-lived plant species and is in balance with its environment. This equilibrium model of succession is what ecologists once meant when they talked about the *balance of nature.*

Over the last several decades, many ecologists have changed their views about balance and equilibrium in nature. Under the old balance-of-nature view, a large terrestrial community undergoing succession eventually became covered with an expected type of climax vegetation such as a mature forest (Figures 7-11 and 7-12). But a close look at almost any community reveals that it consists of an ever-changing mosaic of vegetation patches at different stages of succession. And most ecologists now recognize that mature late-successional communities are not in a state of permanent equilibrium. Instead, they are in a state of continual disturbance and change.

The modern view is that we cannot predict the course of a given succession or view it as preordained progress toward an ideally adapted climax community. Rather, succession reflects the ongoing struggle by different species for enough light, nutrients, food, and space. This competition allows them to survive and gain reproductive advantages over other species.

ECOLOGICAL STABILITY AND SUSTAINABILITY

Stability of Living Systems: Surviving by Changing

Living systems maintain some degree of stability through constant change in response to changing environmental conditions.

All living systems from a cell to the biosphere are dynamic systems that are constantly changing in response to changing environmental conditions. These systems contain complex networks of negative and positive feedback loops that interact to provide some degree of stability over each system's expected life span.

This stability is maintained only by constant change in response to changing environmental conditions or disturbances. For example, in a mature tropical rain forest, some trees die and others take their places. However, unless the forest is cut, burned, or otherwise destroyed, you would still recognize it as a tropical rain forest 50 or 100 years from now.

It is useful to distinguish among three aspects of stability or sustainability in living systems. One is **inertia,** or **persistence:** the ability of a living system to resist being disturbed or altered. A second is **constancy:** the ability of a living system such as a population of plants to keep its numbers within the limits imposed by available resources. A third factor is **resilience:** the ability of a living system to bounce back and repair damage after a disturbance that is not too drastic.

Community Productivity and Sustainability

Having many different species appears to increase the sustainability of many communities.

Do diverse communities tend to produce more biomass than simple communities? The answer seems to be yes. Research by ecologists David Tilman and John Downing suggests that communities with more species tend to have a higher net primary productivity (NPP) and can be more resilient than simpler ones.

The reason for this may be that in diverse communities each species is able to exploit a different portion of the available resources. For example, some plants need lots of water and others need less and some plants bloom early in the growing season while others bloom late.

A second question is whether species diversity in a community leads to greater stability or sustainability. In the 1960s, most ecologists believed the greater the species diversity and the accompanying web of feeding and biotic interactions in a biological community, the greater its stability or ability to withstand environmental disturbances. According to this hypothesis, a complex community with a diversity of species and feeding paths has more ways to respond to most environmental stresses because it does not have "all its eggs in one basket."

Is this a valid hypothesis? Because no community can function without some producers and decomposers, there is a minimum threshold of species diversity below which communities and ecosystems cannot function. Many studies support the idea that some level of biodiversity provides insurance against catastrophe. But how much biodiversity is needed to help sustain various communities remains uncertain.

For example, some recent research suggests that the average annual net primary productivity of an ecosystem reaches a peak with 10–40 producer species. Many ecosystems contain more producer species than this, but it is difficult to distinguish among those that

Ecological Stability: A Closer Look

Part of the problem is that ecologists disagree on how to define *stability*. For example, does a community need both high inertia and high resilience to be considered stable?

Evidence suggests that some communities have one of these properties but not the other. Tropical rain forests have high species diversity and high inertia and thus are resistant to significant alteration or destruction. But once a large tract of tropical forest is severely degraded, the community's

resilience may be so low that the forest may not be restored. Nutrients (stored primarily in the vegetation, not in the soil), and other factors needed for recovery may no longer be present. Such a large-scale loss of forest cover may also change the local or regional climate so that forests can no longer be supported.

By contrast, grasslands are much less diverse than most forests and have low inertia because they burn easily. However, because most of their plant matter is stored in underground roots, these ecosystems have high resilience and recover quickly. Grassland can be de-

stroyed only if its roots are plowed up and something else is planted in its place, or if it is severely overgrazed by livestock or other herbivores. We have been doing both in some grassland areas for many decades.

Another difficulty is that populations, communities, and ecosystems are rarely, if ever, at equilibrium. Instead, nature is in a continuing state of disturbance, fluctuation, and change.

Critical Thinking

Are deserts fairly stable communities? Explain.

---

are essential and those that are not. The Spotlight above sheds more light on this issue.

F RESEARCH FRONTIER Learning more about how biodiversity is related to ecosystem stability and sustainability

Why Should We Bother to Protect Natural Systems?

Human activities are disrupting ecosystem services that support and sustain all life and all economies.

Some developers argue that if biodiversity does not necessarily lead to increased ecological stability, there is no point in trying to preserve and manage old-growth forests and other communities. They conclude that we should cut down diverse old-growth forests, use the timber resources, and replace the forests with tree plantations of single tree species. Furthermore, they say, we should convert most of the world's grasslands to crop fields, drain and develop inland wetlands, dump our toxic and radioactive wastes into the deep ocean, and not worry about the premature extinction of species by our activities.

Ecologists and other environmental scientists strongly disagree. They point out that just because natural, undisturbed systems are not always in equilibrium or balance does not mean they are unimportant parts of the earth's natural capital that help promote ecosystem sustainability.

These scientists point to overwhelming evidence that human disturbances are disrupting, destroying, degrading, and simplifying many of the world's ecosystems (see Figures 3 and 4 on pp. S12–S15 in Supplement 4). Evidence indicates that this threatens

ecosystem services that support and sustain all life and all economies. Thus, we need to use great caution in making potentially harmful changes to communities and ecosystems.

↩ Revisiting the American Alligator and Sustainability

The core case study of the American alligator at the beginning of this chapter illustrates the power humans have over the environment—the power both to do harm and to make amends. As most American alligators were eliminated from their natural areas in the 1950s, scientists began pointing out the ecological benefits these animals had been providing to their habitats (building water holes, nesting mounds, and feeding sites for other species). Understanding this led to protecting this species and allowing it to recover.

In this chapter, we have seen how interactions among organisms in a community determine their abundances and distributions, help limit population size, and influence evolutionary change. And we have seen how communities respond to changes in environmental conditions by undergoing ecological succession.

Biological communities are functioning examples of the four scientific principles of sustainability (Figure 1-16, p. 24) in action. They depend on solar energy, participate in the chemical cycling of nutrients, and have a diversity of types and species, and their populations are controlled by interactions among their species.

No part of the world is what it was before there were humans.

LAWRENCE B. SLOBODKIN

CRITICAL THINKING

1. Some homeowners in the U.S. state of Florida believe they should have the right to kill any alligator found on their property. Others argue against this notion, saying alligators are a threatened species, and that housing developments have invaded the habitats of alligators, not the other way around. Some would say the American alligator has an inherent right to exist, regardless of how we feel about it. What is your opinion on this issue? Explain. What would happen in the areas where alligators live if they were all killed or removed from those areas?

2. How would you experimentally determine whether **(a)** an organism is a keystone species and **(b)** two bird species feeding on the same plant are competing for the same resources or are engaged in resource partitioning?

3. How would you respond to someone who claims it is not important to protect areas of temperate and polar biomes because most of the world's biodiversity is in the tropics?

4. Use the second law of thermodynamics to help explain why predators are generally less abundant than their prey.

5. Describe how natural selection can affect predator–prey relationships and how predator–prey relationships can affect natural selection.

6. How would you reply to someone who argues that **(a)** we should not worry about our effects on natural systems because succession will heal the wounds of human activities and restore the balance of nature, **(b)** efforts to preserve natural systems are not worthwhile because nature is largely unpredictable, and **(c)** because there is no balance in nature we should cut down diverse old-growth forests and replace them with tree farms?

7. Develop three guidelines for sustaining the earth's biological communities based on the four scientific principles of sustainability (Figure 1-16, p. 24).

8. Congratulations! You are in charge of the world. What are the three most important features of your plan to help sustain the earth's biological communities?

9. List two questions that you would like to have answered as a result of reading this chapter.

PROJECTS

1. Use the library or Internet to find and describe two species not discussed in this textbook that are engaged in a **(a)** commensalistic interaction, **(b)** mutualistic interaction, and **(c)** parasite–host relationship.

2. Visit a nearby natural area and try to identify examples of **(a)** mutualism and **(b)** resource partitioning.

3. Do some research to identify the parasites likely to be found in your body.

4. Visit a nearby land area such as a partially cleared or burned forest or grassland or an abandoned crop field and record signs of secondary ecological succession. Study the area carefully to see whether you can find patches that are at different stages of succession because of various disturbances.

5. Make a concept map of this chapter's major ideas, using the section heads, subheads, and key terms (in boldface). Look on the website for this book for information about making concept maps.

LEARNING ONLINE

The website for this book contains study aids and many ideas for further reading and research. They include a chapter summary, review questions for the entire chapter, flash cards for key terms and concepts, a multiple-choice practice quiz, interesting Internet sites, references, information about green careers, and a guide for accessing thousands of InfoTrac® College Edition articles. Log into

www.thomsonedu.com/biology/miller

Then choose Chapter 7, and select a learning resource. For access to animations, additional quizzes, chapter outlines and summaries, register and log into

at **www.thomsonedu.com** using the access code card in the front of your book.

Population Ecology

CORE CASE STUDY
Southern Sea Otters: Are They Back from the Brink of Extinction?

Southern sea otters (Figure 8-1, right) live in kelp forests (Figure 8-1, left) in shallow waters along much of the Pacific coast of North America. These tool-using marine mammals use stones to pry shellfish off rocks under water. Then they break open the shells while swimming on their backs, using their bellies as a table (Figure 8-1, right).

Before European settlers arrived, about 1 million southern sea otters lived along the Pacific coastline of North America. By the early 1900s, the species was almost extinct in this region because of overhunting for their thick and luxurious fur and because they competed with fishers for valuable abalone and other shellfish.

Between 1938 and 2005 the population of southern sea otters off California's coast increased from about 300 to 2,600. This partial recovery was helped when in 1977 the U.S. Fish and Wildlife Service (FWS) declared the species endangered in most of its range. The FWS says that the sea otter population would have to reach about 8,400 animals before it can be removed from the endangered species list.

Biologists classify the southern sea otters as *keystone species*. They help keep sea urchins and other kelp-eating species from depleting highly productive kelp forests (Figure 8-1, left) that provide habitats for a number of species in offshore coastal waters. Kelp also help reduce shore erosion and are used for hundreds of products including toothpaste, beer, and ice cream. Without southern sea otters, many kelp-dependent species would decline or disappear and some of the ecological and economic services provided by kelp forests would be lost as kelp-consuming sea urchin populations would take over.

Wherever southern sea otters have returned or have been reintroduced, formerly deforested kelp areas recover within a few years and fish populations increase. This upsets many commercial and recreational fishers, who argue that sea otters consume too many shellfish, especially dwindling stocks of abalone.

Population dynamics, the subject of this chapter, is a study of how and why populations change in their *distribution, numbers, age structure* and *density* in response to changes in environmental conditions. Studying the population dynamics of southern sea otter populations and their interactions with other species has helped us to better understand the ecological importance of this keystone species.

Figure 8-1 Natural capital: an endangered southern sea otter in Monterey Bay, California (USA) (above, using a stone to crack the shell of a clam) and a giant kelp bed near San Clemente, Island, California (left). Scientific studies indicate that the otters act as a keystone species in a kelp forest system by helping control the populations of sea urchins and other kelp-eating species.

In looking at nature, never forget that every single organic being around us may be said to be striving to increase its numbers.

CHARLES DARWIN, 1859

This chapter addresses the following questions:

- What are the major characteristics of populations?

- How do populations respond to changes in environmental conditions?

- How do species differ in their reproductive patterns?

POPULATION DYNAMICS AND CARRYING CAPACITY

Population Distribution

Most populations live in clumps.

In this section, we will look at how and why populations change in their *distribution, numbers, age structure,* and *density* in response to changes in environmental conditions.

Let's begin with how individuals in populations are distributed or dispersed within a particular area or volume. Three general patterns of *population distribution* or *dispersion* in a habitat are *clumping, uniform dispersion,* and *random dispersion* (Figure 8-2).

The individuals in the populations of most species live in clumps or groups (Figure 8-2a). Examples are patches of desert vegetation around springs, cottonwood trees clustered along streams, wolf packs, flocks of geese, and schools of fish. The location and size of these clumps vary with the availability of resources.

Why clumping? Four reasons. *First,* the resources a species needs vary greatly in availability from place to place. *Second,* living in herds, flocks, and schools can provide some animals with better protection from predators and population declines. *Third,* living in packs gives some predator species such as wolves a

better chance of getting a meal. *Fourth,* some animal species form temporary groups for mating and caring for their young.

Some species maintain a fairly constant distance between individuals. By having this pattern creosote bushes in a desert (Figure 8-2b) have better access to scarce water resources. Organisms with a random distribution (Figure 8-2c) are fairly rare. The world is mostly clumpy.

Changes in Population Size: Entrances and Exits

Populations increase through births and immigration and decrease through deaths and emigration.

Over time the number of individuals in a population may increase, decrease, remain about the same, or go up and down in cycles in response to changes in environmental conditions. Four variables—*births, deaths, immigration,* and *emigration*—govern such changes in population size. A population increases by birth and immigration (arrival of individuals from outside the population) and decreases by death and emigration (departure of individuals from the population):

$$\text{Population change} = (\text{Births} + \text{Immigration}) - (\text{Deaths} + \text{Emigration})$$

Age Structure: Young Populations Can Grow Fast

How fast a population grows or declines depends on its age structure.

A population's **age structure**—the proportions of individuals at various ages—can have a strong effect on how rapidly its size increases or decreases. Age structures are usually described in terms of organisms not mature enough to reproduce (*prereproductive ages*), those capable of reproduction (*reproductive ages*), and those too old to reproduce (*postreproductive ages*).

Figure 8-2 **Natural capital:** generalized *dispersion patterns* for individuals in a population throughout their habitat. The most common pattern is *clumps* of members of a population throughout their habitat, mostly because resources are usually found in patches.

(a) Clumped (elephants)

(b) Uniform (creosote bush)

(c) Random (dandelions)

The size of a population made up mostly of individuals in their reproductive ages or soon to reach these ages is likely to increase. In contrast, a population dominated by individuals past their reproductive age will tend to decrease over time. The size of a population with a fairly even distribution among these three age groups tends to remain stable because reproduction by younger individuals is roughly balanced by the deaths of older individuals.

Limits on Population Growth: Biotic Potential versus Environmental Resistance

No population can increase its size indefinitely.

Populations vary in their capacity for growth, also known as their **biotic potential**. The **intrinsic rate of increase (r)** is the rate at which a population would grow if it had unlimited resources.

Individuals in populations with a high intrinsic rate of growth typically *reproduce early in life, have short generation times* (the time between successive generations), *can reproduce many times* (have a long reproductive life), and *have many offspring each time they reproduce*.

Some species have an astounding biotic potential. Without any controls on population growth, the descendants of a single female housefly could total about 5.6 trillion houseflies within about 13 months. If this rapid exponential growth kept up, within a few years these houseflies would cover the earth's entire surface! Bacteria are even more prolific. Under ideal conditions, a bacterium that reproduced itself every 20 minutes would form a layer 0.3 meter (1 foot) deep over the entire earth's surface in only 36 hours! Exponential growth is amazing.

Fortunately, this is not realistic because *no population can grow indefinitely*. In the real world, a rapidly growing population reaches some size limit imposed by a shortage of one or more *limiting factors* such as light, water, living space, or nutrients, or by exposure to too many competitors, predators, or infectious diseases. *There are always limits to population growth in nature*. This important lesson is one of nature's four principles of sustainability (Figure 1-16, p. 24).

For example, the population size of southern sea otters has fluctuated in response to changes in environmental conditions. And in recent decades, humans have played a major role in their decline, despite current attempts to protect and rebuild the population of this endangered species (Core Case Study, p. 161).

Environmental resistance consists of all factors that act to limit the growth of a population. It is an excellent example of negative or corrective feedback (p. 33). Together biotic potential and environmental resistance determine the **carrying capacity (K):** the maximum population of a given species that a partic-

ular habitat can sustain indefinitely without degrading the habitat. The growth rate of a population decreases as its size nears the carrying capacity of its environment because resources such as food and water begin to dwindle.

F **RESEARCH FRONTIER** Estimating carrying capacity more exactly for various species and ecosystems and for the earth

---

Exponential and Logistic Population Growth: J-Curves and S-Curves

With ample resources a population can grow rapidly, but as resources become limited, its growth rate slows and levels off.

A population with few, if any, resource limitations grows exponentially at a fixed rate such as 1% or 2% a year. *Exponential* or *geometric growth* starts slowly but then accelerates as the population increases because the base size of the population is increasing. Plotting the number of individuals against time yields a J-shaped growth curve (Figure 8-3, bottom half of curve). Whether an exponential growth curve looks steep or "fast" depends on the time period of observation.

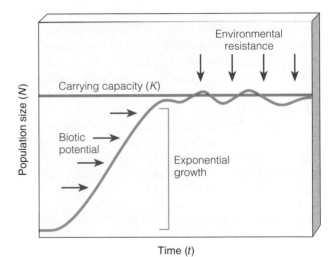

ThomsonNOW **Active Figure 8-3** **Natural capital:** no population can continue to increase in size indefinitely. *Exponential growth* (lower part of the curve) occurs when resources are not limited and a population can grow at its *intrinsic rate of increase (r)* or *biotic potential.* Such exponential growth is converted to *logistic growth*, in which the growth rate decreases as the population becomes larger and faces environmental resistance. Over time, the population size stabilizes at or near the *carrying capacity (K)* of its environment, which results in a sigmoid (S-shaped) population growth curve. Depending on resource availability, the size of a population often fluctuates around its carrying capacity, although a population may temporarily exceed its carrying capacity and suffer a sharp decline or crash in its numbers. *See an animation based on this figure and take a short quiz on the concept.*

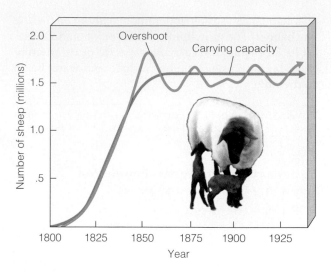

Figure 8-4 Boom and bust: *logistic growth* of a sheep population on the island of Tasmania between 1800 and 1925. After sheep were introduced in 1800, their population grew exponentially thanks to an ample food supply. By 1855, they had overshot the land's carrying capacity. Their numbers then stabilized and fluctuated around a carrying capacity of about 1.6 million sheep.

Logistic growth involves rapid exponential population growth followed by a steady decrease in population growth with time until the population size levels off (Figure 8-3, top half of curve). This slowdown occurs as the population encounters environmental resistance and approaches the carrying capacity of its environment. After leveling off, a population with this type of growth typically fluctuates slightly above and below the carrying capacity.

A plot of the number of individuals against time yields a sigmoid, or S-shaped, logistic growth curve (the whole curve in Figure 8-3). Figure 8-4 depicts such a curve for sheep on the island of Tasmania, south of Australia, in the early 19th century.

Thomson NOW! Learn how to estimate a population of butterflies and see a mouse population growing exponentially at ThomsonNOW.

The *brown tree snake* (Figure 8-5) is native to the Solomon Islands, New Guinea, and Australia. After World War II, a few of these snakes stowed away on military planes going to the island of Guam. With no enemies or rivals in Guam, they have multiplied exponentially for several decades reaching densities of up to 5,000 snakes per square kilometer (13,000 per square mile). Their venomous bites have sent large numbers of people to emergency rooms and their climbing habits have caused more than 2,000 electrical outages. They have also wiped out 8 of Guam's 11 native forest bird species. Sooner or later the brown tree snake will use up its food supply in Guam and decline in numbers, but meanwhile they are causing serious ecologi-

cal and economic damage. And they may end up on islands such as those in Hawaii.

Changes in the population sizes of keystone species such as southern sea otters (Core Case Study, p. 161) and the American alligator (Core Case Study, p. 143) can influence the composition of communities by causing decreases in populations of species dependent on such keystone species and increases in the populations of species that move in to occupy part or all of their ecological niches.

Exceeding Carrying Capacity: Move, Change Habits, or Decline in Size

When a population exceeds its resource supplies, many of its members will die unless they can switch to new resources or move to an area with more resources.

Some species do not make a smooth transition from exponential growth to logistic growth. Some populations use up their resource supplies and temporarily *overshoot*, or exceed, the carrying capacity of their environment. This occurs because of a *reproductive time lag:* the period needed for the birth rate to fall and the death rate to rise in response to resource overconsumption.

In such cases, the population suffers a *dieback,* or *crash,* unless the excess individuals can switch to new resources or move to an area with more resources. Such a crash occurred when reindeer were introduced onto a small island in the Bering Sea (Figure 8-6).

The carrying capacity of an area or volume is not fixed. Sometimes when a population exceeds the carrying capacity of an area, it causes damage that reduces the area's carrying capacity. For example, overgrazing by cattle on dry western lands in the United States has reduced grass cover in some areas. This has allowed sagebrush—which cattle cannot eat—to move

Figure 8-5 Natural capital degradation: the brown tree snake was accidentally introduced onto the island of Guam several decades ago. Since then its numbers have increased exponentially and it has exterminated many of the island's bird species.

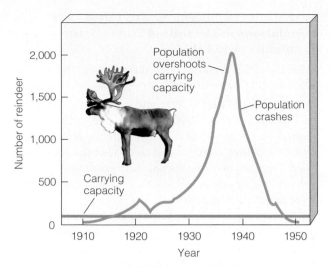

Figure 8-6 Exponential growth, overshoot, and population crash of reindeer introduced to the small Bering Sea island of St. Paul. When 26 reindeer (24 of them female) were introduced in 1910, lichens, mosses, and other food sources were plentiful. By 1935, the herd size had soared to 2,000, overshooting the island's carrying capacity. This led to a population crash, with the herd size plummeting to only 8 reindeer by 1950.

in, thrive, and replace grasses, reducing the land's carrying capacity for cattle.

Humans are not exempt from population overshoot and dieback. Ireland experienced a population crash after a fungus destroyed the potato crop in 1845. About 1 million people died, and 3 million people migrated to other countries. Polynesians on Easter Island experienced a population crash after using up most of the island's trees that helped keep them alive (Core Case Study, p. 28).

Over time some species may increase their carrying capacity by developing adaptive traits through natural selection that reduce environmental resistance to their population growth. In addition, carrying capacity can increase or decrease seasonally and from year-to-year because of variations in weather, climate, and other factors. And some species nearing their carrying capacity may be able to keep growing in size by migrating to other areas with more resources.

So far technological, social, and other cultural changes have extended the earth's carrying capacity for the human species. We have increased food production and used large amounts of energy and matter resources to make normally uninhabitable areas habitable. Some say we can keep doing this indefinitely mostly because of our technological ingenuity. Others say that sooner or later we will reach the limits that nature always imposes on populations.

✗ HOW WOULD YOU VOTE? Can we continue to expand the earth's carrying capacity for humans? Cast your vote online at www.thomsonedu.com/biology/miller.

? THINKING ABOUT THE HUMAN SPECIES If the human species suffered a sharp population decline, name three types of species that might move in to occupy part of our ecological niche.

Population Density and Population Change: Effects of Crowding

A population's density can affect how rapidly it can grow or decline, but some population control factors are not affected by population density.

Population density is the number of individuals in a population found in a particular area or volume. Some factors that limit population growth have a greater effect as a population's density increases. Examples of such *density-dependent population controls* include competition for resources, predation, parasitism, and infectious disease.

Higher population densities may help sexually reproducing individuals find mates but can also lead to increased competition for mates, food, living space, water, sunlight, and other resources. High densities can help shield some members from predators but can also make large groups vulnerable to predators such as humans. And close contact among individuals in dense populations can increase the transmission of infectious disease. When population density decreases, the opposite effects occur. Density-dependent factors tend to regulate a population at a fairly constant size, often near the carrying capacity of its environment.

Infectious disease is a classic type of density-dependent population control. An example is the *bubonic plague,* which swept through densely populated European cities during the fourteenth century. The bacterium causing this disease normally lives in rodents. It was transferred to humans by fleas that fed on infected rodents and then bit humans. The disease spread like wildfire through crowded cities, where sanitary conditions were poor and rats were abundant. At least 25 million people in European cities died from the disease. There is growing concern that a global flu epidemic may kill hundreds of millions of people.

Some factors—mostly abiotic—that can kill members of a population are *density independent.* In other words, their effect is not dependent on the density of the population. For example, a severe freeze in late spring can kill many individuals in a plant population or a population of monarch butterflies (Figure 3-1, left, p. 50) that fly each year from Canada to Mexico's forested mountains to spend the winter, regardless of their density. Other such factors include floods, hurricanes, fire, pollution, and habitat destruction, such as clearing a forest of its trees or filling in a wetland.

Types of Population Change Curves in Nature

Population sizes may stay about the same, suddenly increase and then decrease, vary in regular cycles, or change erratically.

In nature we find four general patterns of variation in population size: *stable, irruptive, cyclic,* and *irregular.* A species whose population size fluctuates slightly above and below its carrying capacity is said to have a fairly stable population size (Figure 8-4). Such stability is characteristic of many species found in undisturbed tropical rain forests, where average temperature and rainfall vary little from year to year.

For some species, population growth may occasionally explode, or *irrupt,* to a high peak and then crash to a more stable lower level or in some cases to a very low level. Many short-lived, rapidly reproducing species such as algae and many insects have irruptive population cycles that are linked to seasonal changes in weather or nutrient availability. For example, in temperate climates, insect populations grow rapidly during the spring and summer and then crash during the hard frosts of winter.

A third type of fluctuation consists of regular *cyclic fluctuations,* or *boom-and-bust cycles,* of population size over a time period (Figure 8-7). Examples are lemmings, whose populations rise and fall every 3–4 years, and the lynx and snowshoe hare, whose populations generally rise and fall in a 10-year cycle. Ecologists distinguish between *top-down population regulation* by predation and *bottom-up population regulation* by the scarcity of one or more resources. Supplement 12 on p. S46 discusses wolf–moose interactions.

Finally, some populations appear to have *irregular behavior* in their changes in population size, with no recurring pattern. Some scientists attribute this behavior to chaos in such systems. Others scientists contend that it may represent fluctuations in response to periodic catastrophic population crashes due to severe winter weather.

F RESEARCH FRONTIER Learning how the premature extinction of some species can affect the populations of other species and the ecological and economic services provided by their ecosystems

Case Study: Exploding White-Tailed Deer Populations in the United States (Science, Economics, and Ethics)

Since the 1930s the white-tailed deer population in the United States has exploded.

By 1900, habitat destruction and uncontrolled hunting had reduced the white-tailed deer population in the United States to about 500,000 animals. In the 1920s and 1930s, laws were passed to protect the remaining deer. Hunting was restricted and predators such as wolves and mountain lions that preyed on the deer were nearly eliminated.

It worked, and to some suburbanites and farmers, perhaps too well. Today there are 25–30 million white-tailed deer in the United States. During the last 50 years, large numbers of Americans have moved into the wooded habitat of deer and provided them with flowers, garden crops, and other plants they favor.

Deer are edge species that like to live in the woods for security and go to nearby fields, orchards, lawns, and gardens for food. Suburbanization has created an all-you-can-eat edge paradise for deer and their populations in such areas have soared. In some forests, they are consuming native ground cover vegetation, allowing nonnative weed species to take over. And deer spread Lyme disease (carried by deer ticks) to humans.

In addition, deer accidentally kill and injure more people in the United States than do any other wild species each year in 1.5 million deer–vehicle collisions. These accidents annually injure at least 14,000 people, kill at least 200 (up from 101 deaths in 1993), and cause more than $1.1 billion in damages.

There are no easy answers to the deer population problem in the suburbs. Changing hunting regulations to allow killing of more female deer (does) can cut down the overall deer population. But these actions have little effect on deer in suburban areas because it is too dangerous to allow widespread hunting there. And animal activists strongly oppose killing deer on the ethical grounds that this is cruel and inhumane treatment.

Deer could also be trapped and moved somewhere else, but this is expensive and must be repeated

Figure 8-7 Population cycles for the snowshoe hare and Canadian lynx. At one time scientists believed these curves provided circumstantial evidence that these predator and prey populations regulated one another. More recent research suggests that the periodic swings in the hare population are caused by a combination of *top-down population control*—predation by lynx and other predators—and *bottom-up population control.* In the latter, changes in the availability of the food supply for hares help determine hare population size, which in turn helps determine the lynx population size. (Data from D. A. MacLulich)

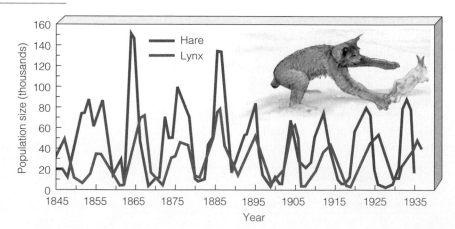

every few years. Where are we going to take them and who pays for this?

Should we put Bambi on birth control? Darts loaded with deer contraceptive could be fired into does each year to hold down the birth rate. But this is expensive and must be repeated each year. One possibility is an experimental single-shot contraceptive vaccine that causes does to stop producing eggs for several years. Another approach, being tested by state biologists in Connecticut, is to trap dominant males and use chemical injections to sterilize them. Both these approaches will require years of testing.

Meanwhile, if you live in the suburbs, expect deer to chow down on your shrubs, flowers, and garden plants. They have to eat every day like you do. You might consider not planting their favorite plants around your house.

> ⁇ *THINKING ABOUT DEER POPULATIONS* Some blame the deer for invading farms and suburban yards and gardens to eat food that we have made easily available to them. Others say we are mostly to blame because we have invaded deer territory, eliminated most of the predators that kept their populations under control, and fed them from our lawns and gardens. Which view do you hold? Do you see a way out of this dilemma?

REPRODUCTIVE PATTERNS

Ways to Reproduce: Sexual Partners Not Always Needed

Some species reproduce without having sex and others reproduce by having sex.

Two types of reproduction pass genes on to offspring. One is **asexual reproduction,** in which offspring are generally exact genetic copies (clones) of a single parent. This is common in species such as single-celled bacteria. Each cell can divide to produce two identical cells that are genetic clones, or replicas of the original. Many plants and animals such as corals reproduce this way.

The second type is **sexual reproduction,** which mixes the genetic material of two individuals and produces offspring with combinations of genetic traits from each parent.

Sexual reproduction has three disadvantages. *First,* males do not give birth. This means that females have to produce twice as many offspring as an asexually reproducing organism does to maintain the same number of young in the next generation.

Second, there is an increased chance of genetic errors and defects during the splitting and recombination of chromosomes. *Third,* courtship and mating rituals (Figure 8-8) consume time and energy, can transmit disease, and can inflict injury on males of some species as they compete for sexual partners.

Figure 8-8 Courtship ritual. A male peacock displays his elaborate and beautiful tail to attract the larger female.

If sexual reproduction is so costly, why do 97% of the earth's species use it? According to biologists, this happens because of two important advantages of sexual reproduction. One is that it provides a greater genetic diversity in offspring. A population with many different genetic possibilities has a greater chance of reproducing when environmental conditions change than does a population of genetically identical clones. Second, males of some species can gather food for the female and the young and protect and help train the young.

Reproductive Patterns: Opportunists and Competitors

While some species have a large number of small offspring and give them little parental care, other species have a few larger offspring and take care of them until they can reproduce.

Species use different reproductive patterns to help ensure their survival. Species with a capacity for a high rate of population increase (*r*) are called **r-selected species** (Figure 8-9, p. 168, and Figure 8-10, p. 168, left). These species have many, usually small, offspring and give them little or no parental care or protection. They overcome the massive loss of their offspring by producing so many that a few will survive to reproduce many more offspring to begin this reproductive pattern again, much like buying a large number of lottery tickets to increase one's chance of winning a big prize (in this case the survival of a species). Examples include algae, bacteria, rodents, annual plants (such as dandelions), and most insects.

Such species tend to be *opportunists.* They reproduce and disperse rapidly when conditions are favorable or when a disturbance opens up a new habitat or niche for invasion, as in the early stages of ecological succession.

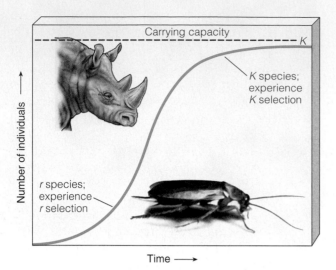

Figure 8-9 Positions of *r-selected* and *K-selected* species on the sigmoid (S-shaped) population growth curve.

Environmental changes caused by disturbances can allow opportunist species to gain a foothold. However, once established, their populations may crash because of unfavorable changes in environmental conditions or invasion by more competitive species. This helps explain why most opportunist species go through irregular and unstable boom-and-bust cycles in their population size.

At the other extreme are *competitor* or **K-selected species** (Figure 8-9 and Figure 8-10, right). They tend to reproduce later in life and have a small number of offspring with fairly long life spans. Typically the offspring of such species develop inside their mothers (where they are safe), are born fairly large, mature slowly, and are cared for and protected by one or both parents until they reach reproductive age. This reproductive pattern results in a few big and strong individuals that can compete for resources and reproduce a few young to begin their life cycle again.

Such species are called K-selected species because they tend to do well in competitive conditions when their population size is near the carrying capacity (*K*) of their environment. Their populations typically follow a logistic growth curve.

Most large mammals (such as elephants, whales, and humans), birds of prey, and large and long-lived plants (such as the saguaro cactus, and most tropical rain forest trees) are K-selected species. Many K-selected species—especially those with long generation times and low reproductive rates like elephants, rhinoceroses, and sharks—are prone to extinction.

Most organisms have reproductive patterns between the extremes of r-selected species and K-selected species. In agriculture we raise both r-selected species (crops) and K-selected species (livestock).

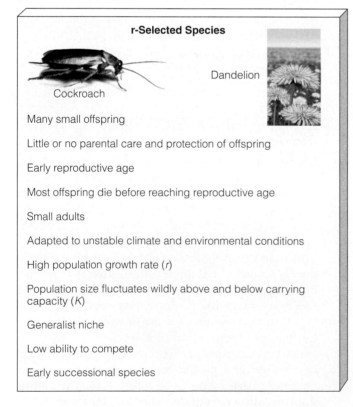

r-Selected Species

Cockroach   Dandelion

Many small offspring

Little or no parental care and protection of offspring

Early reproductive age

Most offspring die before reaching reproductive age

Small adults

Adapted to unstable climate and environmental conditions

High population growth rate (*r*)

Population size fluctuates wildly above and below carrying capacity (*K*)

Generalist niche

Low ability to compete

Early successional species

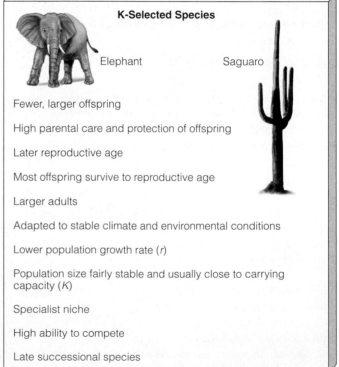

K-Selected Species

Elephant   Saguaro

Fewer, larger offspring

High parental care and protection of offspring

Later reproductive age

Most offspring survive to reproductive age

Larger adults

Adapted to stable climate and environmental conditions

Lower population growth rate (*r*)

Population size fairly stable and usually close to carrying capacity (*K*)

Specialist niche

High ability to compete

Late successional species

Figure 8-10 Natural capital: generalized characteristics of *r-selected* (opportunist) species and *K-selected* (competitor) species. Many species have characteristics between these two extremes.

THINKING ABOUT REPRODUCTIVE BEHAVIOR Is the southern sea otter (Core Case Study, p. 161) an r-selected or a K-selected species?

The reproductive pattern of a species may give it a temporary advantage, but *the availability of a suitable habitat for individuals of a population in a particular area determines its ultimate population size.* No matter how fast a species can reproduce, there can be no more dandelions than there is dandelion habitat and no more zebras than there is zebra habitat in a particular area.

? THINKING ABOUT r-SELECTED AND K-SELECTED SPECIES If the earth experiences significant warming during this century as projected, is this likely to favor r-selected or K-selected species? Explain.

Survivorship Curves: Short to Long Lives

The populations of different species vary in how long individual members typically live.

Individuals of species with different reproductive strategies tend to have different *life expectancies*, or expected lengths of life. One way to represent the age structure of a population is with a **survivorship curve**, which shows the percentages of the members of a population surviving at different ages. There are three generalized types of survivorship curves: *late loss, early loss*, and *constant loss* (Figure 8-11).

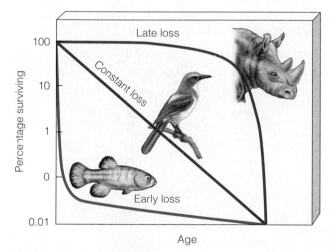

Figure 8-11 When does death come? Survivorship curves for populations of different species, show the percentages of the members of a population surviving at different ages. Most members of a *late loss* population (such as elephants, rhinoceroses, and humans) live to an old age. Members of a *constant loss* population (such as many songbirds) die at all ages. In an *early loss* population (such as annual plants and many bony fish species), most members die at a young age. These generalized survivorship curves only approximate the realities of nature.

 Explore survivorship curves, showing how certain species differ in their local lifespans, at ThomsonNOW.

A *life table* shows the projected life expectancy and probability of death for individuals at each age in a survivorship curve. Insurance companies use life tables of human populations to determine policy costs for customers. These tables show that women in the United States survive an average of 6 years longer than men. This explains why a 65-year-old American man normally pays more for life insurance than a 65-year-old American woman. Supplement 13 on p. S47 describes how various genetic variations can affect population size.

Revisiting Sea Otters and Sustainability

Before the arrival of European settlers on the North American west coast, the sea otter population was part of a complex ecosystem made up of bottom-dwelling creatures, kelp, otters, whales, and other species all depending on each other for survival. Kelp forests served as food and shelter for sea urchins. Otters ate the sea urchins and other kelp eaters. Some whales ate the otters. And detritus from all these species helped maintain the kelp forests. Each of these interacting populations was kept in check by, and helped to sustain, all others.

When humans arrived and began hunting the otters for their pelts, they probably didn't know much about the intricate web of life beneath the ocean surface. But with the effects of overhunting, people realized they had done more than simply take otters. They had torn the web, disrupted an entire ecosystem, and triggered a loss of valuable natural services.

Populations of most plants and animals depend directly or indirectly on solar energy and each population plays a role in the recycling of nutrients in the ecosystems where they live. And the biodiversity found in the variety of species in different terrestrial and aquatic ecosystems provides alternative paths for energy flow and nutrient cycling and better opportunities for natural selection as environmental conditions change.

In this chapter we looked more closely at the sustainability principle that *there are always limits to population growth in nature.* Chapter 9 applies the principles of population dynamics discussed in this chapter to the growth of the human population and its environmental impact. The principles of population dynamics are also used to help us harvest fish and other wildlife resources more sustainably, as discussed in Chapters 11 and 12.

We cannot command nature except by obeying her.

SIR FRANCIS BACON

CRITICAL THINKING

1. What difference would it make if the southern sea otter (Core Case Study, p. 161) became prematurely extinct because of human activities? What three things would you do to help prevent the premature extinction of this species?

2. (a) Why do biotic factors that regulate population growth tend to depend on population density, and **(b)** why do abiotic factors that regulate population tend to be independent of population density?

3. Explain why most species with a high capacity for population growth (high biotic potential) tend to have a small size (such as bacteria and flies) while those with a low capacity for population growth tend to be large (such as humans, elephants, and whales)?

4. Why are pest species likely to be extreme r-selected species? Why are many endangered species likely to be extreme K-selected species?

5. Given current environmental conditions, if you had a choice, would you rather be an r-strategist or a K-strategist? Explain your answer.

6. List the type of survivorship curve you would expect given descriptions of the following organisms:
   a. This organism is an annual plant. It lives only 1 year. During that time, it sprouts, reaches maturity, produces many wind-dispersed seeds, and dies.
   b. This organism is a mammal. It reaches maturity after 10 years. It bears one young every 2 years. The parents and the rest of the herd protect the young.

7. Explain why a simplified ecosystem such as a cornfield usually is much more vulnerable to harm from insects and plant diseases than a more complex, natural ecosystem such as a grassland. Does this mean that we should never convert a grassland to a cornfield? Explain. What restrictions, if any, would you put on such conversions?

8. In your own words, restate this chapter's closing quotation by Sir Francis Bacon. Do you agree with this notion? Why or why not?

9. List two questions that you would like to have answered as a result of reading this chapter.

PROJECTS

1. Using the library or the Internet, choose one wild plant species and one animal species and analyze the factors that are likely to limit the population of each species.

2. Make a concept map of this chapter's major ideas, using the section heads, subheads, and key terms (in bold-face). Look on the website for this book for information about making concept maps.

LEARNING ONLINE

The website for this book contains study aids and many ideas for further reading and research. They include a chapter summary, review questions for the entire chapter, flash cards for key terms and concepts, a multiple-choice practice quiz, interesting Internet sites, references, information about green careers, and a guide for accessing thousands of InfoTrac® College Edition articles. Log into

www.thomsonedu.com/biology/miller

Then choose Chapter 8, and select a learning resource. For access to animations, additional quizzes, chapter outlines and summaries, register and log into

at **www.thomsonedu.com** using the access code card in the front of your book.

Active Graphing

Log into ThomsonNow at www.thomsonedu.com to explore the graphing exercise for this chapter.

9 Applying Population Ecology: The Human Population and Its Impact

CORE CASE STUDY
Is the World Overpopulated?

The world's human population is projected to increase from 6.6 to 9.2 billion or more between 2006 and 2050 (Figure 1-1, p. 6), with much of this growth occurring in several rapidly developing countries such as India and China (Figure 9-1). Are there too many people on the earth?

Some argue that the planet has too many people collectively degrading the earth's natural capital. To some the problem is the sheer number of people in developing countries (Figure 1-14, top, p. 20). To others it is the number of people in developed countries such as the United States where high resource consumption rates magnify the environmental impact of each person (Figure 1-7, p. 13 and Figure 1-14, bottom, p. 20).

Those who do not believe the world is overpopulated point out that the average life span of the world's 6.6 billion people is longer today than at any time in the past and is projected to increase. According to them, the world can support billions more people as a result of human technological ingenuity in providing food and other resources. They also see more people as the world's most valuable resource for solving environmental and other problems and for stimulating economic growth by increasing the number of consumers.

Some view any form of population regulation as a violation of their religious or moral beliefs. Others see it as an intrusion into their privacy and personal freedom to have as many children as they want. Some developing countries and some members of minorities in developed countries regard population control as a form of genocide to keep their numbers and political power from growing.

Proponents of slowing and eventually stopping population growth ask if we cannot or will not provide basic support for about one of every five people—about 1.4 billion people today, how will we be able to do so for the projected 2.6 billion more people by 2050?

They also warn that if we do not sharply lower birth rates, the death rate may increase because of declining health and environmental conditions in some areas, as is already happening in parts of Africa. They also warn that resource use and environmental degradation may intensify as more consumers increase their already large ecological footprint in developed countries and in rapidly developing countries, such as China and India.

This debate over interactions among population growth, economic growth, politics, and moral beliefs is one of the most important and controversial issues in environmental science.

Figure 9-1 Crowded street in China. Together, China and India have 37% of the world's population and the resource use per person in these countries is projected to grow rapidly as they become more modernized (Case Study, p. 15).

L. Yong/UNEP/Peter Arnold, Inc.

X *How Would You Vote?* Should the population of the country where you live be stabilized as soon as possible? Cast your vote online **www.thomsonedu.com/biology/miller**.

This chapter looks at the factors that affect the growth and decline of the human population. It addresses the following questions:

- What is the history of human population growth, and how many people are likely to be here by 2050?

- How is population size affected by birth, death, fertility, and migration rates?

- How is population size affected by the percentages of males and females at each age level?

- How can we slow population growth?

- What success have India and China had in slowing population growth?

- What are the major impacts of human activities on the world's natural ecosystems?

HUMAN POPULATION GROWTH: A BRIEF HISTORY

Population Growth in the Past (Science and Economics)

We have kept the human population growing by expanding into ecosystems throughout the world and using technological innovations to expand the food supply and lower death rates.

For most of history, the human population grew slowly (Figure 1-1, left part of curve, p. 6). But for the past 200 years, the human population has experienced rapid exponential growth reflected in the characteristic J-curve (Figure 8-3, left side, p. 163 and Figure 1-1, right part of curve, p. 6).

Three major factors explain this population increase. *First,* humans developed the ability to expand into diverse new habitats and different climate zones. *Second,* the emergence of early and modern agriculture allowed more people to be fed per unit of land area.

Third, we developed sanitation systems, antibiotics, and vaccines to help control infectious disease agents, and we tapped into concentrated sources of energy (mostly fossil fuels, Case Study, p. 42). As a result, death rates dropped sharply below birth rates.

About 10,000 years ago when agriculture began there were about 5 million humans on the planet; now there are 6.6 billion of us. It took from the time we arrived until about 1927 to add the first 2 billion people to the planet; less than 50 years to add the next 2 billion (by 1974); and just 25 years to add the next 2 billion (by 1999)—an illustration of the power of exponential growth. Such growth raises the question of whether the earth is overpopulated (Core Case Study, p. 171).

Population Growth Today: Slowing but Still Growing (Science and Economics)

The rate of population growth has slowed but is still growing rapidly.

During 2006, about 81 million people were added to the world's population—79.5 million in developing countries and 1.2 million in developed countries. At this exponential growth rate of 1.23% a year, we share the earth and its resources with about 222,000 more people each day and 2.4 more people every time your heart beats.

An exponential growth rate of 1.23% may seem small, but compare the 81 million people added to the world's population in 2006 to the 69 million added in 1963, when the world's population growth reached its peak. This increase of 81 million people per year is roughly equal to adding another New York City (USA) every month or another Germany every year.

Also, there is a big difference between population growth rates in developed and developing countries. In 2006, the population of developed countries was growing exponentially at a rate of 0.1% per year. That of the developing countries was 1.5% per year—15 times faster.

Thus, population growth in recent decades is a result of keeping more people alive by increasing life expectancy and reducing death rates. This is a good thing. But coupled with increased resource consumption per person this can to lead increased environmental degradation as the human ecological footprint spreads across the planet (Figure 1-7, p. 13, and Figure 3 on pp. S12–S13 in Supplement 4). To those arguing that both developed and developing countries are overpopulated, the solution is to focus on sharply reducing birth rates.

Where Are We Headed? (Science, Economics, and Politics)

We do not know how long we can continue increasing the earth's carrying capacity for humans.

Scientific studies of populations of other species (Chapter 8), tell us that *no population can continue growing indefinitely.* How long can we continue increasing the earth's carrying capacity for our species by side-stepping many of the factors that sooner or later limit the growth of any population?

The debate over this important question has been going on since 1798 when Thomas Malthus, a British economist, hypothesized that the human population tends to increase exponentially, while food supplies tend to increase more slowly at a linear rate. So far, Malthus has been proven wrong. Food production has grown at an exponential rate instead of at a linear rate because of genetic and technological advances in industrialized food production.

No one knows how close we are to the environmental limits that sooner or later will control the population size of the human species, but the evidence is growing that we are steadily degrading the natural capital that keeps us and other species alive and supports our economies—one of the arguments supporting the idea that the world is overpopulated (Core Case Study, p. 171).

How many of us are likely to be here in 2050? Answer: 7.2–10.6 billion people, depending mostly on projections about the average number of babies women are likely to have. The medium projection is 8.9 billion people (Figure 9-2). About 97% of this growth is projected to take place in developing countries, where acute poverty (living on less than $1 per day) is a way of life for about 1.4 billion people.

During this century the human population may level off as it moves from a J-shaped curve of exponential growth to an S-shaped curve of logistic growth (Figure 8-3, p. 163).

The key question, as raised in the Core Case Study at the beginning of this chapter, is, *"Can the world pro-vide an adequate standard of living for the medium projection of 8.9 billion people in 2050 without causing widespread environmental damage?"* In other words, how many people can the earth support indefinitely? Some say about 2 billion. Others say as many as 30 billion.

Some analysts believe this is the wrong question. Instead, they say, we should ask what the *optimum sustainable population* of the earth might be, based on the planet's *cultural carrying capacity*. Such an optimum level would allow most people to live in reasonable comfort and freedom without impairing the ability of the planet to sustain future generations. (See the Guest Essay by Garrett Hardin on this topic on the website for this chapter.) There is disagreement on what the optimum sustainable population size for the world is.

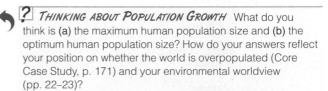

? **THINKING ABOUT POPULATION GROWTH** What do you think is (a) the maximum human population size and (b) the optimum human population size? How do your answers reflect your position on whether the world is overpopulated (Core Case Study, p. 171) and your environmental worldview (pp. 22–23)?

F **RESEARCH FRONTIER** Determining the optimum sustainable population size for the earth and various parts of the earth

FACTORS AFFECTING HUMAN POPULATION SIZE

Birth Rates and Death Rates: Entrances and Exits (Science)

Population increases because of births and immigration and decreases through deaths and emigration.

Human populations grow or decline through the interplay of three factors: *births (fertility), deaths (mortality),* and *migration.* As pointed out in Chapter 8, we can calculate **population change** by subtracting the number of people leaving a population (through death and emigration) from the number entering it (through birth and immigration) during a specified period of time (usually one year):

$$\text{Population change} = (\text{Births} + \text{Immigration}) - (\text{Deaths} + \text{Emigration})$$

When births plus immigration exceed deaths plus emigration, population increases; when the reverse is true, population declines.

Instead of using the total numbers of births and deaths per year, population experts (demographers) use the **birth rate,** or **crude birth rate** (the number of live births per 1,000 people in a population in a given

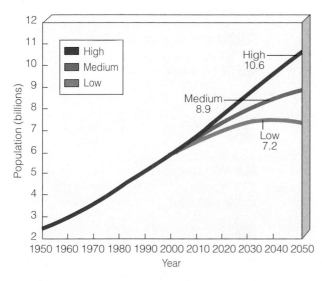

Figure 9-2 Global connections: UN world population projections, assuming that by 2050 women have an average of 2.5 children (high), 2.0 children (medium), or 1.5 children (low). The most likely projection is the medium one—8.9 billion by 2050. (Data from United Nations, *World Population Prospects: The 2001 Revision,* 2002)

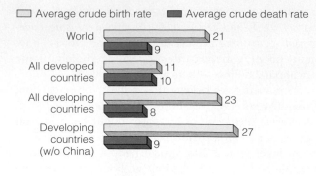

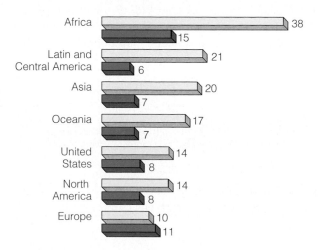

Figure 9-3 Global connections: average crude birth and death rates for various groupings of countries in 2006. (Data from Population Reference Bureau)

year), and the **death rate,** or **crude death rate** (the number of deaths per 1,000 people in a population in a given year). Figure 9-3 shows the crude birth and death rates for various groupings of countries in 2006.

What five countries had the largest numbers of people in 2006? Number 1 is China with 1.3 billion people, or one of every five people in the world. Number 2 is India with 1.1 billion people, or one of every six people. Together China and India have 37% of the world's population. The United States, with 300 million people in 2006—has the world's third largest population but only 4.5% of its people.

Can you guess the next two most populous countries? What three countries are expected to have the most people in 2025? Look at Figure 9-4 to see if your answers are correct.

Declining Fertility Rates: Fewer Babies per Woman (Science and Economics)

The average number of children that a woman bears has dropped sharply, but is not low enough to stabilize the world's population in the near future.

Fertility is the number of children born to a woman during her lifetime. Two types of fertility rates affect a country's population size and growth rate.

The first type, **replacement-level fertility,** is the number of children a couple must bear to replace themselves. It is slightly higher than two children per couple (2.1 in developed countries and as high as 2.5 in some developing countries), mostly because some children die before reaching their reproductive years.

Does reaching replacement-level fertility bring an immediate halt in population growth? No, because so many future parents are alive. If each of today's couples had an average of 2.1 children and their children also had 2.1 children, the world's population would continue to grow for 50 years or more (assuming death rates do not rise).

The second type of fertility rate, the **total fertility rate (TFR)** is the average number of children a woman typically has during her reproductive years. Today, the average woman in the world has half as many children as her counterpart in 1972. In 2006, the average global TFR was 2.7 children per woman: 1.6 in developed countries (down from 2.5 in 1950) and 2.9 in developing countries (down from 6.5 in 1950). Although the decline in TFR in developing countries is impressive, it is still far above the replacement level of 2.1.

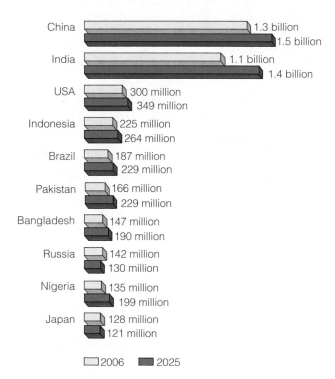

Figure 9-4 Global connections: the world's 10 most populous countries in 2006, with projections of their population size in 2025. In 2006, more people lived in China than in all of Europe, Russia, North America, Japan, and Australia combined. By 2050, India is expected to have a larger population than China. (Data from World Bank and Population Reference Bureau)

Case Study: Fertility and Birth Rates in the United States (Science and Economics)

Population growth in the United States has slowed but is not close to leveling off.

The population of the United States has grown from 76 million in 1900 to 299 million in 2006, despite oscillations in the country's TFR (Figure 9-5) and birth rate (Figure 9-6). The period of high birth rates between 1946 and 1964 is known as the *baby-boom period*. This added 79 million people to the U.S. population. In 1957, the peak of the baby boom, the TFR reached 3.7 children per woman. Since then it has generally declined, remaining at or below replacement level since 1972. The *baby bust* that followed the *baby boom* (Figure 9-6) was due in large part to delayed marriage, widespread contraceptive use, and abortion.

The drop in the TFR has led to a decline in the rate of population growth in the United States. But the country's population is still growing faster than that of any other developed country and is not close to leveling off.

Nearly 3.0 million people were added to the U.S. population in 2006. About 56% of this growth occurred because births outnumbered deaths; the rest came from legal and illegal immigration.

According to the U.S. Census Bureau, the U.S. population is likely to increase from 300 million in 2006 to 420 million by 2050 and then to 571 million by 2100. In contrast, population growth has slowed in other major developed countries since 1950, most of which are expected to have declining populations after 2010. Because of a high per capita rate of resource use, each addition to the U.S. population has an enormous environmental impact (Figure 1-7, p. 13, and Figure 4 on pp. S14–S15 in Supplement 4 at the end of this book).

In addition to the almost fourfold increase in population growth, some amazing changes in lifestyles took place in the United States during the twentieth century (Figure 9-7, p. 176), which led to dramatic increases in per capita resource use and a much larger U.S. ecological footprint.

Here are a few more changes during the last century. In 1905, the three leading causes of death in the United States were pneumonia, tuberculosis, and diarrhea; 90% of U.S. doctors had no college education; one out of five adults could not read or write; the average U.S. worker earned $200–400 per year and the average daily wage was 22 cents per hour; there were only 9,000 cars in the U.S., and only 232 kilometers (144 miles) of paved roads; a 3-minute phone call from Denver, Colorado, to New York City cost $11; only 30 people lived in Las Vegas, Nevada; and most women washed their hair only once a month.

? THINKING ABOUT OVERPOPULATION Do you think the United States, or the country where you live, is overpopulated? Explain.

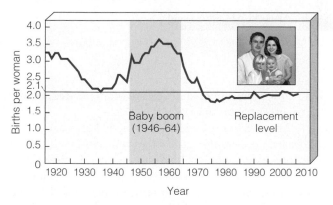

Figure 9-5 Total fertility rates for the United States between 1917 and 2006. Use this figure to trace changes in total fertility rates during your lifetime. QUESTION: *How many children do you plan to have?* (Data from Population Reference Bureau and U.S. Census Bureau)

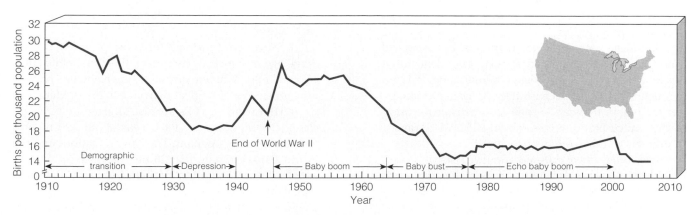

Figure 9-6 Birth rates in the United States, 1910–2006. Use this figure to trace changes in crude birth rates during your lifetime. (Data from U.S. Bureau of Census and U.S. Commerce Department)

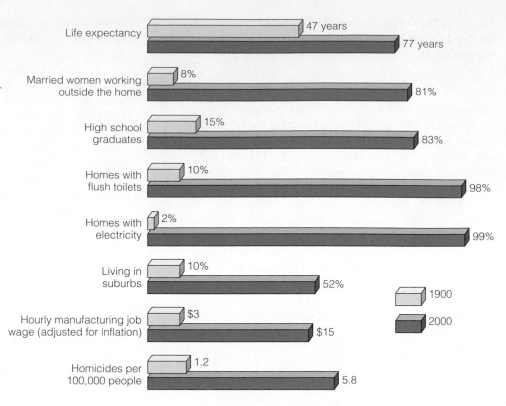

Figure 9-7 Some major changes that took place in the United States between 1900 and 2000. QUESTION: *Which two of these changes do you think were the most important?* (Data from U.S. Census Bureau and Department of Commerce)

Factors Affecting Birth Rates and Fertility Rates (Science and Economics)

The number of children women have is affected by the cost of raising and educating children, availability of pensions, urbanization, educational and employment opportunities for women, infant deaths, marriage age, and availability of contraceptives and abortions.

Many factors affect a country's average birth rate and TFR. One is the *importance of children as a part of the labor force.* Proportions of children working tend to be higher in developing countries—especially in rural areas, where children begin working to help raise crops at an early age.

Another economic factor is the *cost of raising and educating children.* Birth and fertility rates tend to be lower in developed countries, where raising children is much more costly because they do not enter the labor force until they are in their late teens or twenties. In the United States, it costs about $250,000 to raise a middle-class child from birth to age 18.

The *availability of private and public pension systems* can affect a couple's decision on how many children to have. Pensions reduce a couple's need to have many children to help support them in old age.

Urbanization plays a role. People living in urban areas usually have better access to family planning services and tend to have fewer children than those living in rural areas, where children are often needed to perform essential tasks.

Another important factor is *the educational and employment opportunities available for women.* TFRs tend to be low when women have access to education and paid employment outside the home. In developing countries, women with no education generally have more children than women with a secondary school education. In Brazil, for example, illiterate women on average have more than six children compared to two children for literate women.

Another factor is the *infant mortality rate.* In areas with low infant mortality rates, people tend to have fewer children because not as many children die at an early age.

Average age at marriage (or, more precisely, the average age at which women have their first child) also plays a role. Women normally have fewer children when their average age at marriage is 25 or older.

Birth rates and TFRs are also affected by the *availability of legal abortions.* Each year about 190 million women become pregnant. The United Nations and the World Bank estimate that 46 million of these women get abortions—26 million of them legal and 20 million illegal (and often unsafe).

The *availability of reliable birth control methods* allows women to control the number and spacing of the children they have. *Religious beliefs, traditions,* and *cultural norms* also play a role. In some countries, these factors favor large families and strongly oppose abortion and some forms of birth control.

E RESEARCH FRONTIER Learning more about how these and other factors interact and affect birth and fertility rates

Factors Affecting Death Rates (Science and Economics)

Death rates have declined because of increased food supplies, better nutrition, advances in medicine, improved sanitation and personal hygiene, and safer water supplies.

The rapid growth of the world's population over the past 100 years is not the result of a rise in the crude birth rate. Instead, it has been caused largely by a decline in crude death rates, especially in developing countries.

More people started living longer and fewer infants died because of increased food supplies and distribution, better nutrition, medical advances such as immunizations and antibiotics, improved sanitation, and safer water supplies (which curtailed the spread of many infectious diseases).

Two useful indicators of overall health of people in a country or region are **life expectancy** (the average number of years a newborn infant can expect to live) and the **infant mortality rate** (the number of babies out of every 1,000 born who die before their first birthday).

Great news. Between 1955 and 2006, the global life expectancy increased from 48 years to 67 years (77 years in developed countries and 65 years in developing countries). It ranges from a low of 34 in Botswana to a high of 82 in Iceland. Average life expectancy is projected to reach 74 in developing countries by 2050. Between 1900 and 2006, life expectancy in the United States increased from 47 to 78 years and is projected to reach 82 years by 2050.

Bad news. In the world's poorest countries, life expectancy is 49 years or less. In many African countries, life expectancy is expected to fall further because of rising numbers of deaths from AIDS.

Infant mortality is viewed as the best single measure of a society's quality of life because it reflects a country's general level of nutrition and health care. A high infant mortality rate usually indicates insufficient food (undernutrition), poor nutrition (malnutrition), and a high incidence of infectious disease (usually from contaminated drinking water and weakened disease resistance from undernutrition and malnutrition).

Good news. Between 1965 and 2006, the world's infant mortality rate dropped from 20 (per 1,000 live births) to 6.3 in developed countries and from 118 to 59 in developing countries.

Bad news. Annually, at least 7.6 million infants (most in developing countries) die of preventable causes during their first year of life—an average of 21,000 mostly unnecessary infant deaths per day. This is equivalent to 55 jumbo jets, each loaded with 400 infants younger than age 1, crashing each day with no survivors!

The U.S. infant mortality rate declined from 165 in 1900 to 6.7 in 2006. This sharp decline was a major factor in the marked increase in U.S. average life expectancy during this period. Still, some 46 countries (most in Europe) had lower infant mortality rates than the United States had in 2006. If the U.S. infant mortality rate was as low as that of Singapore (ranked no. 1) in 2006, this would have saved the lives of 18,900 American children.

Three factors have helped keep the U.S. infant mortality rate higher than it could be: *inadequate health care for poor women during pregnancy and for their babies after birth, drug addiction among pregnant women,* and *a high birth rate among teenagers.*

Case Study: U.S. Immigration (Economics and Politics)

Immigration has played, and continues to play, a major role in the growth and cultural diversity of the U.S. population.

The third factor in population change is **migration:** the movement of people into (*immigration*) and out of (*emigration*) a specific geographic area. Only five countries—the United States, Canada, Australia, New Zealand, and Israel—have official policies to encourage immigration.

Most migrants seek jobs and economic improvement. But some are driven by religious persecution, ethnic conflicts, political oppression, wars, and environmental degradations such as water shortages, soil erosion, deforestation, desertification, population pressures, and severe poverty. According to a UN study, there were about 19.2 million *environmental refugees* in 2005 and the number could reach 50 million by 2010. In a globally warmed world the number could soar to 250 million or more.

Since 1820, the United States has admitted almost twice as many immigrants and refugees as all other countries combined! However, the number of legal immigrants (including refugees) has varied during different periods because of changes in immigration laws and rates of economic growth (Figure 9-8, p. 178). Currently, legal and illegal immigration account for about 44% of the country's annual population growth.

Between 1820 and 1960, most legal immigrants to the United States came from Europe. Since 1960, most have come from Latin America (53%) and Asia (25%), followed by Europe (14%). In 2006, Latinos (67% of them from Mexico) made up 14% of the U.S. population and by 2050 are projected to make up 25% of the population.

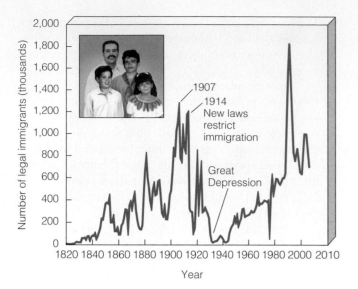

Figure 9-8 Legal immigration to the United States, 1820–2003. The large increase in immigration since 1989 resulted mostly from the Immigration Reform and Control Act of 1986, which granted legal status to illegal immigrants who could show they had been living in the country for several years. (Data from U.S. Immigration and Naturalization Service and the Pew Hispanic Center)

In 1995, the U.S. Commission on Immigration Reform recommended reducing the number of legal immigrants from about 1,000,000 to 700,000 per year for a transition period and then to 550,000 per year. Some analysts want to limit legal immigration to 20% of the country's annual population growth. They would accept new entrants only if they can support themselves, arguing that providing immigrants with public services makes the United States a magnet for the world's poor.

In 2006, there were about 35.2 million foreign-born legal immigrants—roughly one of every eight Americans—living in the United States. Add to this about 11–12 million illegal immigrants—81% of them from Latin America and 56% from Mexico. According to a 2006 poll, 75% of Americans believe that illegal immigrants should be able to earn citizenship and favor guest-worker registration for those already here.

Proponents of reducing legal immigration argue that it would allow the United States to stabilize its population sooner and help reduce the country's enormous environmental impact. Polls show that almost 60% of the U.S. public strongly supports reducing legal immigration.

There is also strong public support for sharply reducing illegal immigration. But some people are concerned that a crackdown on the country's illegal immigrants can also lead to discrimination against some of the country's legal immigrants.

Others oppose reducing current levels of legal immigration. They argue that it would diminish the historical role of the U.S. as a place of opportunity for the world's poor and oppressed. In addition, legal and illegal immigrants pay taxes, take many menial and low-paying jobs that most other Americans shun, open businesses, and create jobs. And according to the U.S. Census Bureau, after 2020 higher immigration levels will be needed to supply enough workers as baby boomers retire.

✗ *HOW WOULD YOU VOTE?* Should legal immigration into the United States (or the country where you live) be reduced? Cast your vote online at **www.thomsonedu.com/biology/miller**.

POPULATION AGE STRUCTURE

Age Structure Diagrams (Science and Economics)

The number of people in young, middle, and older age groups determines how fast populations grow or decline.

As mentioned earlier, even if the replacement-level fertility rate of 2.1 children per woman were magically achieved globally tomorrow, the world's population would keep growing for at least another 50 years (assuming no large increase in death rates). This results mostly from the **age structure:** the distribution of males and females in each age group in the world's population.

Demographers construct a population age structure diagram by plotting the percentages or numbers of males and females in the total population in each of three age categories: *prereproductive* (ages 0–14), *reproductive* (ages 15–44), and *postreproductive* (ages 45 and up). Figure 9-9 presents generalized age structure diagrams for countries with rapid, slow, zero, and negative population growth rates.

Effects of Age Structure on Population Growth: Teenagers Rule (Science and Economics)

The number of people younger than age 15 is the major factor determining a country's future population growth.

Any country with many people younger than age 15 (represented by a wide base in Figure 9-9, left) has a powerful built-in momentum to increase its population size unless death rates rise sharply. The number of births will rise even if women have only one or two children, because a large number of girls will soon be moving into their reproductive years.

What is perhaps the world's most important population statistic? *Twenty-nine percent of the people on the planet were under 15 years old in 2006. These 1.9 billion*

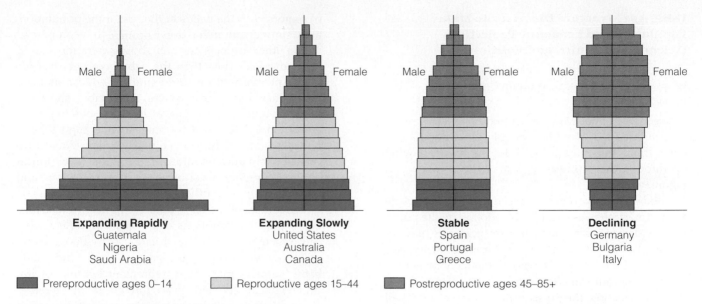

Expanding Rapidly	Expanding Slowly	Stable	Declining
Guatemala	United States	Spain	Germany
Nigeria	Australia	Portugal	Bulgaria
Saudi Arabia	Canada	Greece	Italy

■ Prereproductive ages 0–14   □ Reproductive ages 15–44   ■ Postreproductive ages 45–85+

ThomsonNOW™ Active Figure 9-9 Generalized population age structure diagrams for countries with rapid (1.5–3%), slow (0.3–1.4%), zero (0–0.2%), and negative population growth rates (a declining population). Populations with a large proportion of its people in the prereproductive ages of 1–14 (at left) have a large potential for rapid population growth. **QUESTION:** *Which of these diagrams best represents the country where you live? See an animation based on this figure and take a short quiz on the concept.* (Data from Population Reference Bureau)

young people are poised to move into their prime reproductive years. In developing countries the number is even higher: 32% (with 42% in Africa), compared with 17% in developed countries. We live in *a demographically divided world*, as shown in Figure 9-10.

The youthful age structure of most developing countries contributes to an *unemployment crisis.*

In parts of Asia, Africa, and Latin America with rapid population growth and youthful age structures 20–50% of the 15–24 age group is unemployed. This provides a tinderbox for social unrest and recruits for terrorist activities.

Rapid population growth and cultural and religious conflicts can also lead to political tensions and armed conflict within countries such as the African country of Rwanda. Between 1950 and 1993, its population tripled. By 1991, the demand for firewood was twice the sustainable yield of local forests. As trees disappeared, crop residues were burned for heat and cooking. This reduced soil fertility and decreased crop productivity.

Desperation for food and land grew and when a 1994 plane crash killed the country's president, an organized attack by Hutus led to an estimated 800,000 deaths of Tutsi people within 100 days with the Hutus claiming precious plots of land.

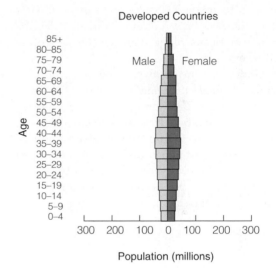

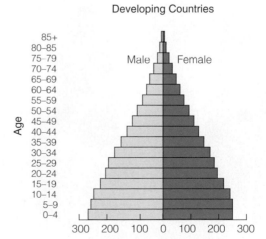

Figure 9-10 Global connections: population structure by age and sex in developing countries and developed countries, 2006. (Data from United Nations Population Division and Population Reference Bureau)

Using Age Structure Diagrams to Make Population and Economic Projections (Science, Economics, and Politics)

Changes in the distribution of a country's age groups have long-lasting economic and social impacts.

Between 1946 and 1964, the United States had a *baby boom* that added 79 million people to its population. Over time, this group looks like a bulge moving up through the country's age structure, as shown in Figure 9-11.

Today baby boomers make up nearly half of all adult Americans. As a result, they dominate the population's demand for goods and services and play increasingly important roles in deciding who gets elected and what laws are passed. Baby boomers who created the youth market in their teens and twenties are now creating the 50-something market and will soon move on to create a 60-something market. After 2011, when the first baby boomers will turn 65, the number of Americans older than age 65 will grow sharply through 2029.

According to some analysts, the retirement of baby boomers is likely to create a shortage of workers in the United States unless immigrant workers replace some of them. Retired baby boomers are also likely to use their political clout to force the smaller number of people in the baby-bust generation (Figure 9-6) to pay higher income, health-care, and Social Security taxes.

An important U.S. public policy question is how the country will balance the needs of growing numbers of seniors with the needs of the rest of the population, at levels of taxation that don't strangle the economy.

In other respects, the baby-bust generation should have an easier time than the baby-boom generation. Fewer people will be competing for educational opportunities, jobs, and services. Also, labor shortages may drive up their wages, at least for jobs requiring education or technical training beyond high school. However, this may not happen if many American-owned companies operating at the global level (multinational companies) continue to export many low- and high-paying jobs to other countries.

Members of the baby-bust group may find it difficult to get job promotions as they reach middle age because members of the much larger baby-boom group will occupy most upper-level positions. And many baby boomers may delay retirement because of improved health, the need to accumulate adequate retirement funds, or extension of the retirement age at which they can collect Social Security.

 Examine how the baby boom affects the U.S. age structure over several decades at ThomsonNOW.

Stable Populations in Some Countries (Science and Economics)

About 13% of the world's people live in countries with stabilized or declining populations.

In 2006, about 884 million people lived in 41 countries with essentially stable (annual growth rates at or be-

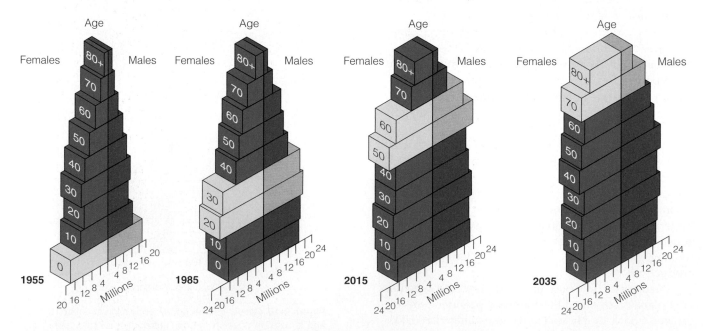

ThomsonNOW **Active Figure 9-11** Tracking the baby-boom generation in the United States. *See an animation based on this figure and take a short quiz on the concept.* (Data from Population Reference Bureau and U.S. Census Bureau)

CASE STUDY

Family Planning in Iran (Science, Politics, and Religion)

When Ayatollah Khomeini assumed power in Iran in 1979, he did away with the family planning programs the Shah of Iran had put into place in 1967. He saw large families as a way to increase the size of his army.

Religious authorities urged couples to have many children and any mention of family planning or birth control for preventing pregnancy was forbidden. And the marriage age for girls was reduced to 9 years old.

Iranians responded and the country's population growth reached 4.4%—one of the world's highest rates. But this rapid growth in numbers began overburdening the country's economy and environment.

In 1989, the government reversed its policy and restored its family planning program. Government agencies were mobilized to raise public awareness of population issues, encourage smaller families, and provide free modern contraception.

Religious leaders helped by mounting a crusade for smaller families and commanding faithful Muslims to participate in family planning. TV stations, billboards, and newspapers advertised the program. The minimum age for marriage was restored to 15 and couples were urged not to begin families until at least the age of 20.

Couples were provided with free contraceptives and female and male sterilization. Indeed, Iran became the world's first country to require couples to take a class on contraception before they could receive a

marriage license. Between 1970 and 2000, the country also increased female literacy from 25% to 70% and female school enrollment from 60% to 90%.

These efforts paid off. Between 1989 and 2006, the country cut its population growth rate from 2.5% to 1.2% and its average family size from 7 children to 2.2—a remarkable change in 17 years. Within another decade, Iran's population could stabilize and perhaps begin a declining rate of growth. However, it is not known whether the new leaders in Iran will continue to support reduced population growth.

Critical Thinking

Do you support such government efforts to slow population growth in the country where you live? Explain.

---

low 0.3%) or declining population sizes. All, except Japan, are in Europe or African countries with high HIV/AIDS rates.

And the rate of population growth is slowing in many countries such as China, India, Bangladesh, and Iran (Case Study, above), mostly because of strong national family planning programs. On the other hand, about 1 billion other people live in countries whose populations are expected to double by 2050 and fuel the harmful effects of rapid population growth (Core Case Study, p. 171).

Rapid Population Decline from Declining Fertility Rates: Demographic Liabilities (Science, Economics, and Politics)

Rapid population decline can lead to long-lasting economic and social problems.

As the age structure of the world's population changes and the percentage of people age 60 or older increases (Figure 9-12), more countries will begin experiencing population declines. If population decline is gradual, its harmful effects usually can be managed.

But rapid population decline, like rapid population growth, can lead to severe economic and social problems. A country that experiences a fairly rapid "baby bust" or a "birth dearth" has a sharp rise in the proportion of older people. This puts severe strains on government budgets because these older people con-

sume an increasingly larger share of medical care, social security funds, and other costly public services funded by a decreasing number of working taxpayers. Such countries can also face labor shortages unless

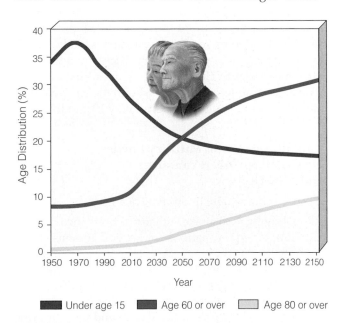

Figure 9-12 *Global aging.* Projected percentage of world population under age 15, age 60 or over, and age 80 or over, 1950–2150, assuming the medium fertility projection shown in Figure 9-2. The cost of supporting a much larger elderly population will place enormous strains on many nations and the global economy. (Data from the United Nations)

• Can threaten economic growth

• Less government revenues with fewer workers

• Less entrepreneurship and new business formation

• Less likelihood for new technology development

• Increasing public deficits to fund higher pension and healthcare costs

Figure 9-13 Some problems with rapid population decline. QUESTION: *Which three of these problems do you believe are the most important?*

they rely more heavily on automation or immigration of foreign workers. Figure 9-13 lists some of the problems associated with rapid population decline.

According to a recent study by the UN Population Division, if the United States wants to maintain its current ratio of workers to retirees, it will need to absorb an average of 10.8 million immigrants each year—more than 13 times the current immigration level—through 2050. At that point, the U.S. population would total 1.1 billion people, 73% of them fairly recent immigrants or their descendants. Housing this annual influx of almost 11 million immigrants would require the equivalent of building another New York City every 10 months.

Global Aging and Military Power: Peace Dividends (Economics and Politics)

Global aging may help promote peace.

Global aging will make military actions increasingly difficult for many nations for several reasons. *First*, there will be fewer young people available for military service. *Second*, parents with only one or two children will be increasingly reluctant to support military ventures that could wipe out their offspring.

Third, as the costs of health care and pensions rise and as the labor force stops growing, there will be intense competition between the general populace and the military for limited and perhaps decreasing government funds.

Population Decline from a Rising Death Rate: The AIDS Tragedy (Science and Economics)

Deaths from AIDS can disrupt a country's social and economic structure by removing significant numbers of young adults.

Between 2000 and 2050, AIDS is projected to cause the premature deaths of 278 million people in 53 countries—including 38 countries in Africa. These premature deaths are almost equal to the entire current population of the United States. Read this paragraph again, and think hard about the enormity of this tragedy.

Unlike hunger and malnutrition, which kill mostly infants and children, AIDS kills many young adults. This change in the young-adult age structure of a country has a number of harmful effects. One is a sharp drop in average life expectancy. In 8 African countries, where 16–39% of the adult population is infected with HIV, life expectancy could drop to 34–40 years.

Another effect is a loss of a country's most productive young adult workers and trained personnel such as scientists, farmers, engineers, teachers, and government, business, and health-care workers. This causes a sharp drop in the number of productive adults needed to support the young and the elderly and to grow food and provide essential services.

Analysts call for the international community—especially developed countries—to create and fund a massive program to help countries ravaged by AIDS in Africa and elsewhere. This program would have two major goals. *First*, reduce the spread of HIV through a combination of improved education and health care. *Second*, provide financial assistance for education and health care as well as volunteer teachers and health care and social workers to help compensate for the missing young-adult generation.

? THINKING ABOUT AIDS Should government and private interests in developed countries fund a massive program to help countries ravaged by AIDS prevent HIV infections and rebuild their work forces of younger adults? Explain.

SOLUTIONS: INFLUENCING POPULATION SIZE

The Demographic Transition (Economics)

As countries become economically developed, their birth and death rates tend to decline.

Demographers examining the birth and death rates of western European countries that became industrialized during the nineteenth century developed a hypothesis of population change known as the **demographic tran-**

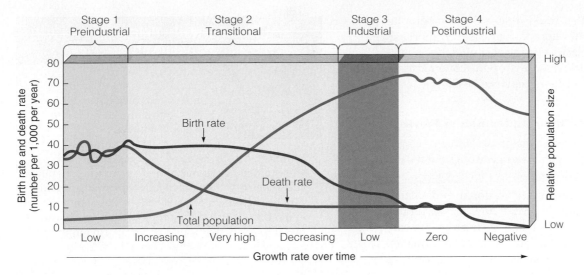

Stage 1 Preindustrial | Stage 2 Transitional | Stage 3 Industrial | Stage 4 Postindustrial

Birth rate

Death rate

Total population

Low — Increasing — Very high — Decreasing — Low — Zero — Negative

⟶ Growth rate over time ⟶

ThomsonNOW™ Active Figure 9-14 Generalized model of the *demographic transition*. There is uncertainty over whether this model will apply to some of today's developing countries. QUESTION: *At what stage is the country where you live? See an animation based on this figure and take a short quiz on the concept.*

sition: as countries become industrialized, first their death rates and then their birth rates decline. According to this hypothesis, this transition takes place in four distinct stages (Figure 9-14).

First is the *preindustrial stage,* when there is little population growth because harsh living conditions lead to both a high birth rate (to compensate for high infant mortality) and a high death rate.

Next is the *transitional stage,* when industrialization begins, food production rises, and health care improves. Death rates drop and birth rates remain high, so the population grows rapidly (typically 2.5–3% a year).

During the third phase, called the *industrial stage,* the birth rate drops and eventually approaches the death rate as industrialization, medical advances, and modernization become widespread. Population growth continues, but at a slower and perhaps fluctuating rate, depending on economic conditions. Most developed countries and a few developing countries are in this third stage.

The last phase is the *postindustrial stage,* when the birth rate declines further, equaling the death rate and reaching zero population growth. If the birth rate falls below the death rate, population size decreases slowly. Forty-two countries containing about 13% of the world's population have entered this stage and more of the world's developed countries are expected to enter this phase by 2050.

 ? *THINKING ABOUT OVERPOPULATION AND THE DEMOGRAPHIC TRANSITION* How does the demographic transition affect the debate over whether the world is over-populated (Core Case Study, p. 171)?

Thomson NOW! Explore the effects of economic development on birth and death rates and population growth at ThomsonNOW.

Can the World's Developing Countries Make a Demographic Transition? (Science, Economics, and Politics)

Some developing countries may have difficulty making the demographic transition.

In most developing countries today, death rates have fallen much more than birth rates. In other words, these developing countries are still in the transitional stage, halfway up the economic development ladder, with fairly high population growth rates. The question is, can they get to stages 3 and 4? Experts disagree on this issue.

Some analysts believe that most of the world's developing countries can make a demographic transition within the next few decades mostly because modern technology can bring economic development and birth control to such countries.

However, some population analysts fear that the still-rapid population growth in a number of developing countries may outstrip economic growth and overwhelm some local life-support systems. As a consequence, some of these countries could become caught in a *demographic trap* at stage 2. This is now happening as death rates rise in a number of developing countries, especially in Africa. Indeed, countries in Africa being ravaged by the HIV/AIDS epidemic are falling back to stage 1.

Other factors that could hinder the demographic transition in some developing countries are shortages of skilled workers, lack of financial capital, large debts to developed countries, and a drop in economic assistance from developed countries since 1985.

Family Planning: Planning for Babies Works

Family planning has been a major factor in reducing the number of births and abortions throughout most of the world.

Family planning provides educational and clinical services that help couples choose how many children to have and when to have them. Such programs vary from culture to culture, but most provide information on birth spacing, birth control, and health care for pregnant women and infants.

Family planning has helped increase the proportion of married women in developing countries who use modern forms of contraception from 10% of married women of reproductive age in the 1960s to 51% of these women in 2006. Family planning has also reduced the number of legal and illegal abortions performed each year and lowered the risk of maternal and fetal death from pregnancy.

Studies also show that family planning has been responsible for at least 55% of the drop in TFRs in developing countries, from 6.0 in 1960 to 3.0 in 2006. Between 1971 and 2005, Thailand used family planning to cut its annual population growth rate from 3.2% to 0.8% and its TFR from 6.4 to 1.7 children per family. Another success story for family planning is Iran (Case Study, p. 181) and Bangladesh (Case Study, at right).

Despite the successes of family planning in many countries, two major problems remain. *First,* according to the United Nations Population Fund, 42% of all pregnancies in developing countries are unplanned and 26% end with abortion. *Second,* an estimated 201 million couples in developing countries want to limit the number and determine the spacing of their children, but they lack access to family planning services. According to a recent study by the United Nations Population Fund and the Alan Guttmacher Institute, meeting women's current unmet needs for family planning and contraception could *each year* prevent 52 million unwanted pregnancies, 22 million induced abortions, 1.4 million infant deaths, and 142,000 pregnancy-related deaths.

Some analysts call for expanding family planning programs to include teenagers and sexually active unmarried women, who are excluded from many existing programs. For teenagers, many advocate much greater emphasis on abstinence.

Another suggestion is to develop programs that educate men about the importance of having fewer children and taking more responsibility for raising

Family Planning in Bangladesh

CASE STUDY

Bangladesh has made striking progress in reducing its population growth since 1965, despite being one of the world's poorest and most densely populated countries. Mostly because of an aggressive and well-funded government family planning program and aid from international family planning organizations such as the United Nations Population Fund, the country's TFR dropped from 6.5 children to 3 children between 1965 and 1990. An important factor was an outreach program where families received biweekly visits from local women who provided education, counseling, and free contraceptives. Studies indicate that each $62 spent by the government to prevent an unwanted birth saved $615 in expenditures on other social services. However, since 1990, TFRs in Bangladesh have leveled off at 3 children per woman and thus the country still faces fairly rapid population growth.

Critical Thinking

What two things would you do to slow population growth further in Bangladesh?

them. Proponents also call for greatly increased research on developing more effective and more acceptable birth control methods for men.

Finally, a number of analysts urge polarized pro-choice and pro-life groups to join forces in greatly reducing unplanned births and abortions, especially among teenagers.

Empowering Women: Ensuring Education, Jobs, and Human Rights

Women tend to have fewer children if they are educated, hold a paying job outside the home, and do not have their human rights suppressed.

Three key factors lead women to have fewer and healthier children: education, employment outside the home, and living in societies where their rights are not suppressed

Women make up roughly half of the world's population. They do almost all of the world's domestic work and childcare for little or no pay and provide more unpaid health care than all of the world's organized health services combined.

They also do 60–80% of the work associated with growing food, gathering fuelwood, and hauling water (photo 4 in the Detailed Contents) in rural areas of

Africa, Latin America, and Asia. As one Brazilian woman put it, "For poor women the only holiday is when you are asleep."

Globally, women account for two-thirds of all hours worked but receive only 10% of the world's income, and they own less than 2% of the world's land. In most developing countries, women do not have the legal right to own land or to borrow money. Women also make up 70% of the world's poor and 64% of the world's 800 million illiterate adults.

According to Thorya Obaid, executive director of the United Nations Population Fund, "Many women in the developing world are trapped in poverty by illiteracy, poor health, and unwanted high fertility. All of these contribute to environmental degradation and tighten the grip of poverty. If we are serious about sustainable development, we must break this vicious cycle."

That means giving women everywhere full legal rights and the opportunity to become educated and earn income outside the home. Achieving these goals would slow population growth, promote human rights and freedom, reduce poverty, and slow environmental degradation—a win–win result.

Empowering women by seeking gender equality will require some major social changes. Although it will be difficult to achieve in male-dominated societies, it can be done.

An increasing number of women in developing countries are taking charge of their lives and reproductive behavior. They are not waiting for the slow processes of education and cultural change. As it expands, such bottom-up change by individual women will play an important role in stabilizing populations and providing more women with basic human rights.

Solutions: Reducing Population Growth

The best way to slow population growth is a combination of investing in family planning, reducing poverty, and elevating the status of women.

In 1994, the United Nations held its third Conference on Population and Development in Cairo, Egypt. One of the conference's goals was to encourage action to stabilize the world's population at 7.8 billion by 2050 instead of the projected 8.9 billion.

The major goals of the resulting population plan, endorsed by 180 governments, are to do the following by 2015:

- Help countries develop and implement national population polices.

- Provide universal access to family planning services and reproductive health care.

- Sharply reduce poverty.

- Provide better health care and nutritious food supplements for infants, children, pregnant women, and nursing mothers.

- Provide universal education with special emphasis on girls and women and adult literacy programs.

- Improve the status of women and expand educational and job opportunities for young women. Educated women tend to have smaller families and provide better child nutrition and health care.

- Launch nutritious school lunch programs in low-income countries to increase school attendance and improve academic performance and health. Girls drawn to school for the lunches tend to stay in school longer, marry later, and have fewer children.

- Increase the involvement of men in child-rearing responsibilities and family planning.

- Sharply reduce unsustainable patterns of production and consumption.

The experiences of countries such as Japan, Thailand, South Korea, Taiwan, Iran, and China indicate that a country can achieve or come close to replacement-level fertility within a decade or two. Such experiences also suggest that the best way to slow population growth is through the combination of *investing in family planning, reducing poverty,* and *elevating the status of women.*

SLOWING POPULATION GROWTH IN INDIA AND CHINA

Case Study: India (Science, Economics, and Politics)

For more than five decades, India has tried to control its population growth with only modest success.

The world's first national family planning program began in India in 1952, when its population was nearly 400 million. In 2006, after 54 years of population control efforts, India had 1.1 billion people.

In 1952, India added 5 million people to its population. In 2006, it added 18.6 million people—more than any other country. By 2035, India is projected to be the world's most populous country, with its population projected to peak at 1.6 billion around 2065. Figure 9-15 (p. 186) compares demographic data for India and China.

India faces a number of already serious poverty, malnutrition, and environmental problems that could worsen as its population continues to grow rapidly. By global standards, one of every four people in India is poor. And nearly half of the country's labor force is unemployed or can find only occasional work.

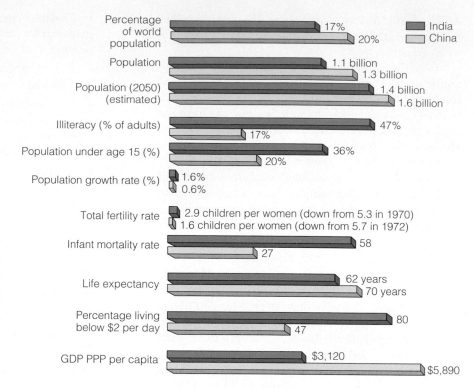

Figure 9-15 Global connection: basic demographic data for India and China in 2006. (Data from United Nations and Population Reference Bureau)

	India	China
Percentage of world population	17%	20%
Population	1.1 billion	1.3 billion
Population (2050) (estimated)	1.4 billion	1.6 billion
Illiteracy (% of adults)	47%	17%
Population under age 15 (%)	36%	20%
Population growth rate (%)	1.6%	0.6%
Total fertility rate	2.9 children per women (down from 5.3 in 1970)	1.6 children per women (down from 5.7 in 1972)
Infant mortality rate	58	27
Life expectancy	62 years	70 years
Percentage living below $2 per day	80	47
GDP PPP per capita	$3,120	$5,890

Although India currently is self-sufficient in food grain production, about 40% of its people and more than half of its children suffer from malnutrition, mostly because of poverty (Photo 3 in the Detailed Contents). Furthermore, India faces critical resource and environmental problems. With 17% of the world's people, it has just 2.3% of the world's land resources and 2% of the world's forests. About half of the country's cropland is degraded as a result of soil erosion, waterlogging, salinization, overgrazing, and deforestation. More than two-thirds of its water is seriously polluted, sanitation services often are inadequate, and many of its major cities suffer from serious air pollution (Photo 2 in the Detailed Contents).

India is undergoing rapid economic growth, which is expected to accelerate. its huge and growing middle class (Case Study, p. 15). As these individuals increase their resource use per person, India's ecological footprint will expand and increase the pressure on the country's and the earth's natural capital. On the other hand, economic growth may help slow population growth by accelerating India's ability to make the demographic transition. By 2050, India—the largest democracy the world has ever seen—could become the world's leading economic power.

Without its long-standing family planning program, India's population and environmental problems would be growing even faster. Still, the results of the program have been disappointing because of poor planning, bureaucratic inefficiency, the low status of women (despite constitutional guarantees of equality), extreme poverty, lack of administrative and financial support, and disagreement over the best ways to slow population growth.

The government has provided information about the advantages of small families for years. Even so, Indian women have an average of 2.9 children. Most poor couples still believe they need many children to do work and care for them in old age. The strong cultural preference for male children also means some couples keep having children until they produce one or more boys. One result is of this cultural preference is an avoidance of birth control; even though 90% of Indian couples know of at least one modern birth control method, only 46% actually use one.

Some urge the government to implement a much more aggressive national population control policy. However, the government has decided to let each state deal with this issue on its own. Top-down control by India's central government is difficult in this huge democratic country consisting of many different states, cultures, religions, and languages.

The state of Andra Pradesh has stabilized its population size by implementing a strongly enforced sterilization program. The poor are strongly encouraged—some say coerced—to be sterilized after having one or two children. Those (mostly women) that agree to sterilization receive a small cash payment (equivalent to $11 U.S.) equal to about a month's wages and have better access to land, water wells, housing, and subsidized loans.

Since the mid-1980s, the state of Kerala, has taken a different approach to stabilizing its population by emphasizing social justice based on economic redistribution. Although it is one of the poorest and most crowded places on the earth in terms of quality of life it has some of the highest scores among developing countries. Its life expectancy is 70 years, compared to 62 years for India as a whole and its infant mortality rate is four times lower than the country as a whole.

And its adult literacy rate is essentially 100% compared to a nationwide rate of 47%.

Kerala also leads India in the quality of its roads, schools, hospitals, public housing, drinking water, sanitation, immunization programs, women's rights, and nutrition programs for infants and for pregnant and nursing women. All its people have access to free or inexpensive medical care, and all households receive ration cards that allow them to buy rice and certain basic commodities at subsidized prices. Since 1960, a land reform program has distributed small plots of land to more than 3 million tenants and landless poor. Kerala provides an example of how to improve life quality and stabilize population without depending on ever-increasing economic growth and a widening gap between the rich and poor. A number of other Indian states have population growth rates of 2–2.5% a year.

 ? *THINKING ABOUT INDIA, THE UNITED STATES, AND OVERPOPULATION* Based on population size and resource use per person (Figure 1-7, p. 13, and Figure 1-14, p. 20) is the United States more overpopulated than India? Explain.

Case Study: China

Since 1970, China has used a government-enforced program to cut its birth rate in half and sharply reduce its fertility rate.

Since 1970, China has made impressive efforts to feed its people, bring its population growth under control, and encourage economic growth. Between 1972 and 2006, the country cut its crude birth rate in half and trimmed its TFR from 5.7 to 1.6 children per woman (Figure 9-15). If current trends continue, China's population should peak around 2040 and then begin a slow decline.

In addition, since 1980, China has moved 300 million people—more than the number of people in the U.S. population—out of poverty and quadrupled the average person's income (Case Study, p. 15). China also has a literacy rate of 91% and has boosted life expectancy to 72 years. It now has more than 300 million new middle-class consumers and is likely to have 600 million such consumers by 2010. By 2020, China could become the world's leading economic power.

In the 1960s, government officials concluded that the only alternative to strict population control was mass starvation. Thus, to achieve its sharp drop in fertility, China has established the world's most extensive, intrusive, and strict family planning and population control program.

It discourages premarital sex and urges people to delay marriage and limit families to no more than one child. Married couples who pledge to have no more than one child receive more food, larger pensions, better housing, free health care, salary bonuses, free school tuition for their child, and preferential employment opportunities when their child enters the job market. Couples who break their pledge lose such benefits.

The government also provides married couples with free sterilization, contraceptives, and abortion. As a consequence, 86% of married women in China use modern contraception. Reports of forced abortions and other coercive actions have brought condemnation from the United States and other national governments.

In China (as in India), there is a strong preference for male children. A folk saying goes, "Rear a son, and protect yourself in old age." Many pregnant women use ultrasound to determine the gender of their fetus and some get an abortion if it is female because unlike sons they are likely to marry and leave their parents. The result: a rapidly growing *gender imbalance* or "bride shortage" in China's population, with a projected 30–40 million surplus of men expected by 2020. Because of this skewed sex ratio, teen-age girls in some parts of rural China are being kidnapped and sold as brides for single men in other parts of the country.

With fewer children, the average age of China's population is increasing rapidly, which could result in a declining work force, lack of funds for supporting continuing economic development, and fewer children and grandchildren to care for the growing number of elderly people. These concerns and other factors may lead to some relaxation of China's one-child population control policy in the future.

Another problem is environmental degradation because supplies of some renewable resources are exceeding the rising demand, and increased resource use is polluting the air and water. China has 20% of the world's population. But it has only 7% of the world's freshwater and cropland, 4% of its forests, and 2% of its oil. Soil erosion in China is serious and apparently getting worse. In 2005, China's deputy minister of the environment summarized the country's environmental problems: "Our raw materials are scarce, we don't have enough land, and our population is constantly growing. . . . Half of the water in our seven largest rivers is completely useless. One-third of the urban population is breathing polluted air."

China's economy is growing at one of the world's highest rates as the country undergoes rapid industrialization. More middle-class Chinese will consume more resources per person, increasing the ecological footprint of China in its own country and in other parts of the world providing it with resources. This will put a strain on the earth's natural capital.

 ? *THINKING ABOUT CHINA, THE UNITED STATES AND OVERPOPULATION* Based on population size and resource use per person (Figure 1-7, p. 13 and Figure 1-14, p. 20) is the United States more overpopulated than China? Explain.

Figure 9-16 Some typical characteristics of natural and human-dominated systems. Many human activities threaten local ecological processes and some bring about harmful regional and global changes.

HUMAN IMPACTS ON NATURAL SYSTEMS

Effects of Humans on Natural Ecosystems: Our Big Footprints

We have used technology to alter much of the rest of nature in ways that threaten the survival of many other species and could reduce the quality of life for our own species.

To survive and provide resources for growing numbers of people, humans have modified, cultivated, built on, or degraded a large and increasing area of the earth's natural systems. Excluding Antarctica, our activities have directly affected to some degree about 83% of the

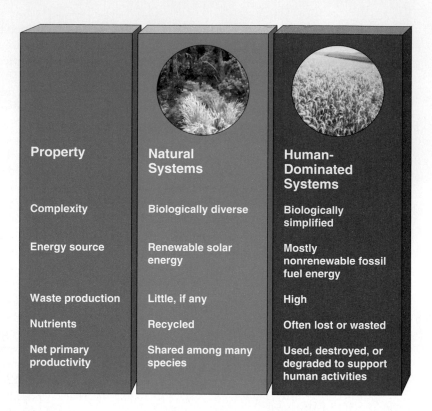

Property	Natural Systems	Human-Dominated Systems
Complexity	Biologically diverse	Biologically simplified
Energy source	Renewable solar energy	Mostly nonrenewable fossil fuel energy
Waste production	Little, if any	High
Nutrients	Recycled	Often lost or wasted
Net primary productivity	Shared among many species	Used, destroyed, or degraded to support human activities

Natural Capital Degradation

Altering Nature to Meet Our Needs

Reduction of biodiversity

Increasing use of the earth's net primary productivity

Increasing genetic resistance of pest species and disease-causing bacteria

Elimination of many natural predators

Deliberate or accidental introduction of potentially harmful species into communities

Using some renewable resources faster than they can be replenished

Interfering with the earth's chemical cycling and energy flow processes

Relying mostly on polluting fossil fuels

earth's land surface (see Figure 3 on pp. S12–S13 in Supplement 4). Figure 9-16 compares some of the characteristics of natural and human-dominated systems.

We have used technology to alter much of the rest of nature to meet our growing needs and wants in eight major ways (Figure 9-17).

 Thomson NOW! Examine how resources have been depleted or degraded around the world at ThomsonNOW.

To survive, we must exploit and modify parts of nature. However, we are beginning to understand that doing so has multiple effects, most of them unintended and unpredictable (Spotlight, p. 189).

We face two major challenges. *First*, we need to maintain a balance between simplified, human-altered communities and the more complex natural communities on which we and other species depend. *Second*, we

ThomsonNOW **Active Figure 9-17** Natural capital degradation: major ways humans have altered the rest of nature to meet our growing population, needs, and wants. QUESTIONS: *Which three of these items do you believe have been the most harmful? How does your lifestyle contribute directly or indirectly to each of these items? See an animation based on this figure and take a short quiz on the concept.*

need to slow down the rates at which we are simplifying, homogenizing, and degrading nature for our purposes. Otherwise, what is at risk is not the resilient earth (Core Case Study, p. 82) but rather the quality of life for our own species and the very existence of the increasing number of species our activities are driving to premature extinction.

We cannot save the earth because it can get along very nicely without us, just as it has done for 3.7 billion years. But we can learn how the earth works and work with its natural processes (Figure 1-16, p. 24).

Ecological Surprises

SPOTLIGHT

Malaria once infected nine of ten people in North Borneo, now known as Sabah. In 1955, the World Health Organization (WHO) began spraying the island with dieldrin (a DDT relative) to kill malaria-carrying mosquitoes. The program was so successful that the dreaded disease was nearly eliminated.

Then unexpected things began to happen. The dieldrin also killed other insects, including flies and cockroaches living in houses. The islanders applauded. Next, small insect-eating lizards that also lived in the houses died after gorging themselves on dieldrin-contaminated insects.

Cats began dying after feeding on the lizards. In the absence of cats, rats flourished and overran the villages. When the people became threatened by sylvatic plague carried by rat fleas, the WHO parachuted healthy cats onto the island to help control the rats. Operation Cat Drop worked.

But then the villagers' roofs began to fall in. The dieldrin had killed wasps and other insects that fed on a type of caterpillar that either avoided or was not affected by the insecticide. With most of its predators eliminated, the caterpillar population exploded, munching its way through its favorite food: the leaves used to thatch roofs.

Ultimately, this episode ended happily: Both malaria and the unexpected effects of the spraying program were brought under control. Nevertheless, this chain of unintended and unforeseen events emphasizes the unpredictability of interfering with a community. It reminds us that when we intervene in nature, we need to ask, "Now what will happen?"

Critical Thinking

Do you believe the beneficial effects of spraying pesticides on Sabah outweighed the resulting unexpected and harmful effects? Explain.

 ## Revisiting Population Growth and Sustainability

This chapter began by discussing the issue of whether the world is overpopulated (Core Case Study, p. 171). As we have seen, some experts say this is the wrong question. Instead, they believe we ought to ask what is the optimal level of human population that the planet can support *sustainably*. In other words, at what level could the maximum number of people live comfortably and freely without jeopardizing the earth's ability to provide the same comforts and freedoms for future generations and for other species?

In the first eight chapters of this book, you have learned how ecosystems and species have sustained themselves through long periods of history by use of four scientific principles of sustainability—reliance on solar energy, biodiversity, population control, and nutrient recycling (Figure 1-16, p. 24). In this chapter, you may have gained a feel for the need for humans to apply these sustainability principles to their lifestyles and economies.

In the next three chapters, you will learn how various principles of ecology and the four sustainability principles can be applied to help preserve the earth's biodiversity.

Our numbers expand but Earth's natural systems do not.
LESTER R. BROWN

CRITICAL THINKING

 1. Do you believe the population of **(a)** the world (Core Case Study, p. 171), **(b)** your own country and **(c)** the area where you live is too high? Explain.

2. Which of the three major environmental worldviews summarized on pp. 22–23 do you believe underlie the two major positions on whether the world is overpopulated (Core Case Study, p. 171)?

3. Why is it rational for a poor couple in a developing country such as India to have four or five children? What changes might induce such a couple to consider their behavior irrational?

4. Identify a major local environmental problem and describe the role population growth plays in this problem.

5. Should everyone have the right to have as many children as they want? Explain.

6. Some people have proposed that the earth could solve its population problem by shipping people off to space colonies, each containing about 10,000 people. Assuming

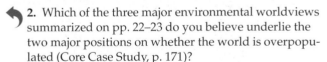

we could build such large-scale, self-sustaining space stations (a big assumption), how many people would we have to ship off each day to provide living spaces for the 81 million people added to the earth's population this year? Assuming a space shuttle could carry 100 passengers, how many shuttles would have to be launched per day to offset the 79 million people added this year? According to your calculations, determine whether this proposal is a logical solution to the earth's population problem.

7. Some people believe the most important goal is to sharply reduce the rate of population growth in developing countries. Others would reduce the high levels of resource consumption per person in developed countries. What is your view on this issue? Explain.

8. Do you agree with China's strict population control policy? If you disagree, what three things do you believe China should have done and should continue doing to help feed its people and support its rapid economic growth?

9. Congratulations! You are in charge of the world. List the three most important features of your population policy.

10. List two questions that you would like to have answered as a result of reading this chapter.

PROJECTS

1. Assume your entire class (or manageable groups of your class) is charged with coming up with a plan for halving the world's population growth rate within the next 20 years. Develop a detailed plan that would achieve this goal, including any differences between policies in developing countries and those in developed countries. Justify each part of your plan. Try to anticipate what problems you might face in implementing the plan, and devise strategies for dealing with these problems.

2. Prepare an age structure diagram for your community. Use the diagram to project future population growth and economic and social problems.

3. Do some research to find information on Generations X and Y. Write a brief report comparing them in terms of where they are in the world, when they were born, the sizes of each group, and the issues that might arise for each group relating to population size and structure.

4. Make a concept map of this chapter's major ideas, using the section heads, subheads, and key terms (in boldface). See material on the website for this book about how to prepare concept maps.

LEARNING ONLINE

The website for this book contains study aids and many ideas for further reading and research. They include a chapter summary, review questions for the entire chapter, flash cards for key terms and concepts, a multiple-choice practice quiz, interesting Internet sites, references, information about green careers, and a guide for accessing thousands of InfoTrac® College Edition articles. Log into

www.thomsonedu.com/biology/miller

Then choose Chapter 9, and select a learning resource. For access to animations, additional quizzes, chapter outlines and summaries, register and log into

at **www.thomsonedu.com** using the access code card in the front of your book.

Active Graphing

Log into ThomsonNow at www.thomsonedu.com to explore the graphing exercise for this chapter.

10 Sustaining Terrestrial Biodiversity: The Ecosystem Approach

CORE CASE STUDY

Reintroducing Wolves to Yellowstone

At one time, the gray wolf, also known as the eastern timber wolf (Figure 10-1), roamed over most of North America. But between 1850 and 1900, an estimated 2 million wolves were shot, trapped, and poisoned by ranchers, hunters, and government employees. The idea was to make the West and the Great Plains safe for livestock and for big-game animals prized by hunters.

It worked. When Congress passed the U.S. Endangered Species Act in 1973, only a few hundred gray wolves remained outside of Alaska, primarily in Minnesota and Michigan.

Ecologists recognize the important role this keystone predator species once played in parts of the West and the Great Plains. These wolves culled herds of bison, elk, caribou, and mule deer, and kept down coyote populations. They also provided uneaten meat for scavengers such as ravens, bald eagles, ermines, grizzly bears, and foxes.

In recent years, herds of elk, moose, deer, and antelope have expanded. Their larger numbers have devastated some vegetation such as willow and aspen trees, increased erosion, and threatened the niches of other wildlife species such as beavers that help create wetlands. Reintroducing a keystone species such as the gray wolf into a terrestrial ecosystem is one way to help sustain the biodiversity of the ecosystem and prevent further environmental degradation.

In 1987, the U.S. Fish and Wildlife Service (USFWS) proposed reintroducing gray wolves into the Yellowstone National Park ecosystem. This brought angry protests. Some objections came from ranchers who feared the wolves would leave the park

Figure 10-1 Natural capital restoration: the *gray wolf*. Ranchers, hunters, miners, and loggers have vigorously opposed efforts to return this keystone species to its former habitat in the Yellowstone National Park. Wolves were reintroduced beginning in 1995 and now number around 118.

Tom Kitchin/Tom Stack & Associates

and attack their cattle and sheep; one enraged rancher said that the idea was "like reintroducing smallpox." Other protests came from hunters who feared the wolves would kill too many big-game animals, and from mining and logging companies that feared the government would halt their operations on wolf-populated federal lands.

Since 1995, federal wildlife officials have caught gray wolves in Canada and relocated them in Yellowstone National Park. Scientists estimate that the long-term carrying capacity of the park is 110 to 150 gray wolves. In 2005, the park had 118 gray wolves.

With wolves around, elk are gathering less near streams and rivers. This has spurred the growth of aspen, cottonwoods, and willow trees. This helps stabilize stream banks, which lowers the water temperature and makes the habitat better for trout. Beavers seeking willow and aspen have returned. And leftovers of elk killed by wolves are an important food source for grizzly bears and other scavengers.

The wolves have also cut coyote populations in half. This has increased populations of smaller animals such as ground squirrels, mice, and gophers preyed upon by coyotes. This provides more food for eagles and hawks.

Since 1980, *biodiversity* (Figure 3-15, p. 62) has emerged as one of the most important integrative principles of biology and as one of the four principles of sustainability (Figure 1-16, p. 24). Biologists warn that population growth, economic development, and poverty are exerting increasing pressure on the world's forests, grasslands, parks, wilderness, and other terrestrial storehouses of biodiversity. This chapter and the two that follow are devoted to helping us understand and sustain the earth's terrestrial and aquatic biodiversity.

Forests precede civilizations; deserts follow them.

FRANCOIS-AUGUSTE-RENÉ DE CHATEAUBRIAND

This chapter discusses how we can help sustain the earth's terrestrial biodiversity by protecting and restoring places where wild species live and curbing the size and force of our ecological footprints by learning to walk more lightly on the earth. It addresses the following questions:

- How have human activities affected the earth's biodiversity?

- How should forest resources be used, managed, and sustained globally and in the United States?

- How serious is tropical deforestation, and how can we help sustain tropical forests?

- How should rangeland resources be used, managed, and sustained?

- What problems do parks face, and how should we manage them?

- How should we establish, design, protect, and manage terrestrial nature reserves?

- What is wilderness, and why is it important?

- What is ecological restoration, and why is it important?

- What can we do to help sustain the earth's terrestrial biodiversity?

HUMAN IMPACTS ON TERRESTRIAL BIODIVERSITY

Effects of Human Activities on Global Biodiversity (Science and Economics)

We have depleted and degraded some of the earth's biodiversity and these threats are expected to increase.

Figure 10-2 summarizes how many human activities decrease biodiversity. You can also get an idea of our impact on the earth's natural terrestrial systems in Supplement 4 at the end of this book by comparing the two-page satellite map of the earth's natural terrestrial systems (Figure 1, on pp. S8–S9), Figure 5-9 (p. 106) of the earth's major biomes, the two-page map of our large and growing ecological footprint on these natural systems (Figure 3, pp. S12–S13), and the map of our ecological footprint in the United States (Figure 4, pp. S14–S15). According to biodiversity expert Edward O. Wilson, "The natural world is everywhere disappearing before our eyes—cut to pieces, mowed down, plowed under, gobbled up, replaced by human artifacts."

Consider a few examples of how human activities have decreased and degraded the earth's biodiversity. According a 2002 study on the impact of the human ecological footprint on the earth's land and the 2005

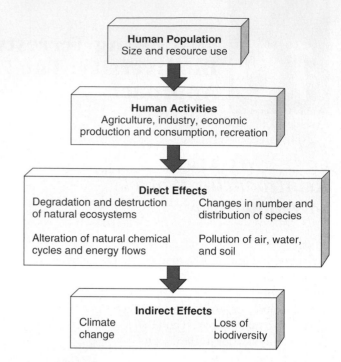

Figure 10-2 Natural capital degradation: major connections between human activities and the earth's biodiversity.

Millennium Ecosystem Assessment, we have disturbed to some extent at least half and probably about 83% of the earth's land surface (excluding Antarctica and Greenland). Most of this is the result of filling in wetlands and converting grasslands and forests to crop fields and urban areas.

In the United States, at least 95% of the virgin forests in the lower 48 states have been logged for lumber and to make room for agriculture, housing, and industry. In addition, 98% of tallgrass prairie in the Midwest and Great Plains has disappeared, and 99% of California's native grassland and 85% of its original redwood forests are gone.

From this brief overview, you can see why protecting and sustaining the genes, species, ecosystems, and ecological functions that make up the world's biodiversity (Figure 3-15, p. 62) is one of the most important and urgent global issues we face. Figure 3-16 (p. 63) outlines the goals, strategies, and tactics for preserving and restoring the earth's terrestrial ecosystems (as discussed in this chapter) and preventing the premature extinction of species (as discussed in Chapter 11). Sustaining aquatic diversity is discussed in Chapter 12.

Why Should We Care about Biodiversity? (Science, Economics, and Ethics)

Biodiversity should be protected from degradation by human activities because it exists and because of its usefulness to us and other species.

Biodiversity researchers contend that we should act to preserve the earth's overall biodiversity because its genes, species, ecosystems, and ecological services have two types of value. One is **intrinsic value** because these components exist, regardless of their use to us. Protecting biodiversity on this basis is basically an ethical decision.

The other is **instrumental value** because of their usefulness to us in the form of numerous economic and ecological services (Figure 1-4, p. 9). For example, more than half the world's people depend directly on forests, rangelands, croplands, and fisheries for their livelihoods. In addition, the jobs of many more people depend on the processing of food, paper, textiles, and other resources these systems provide. Biodiversity also provides economic benefits and pleasure from recreation and tourism.

And biodiversity helps maintain the structure and function of ecosystems (Figure 3-14, p. 61) and control populations of pests and other species, provides a variety of options for nature to adapt to environmental change, and supplies us and other species with food and a variety of medicines and drugs. In other words, biodiversity is one of the most important forms of natural capital.

Instrumental values take two forms. One is a *use value* that benefits us in the form of economic goods and services, ecological services, recreation, scientific information, and preservation of options for such uses in the future.

The other form is a *nonuse value*. For example, there is *existence value*—the satisfaction of knowing that a redwood forest, wilderness, orangutans (Figure 10-3), or wolf pack (Core Case Study, p. 191) exists, even if we will never see it or get direct use from it. *Aesthetic value* is another nonuse value because many people appreciate a tree, a forest, a wild species such as a parrot (see photo 7 in the Detailed Contents), or a vista because of its beauty. *Bequest value,* a third type of nonuse value, is based on the willingness of some people to pay to protect some forms of natural capital for use by future generations.

MANAGING AND SUSTAINING FORESTS

Economic and Ecological Services Provided by Forests (Science and Economics)

Forests provide a number of ecological and economic services and researchers have attempted to estimate their total monetary value.

Forests with at least 10% tree cover occupy about 30% of the earth's land surface (excluding Greenland and Antarctica). Figure 5-9 (p. 106) shows the distribution of the world's boreal, temperate, and tropical forests, which provide many important ecological and economic services (Figure 10-4).

© age fotostock/SuperStock

Figure 10-3 Natural capital degradation: endangered orangutans in a tropical forest. In 1900, there were over 315,000 wild orangutans. Now there are less than 20,000 and they are disappearing at a rate of over 2,000 individuals a year. An illegally smuggled live orangutan typically sells for a street price of $10,000.

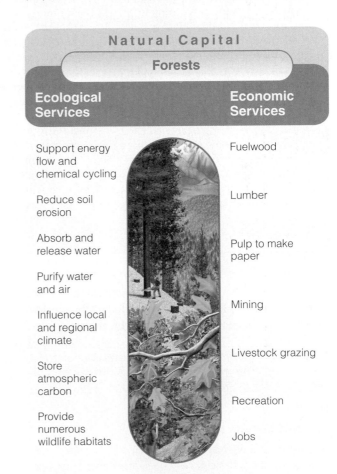

Natural Capital

Forests

Ecological Services	Economic Services
Support energy flow and chemical cycling	Fuelwood
Reduce soil erosion	Lumber
Absorb and release water	Pulp to make paper
Purify water and air	Mining
Influence local and regional climate	Livestock grazing
Store atmospheric carbon	Recreation
Provide numerous wildlife habitats	Jobs

Figure 10-4 Natural capital: major ecological and economic services provided by forests. QUESTION: *Which two ecological services and which two economic services do you think are the most important?*

Putting a Price Tag on Nature's Ecological Services (Science and Economics)

The long-term health of an economy cannot be separated from the health of the natural systems that support it. Currently, forests and other ecosystems are valued mostly for their economic services (Figure 10-4, right). But suppose we took into account the monetary value of the ecological services provided by forests (Figure 10-4, left).

In 1997, a team of ecologists, economists, and geographers—led by ecological economist Robert Costanza of the University of Vermont—estimated the monetary worth of the earth's natural ecological services and the biological income they provide. According to this appraisal, the estimated economic value of the biological income from the earth's ecological services is at least $33.2 trillion per year—fairly close to the economic value of all of the goods and services produced throughout the world each year. To provide an annual natural income of $33.2 trillion per year, the world's natural capital would have a value of at least $500 trillion—an average of about $82,000 for each person on earth!

According to this study, the world's forests provide us with eco-logical services worth at least $4.7 trillion per year—hundreds of times more then the economic value of forests. And these are very conservative estimates. In 2005, a study by the Canadian Boreal Initiative estimated that the ecological services provided by Canada's boreal forest alone were worth $3.7 trillion a year—2.5 times the annual value of timber harvesting, mining, and hydroelectric power production in this region.

The authors of such studies warn that unless we include the financial value of their ecological services in evaluating how we use forests and other ecological resources they will be used unsustainably and destroyed or degraded for short-term economic gain.

These researchers hope their estimates will alert people to three important facts. The earth's ecosystem services are essential for all humans and their economies; their economic value is huge; and they are an ongoing source of ecological income as long as they are used sustainably.

According to ecological economist Robert Costanza, "We have been cooking the books for a long time by leaving out the worth of nature." Biologist David Suzuki warns, "Our economic system has been constructed under the premise that natural services are free. We can't afford that luxury any more."

Why haven't we changed our accounting systems to reflect the values of these sources and the losses from destroying or degrading these ecological services? One reason is that economic savings provided by conserving natural resources benefit everyone now and in the future, whereas the profits made by exploiting these resources are immediate and benefit a relatively small group of people who have the motivation and means to develop them.

A second reason is that many current government subsidies and tax incentives support destruction and degradation of forests and other ecosystems for short-term economic gain. Also, most people are unaware of the value of the ecological services and biological income provided by forests and other parts of nature.

Critical Thinking

Some analysts believe that we should not try to put economic values on the world's irreplaceable ecological services because their value is infinite. Do you agree with this view? Explain.

There have been efforts to estimate the economic value of the ecological services provided by the world's forests and other ecosystems (Spotlight, above).

F **RESEARCH FRONTIER** Improving estimates of the economic values of the ecological services provided by forests and other major ecosystems

Types of Forests (Science)

Some forests have not been disturbed by human activities, others have grown back after being cut, and some consist of planted stands of a particular tree species.

Forest managers and ecologists classify forests into three major types based on their age and structure. The first type is an **old-growth forest:** an uncut or regenerated forest that has not been seriously disturbed by human activities or natural disasters for at least several hundred years (Figure 10-5 and Figure 5-19, top photo, p.116). Old-growth forests are storehouses of biodiversity because they provide ecological niches for a multitude of wildlife species (Figure 5-21, p. 118).

The second type is a **second-growth forest:** a stand of trees resulting from natural secondary ecological succession (Figure 7-12, p. 157). These forests develop after the trees in an area have been removed by *human activities* (such as clear-cutting for timber or conversion to cropland) or by *natural forces* (such as fire, hurricanes, or volcanic eruption).

A **tree plantation,** also called a **tree farm,** is a third type (see Figure 10-6 and photo 1 in the Detailed Con-

Figure 10-5 **Natural capital:** old-growth forest in Washington State's Olympic National Forest (USA).

and 5% are tree plantations (that produce about one-fourth of the world's commercial wood). Five countries—Russia, Canada, Brazil, Indonesia, and Papua, New Guinea—have more than three-fourths of the world's remaining old-growth forests. In order, China (which has little original forest left), Russia, the United States, India, and Japan have two-thirds of the world's tree plantations. The fate of the world's forests will be decided mostly by governments, which own about 84% of the world's remaining forests.

Some analysts believe that establishing tree plantations on much of the earth's deforested and degraded land could conceivably meet most of the world's future needs for wood, help reduce soil erosion, and slow global warming by removing more CO_2 from the troposphere. However, they oppose clearing existing old-growth and secondary forests to establish tree plantations. This trade-off would help protect most of the world's remaining old-growth forests and their extraordinary biodiversity.

X *How Would You Vote?* Should there be a global effort to sharply reduce the cutting of old-growth forests? Cast your vote online at **www.thomsonedu.com/biology/miller**.

tents). It is a managed tract with uniformly aged trees of one or two genetically uniform species that are harvested by clear-cutting as soon as they become commercially valuable. The land is then replanted and clear-cut again in a regular cycle (Figure 10-6).

Currently, about 63% of the world's forests are secondary-growth forests, 22% are old-growth forests,

Global Outlook: Extent of Deforestation (Science)

Human activities have reduced the earth's forest cover by as much as half.

Surveys by the World Resources Institute (WRI) indicate that over the past 8,000 years human activities have reduced the earth's original forest cover by as much as 50%, with an estimated 22% loss since the beginning of

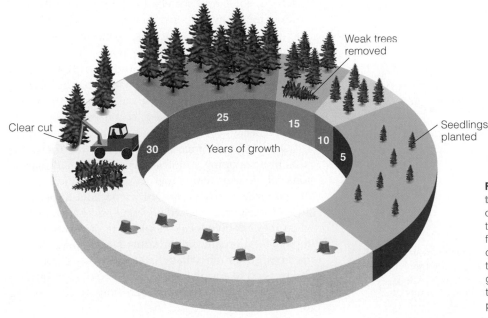

Clear cut

Weak trees removed

Seedlings planted

30 25 15 10 5

Years of growth

Figure 10-6 **Natural capital degradation:** Short (25- to 30-year) rotation cycle of cutting and regrowth of a monoculture tree plantation in modern industrial forestry. In tropical countries, where trees can grow more rapidly year-round, the rotation cycle can be 6–10 years. Old-growth or secondary forests are clear-cut to provide land for growing most tree plantations.

the twentieth century. Surveys by the UN Food and Agricultural Organization (FAO) and the WRI in 2005 indicate that the global rate of forest cover loss between 1990 and 2005 was between 0.2% and 0.5% per year, and that at least another 0.1–0.3% of the world's forests were degraded.

If these estimates are correct, the world's forests are being cleared or degraded exponentially at a rate of 0.3–0.8% per year, with much higher rates in some areas. These losses are concentrated in developing countries, especially those in the tropics. Since 1990, an area of forest about the size of the U.S. state of Kansas has been lost in developing countries each year—amounting to a 6% forest loss each decade. According to the WRI, if current deforestation rates continue, about 40% of the world's remaining intact forests will have been logged or converted to other uses within two decades, if not sooner.

Cutting down large areas of forests, especially old-growth forests, has important short-term economic benefits (Figure 10-4, right), but it also has a number of other harmful environmental effects (Figure 10-7). See Supplement 9 on p. S36 for details of an experiment showing the effects of deforestation on water loss and nutrient cycling.

 Learn more about how deforestation can affect the drainage of a watershed and disturb its ecosystem at ThomsonNOW.

Natural Capital Degradation

Deforestation

- Decreased soil fertility from erosion

- Runoff of eroded soil into aquatic systems

- Premature extinction of species with specialized niches

- Loss of habitat for native species and migratory species such as birds and butterflies

- Regional climate change from extensive clearing

- Release of CO_2 into atmosphere

- Acceleration of flooding

Figure 10-7 Natural capital degradation: harmful environmental effects of deforestation that can reduce biodiversity and the ecological services provided by forests (Figure 10-4, left). QUESTION: *What are the two direct and two indirect effects of your lifestyle on deforestation?*

There are also two encouraging trends. *First,* according to the 2005 Global Forest Resources Assessment by the FAO, the total area occupied by temperate forests in North America, Europe, and China increased slightly between 1990 and 2005 because of the spread of commercial tree plantations and natural reforestation from secondary ecological succession on cleared forest areas and abandoned croplands. *Second,* some of the cut areas in tropical forests have increased tree cover from regrowth and planting of tree plantations. But pressures on the world's remaining forests are increasing as exponentially increasing population, economic growth, and resource consumption per person are increasing the demand for paper, lumber, and fuelwood.

Case Study: Deforestation and the Fuelwood Crisis (Economics and Poverty)

Almost half the people in the developing world face a shortage of fuelwood and charcoal and this shortage is expected to grow.

About half of the wood harvested each year and three-fourths of that in developing countries is used for fuel. Fuelwood and charcoal made from fuelwood are used for heating and cooking for more than 2 million people in developing countries. This is leading to unsustainable cutting of trees in many areas. As the demand for fuelwood in urban areas exceeds the sustainable yield of nearby forests, expanding rings of deforested land encircle such cities. By 2050, the demand for fuelwood could easily be 50% greater than the amount that can be sustainably supplied.

Haiti, a country with 8.5 million people, was once a tropical paradise covered largely with forests. Now it is an ecological basket case because of the collapse of its forests, soils, and society. Largely because its trees were cut for fuelwood, only about 2% of its land is covered with trees. With the trees gone, its soils have eroded away and decreased the ability to grow crops. Like Easter Island (Core Case Study, p. 28), this is an example of how unsustainable use of natural capital can lead to a downward spiral of environmental degradation, poverty, disease, social injustice, crime, and violence.

One way to reduce the severity of the fuelwood crisis in developing countries is to plant small plantations of fast-growing fuelwood trees and shrubs around farms and in community woodlots. Experience shows that such *community forestry* projects work best when they involve local people in their planning and implementation and when farmers and villagers own the trees grown on village land. Tree plantations can also be used to supply nuts as a source of high-quality protein to help reduce malnutrition in developing countries.

A second way to reduce unsustainable harvesting of fuelwood trees is to burn wood more efficiently by providing villagers with cheap, more efficient, and less-polluting wood stoves or solar ovens and in the future electric hotplates powered by wind-generated electricity. This will also greatly reduce premature deaths from indoor air pollution from open fires and poorly designed stoves.

In addition, villagers can switch to burning the renewable sun-dried roots of various gourds and squash plants. Scientists are also looking for ways to produce charcoal used for heating and cooking without cutting down trees (Individuals Matter, at right).

Countries such as South Korea, China, Nepal, and Senegal have used such methods to reduce fuelwood shortages, sustain biodiversity by reforestation, and reduce soil erosion. Indeed, the mountainous country of South Korea is a global model for successful reforestation. When the war between North and South Korea ended in 1953, most of the country's mountains were deforested. Around 1960 the government launched a national reforestation program based on village cooperatives. Today forests cover almost two-thirds of the country and tree plantations near villages supply fuelwood on a sustainable basis. However, most countries suffering from fuelwood shortages are cutting trees for fuelwood and forest products 10–20 times faster than new trees are being planted.

Harvesting Trees (Science and Economics)

Trees can be harvested individually from diverse forests, or an entire forest stand can be cut down.

The first step in forest management is to build roads for access and timber removal. Even carefully designed logging roads have a number of harmful effects (Figure 10-8)—namely, increased erosion and sediment runoff into waterways, habitat fragmentation, and biodiversity loss. Logging roads also expose forests to invasion by nonnative pests, diseases, and wildlife species. And they open once inaccessible

Making Charcoal from Agricultural Wastes in Haiti (Science)

INDIVIDUALS MATTER

Amy Smith teaches at MIT in Cambridge, Massachusetts (USA). In 2004, she was given a McArthur "genius award" for her work in developing appropriate technologies for developing countries.

One of her projects involves Haiti, where about 98% of the country has been deforested, mostly because rural Haitians cook with charcoal made from wood. Smith is developing a way to make charcoal (which is essentially carbon) from the fibers that are left as a waste product (called bagasse) when the juice is squeezed from sugar cane

A benefit of this sugar-cane charcoal is that it burns cleaner than wood charcoal. In addition to saving trees, this new fuel can reduce lung diseases and premature deaths among rural Haitians breathing smoke from indoor cooking fires.

Smith says, "If you see a problem and there's something you can do about it, then you do something about it." And in designing technologies for use in developing countries, "the simpler you can make it, the better."

forests to farmers, miners, ranchers, hunters, and off-road vehicle users. In addition, logging roads on public lands in the United States disqualify the land for protection as wilderness.

Once loggers can reach a forest, they use various methods to harvest the trees (Figure 10-9, p. 198). With **selective cutting**, intermediate-aged or mature trees in an uneven-aged forest are cut singly or in small groups (Figure 10-9a). Selective cutting reduces crowding, removes diseased trees, encourages growth of younger trees, maintains a stand of trees of different species

Figure 10-8 Natural capital degradation: Building roads into previously inaccessible forests paves the way to fragmentation, destruction, and degradation.

and ages, and allows a forest to be used for multiple purposes. However, a form of selective cutting in which most or all of the largest trees are removed (a process called *creaming*) leads to environmental degradation and loss of biodiversity

Some tree species that grow best in full or moderate sunlight are all removed in a single *clear-cut* (Figures 10-9b and 10-10). Figure 10-11 lists the advantages and disadvantages of clear-cutting.

(a) Selective cutting

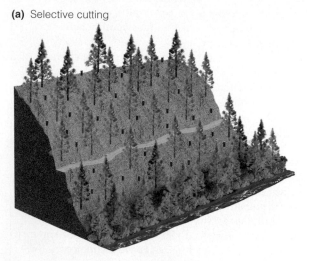

(b) Clear-cutting

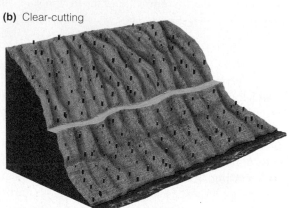

(c) Strip cutting

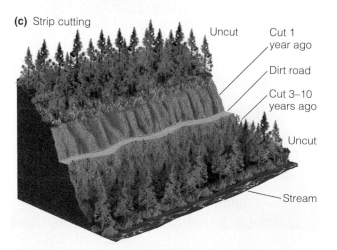

Uncut
Cut 1 year ago
Dirt road
Cut 3–10 years ago
Uncut
Stream

Figure 10-9 Major tree harvesting methods.

Figure 10-10 Natural capital degradation: clear-cut logging in Washington State (USA).

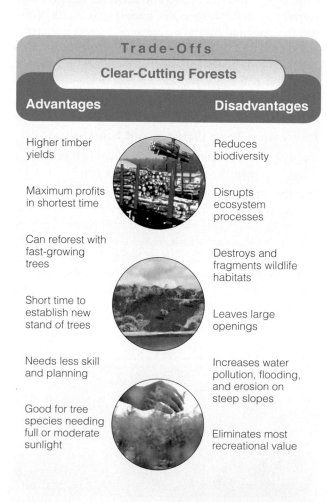

Trade-Offs	
Clear-Cutting Forests	
Advantages	**Disadvantages**
Higher timber yields	Reduces biodiversity
Maximum profits in shortest time	Disrupts ecosystem processes
Can reforest with fast-growing trees	Destroys and fragments wildlife habitats
Short time to establish new stand of trees	Leaves large openings
Needs less skill and planning	Increases water pollution, flooding, and erosion on steep slopes
Good for tree species needing full or moderate sunlight	Eliminates most recreational value

Figure 10-11 Trade-offs: advantages and disadvantages of clear-cutting forests. QUESTION: *Which single advantage and which single disadvantage do you think are the most important?*

- Identify and protect forest areas high in biodiversity

- Grow more timber on long rotations

- Rely more on selective cutting and strip cutting

- Stop clear-cutting on steep slopes

- Cease logging of old-growth forests

- Prohibit fragmentation of remaining large blocks of forest

- Sharply reduce road building into uncut forest areas

- Leave most standing dead trees and fallen timber for wildlife habitat and nutrient recycling

- Certify timber grown by sustainable methods

- Include ecological services of forests in estimating their economic value

- Plant tree plantations on deforested and degraded land

- Shift government subsidies from harvesting trees to planting trees

Figure 10-12 Solutions: ways to manage forests more sustainably. QUESTION: *Which three of these solutions do you think are the most important?*

A clear-cutting variation that can allow a more sustainable timber yield without widespread destruction is *strip-cutting* (Figure 10-9c). It involves clear-cutting a strip of trees along the contour of the land, with the corridor narrow enough to allow natural regeneration within a few years. After regeneration, loggers cut another strip above the first, and so on.

Solutions: Managing Forests More Sustainably (Science and Economics)

We can use forests more sustainably by emphasizing the economic value of their ecological services, harvesting trees no faster than they are replenished, and protecting old-growth and vulnerable areas.

Figure 10-12 lists ways that biodiversity researchers and foresters have suggested for managing the world's forests more sustainably. This includes certifying sustainably grown timber (see Solutions below). *Green Career:* Sustainable forestry

 THINKING ABOUT GRAY WOLVES AND MORE SUSTAINABLE FORESTS What is the connection between introducing gray wolves into the Yellowstone ecosystem (Core Case Study, p. 191) and the sustainability of some of its forests?

CASE STUDY: FOREST RESOURCES AND MANAGEMENT IN THE UNITED STATES

U.S. Forests: Encouraging News (Science, Economics, and Stewardship)

U.S. forests cover more area than they did in 1920, more wood is grown than cut, and the country has set aside large areas of protected forests.

Certifying Sustainably Grown Timber (Science and Stewardship)

SOLUTIONS

Collins Pine owns and manages a large area of productive timberland in northeastern California (USA). Since 1940, the company has used selective cutting to help maintain the ecological, economic, and social sustainability of its timberland.

Since 1993, Scientific Certification Systems (SCS) has evaluated the company's timber production. SCS, which is part of the nonprofit Forest Stewardship Council (FSC), was formed in 1993 to develop a list of environmentally sound practices for use in certifying timber and products made from certified timber.

Each year, SCS evaluates Collins Pine's landholdings to ensure that: cutting has not exceeded long-term forest regeneration; roads and harvesting systems have not caused unreasonable ecological damage; soils are not damaged; downed wood (boles) and standing dead trees (snags) are left to provide wildlife habitat; and the company is a good employer and a good steward of its land and water resources.

According to the FSC, between 1995 and 2005 the forested area in 65 countries that meets its international certification standards grew tenfold. The countries with the largest areas of FSC-certified forests are, in order, Sweden, Poland, the United States, and Canada. Despite this progress, by 2005 only about 6% of the world's forested area was certified.

Critical Thinking

Should governments provide tax breaks for timber that has been grown sustainably to encourage this practice? Explain.

Butterfly in a Redwood Tree (Stewardship)

"Butterfly" is the nickname given to Julia Hill. This young woman spent two years of her life on a small platform near the top of a giant redwood tree in California to protest the clear-cutting of a forest of these ancient trees, some of them more than 1,000 years old.

She and other protesters were illegally occupying these trees as a form of *nonviolent civil disobedience,* similar to that used decades ago by Mahatma Gandhi in his efforts to end the British occupation of India and by Martin Luther King in the U.S. civil rights movement.

Butterfly had never before participated in any environmental protest. She went to the site to express her belief that it was wrong to cut down these ancient giants for short-term economic gain. She planned to stay for only a few days. But after seeing the destruction and climbing one of these magnificent trees, she ended up staying in the tree for two years to publicize what was happening and to help save the surrounding trees. She became a symbol of the protest and during her stay used a cell phone to communicate with members of the mass media throughout the world to help develop public support for saving the trees.

Can you imagine spending two years of your life in a tree on a platform not much bigger than a king-sized bed, hovering 55 meters (180 feet) above the ground, and enduring high winds, intense rainstorms, snow, and ice? And Butterfly was not living in a quiet, pristine forest. All round her was noise from trucks, chainsaws, and helicopters trying to scare her into returning to the ground.

Although Butterfly lost her courageous battle to save the surrounding forest, she persuaded Pacific Lumber MAXXAM to save her tree (called Luna) and a 60-meter (200-foot) buffer zone around it. Not too long after she descended from her perch, someone used a chainsaw to seriously damage the tree. Cables and steel plates are now used to preserve it.

But maybe Butterfly and the earth did not lose. A book she wrote about her ethical stand, and her subsequent travels to campuses all over the world, have inspired a number of young people to stand up for protecting biodiversity and other environmental causes.

Butterfly led others by following in the tradition of Gandhi, who said, "My life is my message." Would you spend a day or a week of your life protesting something that you believed to be wrong?

Forests cover about 30% of U.S. land area, providing habitats for more than 80% of the country's wildlife species, and serve as sources for about two-thirds of the nation's total surface water.

Forests (including tree plantations) in the United States cover more area than they did in 1920. Many of the old-growth forests that were cleared or partially cleared between 1620 and 1960 have grown back naturally through secondary ecological succession as fairly diverse second-growth forests in every region of the United States, except much of the West. In 1995, environmental writer Bill McKibben cited forest regrowth in the United States—especially in the East—as "the great environmental story of the United States, and in some ways the whole world." And U.S. timber companies and conservation organizations are working together to protect large areas of forest land from development.

Every year more wood is grown in the United States than is cut, and each year the total area planted with trees increases. And protected forests make up about 40% of the country's total forest area, mostly in the *National Forest System,* which contains 155 national forests and 22 national grasslands, all managed by the U.S. Forest Service (USFS).

On the other hand, since the mid-1960s, an increasing area of the nation's remaining old-growth and fairly diverse second-growth forests has been clear-cut (Figure 10-10) and often replaced with biologically simplified tree plantations. According to biodiversity researchers, this reduces overall forest biodiversity and disrupts ecosystem processes such as energy flow and chemical cycling (see Supplement 9 on pp. S36–S37). Some environmentally concerned citizens have protested the cutting of ancient trees and forests (Individuals Matter, at left).

Types and Effects of Forest Fires (Science)

Depending on their intensity, fires can benefit or harm forests

Three types of fires can affect forest ecosystems. *Surface fires* (Figure 10-13, left) usually burn only undergrowth and leaf litter on the forest floor. They may kill seedlings and small trees but spare most mature trees and allow most wild animals to escape.

Occasional surface fires have a number of ecological benefits. They burn away flammable ground material and help prevent more destructive fires. They also release valuable mineral nutrients (tied up in slowly decomposing litter and undergrowth), release seeds from the cones of lodgepole pines, stimulate the germination of certain tree seeds (such as those of the giant sequoia and jack pine), and help control pathogens and insects. In addition, wildlife species such as deer, moose, elk, muskrat, woodcock, and quail depend on occasional surface fires to maintain their habitats and provide food in the form of vegetation that sprouts after fires.

Figure 10-13 Surface fires, such as this one in Tilton, Georgia (USA) (left), usually burn undergrowth and leaf litter on a forest floor. They can help prevent more destructive crown fires (right) by removing flammable ground material. They also recycle nutrients and thus help maintain the productivity of a variety of forest ecosystems. Sometimes carefully controlled surface fires are deliberately set to prevent buildup of flammable ground material in forests.

Some extremely hot fires, called *crown fires* (Figure 10-13, right), may start on the ground but eventually burn whole trees and leap from treetop to treetop. They usually occur in forests that have not experienced surface fires for several decades, where dead wood, leaves, and other flammable ground litter have accumulated. These rapidly burning fires can destroy most vegetation, kill wildlife, increase soil erosion, and burn or damage human structures in their paths.

Sometimes surface fires go underground and burn partially decayed leaves or peat. Such *ground fires* are common in northern peat bogs. They may smolder for days or weeks and are difficult to detect and extinguish. Forests can also be damaged by pests and diseases, as discussed in Supplement 15 on p. S49.

Solutions: Controversy over Fire Management (Science, Economics, and Politics)

To reduce fire damage, we can set controlled surface fires, allow fires on most public lands to burn unless they threaten human structures and lives, and clear small areas around buildings in areas subject to fire.

In the United States, the Smokey Bear educational campaign undertaken by the Forest Service and the National Advertising Council has prevented countless forest fires. It has also saved many lives and prevented billions of dollars in losses of trees, wildlife, and human structures.

At the same time, this educational program has convinced much of the public that all forest fires are bad and should be prevented or put out. Ecologists warn that trying to prevent all forest fires increases the likelihood of destructive crown fires by allowing accumulation of highly flammable underbrush and smaller trees in some forests.

According to the U.S. Forest Service, severe fires could threaten 40% of all federal forest lands, mainly through fuel buildup from past rigorous fire protection programs (the Smokey Bear era), increased logging in the 1980s that left behind highly flammable logging debris (called *slash*), and greater public use of federal forest lands.

Ecologists and forest fire experts have proposed several strategies for reducing fire-related harm to forests and people. One approach is to set small, contained surface fires or clear out (thin) flammable small trees and underbrush in the highest-risk forest areas. Such *prescribed fires* require careful planning and monitoring to keep them from getting out of control. In parts of fire-prone California near human settlements, local officials use herds of goats (kept in moveable pens) to eat away underbrush as an alternative to prescribed burns.

Another strategy is to allow many fires in public lands to burn and thereby remove flammable underbrush and smaller trees as long as the fires do not threaten human structures and life. A third approach is to protect houses or other buildings in fire-prone areas by thinning a zone of about 46 meters (200 feet) around them and eliminating the use of flammable materials such as wooden roofs.

In 2003, the U.S. Congress passed the *Healthy Forests Restoration Act*. Under this law, timber companies are allowed to cut down economically valuable medium-size and large trees in 71% of the total area of the national forests in return for clearing away smaller, more fire-prone trees and underbrush. This law also exempts most thinning projects from environmental reviews and appeals currently required by forest protection laws.

According to many biologists and forest fire scientists, this law is likely to *increase* the chances of severe forest fires for two reasons. *First,* removing the most fire-resistant large trees—the ones that are valuable to timber companies—encourages dense growth of highly flammable young trees and underbrush. *Second,* removing the large and medium trees leaves behind highly flammable slash. Many of the worst fires in U.S. history—including some of those during the 1990s—burned through cleared forest areas containing slash.

Fire scientists agree that some national forests need thinning to reduce the chances of catastrophic fires, but they believe such efforts should focus on two goals. One is to reduce ground-level fuel and vegetation in dry forest types and leave widely spaced medium and large trees that are the most fire resistant and thus can help forest recovery after a fire. These trees also provide critical wildlife habitat, especially as standing dead trees (snags) and logs where many animals live. The other goal would emphasize clearing of flammable vegetation around individual homes and buildings and near communities that are especially vulnerable to wildfire.

Critics of the Healthy Forests Restoration Act of 2003 say that these goals could be accomplished at a much lower cost to taxpayers by giving communities that seem especially vulnerable to wildfires grants to create a buffer around their communities to protect homes and buildings in their areas. They call for citizens to exert political pressure to repeal or drastically change this act as soon as possible before most of the valuable trees owned by the U.S. public are cut and fire hazards in some areas increase.

X *HOW WOULD YOU VOTE?* Do you support repealing or modifying the Healthy Forests Restoration Act of 2003? Cast your vote online at www.thomsonedu.com/biology/miller.

Controversy over Logging in U.S. National Forests (Science and Politics)

There has been an ongoing debate over whether U.S. national forests should be managed primarily for timber, their ecological services, recreation, or a mix of these uses.

For decades, controversy has swirled around the use of resources in the national forests. Timber companies have pushed to cut as much of the timber in these forests as possible at low prices. Biodiversity experts and many environmental scientists believe that national forests should be managed primarily to provide recreation and to sustain biodiversity, water resources, and other ecological services.

The U.S. Forest Service's timber-cutting program loses money because the revenues from timber sales do not cover the expenses of road building, timber sale preparation, administration, and other overhead costs. More than 644,000 kilometers (400,000 miles) of roads have been cut through the national forests at taxpayer's expense, primarily to facilitate logging. Because of such government subsidies, timber sales from U.S. federal lands have lost money for taxpayers in 97 of the last 100 years, including a loss of $6.7 billion between 1997 and 2004, according to a 2005 study by the Earth Island Institute.

Figure 10-14 lists the advantages and disadvantages of logging in national forests. According to a 2000 study by the accounting firm ECONorthwest, recreation, hunting, and fishing in national forests add ten times more money to the national economy and provide seven times more jobs than does extraction of timber and other resources.

X *HOW WOULD YOU VOTE?* Should Congress ban logging in U.S. national forests? Cast your vote online at www.thomsonedu.com/biology/miller.

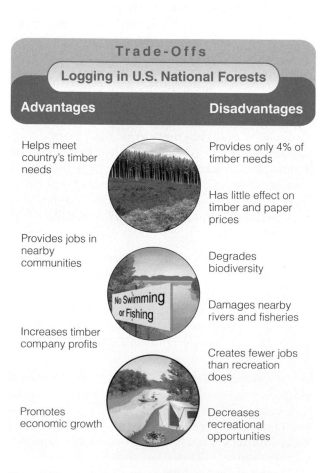

Figure 10-14 **Trade-offs:** advantages and disadvantages of allowing logging in U.S. national forests. QUESTION: *Which two advantages and which two disadvantages do you think are the most important?*

Solutions: Reducing Demand for Harvested Trees (Science)

Tree harvesting can be reduced by wasting less wood and making paper and charcoal fuel from fibers that do not come from trees.

One way to reduce the pressure to harvest trees on public and private land in the United States (and elsewhere) is to improve the efficiency of wood use. According to the Worldwatch Institute and forestry analysts, *up to 60% of the wood consumed in the United States is wasted unnecessarily.* This occurs because of inefficient use of construction materials, excess packaging, overuse of junk mail, inadequate paper recycling, and failure to reuse wooden shipping containers.

Only 4% of the total U.S. production of softwood timber comes from the national forests. Thus, reducing the waste of wood and paper products by only 4% could eliminate the need to remove any timber from national forests. This would allow these lands to be used primarily for recreation and biodiversity protection.

One way to reduce the pressure to harvest trees for paper production is to make more paper by using fiber that does not come from trees. China uses tree-free pulp from rice straw and other agricultural wastes to make almost two-thirds of its paper. Most of the small amount of tree-free paper produced in the United States is made from the fibers of a rapidly growing woody annual plant called *kenaf* (pronounced "kuh-NAHF"; Figure 10-15). Kenaf and many other non-tree fibers yield more paper pulp per hectare than tree farms and require fewer pesticides and herbicides.

American Forests in a Globalized Economy (Economics)

Timber from tree plantations in temperate and tropical countries is decreasing the need for timber production in the United States.

In today's global economy, timber and pulpwood for making paper can be produced rapidly and efficiently in the intensively managed tree plantations found in the temperate and tropical regions of the southern hemisphere.

As a result, the United States will likely play a decreasing role in the global production of wood-based products such as lumber and pulp. This trend could help preserve the nation's biodiversity by decreasing the pressure to clear-cut old-growth and second-growth forests on public and private lands.

Conversely, this shift in timber production to other countries could decrease biodiversity and watershed protection in the United States. Most forested land in the United States is privately owned, and private owners who can no longer make a profit by selling off some of their timber might be tempted to sell their land to housing developers.

Figure 10-15 Pressure to cut trees to make paper could be greatly reduced by planting and harvesting a fast-growing plant known as kenaf. According to the USDA kenaf is "the best option for tree-free papermaking in the United States" and could replace wood-based paper within 20–30 years.

The lower income from harvesting trees also means that both the government and private owners will have less money and fewer incentives for managing forests, forest restoration, and forest thinning projects to reduce damage from fire and insects. As a result, stewardship of public and private forests in the United States may decline unless the country overhauls its forest policy to meet this new reality and the challenges it presents.

CASE STUDY: TROPICAL DEFORESTATION

Global Outlook: Tropical Forests Are Disappearing (Science and Economics)

Large areas of ecologically and economically important tropical forests are being cleared and degraded at a fast rate.

Tropical forests cover about 6% of the earth's land area—roughly the area of the lower 48 U.S. states. Climatic and biological data suggest that mature tropical forests once covered at least twice as much area as they do today. Most of the destruction has occurred since

Figure 10-16 Natural capital degradation: satellite images of Amazon deforestation in the state of Rondônia, Brazil, between 1975 and 2001.

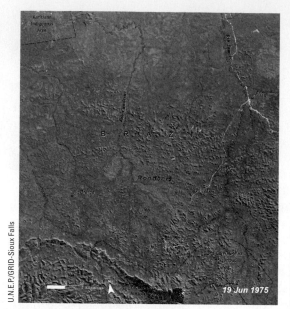

1950. Satellite scans and ground-level surveys indicate that large areas of tropical rainforests and tropical dry forests are being cut rapidly in parts of South America (especially Brazil), Africa, and Southeast Asia.

Studies indicate that at least half of the world's species of terrestrial plants and animals live in tropical rain forests. Because of their specialized niches, these species are highly vulnerable to extinction when their forest habitats are cleared or degraded. Each time a tract of tropical rain forest is cleared, several species—some with possible medical benefits to humans—may be lost forever.

Brazil has about 40% of the world's remaining tropical rain forest and an estimated 30% of the world's terrestrial plant and animal species in its vast Amazon basin, which is roughly twice the size of Europe. In 1970, only 1% of the Amazon basin had been deforested. By 2005, Brazil's government estimated that 16% had been deforested or degraded and converted mostly to scrub forest or tropical grassland (savanna)—an area roughly equal to the combined areas of Germany and Poland. Figure 10-16 shows deforestation in Brazil's state of Rondônia between 1975 and 2001. Between 2002 and 2005, the rate of deforestation increased sharply—mostly to make way for cattle ranching and large plantations for growing crops such as soybeans used for cattle feed.

In 2004, researchers at the Smithsonian Tropical Research Institute estimated that loggers, ranchers, and farmers in Brazil cleared and burned an area equivalent to a loss of 11 football fields a minute! The large-scale burning of the Amazon (Figure 10-17) accounts for three-fourths of Brazil's greenhouse gas emissions.

In 2004, Imazon, a leading Brazilian environmental group, said that satellite photos show that land oc-cupation and deforestation covers some 47% of Brazil's Amazon basin—almost three times the government estimate of 16%. This destruction is likely to accelerate because in 2000 Brazil announced plans to invest $40 billion in new highways and roads, railroads, gas lines, power lines, and hydroelectric dams

Figure 10-17 Natural capital degradation: each year large areas of tropical forest in Brazil's Amazon basin are burned to make way for cattle ranches and plantation crops. According to a 2003 study by NASA, the Amazon is slowly getting drier due to this practice. If this trend continues, it will prevent the restoration of forest by secondary ecological succession and convert a large area to tropical grasslands and scrub forest.

and reservoirs in the Amazon basin. Some hopeful news is that 23% of the Brazilian Amazon is listed as being protected. But many of these areas are protected on paper only.

Between 1970 and 2005, about 93% of Brazil's coastal rain forest that was once the size of Europe was cleared. This has caused a major loss of biodiversity because an area in this forest a little larger than two typical suburban house lots in the United States has 450 tree species! The entire United States has only about 865 native tree species.

Estimates of global tropical forest loss vary because of the difficulty of interpreting satellite images and different definitions of forest and deforestation. Also, some countries hide or exaggerate deforestation rates for political and economic reasons.

Because of these factors, estimates of global tropical forest loss vary from 50,000 square kilometers (19,300 square miles) to 170,000 square kilometers (65,600 square miles) per year. This rate is high enough to lose or degrade half of the world's remaining tropical forests in 35–117 years. A 2006 survey found that less than 5% of the world's tropical forest are managed sustainably.

F **RESEARCH FRONTIER** Get more exact figures on rates of deforestation

Why Should We Care about the Loss of Tropical Forests? (Science and Economics)

Cutting and degrading old-growth tropical forests reduces the important economic and ecological services they provide.

Most biologists believe that cutting and degrading most remaining old-growth tropical forests is a serious global environmental problem because of the important ecological and economic services they provide (Figure 10-4). For example, the World Health Organization estimates that 80% of the world's people rely on traditional medicines mostly derived from natural plants in forests.

Tropical forest plants also provide chemicals used as blueprints for making most of the world's prescription drugs (Figure 10-18). For example, each year anticancer drugs derived from chemicals in tropical plants save 30,000 lives in the United States and provide economic benefits of at least $350 billion.

The fast-growing neem tree shown in Figure 10-18 has a remarkable number of benefits. Villagers call the tree a "village pharmacy" because its chemicals can fight bacterial, viral, and fungal infections and its oil acts as a strong spermicide. Its leaves and seed contain natural pesticides that can repel or kill more than 200 insect species. People also use the tree's twigs as an

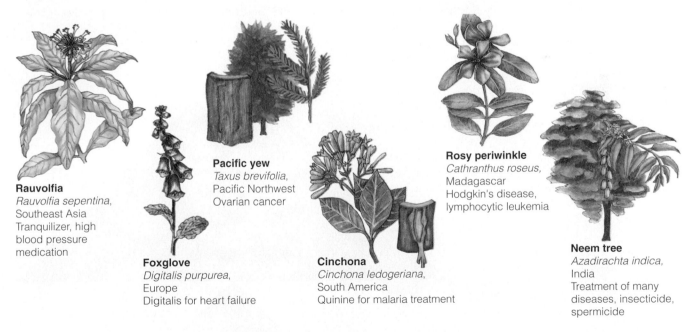

Rauvolfia
Rauvolfia sepentina,
Southeast Asia
Tranquilizer, high
blood pressure
medication

Pacific yew
Taxus brevifolia,
Pacific Northwest
Ovarian cancer

Foxglove
Digitalis purpurea,
Europe
Digitalis for heart failure

Cinchona
Cinchona ledogeriana,
South America
Quinine for malaria treatment

Rosy periwinkle
Cathranthus roseus,
Madagascar
Hodgkin's disease,
lymphocytic leukemia

Neem tree
Azadirachta indica,
India
Treatment of many
diseases, insecticide,
spermicide

Figure 10-18 Natural capital: *nature's pharmacy.* Parts of these and a number of other plants and animals (many of them found in tropical forests) are used to treat a variety of human ailments and diseases. Nine of the ten leading prescription drugs originally came from wild organisms. About 2,100 of the 3,000 plants identified by the National Cancer Institute as sources of cancer-fighting chemicals come from tropical forests. Despite their economic and health potential, fewer than 1% of the estimated 125,000 flowering plant species in tropical forests (and a mere 1,100 of the world's 260,000 known plant species) have been examined for their medicinal properties. Once the active ingredients in the plants have been identified, they can usually be produced synthetically. Many of these tropical plant species are likely to become extinct before we can study them.

antiseptic toothbrush and the oil from its seeds to make toothpaste and soap. And it grows well in poor soil in semiarid lands and provides fuelwood and lumber.

Causes of Tropical Deforestation and Degradation (Economics and Politics)

The primary causes of tropical deforestation and degradation are population growth, poverty, environmentally harmful government subsidies, debts owed to developed countries, and failure to value their ecological services.

Tropical deforestation results from a number of interconnected primary and secondary causes (Figure 10-19). Population growth and poverty combine to drive subsistence farmers and the landless poor to tropical forests, where they try to grow enough food to survive. Government subsidies can accelerate deforestation by reducing the costs of timber harvesting and cattle grazing.

Governments in Indonesia, Mexico, and Brazil encourage the poor to colonize tropical forests by giving them title to land they clear. This practice can help reduce poverty but can lead to environmental degradation unless the new settlers are taught how to use such forests more sustainably, which is rarely done. Mostly because of these government inducements, more than 500,000 families have settled in the Amazon Basin and more keep coming in a desperate struggle to survive.

In addition, international lending agencies encourage developing countries to borrow huge sums of money from developed countries to finance projects such as roads, mines, logging operations, oil drilling, and dams in tropical forests. Another cause is failure to value the ecological services of their forests (Figure 10-4, left).

The depletion and degradation of a tropical forest begin when a road is cut deep into the forest interior for logging and settlement (Figures 10-8 and 10-16). Loggers then use selective cutting to remove the best timber. This topples many other trees because of their shallow roots and the network of vines connecting trees in the forest's canopy. Although illegal exports of legally and illegally harvested timber to developed countries contribute significantly to tropical forest depletion and degradation, domestic use accounts for more than 80% of the trees cut in developing countries.

In many cases logging in tropical countries is done by foreign corporations operating under government concession contracts. Once a country's forests are gone, the companies move on to another country, leaving ecological devastation behind. For example, the Philippines and Nigeria have lost most of their once-abundant tropical hardwood forests and now are net importers of forest products. Several other tropical countries are following this unsustainable ecological and economic path.

After the best timber has been removed, timber companies often sell the land to ranchers. Within a few years, they typically overgraze it and sell it to settlers who have migrated to the forest hoping to grow enough food to survive. Then they move their land-degrading ranching operations to another forest area. After a few years of crop growing and rain erosion, the nutrient-poor tropical soil is depleted of nutrients. Then the settlers move on to newly cleared land to repeat this environmentally destructive process.

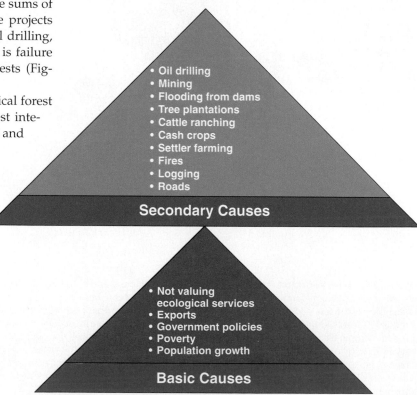

Figure 10-19 Natural capital degradation: major interconnected causes of the destruction and degradation of tropical forests. The importance of specific secondary causes varies in different parts of the world.

According to a 2005 report by the FAO, cattle ranching is the main cause of forest destruction in Brazil and other parts of Central and South America. Raising soybeans to feed Brazil's rapidly growing beef, poultry, and pork industries is also leading to forest destruction.

In some areas—especially Africa and Latin America—large sections of tropical forest are cleared for raising cash crops such as sugarcane, bananas, pineapples, strawberries, soybeans, palm oil for use as a biodiesel fuel, and coffee (see Supplement 14 on p. S48) mostly for export to developed countries. Tropical forests are also cleared for mining and oil drilling and to build dams on rivers that flood large areas of the forest.

Healthy rain forests do not burn. But increased burning (Figure 10-17), logging, settlements, grazing, and farming along roads built in these forests results in patchy fragments of forest (Figure 10-8, right and Figure 10-16) that dry out. In addition to destroying and degrading biodiversity, their combustion releases large amounts of carbon dioxide into the atmosphere.

A 2005 study by forest scientists found that widespread burning of tropical forest areas in the Amazon is changing weather patterns by raising temperatures and reducing rainfall. This is converting deforested areas into tropical grassland (savanna)—a process called *savannization.* Models project that if current burning and deforestation rates continue 20–30% of the Amazon will turn into savanna in the next fifty years.

Political influence and illegal logging also contribute to tropical deforestation. In Indonesia, for example, President Suharto awarded logging concessions covering more than half the country's forested area mostly to a small number of politically powerful companies. And a 2000 study estimated that 65% of the timber in Indonesia is cut illegally.

Solutions: Reducing Tropical Deforestation and Degradation (Science, Economics, and Politics)

There are a number of ways to slow and reduce the deforestation and degradation of tropical forests.

Analysts have suggested various ways to protect tropical forests and use them more sustainably (Figure 10-20).

One method is to help new settlers in tropical forests learn how to practice small-scale sustainable agriculture and forestry. Other methods are to sustainably harvest some of the renewable resources such as fruits and nuts in rain forests and to use strip-cutting (Figure 10-9c) to harvest tropical trees for lumber.

Debt-for-nature swaps can make it financially attractive for countries to protect their tropical forests. In such a swap, participating countries act as custodians of protected forest reserves in return for foreign aid or debt relief. In a similar newer strategy called *conservation concession,* nations are paid for concessions that preserve their resources. In these win-win approaches, the nation gets the money and also does not have to pay off its debts by selling off its resources.

Loggers can also use gentler methods for harvesting trees. For example, cutting canopy vines (lianas) before felling a tree and using the least obstructed paths to remove the logs can sharply

Solutions

Sustaining Tropical Forests

Prevention	Restoration

Protect the most diverse and endangered areas

Educate settlers about sustainable agriculture and forestry

Phase out subsidies that encourage unsustainable forest use

Add subsidies that encourage sustainable forest use

Protect forests with *debt-for-nature* swaps and *conservation easements*

Certify sustainably grown timber

Reduce illegal cutting

Reduce poverty

Slow population growth

Reforestation

Rehabilitate degraded areas

Concentrate farming and ranching on already-cleared areas

Figure 10-20 Solutions: ways to protect tropical forests and use them more sustainably. **QUESTION:** *Which three of these solutions do you think are the most important?*

Kenya's Green Belt Movement

INDIVIDUALS MATTER

Starting with a small tree nursery in her backyard, Wangari Maathai (Figure 10-A) founded the Green Belt Movement in Kenya.

The main goal of this highly regarded women's self-help group is to organize poor women in rural Kenya to plant and protect millions of trees to combat deforestation and provide fuelwood. By 2004, the 50,000 members of this grassroots group had established 6,000 village nurseries and planted and protected more than 30 million trees.

The women are paid a small amount for each seedling they plant that survives. This gives them an income to help break the cycle of poverty. And it improves the environment because trees reduce soil erosion and provide fuel, building materials, fruits, fodder for livestock, shade, and beauty. Such environmental improvement can reduce the distances women and children need to walk to get clean water and fuelwood for cooking and heating.

RADV SIGHETI/Reuters/Corbis

The success of this project has sparked the creation of similar programs in more than 30 other African countries.

This inspiring leader has said,

I don't really know why I care so much. I just have something inside me that tells me that there is a problem and I have got to do something about it. And I'm sure it's the same voice that is speaking to everyone on this planet, at least everybody who

Figure 10-A Wangari Maathai was the first Kenyan woman to earn a Ph.D. (in anatomy) and to head an academic department (veterinary medicine) at the University of Nairobi. In 1977, she organized the internationally acclaimed Green Belt Movement. For her work in protecting the environment, she has received many honors, including the Goldman Prize, the Right Livelihood Award, the UN Africa Prize for Leadership, and the 2004 Nobel Peace Prize. After years of being harassed, beaten, and jailed for opposing government policies, she was elected to Kenya's parliament as a member of the Green Party in 2002.

seems to be concerned about the fate of the world, the fate of this planet.

In 2004, she became the first African woman and first environmentalist to be awarded the Nobel peace prize for her lifelong efforts. Within an hour of learning that she had won the prize, Maathai planted a tree. In her speech accepting the award she urged everyone in the world to plant a tree as a symbol of commitment and hope.

---

reduce damage to neighboring trees. In addition, governments and individuals can mount efforts to reforest and rehabilitate degraded tropical forests and watersheds (Individuals Matter, above). Another suggestion is to clamp down on illegal logging.

MANAGING AND SUSTAINING GRASSLANDS

Rangelands and Overgrazing (Science)

Almost half of the world's livestock graze on natural grasslands (rangelands) and managed grasslands (pastures).

Grasslands provide many important ecological services. Examples are soil formation, erosion control, nutrient cycling, storage of atmospheric carbon, gene pools for crossbreeding grain crops, maintaining biodiversity, and providing habitats and food for a variety of organisms.

After forests, grasslands are the ecosystems most widely used and altered by human activities. **Rangelands** are unfenced grasslands in temperate and tropical climates that supply forage or vegetation for grazing (grass-eating) and browsing (shrub-eating) animals. Cattle, sheep, and goats graze on about 42% of the world's rangeland. The 2005 Millenium Ecosystem Assessment estimated that if we continue on our present course, that percentage will increase to 70% by 2050. Livestock also graze in **pastures:** managed grasslands or enclosed meadows usually planted with domesticated grasses or other forage.

Blades of rangeland grass grow from the base, not the tip. Thus, as long as only its upper half is eaten and its lower half remains, grass is a renewable resource that can be grazed again and again.

Moderate levels of grazing are healthy for grasslands because removal of mature vegetation stimulates rapid regrowth and encourages greater plant diversity. The key is to prevent both overgrazing and undergrazing by domesticated livestock and wild herbivores. **Overgrazing** occurs when too many animals graze for too long and exceed the carrying capacity of a grassland area (Figure 10-21, left). It reduces grass cover, exposes the soil to erosion by water and wind and compacts the soil (which diminishes its capacity to hold water). Overgrazing also enhances invasion of exposed land by species such as sagebrush, mesquite, cactus, and cheatgrass that cattle won't or can't eat. About 200 years ago, grass may have covered nearly half the land in the southwestern United States. Today, it covers only about 20% mostly because of a combination of prolonged droughts, fire, and overgrazing that created footholds for invader species that now cover many former grasslands.

Limited data from surveys in various countries by the FAO indicate that overgrazing by livestock has caused as much as a fifth of the world's rangeland to lose productivity. Some grassland can also suffer from **undergrazing,** where absence of grazing for long periods (at least 5 years) can reduce the net primary productivity of grassland vegetation and grass cover.

Solutions: Managing Rangelands More Sustainably (Science)

We can sustain rangeland productivity by controlling the number and distribution of livestock and by restoring degraded rangeland.

The most widely used method for more sustainable management of rangeland is to control the number of grazing animals and the duration of their grazing in a given area so the carrying capacity of the area is not exceeded. One widely used method is *rotational grazing* in which cattle are confined by portable fencing to

Figure 10-21 Natural capital degradation: overgrazed (left) and lightly grazed (right) rangeland.

one area for a short time (often only 1–2 days) and then moved to a new location.

Livestock tend to aggregate around natural water sources, especially thin strips of lush vegetation along streams or rivers known as *riparian zones* and ponds established to provide water for livestock. Overgrazing by cattle can destroy the vegetation in such areas (Figure 10-22, left). Protecting such land from further grazing by moving livestock around and by fencing off these areas can eventually lead to its natural ecological restoration (Figure 10-22, right). Ranchers can also move cattle around by providing supplemental feed at

Figure 10-22 Natural capital restoration: in the mid-1980s, cattle had degraded the vegetation and soil on this stream bank along the San Pedro River in the U.S. state of Arizona (left). Within 10 years, the area was restored through natural regeneration after banning grazing and off-road vehicles (right).

selected sites and by locating water holes and tanks and salt blocks in strategic places.

A more expensive and less widely used method of rangeland management is to suppress the growth of unwanted invader plants by herbicide spraying, mechanical removal, or controlled burning. A cheaper way to discourage unwanted vegetation in some areas is controlled, short-term trampling by large numbers of livestock.

Replanting barren areas with native grass seeds and applying fertilizer can increase growth of desirable vegetation and reduce soil erosion. But this is an expensive way to restore severely degraded rangeland.

Case Study: Grazing and Urban Development in the American West—Cows or Condos? (Science, Economics, and Politics)

Ecologists, environmentalists, and ranchers are working together to protect grasslands in the western United States from rapidly increasing economic development.

For decades some environmental scientists and environmentalists have pushed to reduce overgrazing on public and nonpublic lands and to reduce or eliminate livestock grazing permits on public lands.

But things are changing in ranch country because large numbers of people have moved to parts of the southwestern United States since 1980. Many ranchers have sold their ranches to developers who convert them to housing developments, condos, and small "ranchettes," creeping out from the edges of many southwestern cities and towns. Most people moving to the southwestern states value the landscape for its scenery and recreational opportunities, not for its traditional extractive industries such as cattle grazing, logging, oil and gas drilling, and mining of coal and minerals. But uncontrolled urban development can degrade the environmental quality they seek.

Now many ranchers, ecologists, and environmentalists are joining together to help preserve many of the cattle ranches as the best hope for helping sustain key remaining grasslands and the habitats they provide for many native species.

They are working together to identify areas that are best for sustainable grazing, areas best for sustainable urban development, and areas that should be neither grazed nor developed. They are forming land trust groups that pay ranchers with key grazing areas for *conservation easements*—restrictions on a deed that bar future owners from developing the land. And they are pressuring local governments to zone the land to prevent large-scale development in ecologically fragile rangeland areas.

Some ranchers are also reducing the harmful environmental impacts of their herds. They are rotating their cattle away from riparian areas (Figure 10-22), giving up most uses of fertilizers and pesticides, and consulting with range and wildlife scientists about ways to make their ranch operations more economically and ecologically sustainable.

NATIONAL PARKS

Global Outlook: Threats to National Parks (Science and Economics)

Countries have established more than 1,100 national parks, but most are threatened by human activities.

Today, more than 1,100 national parks larger than 10 square kilometers (4 square miles) each are located in more than 120 countries. (Examples are shown in Figure 5-14, top, p. 112; Figures 5-24 and 5-25, p. 122; and Figure 10-5.) However, according to a 1999 study by the World Bank and the Worldwide Fund for Nature only 1% of the parks in developing countries receive protection.

Local people invade most parks in search of wood, cropland, game animals, and other natural products for their daily survival. Loggers, miners, and wildlife poachers (who kill animals to obtain and sell items such as rhino horns, elephant tusks, and furs) also operate in many of these parks. Park services in developing countries typically have too little money and too few personnel to fight these invasions, either by force or by education.

Another problem is that most national parks are too small to sustain many large-animal species. Also, many parks suffer from invasions by nonnative species that can reduce the populations of native species and cause ecological disruption.

Case Study: Stresses on U.S. National Parks (Science, Economics, and Politics)

National parks in the United States face many threats.

The U.S. national park system, established in 1912, has 58 national parks, sometimes called the country's *crown jewels*. State, county, and city parks supplement these national parks. Most state parks are located near urban areas and have about twice as many visitors per year as the national parks.

Popularity is one of the biggest problems of many national and state parks in the United States. The Great Smoky Mountains National Park, for example, hosts about 9 million visitor a year. During the summer, users entering the most popular U.S. national and state parks often face hour-long backups and experience noise, congestion, eroded trails, and stress instead of peaceful solitude. In some parks and other public lands, noisy dirt bikes, dune buggies, snowmobiles, and other off-road vehicles degrade the aes-

Figure 10-23 Natural capital degradation: damage from off-road vehicles in a proposed wilderness area near Moab, Utah (USA). Such vehicles damage soils and vegetation, threaten wildlife, and degrade wetlands and streams.

Another problem is *inholdings* consisting of land in private ownership when a park was established. They can be an environmental threat to parks when owners develop hotels, mines, and gas and oil wells deep within some parks.

Many national parks have become threatened islands of biodiversity surrounded by a sea of commercial development. Nearby human activities that threaten wildlife and recreational values in many national parks include mining, logging, livestock grazing, coal-burning power plants, water diversion, and urban development.

Polluted air, drifting hundreds of kilometers, kills ancient trees in California's Sequoia National Park and often degrades the awesome views at Arizona's Grand Canyon. According to the National Park Service, air pollution affects scenic views in national parks more than 90% of the time. In addition, the U.S. General Accounting Office reports that the national parks need at least $6 billion for long overdue repair of trails, buildings, and other infrastructure.

Figure 10-24 lists suggestions that various analysts have made for sustaining and expanding the national park system in the United States. Some analysts also call for requiring private concessionaires who provide

thetic experience for many visitors, destroy or damage fragile vegetation (Figure 10-23), and disturb wildlife.

Many visitors expect parks to have grocery stores, laundries, bars, golf courses, video arcades, and other facilities found in urban areas. U.S. Park Service rangers spend an increasing amount of their time on law enforcement and crowd control instead of conservation management and education. Many overworked and underpaid rangers are leaving for better-paying jobs.

Many parks suffer damage from the migration or deliberate introduction of nonnative species. European wild boars (imported to North Carolina in 1912 for hunting) threaten vegetation in part of the Great Smoky Mountains National Park. Nonnative mountain goats in Washington's Olympic National Park (Figure 10-5) trample native vegetation and accelerate soil erosion. While some nonnative species have moved into parks, some economically valuable native species of animals and plants (including many threatened or endangered species) are killed or removed illegally in almost half of U.S. national parks. On the other hand, an important keystone species, the gray wolf, has been reintroduced into the Yellowstone National Park ecosystem (Core Case Study, p. 191).

[?] *THINKING ABOUT WOLVES AND YELLOWSTONE* Do you agree or disagree with the ecological decision to restore gray wolves to the Yellowstone ecosystem? Explain.

Solutions

National Parks

- Integrate plans for managing parks and nearby federal lands

- Add new parkland near threatened parks

- Buy private land inside parks

- Locate visitor parking outside parks and use shuttle buses for entering and touring heavily used parks

- Increase funds for park maintenance and repairs

- Survey wildlife in parks

- Raise entry fees for visitors and use funds for park management and maintenance

- Limit the number of visitors to crowded park areas

- Increase the number and pay of park rangers

- Encourage volunteers to give visitor lectures and tours

- Seek private donations for park maintenance and repairs

Figure 10-24 Solutions: suggestions for sustaining and expanding the national park system in the United States. QUESTION: *Which two of these solutions do you think are the most important?* (Data from Wilderness Society and National Parks and Conservation Association)

campgrounds, restaurants, hotels, and other services for park visitors to compete for contracts and pay franchise fees equal to 22% of their gross receipts. Currently, these fees average only about 6–7%. And many large concessionaires with long-term contracts pay as little as 0.75% of their gross receipts.

NATURE RESERVES

Protecting Land from Human Exploitation (Science, Economics, and Politics)

Ecologists call for protecting more land to help sustain biodiversity, but powerful economic and political interests oppose doing this.

Most ecologists and conservation biologists believe the best way to preserve biodiversity is to create a worldwide network of protected areas. Currently, 12% of the earth's land area (19% in Western Europe) is protected strictly or partially in nature reserves, parks, wildlife refuges, wilderness, and other areas.

The 12% figure is actually misleading because no more than 5% of the earth's land is strictly protected from potentially harmful human activities. See the map of our ecological footprints (Figures 3 and 4 on pp. S12–S15 in Supplement 4). In other words, *we have reserved 95% of the earth's land for us*, and most of the remaining area consists of ice, tundra, or desert where we do not want to live because it is too cold or too hot.

Conservation biologists call for full protection of at least 20% of the earth's land area in a global system of biodiversity reserves that includes multiple examples of all the earth's biomes. Doing this will require action and funding by national governments (Case Study, at right), private groups, and cooperative ventures involving governments, businesses, and private conservation groups.

Private groups play an important role in establishing wildlife refuges and other reserves to protect biological diversity. For example, since its founding by a group of professional ecologists in 1951, the *Nature Conservancy*—with more than 1 million members worldwide—has created the world's largest system of private natural areas and wildlife sanctuaries in 30 countries.

And eco-philanthrophists are using some of their wealth to buy up wilderness areas in South America and donating the preserved land to the governments of various countries. For example, Douglas and Kris Tompkins have created 11 wilderness parks in Latin America. And in 2005 they donated two new national parks to Chile and Argentina.

In the United States, private, nonprofit *land trust groups* have protected large areas of land. Members pool their financial resources and accept tax-de-

ductible donations to buy and protect farmlands, woodlands, and urban green spaces.

Most developers and resource extractors oppose protecting even the current 12% of the earth's remaining undisturbed ecosystems. They contend that these areas might contain valuable resources that would add to economic growth.

Ecologists and conservation biologists disagree. They view protected areas as islands of biodiversity that help sustain all life and economies, and serve as centers of future evolution. See Norman Myer's Guest Essay on this topic on the website for this chapter.

X *HOW WOULD YOU VOTE?* Should at least 20% of the earth's land area be strictly protected from economic development? Cast your vote online at **www.thomsonedu.com /biology/miller**.

Designing and Connecting Nature Reserves (Science)

Large and medium-sized reserves with buffer zones help protect biodiversity and can be connected by corridors.

Large reserves sustain more species and provide greater habitat diversity than do small reserves. They also minimize the area of outside edges exposed to natural disturbances (such as fires and hurricanes), invading species, and human disturbances from nearby developed areas.

However, research indicates that in some locales, several well-placed, medium-sized reserves may better protect a wider variety of habitats and preserve more biodiversity than a single large reserve of the same area. A mixture of large and medium-sized reserves (Figure 10-B) may be the best way to protect a variety of species and communities against a number of different threats.

Establishing protected *habitat corridors* between isolated reserves can help support more species and allow migration of vertebrates that need large ranges. They also permit migration of individuals and populations when environmental conditions in a reserve deteriorate and help preserve animals that must make seasonal migrations to obtain food. Corridors may also enable some species to shift their ranges if global climate change makes their current ranges uninhabitable.

On the other hand, corridors can threaten isolated populations by allowing movement of pest species, disease, fire, and exotic species between reserves. They also increase exposure of migrating species to natural predators, human hunters, and pollution. In addition, corridors can be costly to acquire, protect, and manage. Geographic information systems (GIS) technology can be used to help make decisions about where to locate corridors (Spotlight, p. 214).

Costa Rica—A Global Conservation Leader (Science, Politics, and Economics)

Tropical forests once completely covered Central America's Costa Rica, which is smaller in area than the U.S. state of West Virginia and about one-tenth the size of France. Between 1963 and 1983, politically powerful ranching families cleared much of the country's forests to graze cattle. They exported most of the beef produced to the United States and Western Europe.

Despite such widespread forest loss, tiny Costa Rica is a superpower of biodiversity, with an estimated 500,000 plant and animal species. A single park in Costa Rica is home to more bird species than all of North America.

In the mid-1970s, Costa Rica established a system of nature reserves and national parks that by 2004 included about a quarter of its land—6% of it in reserves for indigenous peoples. Costa Rica now devotes a larger proportion of land to biodiversity conservation than any other country.

The country's parks and reserves are consolidated into eight *megareserves* designed to sustain about 80% of Costa Rica's biodiversity (Figure 10-B). Each reserve contains a protected inner core surrounded by two buffer zones that local and indigenous people can use for sus-

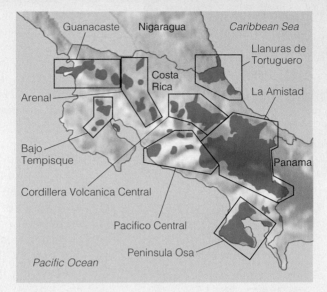

Figure 10-B
Solutions: Costa Rica has consolidated its parks and reserves into eight *megareserves* designed to sustain about 80% of the country's rich biodiversity.

tainable logging, food growing, cattle grazing, hunting, fishing, and ecotourism.

Costa Rica's biodiversity conservation strategy has paid off. Today, the country's largest source of income is its $1-billion-a-year tourism business—almost two-thirds of it from eco-tourists. *Green Career:* Ecotourism guide

To reduce deforestation, the government has eliminated subsidies for converting forest to cattle grazing land. It also pays landowners to maintain or restore tree coverage. The goal is to make sustaining forests profitable. The strategy has

worked: Costa Rica has gone from having one of the world's highest deforestation rates to one of the lowest.

Critical Thinking

At least 1 million tourists visit Costa Rica each year and stimulate the building of hotels, resorts, and other potentially harmful forms of development. How could this threaten some of the biodiversity the country is trying to protect? What two things would you do to prevent this from happening?

Whenever possible, conservation biologists call for using the *buffer zone concept* to design and manage nature reserves. This means protecting an inner core of a reserve by establishing two buffer zones in which local people can extract resources in ways that are sustainable and that do not harm the inner core. Doing this can involve local people as partners in protecting a reserve from unsustainable uses. The United Nations has used this principle in creating its global network of 425 biosphere reserves in 95 countries (Figure 10-25, p. 214).

So far, most biosphere reserves fall short of the ideal and receive too little funding for their protection and management. An international fund to help coun-

tries protect and manage biosphere reserves would cost about $100 million per year—about what the world's nations spend on weapons every 90 minutes.

Connecting Reserves on an Ecoregion Scale: Thinking Big (Science Economics, and Stewardship)

We can establish and connect nature reserves in a large ecoregion.

In 2006, conservation biologist Jonathan S. Adams wrote a book, *The Future of the Wild,* calling for establishing networks of corridors that connect the protected areas in a large *ecoregion.* He points out that the natural

Biosphere Reserve

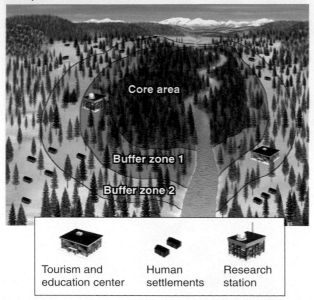

Core area

Buffer zone 1

Buffer zone 2

Tourism and education center

Human settlements

Research station

Figure 10-25 Solutions: a model *biosphere reserve*. Each reserve contains a protected inner core surrounded by two buffer zones that local and indigenous people can use for sustainable logging, food growing, cattle grazing, hunting, fishing, and ecotourism.

order of the wild is connections not fragmentation and that our goal should be to protect existing connections and restore broken ones.

Adams suggests that we do this in a four-step process. First, identify remaining wild areas in an ecoregion. Second, find out what nature tells us about how to protect and connect these wild areas. Third, look at how human development has disrupted key connections. Fourth, get scientists, conservationists, large private landowners, and officials of federal, state, and local land and wildlife management agencies together to devise a strategy for establishing a network of corridors to preserve existing connections between wild areas and to restore severed ones. GIS is a useful tool for making such decisions (Spotlight, below).

The goals of such *ecoregional conservation* are to identify what places should be protected and connected to one another, what places people should continue to use, and how to do this more sustainably by using land well for both people and other species. Accomplishing this over the next several decades will also help sustain biodiversity as climate change from global warming alters the nature and locations of ecosystems and reserves.

This process that some call *rewilding* is based on using a combination of sound conservation science and building the social capital (p. 25) needed to engage peo-

Using GIS Mapping To Understand and Manage Ecosystems (Science)

SPOTLIGHT

In recent decades, environmental planners and resources managers throughout the world have made increasing use of *geographic information systems* (GIS) to make maps that help them understand the components of various ecosystems and how to manage such systems. GIS is a software that makes maps from numerical or remote sensing data about various portions of the earth's surface. You may have used the GIS that produces online mapping services such as MapQuest to get driving directions.

When combined with data from remote sensing satellites GIS has become a powerful and useful technology for helping us use the earth's resources in more sustainable ways.

For example, suppose land-use planners and conservation biologists want to find the best place to establish a nature reserve or a corridor between two nature reserves.

They would get data on the plants and animals, topography (including elevation, forests, grassland, desert, and other features), geology (including soils), water resources, and forms of human development on the area of land involved. Then they would use GIS software to convert each of these sets of data into individual maps and produce a composite map that combines the data in the various layers of maps (Figure 3-33, p. 79). Looking at such a composite map can help them decide where a new nature reserve should and could be established and where to use a corridor to connect it to an existing nature reserve.

GIS is not infallible. As with any scientific modeling tool, GIS maps are no better than the data used to produce them. However, GIS has become an increasingly important tool for helping us manage, protect, and sustain the earth's resources. And GIS maps often spur frontier research by revealing the need for new or improved data. *Green Career:* Geographic Information Systems specialist

Critical Thinking

A developer could use GIS to design a housing development that has the least harmful environmental effects on an area and use such maps to help gain approval from local land-use planning officials. What three things would you include in such a development?

ple in making decisions about the future of the land that shapes their lives, as the people of Chattanooga, Tennessee did for their urban area (Case Study, p. 26).

F **RESEARCH FRONTIER** Learning how to design, locate, connect, and manage networks of more effective nature preserves connected by corridors on an ecoregion basis

In the final analysis, trying to decide how to use land well is an ethical issue. It requires that we see our use of land now and in the future as part of something larger than ourselves, as pointed out decades ago by Aldo Leopold (Individuals Matter, p. 23). Thinking and acting big in this way can give us a sense of the possible and a reason for hope. *Green Career:* Conservation biologist

Protecting Global Biodiversity Hot Spots (Science and Stewardship)

We can prevent or slow down losses of biodiversity by concentrating efforts on protecting global hot spots where significant biodiversity is under immediate threat.

The earth's species are not evenly distributed. Just a few countries, mostly those with large areas of tropical forests, contain most of the world's species. These 17 megadiversity countries contain more than two-thirds of all species. The leading megadiversity country is Brazil, followed by Indonesia, and Colombia.

In reality, few countries are physically, politically, or financially able to set aside and protect large biodiversity reserves. To protect as much of the earth's remaining biodiversity as possible, conservation biologists use an *emergency action* strategy that identifies and quickly protects *biodiversity hot spots,* found mostly in the megadiversity countries. These "ecological arks" are areas especially rich in plant and animal species that are found nowhere else and are in great danger of extinction or serious ecological disruption.

Figure 10-26 (p. 216) shows 34 such hot spots. They cover only a little over 2% of the earth's land surface but contain 52% of the world's plant species and 36% of all terrestrial vertebrates. And they are the only homes for more than one-third of the planet's known terrestrial plant and animal species. According to Norman Myers, "I can think of no other biodiversity initiative that could achieve so much at a comparatively small cost, as the hot spots strategy." Edward O. Wilson, one of the leading authorities on biodiversity, has described the hot spot approach as "the most important contribution to conservation biology of the last century."

 Learn more about hot spots around the world, what is at stake there, and how they are threatened at ThomsonNOW.

F **RESEARCH FRONTIER** Identifying and preserving all the world's terrestrial and aquatic biodiversity hot spots

Identifying and protecting hot spots is very important. But conservation biologists warn that this does little in the long run if we don't work to sustain the entire fabric of biodiversity throughout the world. Everything is connected.

Community-Based Conservation: Thinking and Acting Locally (Science, Economics, and Stewardship)

Conservation biologists are helping people in communities find ways to sustain local biodiversity while providing local economic income.

Many conservation biologists are working with people to help them protect biodiversity in their local communities. With this *community-based conservation* approach, scientists, citizens, and sometimes national and international conservation organizations work together. They seek ways to preserve local biodiversity while allowing people who live in or near protected areas to make sustainable use of some of the resources there.

For example, people learn how protecting local wildlife and ecosystems can help provide economic resources for their communities by encouraging sustainable forms of ecotourism. In the South American country of Belize, conservation biologist Robert Horwich has helped establish a local sanctuary for the black howler monkey. It involved getting local farmers to set aside strips of forest to serve as habitats and traveling corridors for these monkeys. The reserve, run by a local women's cooperative, has attracted ecotourists and biological scientists. The community has built a black howler museum and local residents receive income from housing and guiding visiting ecotourists and biological researchers. They are following the slogan to "Think globally, but act locally."

Natural Capital: Wilderness (Science and Stewardship)

Wilderness is land legally set aside in a large enough area to prevent or minimize harm from human activities.

One way to protect undeveloped lands from human exploitation is by legally setting them aside as undeveloped land called **wilderness.** Hikers and campers can visit such areas but they cannot stay. U.S. President Theodore Roosevelt summarized what we should do with wilderness: "Leave it as it is. You cannot improve it."

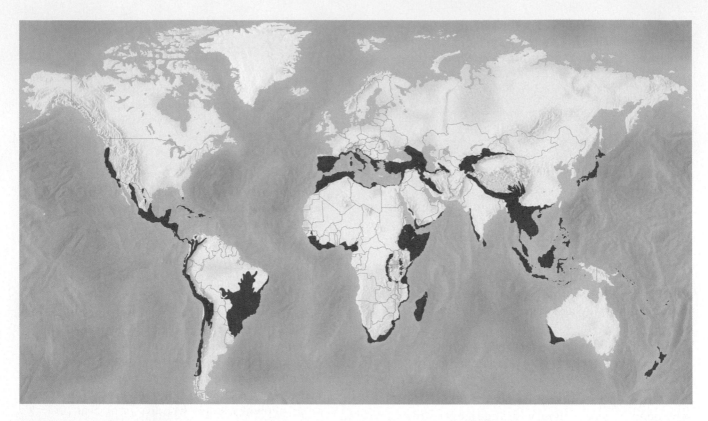

ThomsonNOW **Active Figure 10-26** **Endangered natural capital:** 34 *hot spots* identified by ecologists as important and endangered centers of biodiversity that contain a large number of endemic plant and animal species found nowhere else. Identifying and saving these critical habitats is a vital emergency response. QUESTION: *Are any of these hot spots near where you live? See an animation based on this figure and take a short quiz on the concept.* (Data from Conservation International)

The U.S. Wilderness Society estimates that a wilderness area should contain at least 4,000 square kilometers (1,500 square miles). Otherwise, it can be affected by air, water, and noise pollution from nearby human activities.

Wild places are areas where people can experience the beauty of nature and observe natural biological diversity (Figure 5-19, bottom photo, p. 116). They can also enhance the mental and physical health of visitors by allowing them to get away from noise, stress, development, and large numbers of people.

Wilderness preservationist John Muir advised us:

Climb the mountains and get their good tidings. Nature's peace will flow into you as the sunshine into the trees. The winds will blow their freshness into you, and the storms their energy, while cares will drop off like autumn leaves.

Even those who never use wilderness areas may want to know they are there, a feeling expressed by American novelist Wallace Stegner:

Save a piece of country . . . and it does not matter in the slightest that only a few people every year will go into it. This is precisely its value. . . . We simply need that wild country available to us, even if we never do more than drive to its edge and look in. For it can be a means of reassuring ourselves of our sanity as creatures, a part of the geography of hope.

Some critics oppose protecting wilderness for its scenic and recreational value for a small number of people. They believe this is an outmoded ideal that keeps some areas of the planet from being economically useful to humans.

To most biologists, the most important reasons for protecting wilderness and other areas from exploitation and degradation are to *preserve their biodiversity* as a vital part of the earth's natural capital and to *protect them as centers for evolution* in response to mostly unpredictable changes in environmental conditions. In other words, wilderness is a biodiversity and wildness bank and an eco-insurance policy.

Some analysts also believe wilderness should be preserved because the wild species it contains have a right to exist and play their roles in the earth's ongoing saga of biological evolution and ecological processes, without human interference.

Case Study: Controversy over Wilderness Protection in the United States (Science and Politics)

Only a small percentage of the land area of the United States has been protected as wilderness.

In the United States, conservationists have been trying to save wild areas from development since 1900. Overall, they have fought a losing battle. Not until 1964 did Congress pass the Wilderness Act. This important American idea allowed the government to protect undeveloped tracts of public land from development as part of the National Wilderness Preservation System.

The area of protected wilderness in the United States increased tenfold between 1970 and 2000. Even so, only about 4.6% of U.S. land is protected as wilderness—almost three-fourths of it in Alaska. Only 1.8% of the land area of the lower 48 states is protected, most of it in the West.

In other words, Americans have reserved 98% of the continental United States to be used as they see fit and have protected only 2% as wilderness. According to a 1999 study by the World Conservation Union, the United States ranks 42nd among nations in terms of terrestrial area protected as wilderness, and Canada is in 36th place.

In addition, only 4 of the 413 wilderness areas in the lower 48 U.S. states are larger than 4,000 square kilometers (1,500 square miles). Also, the system includes only 81 of the country's 233 distinct ecosystems. Most wilderness areas in the lower 48 states are threatened habitat islands in a sea of development.

Almost 400,000 square kilometers (150,000 square miles) in scattered blocks of public lands could qualify for designation as wilderness—about 60% of it in the national forests. For over 20 years, these areas have been temporarily protected under the Roadless Rule while they were evaluated for wilderness protection.

For decades, politically powerful oil, gas, mining, and timber industries have sought entry to these areas to develop resources for increased profits and short-term economic growth. Their efforts paid off in 2005 when the Bush administration ceased protecting roadless areas under consideration for classification as wilderness within the national forest system. And the Secretary of the Interior now allows states to claim old cow paths and off-road vehicle trails as roads that would disqualify an area from being protected as wilderness.

ECOLOGICAL RESTORATION

Rehabilitating and Restoring Damaged Ecosystems (Science and Stewardship)

Scientists have developed a number of ways to rehabilitate and restore degraded ecosystems and create artificial ecosystems.

Almost every natural place on the earth has been affected or degraded to some degree by human activities. However, much of the harm we have inflicted on nature is at least partially reversible through **ecological restoration**: the process of repairing damage caused by humans to the biodiversity and dynamics of natural ecosystems. Examples include replanting forests, restoring grasslands, restoring wetlands and stream banks, reclaiming urban industrial areas (brownfields), reintroducing native species (Core Case Study, p. 191), removing invasive species, and freeing river flows by removing dams.

Farmer and philosopher Wendell Berry says we should try to answer three questions in deciding whether and how to modify or rehabilitate a natural ecosystem. *First,* what is here? *Second,* what will nature permit us to do here? *Third,* what will nature help us do here? An important strategy is to mimic nature and natural processes and ideally let nature do most of the work, usually through secondary ecological succession.

By studying how natural ecosystems recover, scientists are learning how to speed up repair operations using a variety of approaches. They include the following measures:

- *Restoration:* trying to return a particular degraded habitat or ecosystem to a condition as similar as possible to its natural state.

- *Rehabilitation:* attempting to turn a degraded ecosystem back into a functional or useful ecosystem without trying to restore it to its original condition. Examples include removing pollutants and replanting areas such as mining sites, landfills, and clear-cut forests to reduce soil erosion.

- *Replacement:* replacing a degraded ecosystem with another type of ecosystem. For example, a productive pasture or tree farm may replace a degraded forest.

- *Creating artificial ecosystems:* an example is the creation of artificial wetlands to help reduce flooding and to treat sewage.

Researchers have suggested five basic science-based principles for carrying out most forms of ecological restoration and rehabilitation.

- Identify what caused the degradation (such as pollution, farming, overgrazing, mining, or invading species).

- Stop the abuse by eliminating or sharply reducing these factors. Examples include removing toxic soil pollutants, adding nutrients to depleted soil, adding new topsoil, preventing fires, and controlling or eliminating disruptive nonnative species (Case Study, below).

- If necessary, reintroduce species, especially pioneer, keystone, and foundation species, to help restore natural ecological processes, as was done with wolves in the Yellowstone area (Core Case Study, p. 191).

- Protect the area from further degradation (Figure 10-22, right).

- Use adaptive management to monitor restoration efforts, assess successes, and modify strategies as needed. *Green Career:* Restoration ecology specialist

Most of the tall-grass prairies in the United States have been plowed up and converted to crop fields. However, these prairies are ideal subjects for ecological restoration for three reasons. *First,* many residual or transplanted native plant species can be established within a few years. *Second,* the technology involved is similar to that in gardening and agriculture. *Third,* the process is well suited for volunteer labor needed to plant native species and weed out invading species until the natural species can take over. There are a number of prairie restoration projects in the United States.

Private enterprise is getting into ecological restoration. In May 2000, the Australian Stock Exchange listed an Australian firm called Earth Sanctuaries, Ltd. This firm buys degraded land, restores it, and earns income from ecotourism and consulting on ecosystem assessment and ecological restoration.

F **RESEARCH FRONTIER** Exploring ways to improve ecological restoration efforts

---

CASE STUDY

Ecological Restoration of a Tropical Dry Forest in Costa Rica (Science and Stewardship)

Costa Rica is the site of one of the world's largest *ecological restoration* projects. In the lowlands of its Guanacaste National Park (Figure 10-B), a small tropical dry deciduous forest has been burned, degraded, and fragmented by large-scale conversion to cattle ranches and farms.

Now it is being restored and relinked to the rain forest on adjacent mountain slopes. The goal is to eliminate damaging nonnative grass and reestablish a tropical dry forest ecosystem over the next 100–300 years.

Daniel Janzen, professor of biology at the University of Pennsylvania and a leader in the field of restoration ecology, has helped galvanize international support for this restoration project. He used his own MacArthur grant money to purchase this Costa Rican land to be set aside as a national park. And he has raised more than $10 million for restoring the park.

He realized that large native animals that ate the fruit of the Guan-

caste tree and spread its seeds in their droppings maintained the original forests. But these animals disappeared about 10,000 years ago. About 500 years ago, horses and cattle introduced by Europeans spread the seeds but farming and ranching took its toll on the forest's trees. Janzen decided to speed up restoration of this tropical dry forest by incorporating horses as seed dispersers in his recovery plan.

He recognizes that ecological restoration and protection of the park will fail unless the people in the surrounding area believe they will benefit from such efforts. Janzen's vision is to make the nearly 40,000 people who live near the park an essential part of the restoration of the degraded forest, a concept he calls *biocultural restoration.*

By actively participating in the project, local residents reap educational, economic, and environmental benefits. Local farmers make money by sowing large areas with tree seeds and planting seedlings started in Janzen's lab. Local grade school, high school, and university students and citizens' groups study

the park's ecology and visit it on field trips. The park's location near the Pan American Highway makes it an ideal area for ecotourism, which stimulates the local economy.

The project also serves as a training ground in tropical forest restoration for scientists from all over the world. Research scientists working on the project give guest classroom lectures and lead some of the field trips.

In a few decades, today's children will be running the park and the local political system. If they understand the ecological importance of their local environment, they are more likely to protect and sustain its biological resources. Janzen believes that education, awareness, and involvement—not guards and fences—are the best ways to restore degraded ecosystems and protect largely intact ecosystems from unsustainable use.

Critical Thinking

Would such an ecological restoration project be possible in the area where you live? Explain.

Will Restoration Encourage Further Destruction? (Science and Stewardship)

There is some concern that ecological restoration could promote further environmental destruction and degradation.

Some analysts worry that ecological restoration could encourage continuing environmental destruction and degradation by suggesting that any ecological harm we do can be undone. Ecologists agree that preventing ecosystem damage in the first place is cheaper and more effective than any form of ecological restoration.

Restoration scientists agree that restoration should not be used as an excuse for environmental destruction. But they point out that so far we have been able to protect or preserve no more than about 5% of the earth's land from the effects of human activities. So ecological restoration is badly needed for many of the world's ecosystems that we have already damaged. They also point out that if a restored ecosystem differs from the original system, it is better than nothing, and that increased experience will improve the effectiveness of ecological restoration. Chapter 12 describes examples of the ecological restoration of aquatic systems such as wetlands and rivers.

X *HOW WOULD YOU VOTE?* Should we mount a massive effort to restore ecosystems we have degraded even though this will be quite costly? Cast your vote online at www.thomsonedu.com/biology/miller.

WHAT CAN WE DO?

Solutions: Establishing Priorities (Science)

Biodiversity expert Edward O. Wilson has proposed eight priorities for protecting most of the world's remaining ecosystems and species.

In 2002, Edward O. Wilson, considered to be one of the world's foremost experts on biodiversity, proposed the following priorities for protecting most of the world's remaining ecosystems and species:

■ Take immediate action to preserve the world's biological hot spots (Figure 10-26).

■ Keep intact the world's remaining old-growth forests and cease all logging of such forests.

■ Complete the mapping of the world's terrestrial and aquatic biodiversity so we know what we have and can make conservation efforts more precise and cost-effective. An important step towards this goal is the *Systematics Agenda 2000*, a carefully designed effort to inventory the world's biodiversity. And in 2005, the National Ecological Observatory Network (NEON) launched a plan to have 17 networks across the United States identify and monitor native and nonnative species and the functioning of ecosystems throughout the country.

■ Determine the world's marine hot spots and assign them the same priority for immediate action as for those on land.

■ Concentrate on protecting and restoring the world's lakes and river systems, which are the most threatened ecosystems of all.

■ Ensure that the full range of the earth's terrestrial and aquatic ecosystems is included in a global conservation strategy.

■ Make conservation profitable. This involves finding ways to raise the income of people who live in or near nature reserves so they can become partners in their protection and sustainable use.

■ Initiate ecological restoration projects worldwide to heal some of the damage we have done and increase the share of the earth's land and water allotted to the rest of nature.

According to Wilson, such a conservation strategy would cost about $30 billion per year—an amount that could be provided by a tax of a penny per cup of coffee.

This strategy for protecting the earth's precious biodiversity will not be implemented without bottom-up political pressure on elected officials from individual citizens and groups. It will also require cooperation among key people in government, the private sector, science, and engineering.

Figure 10-27 lists some ways you can help sustain the earth's terrestrial biodiversity.

What Can You Do?
Sustaining Terrestrial Biodiversity

• Adopt a forest.

• Plant trees and take care of them.

• Recycle paper and buy recycled paper products.

• Buy sustainable wood and wood products.

• Choose wood substitutes such as bamboo furniture and recycled plastic outdoor furniture, decking, and fencing.

• Restore a nearby degraded forest or grassland.

• Landscape your yard with a diversity of plants natural to the area.

• Live in town because suburban sprawl reduces biodiversity.

Figure 10-27 Individuals matter: ways to help sustain terrestrial biodiversity.

Revisiting Wolves and Sustainability

In this chapter we have seen how terrestrial biodiversity is being destroyed and degraded and how we can reduce this by using forests and grasslands more sustainably and by protecting species and ecosystems in parks, wilderness, and other nature reserves.

We have also learned the importance of restoring or rehabilitating some of the ecosystems we have degraded. Reintroducing keystone species such as the gray wolf into ecosystems they once inhabited (Core Case Study, p. 191) is a form of ecological restoration that can reestablish some of the ecological functions and interactions in such systems.

Preserving terrestrial biodiversity involves applying the four scientific principles of sustainability. This means not disrupting the flows of energy from the sun through food webs, the cycling of nutrients in ecosystems, and the species in food webs that help prevent excessive population growth of various species. It also means not prematurely causing the extinction of species and not destroying and degrading critical wildlife habitats.

Shortsighted men ... in their greed and selfishness will, if permitted, rob our country of half its charm by their reckless extermination of all useful and beautiful wild things.

THEODORE ROOSEVELT

CRITICAL THINKING

 1. Do you support the program that reintroduced populations of the gray wolf in the Yellowstone ecosystem in the United States (Core Case Study, p. 191)? Explain. Another keystone species in the Yellowstone ecosystem is the grizzly bear. Would you support a program to reintroduce this species into this ecosystem? Explain.

2. Explain why you agree or disagree with each of the proposals for providing more sustainable use of forests throughout the world, listed in Figure 10-12, p. 199.

3. Should there be severe limitations on the use of off-road motorized vehicles (Figure 10-23, p. 211) and snowmobiles on all public lands in the United States or in the country where you live? Explain.

4. In 2006, Lester R. Brown estimated that reforesting the earth and restoring the earth's degraded rangelands would cost about $15 billion a year. Suppose the United States, as the world's most affluent country, agreed to put up half this money, at an average annual cost of $25 per American. Would you support doing this? Explain. What other part or parts of the federal budget would you decrease to come up with these funds?

5. Should developed countries provide most of the money to help preserve remaining tropical forests in developing countries? Explain.

6. In the early 1990s, Miguel Sanchez, a subsistence farmer in Costa Rica, was offered $600,000 by a hotel developer for a piece of land that he and his family had been using sustainably for many years. The land contained an old-growth rain forest and a black sand beach in an area under rapid development. Sanchez refused the offer. What would you have done if you were a poor subsistence farmer in Miguel Sanchez's position? Explain your decision.

7. Are you in favor of establishing more wilderness areas in the United States, especially in the lower 48 states (or in the country where you live)? Explain. What might be some drawbacks of doing this?

8. If ecosystems are undergoing constant change, why should we **(a)** establish and protect nature reserves and **(b)** carry out ecological restoration?

9. Congratulations! You are in charge of the world. List the three most important features of your policies for using and managing **(a)** forests, **(b)** grasslands, and **(c)** nature reserves such as parks and wildlife refuges.

10. List two questions that you would like to have answered as a result of reading this chapter.

PROJECTS

1. Obtain a topographic map of the region where you live and use it to identify federal-, state-, and local-owned lands designated as parks, rangeland, forests, and wilderness areas. Identify the government agency or agencies responsible for managing each of these areas, and try to evaluate how well these agencies are preserving the natural resources on this public land on your behalf.

2. If possible, try to visit **(a)** a diverse old-growth forest, **(b)** an area that has been recently clear-cut, and **(c)** an area that was clear-cut 5–10 years ago. Compare the biodiversity, soil erosion, and signs of rapid water runoff in each of the three areas.

3. For many decades, New Zealand has had a policy of meeting all its demand for wood and wood products by growing timber on intensively managed tree plantations. Use the library or Internet to evaluate the effectiveness of this approach and its major advantages and disadvantages.

4. Use the library or Internet to find one example of a successful ecological restoration project not discussed in this chapter and one that failed. For your example, describe the strategy used, the ecological principles involved, and why the project succeeded or failed.

5. Make a concept map of this chapter's major ideas, using the section heads, subheads, and key terms (in boldface). Look on the website for this book for information about making concept maps.

LEARNING ONLINE

The website for this book contains study aids and many ideas for further reading and research. They include a chapter summary, review questions for the entire chapter, flash cards for key terms and concepts, a multiple-choice practice quiz, interesting Internet sites, references, information about green careers, and a guide for accessing thousands of InfoTrac® College Edition articles. Log into

www.thomsonedu.com/biology/miller

Then choose Chapter 10, and select a learning resource. For access to animations, additional quizzes, chapter outlines and summaries, register and log into

at **www.thomsonedu.com** using the access code card in the front of your book.

Active Graphing

Log into ThomsonNow at www.thomsonedu.com to explore the graphing exercise for this chapter.

11 Sustaining Biodiversity: The Species Approach

CORE CASE STUDY

The Passenger Pigeon: Gone Forever

In 1813, bird expert John James Audubon saw a single huge flock of passenger pigeons that took three days to fly past him and was so dense that it darkened the skies.

By 1900, North America's passenger pigeon (Figure 11-1), once the most numerous bird on earth, had

Figure 11-1 Lost natural capital: passenger pigeons have been extinct in the wild since 1900 because of human activities. The last known passenger pigeon died in the U.S. state of Ohio's Cincinnati Zoo in 1914.

Michael Sewell/Peter Arnold, Inc.

disappeared from the wild because of a combination of uncontrolled commercial hunting and habitat loss as forests were cleared to make room for farms and cities. These birds were good to eat, their feathers made good pillows, and their bones were widely used for fertilizer. They were easy to kill because they flew in gigantic flocks and nested in long, narrow, densely packed colonies.

Commercial hunters would capture one pigeon alive, sew its eyes shut, and tie it to a perch called a stool. Soon a curious flock would land beside this "stool pigeon"—a term we now use to describe someone who turns in another person for breaking the law. Then the birds would be shot or ensnared by nets that might trap more than 1,000 birds at once.

Beginning in 1858, passenger pigeon hunting became a big business. Shotguns, traps, artillery, and even dynamite were used. People burned grass or sulfur below their roosts to suffocate the birds. Shooting galleries used live birds as targets. In 1878, one professional pigeon trapper made $60,000 by killing 3 million birds at their nesting grounds near Petoskey, Michigan.

By the early 1880s, only a few thousand birds remained. At that point, recovery of the species was doomed because the females laid only one egg per nest each year. On March 24, 1900, a young boy in the U.S. state of Ohio shot the last known wild passenger pigeon.

Eventually all species become extinct or evolve into new species. The archeological record reveals five mass extinctions since life on the earth began (Figure 4-12, p. 93)—each a massive impoverishment of life on the earth. There is considerable evidence that we are now in the early stage of a sixth great extinction. Previous mass extinctions were caused by natural phenomena that drastically changed the earth's environmental conditions. We are causing this one as the human population grows, consumes more resources, disturbs more of the earth's land and aquatic systems, and uses more of the earth's net primary productivity that supports all species. If we keep impoverishing the earth's biodiversity, eventually our species will also be impoverished. And there is no place we can escape to.

The last word in ignorance is the person who says of an animal or plant: "What good is it?". . . . If the land mechanism as a whole is good, then every part of it is good, whether we understand it or not. . . . Harmony with land is like harmony with a friend; you cannot cherish his right hand and chop off his left.

ALDO LEOPOLD

This chapter looks at the problem of premature extinction of species by human activities and ways to reduce this threat to the world's biodiversity. It addresses the following questions:

- How do biologists estimate extinction rates, and how do human activities affect these rates?

- Why should we care about protecting wild species?

- Which human activities endanger wildlife?

- How can we help prevent premature extinction of species?

- What is reconciliation ecology, and how can it help prevent premature extinction of species?

SPECIES EXTINCTION

Three Types of Species Extinction (Science)

Species can become extinct locally, ecologically, or globally.

Biologists distinguish among three levels of species extinction. *Local extinction* occurs when a species is no longer found in an area it once inhabited but is still found elsewhere in the world. Most local extinctions involve losses of one or more populations of species.

Ecological extinction occurs when so few members of a species are left that it can no longer play its ecological roles in the biological communities where it is found.

In *biological extinction*, a species is no longer found anywhere on the earth (Figure 11-2 and Core Case Study, p. 222). Biological extinction is forever.

Endangered and Threatened Species— Ecological Smoke Alarms (Science)

An endangered species could soon become extinct, and a threatened species is likely to become extinct.

Biologists classify species heading toward biological extinction as either *endangered* or *threatened* (Figure 11-3, p. 224). An **endangered species** has so few individual survivors that the species could soon become extinct over all or most of its natural range. Like the passenger pigeon (Figure 11-1), they may soon disappear from the earth. A **threatened species** (also known as a *vulnerable species*) is still abundant in its natural range but because of declining numbers it is likely to become endangered in the near future.

Some species have characteristics that make them especially vulnerable to ecological and biological extinction (Figure 11-4, p. 225). As biodiversity expert Edward O. Wilson puts it, "The first animal species to go are the big, the slow, the tasty, and those with valuable parts such as tusks and skins."

Some species also have *behavioral characteristics* that make them prone to extinction. The passenger pigeon and the Carolina parakeet nested in large flocks. Key deer are "nicotine addicts" that get killed by cars because they forage for cigarette butts along highways.

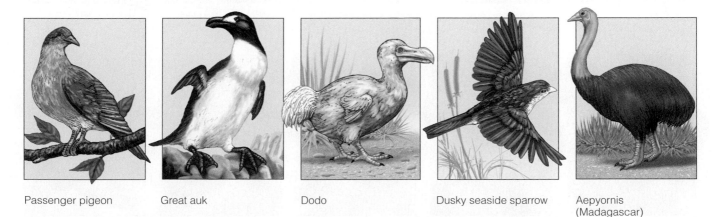

Passenger pigeon   Great auk   Dodo   Dusky seaside sparrow   Aepyornis (Madagascar)

Figure 11-2 Lost natural capital: some animal species that have become prematurely extinct largely because of human activities, mostly habitat destruction and overhunting. The Great Auk became extinct in 1844 from overhunting because of its willingness to march up the boardwalks to ships. QUESTION: *Why do you think birds top this list?*

Grizzly bear   Kirkland's warbler   Knowlton cactus   Florida manatee   African elephant

Utah prairie dog   Swallowtail butterfly   Humpback chub   Golden lion tamarin   Siberian tiger

Giant panda   Black-footed ferret   Whooping crane   Northern spotted owl   Blue whale

Mountain gorilla   Florida panther   California condor   Hawksbill sea turtle   Black rhinoceros

Figure 11-3 Endangered natural capital: species that are endangered or threatened with premature extinction largely because of human activities. Almost 30,000 of the world's species and 1,260 of those in the United States are officially listed as being in danger of becoming extinct. Most biologists believe the actual number of species at risk is much larger.

Characteristic	Examples
Low reproductive rate (K-strategist)	Blue whale, giant panda, rhinoceros
Specialized niche	Blue whale, giant panda, Everglades kite
Narrow distribution	Many island species, elephant seal, desert pupfish
Feeds at high trophic level	Bengal tiger, bald eagle, grizzly bear
Fixed migratory patterns	Blue whale, whooping crane, sea turtles
Rare	Many island species, African violet, some orchids
Commercially valuable	Snow leopard, tiger, elephant, rhinoceros, rare plants and birds
Large territories	California condor, grizzly bear, Florida panther

Figure 11-4 Natural capital loss and degradation: characteristics of species that are prone to ecological and biological extinction. QUESTION: *Which of these characteristics helped lead to the premature extinction of the passenger pigeon within a single human lifetime?*

One study in 2000 found that human activities threaten several types of species with premature extinction (Figure 11-5). Another 2000 survey by the Nature Conservancy and the Association for Biodiversity Information found that about one-third of the 21,000 plant and animal species in the United States are vulnerable to premature extinction.

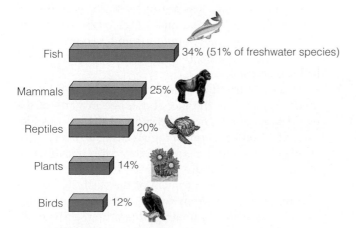

Fish — 34% (51% of freshwater species)
Mammals — 25%
Reptiles — 20%
Plants — 14%
Birds — 12%

Figure 11-5 Endangered natural capital: percentage of various types of species threatened with premature extinction because of human activities. QUESTION: *Why do you think fish top this list?* (Data from World Conservation Union, Conservation International, and World Wildlife Fund)

Estimating Extinction Rates Is a Tough Job (Science)

Scientists use measurements and models to estimate extinction rates.

Biologists trying to catalog extinctions have three problems. *First,* the extinction of a species typically takes such a long time that it is not easy to document. *Second,* scientists have identified only about 1.4 million of the world's 4 million to 100 million species. *Third,* scientists know little about most of the species that have been identified.

Scientists do the best they can with the tools they have to estimate past and projected future extinction rates. One approach is to study records documenting the rate at which mammals and birds have become extinct since humans arrived and to compare this with the fossil records of such extinctions prior to our arrival.

The International Union for the Conservation of Nature and Natural Resources (IUCN)—also known as the World Conservation Union—is a coalition of the world's leading conservation groups. Since the 1960s, it has published annual *Red Lists,* the world standard for listing the world's threatened species.

The 2005 Red List contains more than 16,000 species at risk of extinction. This includes one of every four mammal species, one of every seven plant species, and one of every eight bird species. Those compiling the list say it greatly underestimates the true number of threatened species because only a tiny fraction of the 1.4 million known species have been assessed, much less the estimated 4–100 million additional species that have not been catalogued or studied. You can examine the Red Lists database online at **www.iucnredlist.org.**

In 2006 World Wildlife Fund (WWF) researchers identified 794 species in danger of imminent extinction. According to this study, safeguarding 595 sites around the world that house these species would help stem this global extinction threat. Among mammals other than humans, nearly half of the 240 species of primates, such as the orangutan (Figure 10-3, p. 193) and white ukari (photo 6 in the Detailed Contents) are threatened with extinction, mostly because of habitat loss, the illegal trade in wildlife, and overhunting for their meat (bushmeat).

Another way that biologists project future extinction rates is to observe how the number of species present increases with the size of an area. This *species–area relationship* suggests that on average a 90% loss of habitat causes the extinction of about 50% of the species living in that habitat. This is based on the *theory of island biogeography* (Case Study, p. 146). Scientists are using this model to estimate the number of current and future extinctions in patches or "islands" of shrinking habitat surrounded by degraded habitats or by encroaching human developments.

Scientists also use models to estimate the risk of a particular species becoming endangered or extinct within a certain period of time, based on factors such as trends in population size, changes in habitat availability, interactions with other species, and genetic factors (see Supplement 13 on p. S47).

Researchers know that their estimates of extinction rates are based on inadequate data and sampling and incomplete models. They are continually striving to get better data and improve the models used to estimate extinction rates.

At the same time, they point to clear evidence that human activities have increased the rate of species extinction and that this rate is increasing. According to these biologists, arguing over the numbers and waiting to get better data and models should not be used as excuses for inaction. They agree with the advice of Aldo Leopold (Individuals Matter, p. 23): "To keep every cog and wheel is the first precaution of intelligent tinkering."

F **RESEARCH FRONTIER** Identifying and cataloging the millions of unknown species and improving models for estimating extinction rates

Effects of Human Activities on Extinction Rates (Science)

Biologists estimate that the current rate of extinction is 100 to 10,000 times the rate before humans arrived on earth.

In due time, all species become extinct. Before we came on the scene, the estimated natural extinction rate was roughly one extinct species per million species on earth annually. This amounted to an extinction rate of about 0.0001% per year.

Using the methods just described, biologists conservatively estimate that the current rate of extinction is at least 100 to 1,000—and by some estimates 10,000—times the rate before we arrived on the earth. This amounts to an annual extinction rate of 0.01% to 1% per year.

How many species are we losing prematurely each year? The answer depends on how many species are on the earth and the rate of species extinction. Assuming that the extinction rate is 0.1%, each year we lose 5,000 species per year if there are 5 million species on earth and 14,000 species if there are 14 million species—biologists' current best guess. See Figure 11-6 for more examples.

Most biologists would consider the premature loss of 1 million species over 100–200 years to be an extinction crisis or spasm that, if it continued, would lead to a mass depletion or even a mass extinction.

According to researchers Edward O. Wilson and Stuart Primm, at a 1% extinction rate, at least one-fifth

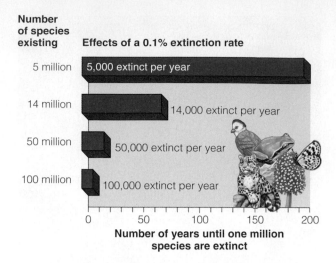

Figure 11-6 Natural capital degradation: effects of a 0.1% extinction rate.

of the world's current animal and plant species could be gone by 2030 and half could vanish by the end of this century. In the words of biodiversity expert Norman Myers, "Within just a few human generations, we shall—in the absence of greatly expanded conservation efforts—impoverish the biosphere to an extent that will persist for at least 200,000 human generations or twenty times longer than the period since humans emerged as a species."

? **THINKING ABOUT EXTINCTION** How might your lifestyle change if our activities cause the premature extinction of up to half of the world's species in your lifetime? List three things in your lifestyle that contribute to this threat.

Most biologists consider extinction rates of 0.01%–1% to be conservative estimates for several reasons. *First*, both the rate of species loss and the extent of biodiversity loss are likely to increase during the next 50–100 years because of the projected growth of the world's human population and resource use per person (Figure 1-7, p. 13 and Figure 3 on pp. S12–S13 in Supplement 4).

Second, current and projected extinction rates are much higher than the global average in parts of the world that are endangered centers of the world's biodiversity. Conservation biologists urge us to focus our efforts on slowing the much higher rates of extinction in such *hot spots* (Figure 10-26, p. 216) as the best and quickest way to protect much of the earth's biodiversity from being lost prematurely.

Third, we are eliminating, degrading, and simplifying many biologically diverse environments—such as tropical forests, tropical coral reefs, wetlands, and estuaries—that serve as potential colonization sites for the emergence of new species. Thus, in addition to increasing the rate of extinction, we may be limiting the long-term recovery of biodiversity by reducing the rate of speciation for some types of species. In other words, we are creating a *speciation crisis*. (See the Guest Essay by Normal Myers on this topic on the website for this chapter.)

Philip Levin, Donald Levin, and other biologists also argue that the increasing fragmentation and disturbance of habitats throughout the world may increase the speciation rate for rapidly reproducing opportunist species such as weeds, rodents, and cockroaches and other insects. Thus, the real threat to biodiversity from current human activities may not be a permanent decline in the number of species but a long-term erosion in the earth's variety of species and habitats. Such a loss of biodiversity would reduce the ability of life to adapt to changing conditions by creating new species.

IMPORTANCE OF WILD SPECIES

Why Should We Preserve Wild Species? (Science and Economics)

We should not cause the premature extinction of species because of the economic and ecological services they provide.

So what is all the fuss about? If all species eventually become extinct, why should we worry about losing a few more because of our activities? Does it matter that the passenger pigeon (Core Case Study, p. 222), the remaining orangutans (Figure 10-3, p. 193), or some unknown plant or insect in a tropical forest becomes prematurely extinct because of our activities?

New species eventually evolve to take the place of those lost through extinction spasms, mass depletions, or mass extinctions (Figure 4-12, p. 93). So why should we care if we speed up the extinction rate over the next 50–100 years? The answer: because it will take at least 5 million years for natural speciation to rebuild the biodiversity we are likely to destroy during this century.

Conservation biologists and ecologists say we should act now to prevent the premature extinction of species because of their *instrumental value*—their usefulness to us in the form of economic and ecological services (Case Study, at right). Many species provide economic value in the form of food crops, fuelwood and lumber, paper, and medicine (Figure 10-18, p. 205).

A 2005 United Nations University report concluded that 62% of all cancer drugs were created by bioprospecting discoveries, including the key ingredient in Taxol, which is used to treat breast cancer. Companies are also looking to nature for industrial applications such as using an enzyme found in deep sea vents to streamline the production of ethanol for use as a biofuel. *Green Career:* Bioprospecting

CASE STUDY

Why Should We Care about Bats? (Science and Economics)

Worldwide there are 950 known species of bats—the only mammals that can fly. But bats have two traits that make them vulnerable to extinction. *First,* they reproduce slowly. *Second,* many bat species live in huge colonies in caves and abandoned mines, which people sometimes block. This prevents them from leaving to get food and can disturb their hibernation.

Bats play important ecological roles. About 70% of all bat species feed on crop-damaging nocturnal insects and other insect pest species such as mosquitoes. This makes them the major nighttime SWAT team for such insects.

In some tropical forests and on many tropical islands, *pollen-eating bats* pollinate flowers, and *fruit-eating bats* distribute plants throughout tropical forests by excreting undigested seeds. As keystone species, such bats are vital for maintaining plant biodiversity and for regenerating large areas of tropical forest cleared by human activities. If you enjoy bananas, cashews, dates, figs, avocados, or mangos, you can thank bats.

Many people mistakenly view bats as fearsome, filthy, aggressive, rabies-carrying bloodsuckers. But most bat species are harmless to people, livestock, and crops. In the United States, only 10 people have died of bat-transmitted disease in more than four decades of record keeping; more Americans die each year from falling coconuts.

Because of unwarranted fears of bats and lack of knowledge about their vital ecological roles, several bat species have been driven to extinction. Currently, about one-fourth of the world's bat species are listed as endangered or threatened. Because of the important ecological and economic roles they play, conservation biologists urge us to view bats as valuable allies, not as enemies to kill.

Critical Thinking

Has reading the case study changed your view of bats? Can you think of two things that could be done to help protect bat species from premature extinction?

Another instrumental value is the *genetic information* in species that allows them to adapt to changing environmental conditions and to form new species. Genetic engineers use this information to produce new types of crops (Figure 4-14, p. 95) and foods. Carelessly eliminating many of the species making up the world's vast genetic library is like burning books before we read them.

Biophilia (Ethics)

Biologist Edward O. Wilson contends that because of the billions of years of biological connections leading to the evolution of the human species, we have an inherent genetic kinship with the natural world. He calls this phenomenon *biophilia* (love of life).

Evidence of this natural and emotional affinity for life is seen in the preference most people have for almost any natural scene over one from an urban environment. Given a choice, most people prefer to live in an area where they can see water, grassland, or a forest. More people visit zoos and aquariums than attend all professional sporting events combined.

In the 1970s, I was touring the space center at Cape Canaveral in Florida. During our bus ride the tour guide pointed out each of the abandoned multimillion-dollar launch sites and gave a brief history of each launch. Most of us were utterly bored. Suddenly people started rushing to the front of the bus and staring out the window with great excitement. What they were looking at was a baby alligator—a dramatic example of how *biophilia* can triumph over *technophilia*.

Not everyone has biophilia. Some have the opposite feeling about many or most forms of life. This fear of many forms of wildlife is called *biophobia*. For example, some movies, books, and TV programs condition us to fear or be repelled by certain species such as alligators (p. 143), cockroaches (p. 90), sharks (p. 149), bats (p. 227), and bacteria (p. 52). Although these species play important ecological roles, as shown many times throughout this book, fear is a difficult emotion to overcome.

Critical Thinking

Do you have an affinity for wildlife and wild ecosystems (biophilia)? If so, how do you display this love of wildlife in your daily actions? List three types of your resource consumption that help destroy and degrade wildlife.

Wild species also provide a way for us to learn how nature works and sustains itself. In addition, the earth's wild plants and animals provide us with *recreational pleasure*. Each year, Americans spend more than three times as many hours watching wildlife—doing nature photography and bird watching, for example—as they spend watching movies or professional sporting events.

Wildlife tourism, or *ecotourism*, generates at least $500 billion per year worldwide, and perhaps twice that much. Conservation biologist Michael Soulé estimates that one male lion living to age 7 generates $515,000 in tourist dollars in Kenya but only $1,000 if killed for its skin. Similarly, over a lifetime of 60 years, a Kenyan elephant is worth about $1 million in ecotourist revenue—many times more than its tusks are worth when they are sold illegally for their ivory. *Green Career:* Ecotourism guide

The upside of ecotourism is that it can inject money into local economies and allows visitors to learn about the natural world and perhaps appreciate its value and fragility. The downside is that large numbers of people cannot visit an ecosystem without disturbing it. Responsible ecotourism limits the number of visitors and strives to minimize ecological damage. It should also provide income for local people who act as guides and provide local lodging for visitors. And a certain percentage of the tour income should be given to local communities for the purchase and maintenance of wildlife reserves and conservation programs. This can motivate local people to help protect and sustain local wildlife.

Much ecotourism does not meet these standards, and excessive and unregulated ecotourism and the building of modern hotels and other tourist facilities can destroy or degrade fragile areas and promote premature extinction of species. The website for this chapter lists some guidelines for evaluating ecotours.

Do We Have an Ethical Obligation to Protect Species from Premature Extinction? (Ethics)

Some believe that each wild species has an inherent right to exist.

Many people with the stewardship and environmental wisdom worldviews (p. 23) believe that each wild species has *intrinsic* or *existence* value based on its inherent right to exist and play its ecological roles, regardless of its usefulness to us. According to this view, we have an ethical responsibility to protect species from becoming prematurely extinct as a result of human activities, and to prevent the degradation of the world's ecosystems and its overall biodiversity.

Biologist Edward O. Wilson believes that deep down most people feel obligated to protect other species and the earth's biodiversity because they seem to have a natural affinity for nature that he calls *biophilia* (Spotlight, left).

Some people distinguish between the survival rights of plants and those of animals, mostly for practical reasons. Poet Alan Watts once said he was a vegetarian "because cows scream louder than carrots."

Other people distinguish among various types of species. For example, they might think little about getting rid of the world's mosquitoes, cockroaches, rats, or disease-causing bacteria. **Question:** Where do you stand on this issue? Explain.

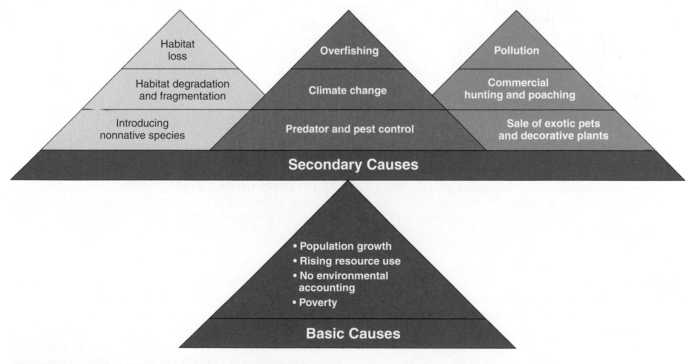
THINKING ABOUT THE PASSENGER PIGEON In earlier times, many people viewed huge flocks of passenger pigeons (Core Case Study, p. 222) as pests that devoured grain and left massive piles of their waste. Do you think this justified their premature extinction? Explain. If you agree that premature extinction of undesirable species is justified, what would be your three top candidates? What might be some harmful ecological effects of such extinctions?

Some biologists caution us not to focus primarily on protecting relatively large organisms—the plants and animals we can see and are familiar with. They remind us that the true foundation of the earth's ecosystems and ecological processes are invisible bacteria and the algae, fungi, and other *microorganisms* that decompose the bodies of larger organisms and recycle the nutrients needed by all life.

HABITAT LOSS, DEGRADATION, AND FRAGMENTATION

Habitat Destruction, Degradation, and Fragmentation: Remember HIPPO (Science)

The greatest threat to a species is the loss, degradation, and fragmenting of the place where it lives.

Figure 11-7 shows the basic and secondary causes of the endangerment and premature extinction of wild species. Conservation biologists summarize the most important causes of premature extinction using the acronym **HIPPO:** **H**abitat destruction, degradation, and fragmentation, **I**nvasive (nonnative) species, **P**opulation growth (too many people consuming too many resources), **P**ollution, and **O**verharvesting.

According to biodiversity researchers, the greatest threat to wild species is habitat loss (Figure 11-8, p. 230), degradation, and fragmentation. Many species have a hard time surviving when we take over their ecological "house" and their food supplies and make them homeless. The passenger pigeon (Figure 11-1) is only one of many species whose extinction was hastened by loss of habitat from forest clearing.

Deforestation of tropical forests is the greatest eliminator of species, followed by the destruction and degradation of coral reefs and wetlands, plowing of grasslands, and pollution of streams, lakes, and oceans. Globally, temperate biomes have been affected more by habitat loss and degradation than have tropical biomes because of widespread economic development in temperate countries over the past 200 years. Economic development is now shifting to many tropical biomes.

Island species—many of them *endemic species* found nowhere else on earth—are especially vulnerable to extinction when their habitats are destroyed, degraded, or fragmented. For example, half of the lemur species on Madagascar, a threatened island jewel of

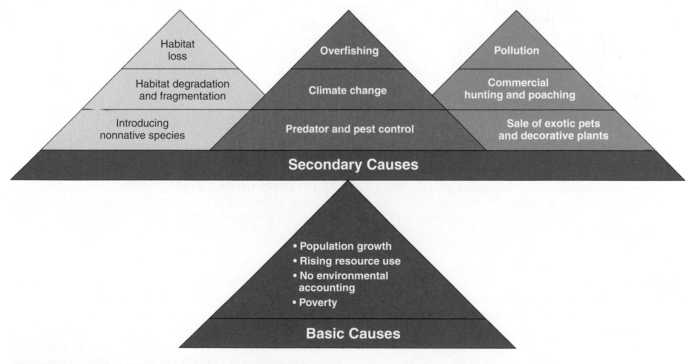

Figure 11-7 Natural capital degradation: underlying and direct causes of depletion and premature extinction of wild species. The major direct cause of wildlife depletion and premature extinction is habitat loss, degradation, and fragmentation. This is followed by the deliberate or accidental introduction of harmful invasive (nonnative) species into ecosystems.

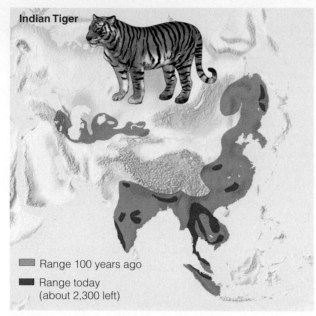

Indian Tiger

Range 100 years ago

Range today
(about 2,300 left)

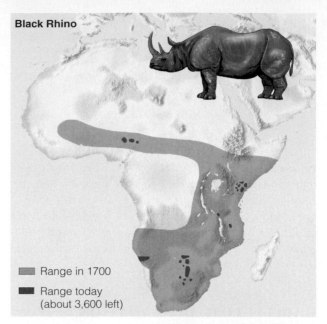

Black Rhino

Range in 1700

Range today
(about 3,600 left)

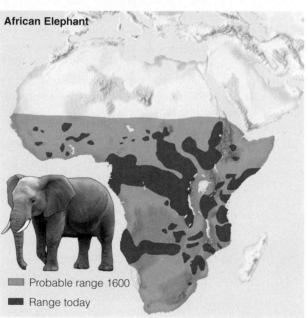

African Elephant

Probable range 1600

Range today

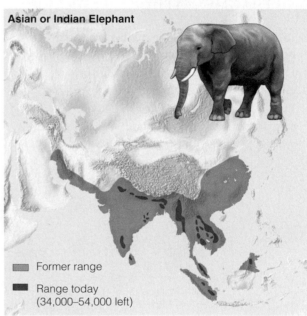

Asian or Indian Elephant

Former range

Range today
(34,000–54,000 left)

ThomsonNOW™ Active Figure 11-8 Natural capital degradation: reductions in the ranges of four wildlife species, mostly as the result of habitat loss and hunting. What will happen to these and millions of other species when the world's human population doubles and per capita resource consumption rises sharply in the next few decades? *See an animation based on this figure and take a short quiz on the concept.* (Data from International Union for the Conservation of Nature and World Wildlife Fund)

biodiversity in the Indian Ocean off the coast of Africa, are threatened (See photo 8 in the Detailed Contents).

Any habitat surrounded by a different one can be viewed as a *habitat island* for most of the species that live there. Most national parks and other nature reserves are habitat islands, many of them encircled by potentially damaging logging, mining, energy extraction, and industrial activities. Freshwater lakes are also habitat islands that are especially vulnerable to the introduction of nonnative species and pollution.

Habitat fragmentation—by roads, logging, agriculture, and urban development—occurs when a large, continuous area of habitat is reduced in area and divided into smaller, more scattered, and isolated patches or "habitat islands." This process can block migration routes and divide populations of a species into smaller and more isolated groups (metapopulations) that are more vulnerable to predators, competitive species, disease, and catastrophic events such as a storm or fire (see Supplement 13 on p. S47). Also, it creates barriers that limit the abilities of some species to disperse and colonize new areas, get enough to eat, and find mates. Highway expansion fragments wildlife habitat and increases the road kill of wild animals. In the United States, vehicles now kill more wildlife than hunters.

Certain types of species are especially vulnerable to local and regional extinction because of habitat fragmentation. They include species that are rare, that need to roam unhindered over large areas, and that cannot rebuild their population because of a low reproductive capacity. Also included are species with specialized niches and species that are sought by people for furs, food, medicines, or other uses.

Scientists use the theory of island biogeography (Case Study, p. 146) to help them understand the effects of fragmentation on species extinction and to develop ways to help prevent such extinction.

 See how serious the habitat fragmentation problem is for elephants, tigers, and rhinos at ThomsonNOW.

Case Study: A Disturbing Message from the Birds (Science)

Human activities are causing serious declines in the populations of many bird species.

Approximately 70% of the world's 9,775 known bird species are declining in numbers, and roughly one of every eight bird species is threatened with extinction, mostly because of habitat loss, degradation, and fragmentation. The majority of the world's bird species are found in South America (Figure 11-9).

About three-fourths of the threatened bird species live in forests. Each year an area of forests about the size of Greece are destroyed. For example, some 40% of Indonesia's moist tropical forests, particularly in Borneo and Sumatra, has been cleared for lumber and palm plantations to supply palm oil used as biofuel, mostly in European nations. As a result, three of every four bird species in Sumatra's lowland forests are on the verge of extinction. And in Brazil, 115 bird species are threatened, mostly because of the burning and clearing of Amazon forests for farms and ranches (Figure 10-16, p. 204), the loss of 93% of Brazil's Atlantic coastal rain forest, and most recently the clearing of the country's savannah-like cerrado area to establish soybean plantations.

A 2004 National Audubon Society study found that 30% of all North American bird species and 70% of those living in grasslands are declining in numbers or are at risk of disappearing. Figure 11-10 (p. 232) shows the 10 most threatened U.S. songbird species.

Occasionally, there is a ray of hope in the gloomy news about birds. In 2005, naturalists reported several

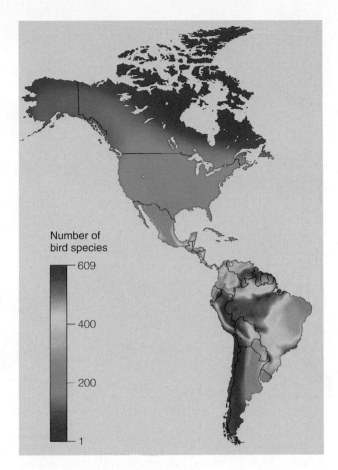

Figure 11-9 Natural capital: distribution of bird species in North America and Latin America. (Data from the Nature Conservancy, Conservation International, World Wildlife Fund, and Environment Canada).

sightings of ivory-billed woodpeckers in a national wildlife refuge in eastern Arkansas (USA). This bird, one of the world's largest woodpeckers, had been listed as extinct since 1944. Scientists are trying to confirm the sightings, and government agencies and private conservation groups are working to create a protected habitat area for this bird and seven other endangered bird species that live in this area. Conservationists hope that the woodpecker won't be loved to death by bird-watchers flocking to the area.

After habitat loss, the intentional or accidental introduction of nonnative species such as bird-eating cats, rats, snakes, and mongooses is the second greatest danger, affecting about 28% of the world's threatened birds. Fifty-two of the world's 388 parrot species (see photo 7 in in the Detailed Contents) are threatened from a combination of habitat loss and capture for the pet trade (often illegal), especially in Europe and the United States.

At least 23 species of seabirds face extinction. Many drown after becoming hooked on one of the miles of baited lines put out by fishing boats. And populations of 40% of the world's waterbirds are in decline because of the global loss of wetlands.

Cerulean warbler   Sprague's pipit   Bichnell's thrush   Black-capped vireo   Golden-cheeked warbler

Florida scrub jay   California gnatcatcher   Kirtland's warbler   Henslow's sparrow   Bachman's warbler

Figure 11-10 Threatened natural capital: the 10 most threatened species of U.S. songbirds, according to a 2002 study by the National Audubon Society. Most of these species are vulnerable because of habitat loss and fragmentation from human activities. An estimated 12% of the world's known bird species may face premature extinction from human activities during this century. (Data from National Audubon Society)

Millions of migrating birds are killed each year when they collide with power lines, communications towers, and skyscrapers that we have erected in the middle of their migration routes. While U.S. hunters kill about 121 million birds a year, as many as 1 billion birds a year in the United States die when they fly into glass windows, especially those in tall buildings in cities that are lit up at night—the number one cause of U.S. avian mortality. Other threats to birds are oil spills, exposure to pesticides, herbicides that destroy their habitats, and swallowing toxic lead shotgun pellets falling into wetlands and lead sinkers left by anglers.

The greatest new threat to birds is climate change. Birds that spend all or part of their lives at the rapidly warming earth's poles are especially vulnerable to rising temperatures. And populations of some migrating birds are declining because climate change has shifted the location of the species they normally feed on.

One reason is that birds are excellent *environmental indicators* because they live in every climate and biome, respond quickly to environmental changes in their habitats, and are relatively easy to track and count.

In addition, birds perform a number of economically and ecologically vital services in ecosystems throughout the world. They help control populations of rodents and insects (which decimate many tree species), clean up dead animal carcasses by eating them, pollinate flowers, and spread plants throughout their habitats by consuming and excreting plant seeds.

Extinctions of birds that play key and specialized roles in pollination and seed dispersal, especially in tropical areas, may lead to extinctions of plants dependent on these ecological services. Then some specialized animals that feed on these plants may become extinct. Everything is connected.

The collapse of bird populations can have other unexpected effects. Since the 1980s, there has been a 95% drop in India's Gyp vulture population, poisoned by a medicine used to treat the livestock they feed on. This decline was followed by a sharp rise in the number of feral dogs and rats that vultures helped control. This put humans at risk. In 1997, alone, more than 30,000 people in India died of rabies—more than half the world's total number of rabies deaths that year.

Conservation biologists urge us to listen more carefully to what birds are telling us about the state of the environment for them and for us.

F **RESEARCH FRONTIER** Learning why birds are declining, what it implies for the biosphere, and what can be done about it

? *THINKING ABOUT BIRD EXTINCTIONS* How does your lifestyle directly or indirectly contribute to the premature extinction of some bird species? What three things do you think should be done to reduce the premature extinction of birds?

INVASIVE SPECIES

Deliberately Introduced Species (Science and Economics)

Many nonnative species provide us with food, medicine, and other benefits but a few can wipe out some native species, disrupt ecosystems, and cause large economic losses.

After habitat loss and degradation, the deliberate or accidental introduction of harmful invasive species into ecosystems is the biggest cause of animal and plant extinctions.

However, most species introductions are beneficial. We depend heavily on introduced species for ecosystem services, food, shelter, medicine, and aesthetic enjoyment. According to a 2000 study by biologist David Pimentel, introduced species such as corn, wheat, rice, other food crops, cattle, poultry, and other livestock provide more than 98% of the U.S. food sup-

ply. Similarly, nonnative tree species are grown in about 85% of the world's tree plantations. Some deliberately introduced species have also helped control pests.

About 50,000 nonnative species now live in the United States and about one in seven of them are harmful invasive species. The problem is that many of these invasive species have no natural predators, competitors, parasites, or pathogens to help control their numbers in their new habitats. Such species can reduce or wipe out populations of many native species and trigger ecological disruptions. Some ecologists call this **biotic pollution.** Figure 11-11 (p. 234) shows some of the estimated 7,100 harmful invasive species that, after being deliberately or accidentally introduced into the United States, have caused ecological and economic harm. According to biologist Thomas Lovejoy, harmful invasive species cost the U.S. public more than $137 billion each year—an average of $16 million per hour. The situation in China is much worse.

Nonnative species threaten almost half of the more than 1,260 endangered and threatened species in the United States and 95% of those in the state of Hawaii, according to the U.S. Fish and Wildlife Service. One example of a deliberately introduced plant species is the *kudzu* ("CUD-zoo") *vine*, which grows rampant in the southeastern United States (Case Study, below).

CASE STUDY

The Kudzu Vine (Science)

In the 1930s, the *kudzu vine* was imported from Japan and planted in the southeastern United States in an attempt to control soil erosion. It does control erosion. Unfortunately, it is so prolific and difficult to kill that it engulfs hillsides, gardens, trees, abandoned houses and cars, stream banks, patches of forest, and anything else in its path (Figure 11-A).

This plant, which is sometimes called "the vine that ate the South," has spread throughout much of the southeastern United States. It could spread as far north as the Great Lakes by 2040 if global warming occurs as projected.

Kudzu is considered a menace in the United States but Asians use a powdered kudzu starch in beverages, gourmet confections, and herbal remedies for a range of diseases. A Japanese firm has built a large kudzu farm and processing

Bruce Coleman USA

plant in the U.S. state of Alabama and ships the extracted starch to Japan.

And almost every part of the kudzu plant is edible. Its deep-fried leaves are delicious and contain high levels of vitamins A and C. Stuffed kudzu leaves, anyone?

Although kudzu can engulf and kill trees, it might eventually help save trees from loggers. Researchers at the Georgia Institute of Technology indicate that it could be

Figure 11-A Natural capital degradation: kudzu taking over an abandoned house in Mississippi. This vine, which can grow 5 centimeters (2 inches) per hour, was deliberately introduced into the United States for erosion control. It cannot be stopped by being dug up or burned. Grazing by goats and repeated doses of herbicides can destroy it, but goats and herbicides also destroy other plants, and herbicides can contaminate water supplies. Recently, scientists have found a common fungus that can kill kudzu within a few hours, apparently without harming other plants. Stay tuned.

used as a source of tree-free paper. And a preliminary 2005 study indicated that kudzu powder could reduce alcoholism and binge drinking. Ingesting small amounts of the powder can lessen one's desire for alcohol.

Critical Thinking

Do the advantages of kudzu in reducing erosion and perhaps as a source of tree-free paper outweigh its disadvantages?

Deliberately Introduced Species

Purple loosestrife

European starling

African honeybee ("Killer bee")

Nutria

Salt cedar (Tamarisk)

Marine toad (Giant toad)

Water hyacinth

Japanese beetle

Hydrilla

European wild boar (Feral pig)

Accidentally Introduced Species

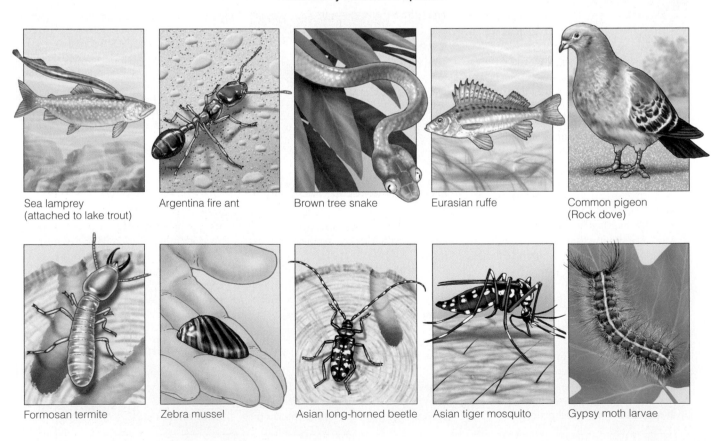

Sea lamprey (attached to lake trout)

Argentina fire ant

Brown tree snake

Eurasian ruffe

Common pigeon (Rock dove)

Formosan termite

Zebra mussel

Asian long-horned beetle

Asian tiger mosquito

Gypsy moth larvae

Deliberately introduced animal species have also caused ecological and economic damage. Consider the estimated 1 million *European wild (feral) boars* (Figure 11-11) found in parts of Florida and other states. They compete for food with endangered animals, root up farm fields, and cause traffic accidents. Game and wildlife officials have failed to control their numbers through hunting and trapping and say there is no way to stop them. Another example is the estimated 30 million *feral cats* and 41 million *outdoor pet cats* introduced into the United States; they kill about 568 million birds per year.

Accidentally Introduced Invasive Species (Science)

A growing number of accidentally introduced species cause serious economic and ecological damage.

Welcome to one of the downsides of global trade. Many unwanted nonnative invaders arrive from other continents as stowaways on aircraft, in the ballast water of tankers and cargo ships, and as hitchhikers on imported products such as wooden packing crates in today's increasingly globalized economy. Cars and trucks can spread seeds of nonnative species embedded in tire treads. Many tourists return home with living plants that may multiply and become invasive or harbor insects that can escape, multiply rapidly, and threaten crops.

The 2005 Millennium Ecosystem Assessment called this effect a "globalization of nature" that has profoundly affected ecosystems around the world. For example, due to the release of ballast water from cargo ships in the Black Sea, the arrival of the American comb jellyfish has led to destruction of 26 commercially valuable stocks of fish, while the nonnative zebra mussel has done considerable damage in the North American Great Lakes. The Baltic Sea now has 100 nonnative species, a third of which come from the Great Lakes. Those lakes, in turn, contain 170 alien species, a third of which come from the Baltic.

In the late 1930s, the extremely aggressive Argentina fire ant (Figure 11-12) was introduced accidentally into the United States in Mobile, Alabama. It probably arrived on shiploads of lumber or coffee imported from South America. Without natural predators, fire ants have spread rapidly by land and water (they can float) throughout the South, from Texas to Florida and as far north as Tennessee and Virginia (Figure 11-12).

When these ants invade an area, they can wipe out as much as 90% of native ant populations. Bother them, and 100,000 fire ants may swarm out of their nest to attack you with painful and burning stings. They have killed deer fawns, birds, livestock, pets, and at least 80 people who were allergic to their venom some of them fragile nursing home residents. They also do an estimated $600 million of economic damage per year to crops and phone and power lines.

Widespread pesticide spraying in the 1950s and 1960s temporarily reduced fire ant populations. But this chemical warfare hastened the advance of the rapidly multiplying fire ants by reducing populations of many native ant species. Even worse, it promoted development of genetic resistance to pesticides in the fire ants through natural selection. In other words, we helped wipe out their competitors and make them genetically stronger.

In the Everglades in the U.S. state of Florida, the population of the huge *Burmese python* is increasing. A native of Southeast Asia, it was imported as a pet and ended up being dumped in the Everglades by people who learned that pythons do not make great pets. They

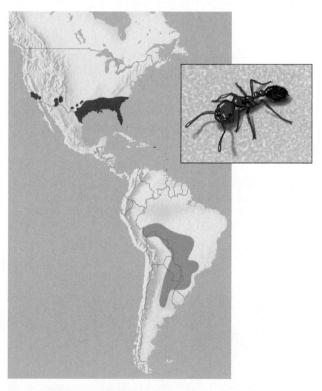

Figure 11-12 Natural capital degradation: the *Argentina fire ant*, introduced accidentally into Mobile, Alabama in 1932 from South America (green area), has spread over much of the southern United States (red area). This invader is also found in Puerto Rico, New Mexico, and California. (Data from S.D. Porter, Agricultural Research Service, U.S. Department of Agriculture)

Figure 11-11 (facing page) Threats to natural capital: some of the more than 7,100 harmful invasive (nonnative) species that have been deliberately or accidentally introduced into the United States.

can live 25 years, reach 6 meters (20 feet) in length, have the girth of a telephone pole, and with their razor-sharp teeth can eat practically anything that moves, including a full-grown deer.

Solutions: Reducing Threats from Invasive Species (Science and International Laws)

Prevention is the best way to reduce the threats from invasive species because once they have arrived it is almost impossible to slow their spread.

Once a harmful nonnative species becomes established in an ecosystem, its wholesale removal is almost impossible—somewhat like trying to get smoke back into a chimney. Thus, the best way to limit the harmful impacts of nonnative species is to prevent them from being introduced and becoming established.

Scientists suggest several ways to do this:

- Fund a massive research program to identify the major characteristics that allow species to become successful invaders and the types of ecosystems that are vulnerable to invaders (Figure 11-13).
- Greatly increase ground surveys and satellite observations to detect and monitor species invasions and develop better models for predicting how they will spread.
- Step up inspection of imported goods and goods carried by travelers that are likely to contain invader species.
- Identify major harmful invader species and pass international laws banning their transfer from one country to another, as is now done for endangered species. Australia and New Zealand no longer assume that a potential invasive species is innocent until proven guilty. With this *precautionary approach,* species that are not on an approved list are denied entry into the country.
- Require cargo ships to discharge their ballast water and replace it with saltwater at sea before entering ports, or require them to sterilize such water or pump nitrogen into the water to displace dissolved oxygen and kill most invader organisms.
- Increase research to find and introduce natural predators, parasites, and disease-causing bacteria and viruses to control populations of established invaders.

Characteristics of Successful Invader Species	Characteristics of Ecosystems Vulnerable to Invader Species
• High reproductive rate, short generation time (r-selected species)	• Climate similar to habitat of invader
• Pioneer species	• Absence of predators on invading species
• Long lived	• Early successional systems
• High dispersal rate	• Low diversity of native species
• Release growth-inhibiting chemicals into soil	• Absence of fire
• Generalists	• Disturbed by human activities
• High genetic variability	

Figure 11-13 **Threats to natural capital:** some general characteristics of successful invasive species and ecosystems vulnerable to invading species.

What Can You Do?

Invasive Species

- Do not allow wild animals to escape.
- Do not spread wild plants to other areas.
- Do not dump the contents of an aquarium into waterways, wetlands, or storm drains.
- When camping use wood near your campsite instead of bringing firewood from somewhere else.
- Do not dump unused bait into the water.
- After dogs visit woods or the water brush them before taking them home.
- After each use clean your vehicle, mountain bike, surfboard, kayaks, canoes, boats, tent, hiking boots, and other gear before heading for home.
- Empty all water from canoes, kayaks, dive gear, and other outdoor equipment before heading home.
- Plant a variety of trees, shrubs, and other plants in your yard to reduce losses from invasive species.
- Do not buy plants from overseas or swap them with others using the Internet.

Figure 11-14 **Individuals matter:** ways to prevent or slow the spread of harmful invasive species. QUESTIONS: *Which two of these actions do you think are the most important? Which of these actions do you plan to take?*

RESEARCH FRONTIER Learning more about invasive species, why they thrive, and how to control them

Figure 11-14 shows some of the things you can do to help prevent or slow the spread of these harmful invaders.

POPULATION GROWTH, POLLUTION, AND CLIMATE CHANGE

Population Growth, Overconsumption, and Pollution

Population growth, affluenza, and pollution have promoted the premature extinction of some species.

Past and projected human population growth (Figure 9-2, p. 173) and excessive and wasteful consumption of resources (affluenza, p. 19) have caused premature extinction of some species. Acting together, these two factors have greatly expanded the human ecological footprint (Figure 1-7, p. 13, and Figures 3 and 4 on pp. S12–S15 in Supplement 4).

An unintended effect of pesticides threatens some species with extinction. According to the U.S. Fish and Wildlife Service, each year pesticides kill about one-fifth of the United States' beneficial honeybee colonies, more than 67 million birds, and 6–14 million fish. They also threaten one-fifth of the country's endangered and threatened species.

During the 1950s and 1960s, populations of fish-eating birds such as the osprey, cormorant, brown pelican, and bald eagle plummeted. A chemical derived from the pesticide DDT, when biologically magnified in food webs (Figure 11-15), made the birds' eggshells so fragile they could not reproduce successfully. Also hard hit were such predatory birds as the prairie falcon, sparrow hawk, and peregrine falcon, which help control rabbits, ground squirrels, and other crop eaters. Since the U.S. ban on DDT in 1972, most of these species have made a comeback.

Climate Change (Science and Economics)

Projected climate change threatens a number of species with premature extinction.

In the past, most natural climate changes have taken place over long periods of time—giving species more time to adapt or evolve into new species to cope with the change. Considerable evidence indicates that human activities such as greenhouse gas emissions and deforestation may bring about rapid climate change during this century, as discussed in Chapter 20.

A 2004 study by Conservation International predicted that climate change could drive more than a quarter of all land animals and plants to extinction by the end of this century. Polar bears and 10 of the world's 17 penguin species are already threatened because of higher temperatures and melting ice in their polar habitats.

OVEREXPLOITATION

Illegal Killing or Sale of Wild Species (Economics and Ethics)

Some protected species are killed for their valuable parts or are sold live to collectors.

The legitimate trade of wildlife products is worth at least $10 billion a year. And the illegal trade in wildlife earns smugglers $6–10 billion a year. Organized crime has moved into illegal wildlife smuggling because of the huge profits involved—surpassed only by the illegal international trade in drugs and weapons. At least

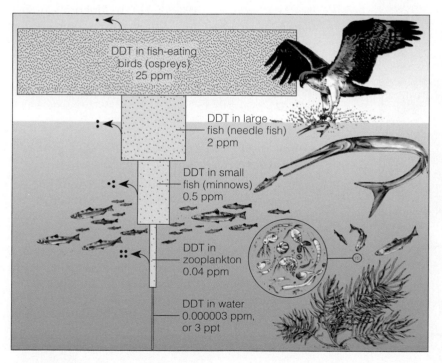

Figure 11-15 **Natural capital degradation:** *bioaccumulation* and *biomagnification*. DDT is a fat-soluble chemical that can accumulate in the fatty tissues of animals. In a food chain or web, the accumulated DDT can be biologically magnified in the bodies of animals at each higher trophic level. The concentration of DDT in the fatty tissues of organisms was biomagnified about 10 million times in this food chain in an estuary near Long Island Sound in New York. If each phytoplankton organism takes up from the water and retains one unit of DDT, a small fish eating thousands of zooplankton (which feed on the phytoplankton) will store thousands of units of DDT in its fatty tissue. Each large fish that eats 10 of the smaller fish will ingest and store tens of thousands of units, and each bird (or human) that eats several large fish will ingest hundreds of thousands of units. Dots represent DDT, and arrows show small losses of DDT through respiration and excretion.

DDT in fish-eating birds (ospreys) 25 ppm

DDT in large fish (needle fish) 2 ppm

DDT in small fish (minnows) 0.5 ppm

DDT in zooplankton 0.04 ppm

DDT in water 0.000003 ppm, or 3 ppt

two-thirds of all live animals smuggled around the world die in transit.

Poor people in areas with rich stores of wildlife may kill or trap such species in an effort to make enough money to survive and feed their families. Professional poachers also prey on these species. To poachers, a live *mountain gorilla* is worth $150,000, a *giant panda* pelt $100,000, a *chimpanzee* $50,000, an *Imperial Amazon macaw* $30,000, and a *Komodo dragon reptile* from Indonesia $30,000. A poached *rhinoceros horn* (Figure 11-16) may be worth as much as $28,600 per kilogram ($13,000 per pound). It is used to make dagger handles in the Middle East and as a fever reducer and alleged aphrodisiac in China and other parts of Asia.

According to a 2005 study by the International Fund for Animal Welfare, the Internet has become a key market for illegal global trade in thousands of live threatened and endangered species and products made from such species. For example, U.S. websites offered chimpanzees dressed as dolls for $60,000–65,000 each and a two-year-old Siberian tiger for $70,000.

In 1950, an estimated 100,000 tigers lived in the world. Despite international protection, only 5,000 to 7,000 tigers remain in the wild, on an ever-shrinking range (Figure 11-8, top left). Today all five tiger subspecies are endangered. The Bengal or Indian tiger is at risk because a coat made from its fur can sell for as much as $100,000 in Tokyo. Some wealthy Chinese and Europeans pay $10,000 or more for a Bengal tiger rug. With the body parts and bones of a single tiger worth as much as $25,000, it is not surprising that illegal

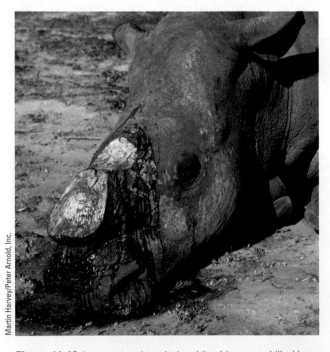

Figure 11-16 Lost natural capital: white rhinoceros killed by a poacher for its horns in South Africa.

hunting has skyrocketed, especially in India where since 1800 the number of tigers has plummeted from about 40,000 to 3,700 today. Without emergency action to curtail poaching and preserve their habitat, few if any tigers may be left in the wild within 20 years.

? *THINKING ABOUT TIGERS* What difference would it make if all the world's tigers disappeared? What three things would you do to help protect the world's remaining tigers from premature extinction?

As commercially valuable species become endangered, their black market demand soars. This increases their chances of premature extinction from poaching. Most poachers are not caught. And the money they can make far outweighs the small risk of being caught, fined, or imprisoned.

Killing Species We Don't Like (Science, Economics, and Ethics)

Killing predators and pests that bother us or cause economic losses threatens some species with premature extinction.

People sometimes try to exterminate species that compete with them for food and game animals or that become pests. African farmers kill large numbers of elephants to keep them from trampling and eating food crops. Each year, U.S. government animal control agents shoot, poison, or trap thousands of coyotes, prairie dogs, wolves, bobcats, and other species that prey on livestock, on species prized by game hunters, or on fish raised in aquaculture ponds.

Since 1929, U.S. ranchers and government agencies have poisoned 99% of North America's prairie dogs because horses and cattle sometimes step into the burrows and break their legs. This has also nearly wiped out the endangered black-footed ferret (Figure 11-3; about 600 are left in the wild) that preyed on the prairie dog—an unintended consequence of a pest control effort.

Collecting Exotic Pets and Plants (Economics and Ethics)

Legal and illegal trade in wildlife species used as pets or for decorative purposes threatens some species with extinction.

The global legal and illegal trade in wild species for use as pets is a huge and very profitable business. However, for every live animal captured and sold in the pet market, an estimated 50 others are killed.

About 25 million U.S. households have exotic birds as pets, 85% of them imported. More than 60 bird species, mostly parrots (see photo 7 in the Detailed Contents), are endangered or threatened because of this wild bird trade. According to the U.S. Fish and Wildlife Service, collectors of exotic birds may pay $10,000 for a threatened hyacinth macaw smuggled out of Brazil. But during its lifetime, a single macaw left in the wild might yield as much as $165,000 in tourist income.

Keeping birds as pets can be dangerous. A 1992 study suggested that keeping a pet bird indoors for more than 10 years doubles a person's chances of getting lung cancer from inhaling tiny particles of bird dander.

Other wild species whose populations are depleted because of the pet trade include amphibians, reptiles, mammals, and tropical fish (taken mostly from the coral reefs of Indonesia and the Philippines). Divers often catch tropical fish by using plastic squeeze bottles of cyanide to stun them. For each fish caught alive, many more die. In addition, the cyanide solution kills the coral animals that create the reef.

Some exotic plants, especially orchids and cacti, are endangered because they are gathered (often illegally) and sold to collectors to decorate houses, offices, and landscapes. The United States imports about 75% of all orchids and 99% of all live cacti sold each year. A collector may pay $5,000 for a single rare orchid, and a single rare mature crested saguaro cactus can earn cactus rustlers as much as $15,000.

THINKING ABOUT COLLECTING WILD SPECIES Some people believe it is unethical to collect wild animals and plants for display and personal pleasure. They believe we should leave most exotic wild species in the wild. Explain why you agree or disagree with this view.

Case Study: Rising Demand for Bushmeat in Africa (Survival and Economics)

Rapid population growth in parts of Africa has increased the number of people hunting wild animals for food or for sale of their meat to restaurants.

Indigenous people in much of West and Central Africa have sustainably hunted wildlife for *bushmeat*, a source of food, for centuries. But in the last two decades bushmeat hunting in some areas has skyrocketed as local people try to provide food for a rapidly growing population and to make a living by supplying restaurants with exotic meat (Figure 11-17). The bushmeat trade is also increasing in Southeast Asia, the Caribbean, and Central and South America.

Killing animals for bushmeat has also increased because logging roads have allowed miners, ranchers, and settlers to move into once inaccessible forests. And a 2004 study showed that people living in coastal areas of West Africa have increased bushmeat hunting because local fish harvests have declined from overfishing by heavily subsidized European Union fishing fleets.

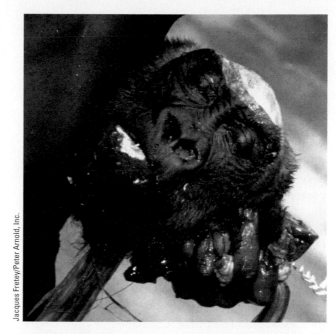

Jacques Fretey/Peter Arnold, Inc.

Figure 11-17 Natural capital degradation: *bushmeat*, such as this severed head of a lowland gorilla in the Congo, is consumed as a source of protein by local people in parts of West Africa and sold in the national and international marketplace. You can find bushmeat on the menu in Cameroon and the Congo in West Africa as well as in Paris, London, Toronto, New York, and Washington, D.C. It is often supplied by illegal poaching. Wealthy patrons of some restaurants regard gorilla meat as a source of status and power.

So what is the big deal? After all, people have to eat. For most of our existence, humans have survived by hunting and gathering wild species.

One problem is that bushmeat hunting has caused the local extinction of many animals in parts of West Africa and has driven one species—Miss Waldron's red colobus monkey—to complete extinction. It is also a factor in reducing gorilla, orangutan (Figure 10-3, p. 193), chimpanzee, elephant, and hippo populations. This practice also threatens forest carnivores such as crowned eagles and leopards by depleting their main prey species.

Some conservationists fear that within one or two decades the Congo Basin's rain forest—the world's second largest remaining tropical forest—will contain few large mammals and most of Africa's great apes will be extinct. Another problem is that butchering and eating some forms of bushmeat has helped spread fatal diseases such as HIV/AIDS and the ebola virus.

 THINKING ABOUT THE PASSENGER PIGEON AND HUMANS Humans exterminated the passenger pigeon within a single lifetime because it was considered a pest and because of its economic value. Suppose a species superior to us arrived and began taking over the earth with the goal of using the planet more sustainably. The first thing they might do is to exterminate us. Do you think such an action would be justified?

PROTECTING WILD SPECIES: LEGAL AND ECONOMIC APPROACHES

Global Outlook: International Treaties (Politics)

International treaties have helped reduce the international trade of endangered and threatened species, but enforcement is difficult.

Several international treaties and conventions help protect endangered or threatened wild species. One of the most far reaching is the 1975 *Convention on International Trade in Endangered Species* (CITES). This treaty, now signed by 169 countries, lists some 900 species that cannot be commercially traded as live specimens or wildlife products because they are in danger of extinction. It also restricts international trade of roughly 5,000 species of animals and 28,000 species of plants because they are at risk of becoming threatened.

CITES has helped reduce international trade in many threatened animals, including elephants, crocodiles, cheetahs, and chimpanzees. But the effects of this treaty are limited because enforcement varies from country to country and convicted violators often pay only small fines. Also, member countries can exempt themselves from protecting any listed species, and much of the highly profitable illegal trade in wildlife and wildlife products goes on in countries that have not signed the treaty.

The *Convention on Biological Diversity* (CBD), ratified by 188 countries, legally commits participating governments to reversing the global decline of biological diversity and equitably sharing the benefits from using the world's genetic resources. This includes efforts to prevent or control the spread of ecologically harmful invasive species.

This Convention is a landmark in international law because of its focus on ecosystems rather than on single species and its linkage of biodiversity protection to issues such as intellectual property rights and the traditional knowledge and rights of indigenous peoples. Its implementation has been slow because some key countries such as the United States have not ratified it. Also, it contains no severe penalties or other enforcement mechanisms.

Case Study: The U.S. Endangered Species Act (Science and Politics)

One of the world's most far-reaching and controversial environmental laws is the 1973 U.S. Endangered Species Act.

The *Endangered Species Act of 1973* (ESA; amended in 1982, 1985, 1988, and 2006) was designed to identify and legally protect endangered species in the United States and abroad. This act is probably the most far-reaching environmental law ever adopted by any nation, and this has made it controversial. Canada and a number of other countries have similar laws.

The National Marine Fisheries Service (NMFS) is responsible for identifying and listing endangered and threatened ocean species. The U.S. Fish and Wildlife Services (USFWS) identifies and lists all other endangered and threatened species. Any decision by either agency to add or remove a species from the list must be based on biological factors alone, without consideration of economic or political factors. However, economic factors can be used in deciding whether and how to protect endangered habitat and in developing recovery plans for listed species.

The ESA forbids federal agencies (except the Defense Department) to carry out, fund, or authorize projects that would jeopardize an endangered or threatened species or destroy or modify the critical habitat it needs to survive. For offenses committed on private lands, fines as high as $100,000 and one year in prison can be imposed to ensure protection of the habitats of endangered species. This part of the act has been controversial because about 80% of the listed species live totally or partially on private land.

The ESA makes it illegal for Americans to sell or buy any product made from an endangered or threatened species or to hunt, kill, collect, or injure such species in the United States.

Between 1973 and 2006, the number of U.S. species on the official endangered and threatened lists increased from 92 to about 1,260 species—60% of them plants and 40% animals. According to a 2000 study by the Nature Conservancy, one-third of the country's species are at risk of extinction, and 15% of all species are at high risk—far more than the 1,260 species on the ESA list. The study also found that many of the country's rarest and most imperiled species are concentrated in a few hot spots (Figure 11-18).

The USFWS or the NMFS is supposed to prepare a plan to help each listed species recover, including designating and protecting its critical habitat. By 2006, only one-fourth of the species on the protected list had active plans and only one-third had designated critical habitats—mostly because of political opposition and limited funds. Examples of successful recovery plans include those for the American alligator (p. 143), the gray wolf (p. 191), the bald eagle, and the peregrine falcon.

The ESA also requires that all commercial shipments of wildlife and wildlife products enter or leave the country through one of nine designated ports. Few illegal shipments are confiscated (Figure 11-19) because the 60 USFWS inspectors can examine less than one-fourth of the approximately 90,000 shipments that enter and leave the United States each year. Even if caught, many violators are not prosecuted, and convicted violators often pay only a small fine.

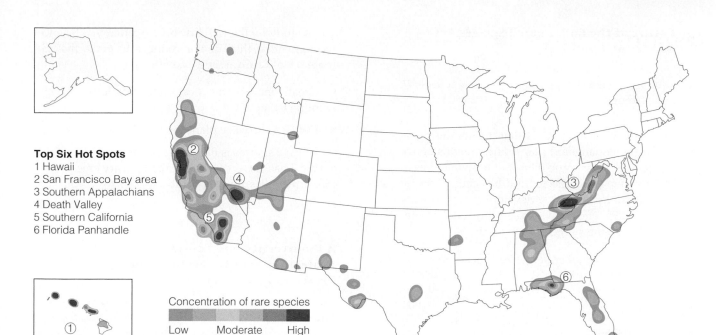

Top Six Hot Spots
1 Hawaii
2 San Francisco Bay area
3 Southern Appalachians
4 Death Valley
5 Southern California
6 Florida Panhandle

Concentration of rare species

Low   Moderate   High

Figure 11-18 Threatened natural capital: biodiversity hot spots in the United States. The shaded areas contain the largest concentrations of rare and potentially endangered species. Compare these areas with those on the map of the human ecological footprint in North America shown in Figure 4 on pp. S14–S15 in Supplement 4. (Data from State Natural Heritage Programs, the Nature Conservancy, and Association for Biodiversity Information)

Encouraging Private Landowners to Protect Endangered Species (Economics and Politics)

Congress has amended the Endangered Species Act to help landowners protect endangered species on their land.

In 1982, Congress amended the ESA to allow the secretary of the interior to use *habitat conservation plans* (HCPs). These are designed to strike a compromise between the interests of private landowners and those of endangered and threatened species.

With an HCP, landowners, developers, or loggers are allowed to destroy some critical habitat in exchange for taking steps to protect members of a species. Such measures might include setting aside a part of the species' habitat as a protected area, paying to relocate the species to another suitable habitat, or paying money to have the government buy suitable habitat elsewhere. Once the plan is approved it cannot be changed, even if new data show that the plan is inadequate to protect a species and help it recover.

In 1999, the USFWS approved two new approaches for encouraging private landowners to protect threatened or endangered species. One is *safe harbor agreements* in which landowners voluntarily agree to take specified steps to restore, improve, or maintain habitat for threatened or endangered species located on their land. In return, landowners get technical help and assurances that the natural resources involved will not face future restrictions once the agreement expires.

Another prevention method is *voluntary candidate conservation agreements* in which landowners agree to take specific steps to help conserve a species whose population is declining but is not yet listed as endangered or threatened. Participating landowners receive technical help, government subsidies, and assurances that no additional resource-use restrictions will be imposed on the land covered by the agreement if the species is listed as endangered or threatened in the future.

Steve Hillebrand/U.S. Fish and Wildlife Service

Figure 11-19 Natural capital degradation: confiscated products made from endangered species. Because of a scarcity of funds and inspectors, probably no more than one-tenth of the illegal wildlife trade in the United States is discovered. The situation is even worse in most other countries.

The Future of the Endangered Species Act (Economics and Politics)

Some believe that the Endangered Species Act should be weakened or repealed and others believe it should be strengthened and modified to focus on protecting ecosystems.

Opponents of the ESA contend that it puts the rights and welfare of endangered plants and animals above those of people. They argue it has not been effective in protecting endangered species and has caused severe economic losses by hindering development on private lands. Since 1995, efforts to weaken the ESA have included the following suggested changes:

- Making protection of endangered species on private land voluntary

- Having the government compensate landowners if it forces them to stop using part of their land to protect endangered species

- Making it harder and more expensive to list newly endangered species by requiring government wildlife officials to navigate through a series of hearings and peer-review panels and requiring hard data instead of computer-based models

- Eliminating the need to designate critical habitats partly because dealing with lawsuits for failure to develop critical habitats takes up most of the limited funds for carrying out the ESA

- Allowing the secretary of the interior to permit a listed species to become extinct without trying to save it and to determine whether a species should be listed

- Allowing the secretary of the interior to give any state, county, or landowner permanent exemption from the law, with no requirement for public notification or comment

By 2006, many of these objectives had been achieved. Other critics would go further and do away with this act. Because this step is politically unpopular with the American public, most efforts are designed to weaken the act and reduce its meager funding.

Most conservation biologists and wildlife scientists agree that the ESA needs to be simplified and streamlined. But they contend that it has not been a failure (Spotlight, at right).

They also contest the charge that the ESA has caused severe economic losses. According to government records, since 1979 only 0.05% of the almost 200,000 projects evaluated by the USFWS have been blocked or canceled as a result of the ESA. And the act authorizes a special cabinet-level panel, nicknamed the "God Squad," to exempt any federal project from having to comply with the act if the economic costs are too high.

A study by the U.S. National Academy of Sciences recommended three major changes to make the ESA more scientifically sound and effective:

- Greatly increase the meager funding for implementing the act

- Develop recovery plans more quickly

- When a species is first listed, establish a core of its survival habitat as critical, as a temporary emergency measure that could support the species for 25–50 years

A Biodiversity and Ecosystem Protection Act—Rethinking Species Protection (Science)

Many scientists believe that we should focus on protecting and sustaining biodiversity and ecosystem functioning as the best way to protect species.

Most biologists and wildlife conservationists believe that the United States needs a new law that emphasizes protecting and sustaining biological diversity and ecosystem functioning rather than focusing mostly on saving individual species. The idea is to prevent species from becoming extinct in the first place by protecting their habitats. This new *ecosystems approach* would follow three principles:

- Find out what species and ecosystems the country has

- Locate and protect the most endangered ecosystems (Figure 11-18) and species

- Make development *biodiversity-friendly* by providing significant financial incentives (tax breaks and write-offs) and technical help to private landowners who agree to help protect specific endangered ecosystem.

X *HOW WOULD YOU VOTE?* Should the Endangered Species Act be modified to protect and sustain the nation's overall biodiversity? Cast your vote online at www.thomsonedu.com/biology/miller.

PROTECTING WILD SPECIES: THE SANCTUARY APPROACH

Wildlife Refuges and Other Protected Areas (Science and Stewardship)

The United States has set aside 544 federal refuges for wildlife, but many refuges are suffering from environmental degradation.

Accomplishments of the Endangered Species Act
(Science, Economics, and Politics)

SPOTLIGHT

Critics of the ESA call it an expensive failure because only 37 species have been removed from the endangered list. Most biologists insist that it has not been a failure, for four reasons.

First, species are listed only when they face serious danger of extinction. This is like setting up a poorly funded hospital emergency room that takes only the most desperate cases, often with little hope for recovery, and saying it should be shut down because it has not saved enough patients.

Second, it takes decades for most species to become endangered or threatened. Not surprisingly, it also takes decades to bring a species in

critical condition back to the point where it can be removed from the list. Expecting the ESA—which has been in existence only since 1973—to quickly repair the biological depletion of many decades is unrealistic.

Third, according to federal data the conditions of more than half of the listed species are stable or improving and 99% of the protected species are still surviving. A hospital emergency room taking only the most desperate cases and then stabilizing or improving the condition of more than half of its patients and not losing 99% of its patients would be considered an astounding success.

Fourth, the ESA budget was only $58 million in 2005—about what the Department of Defense spends

in a little more than an hour or 20¢ per year per U.S. citizen. To its supporters, it is amazing that the ESA has managed to stabilize or improve the conditions of more than half of the listed species on a shoe-string budget.

Yes, the act can be improved and federal regulators have sometimes been too heavy handed in enforcing it. But instead of gutting or doing away with the ESA, biologists call for it to be strengthened and modified to help protect ecosystems and the nation's overall biodiversity.

Critical Thinking

Should the budget for the Endangered Species Act be drastically increased? Explain.

In 1903, President Theodore Roosevelt established the first U.S. federal wildlife refuge at Pelican Island, Florida, to help protect birds such as the brown pelican (Figure 11-20) from extinction. Since then, the National Wildlife Refuge System has grown to include 544 refuges with 81% of the area devoted to refuges in Alaska. More than 35 million Americans visit these refuges each year to hunt, fish, hike, or watch birds and other wildlife.

More than three-fourths of the refuges serve as vital wetland sanctuaries for protecting migratory waterfowl. One-fifth of U.S. endangered and threatened species have habitats in the refuge system, and some refuges have been set aside for specific endangered species. These areas have helped Florida's key deer, the brown pelican (Figure 11-20), and the trumpeter swan to recover.

Conservation biologists call for setting aside more refuges for endangered plants. They also urge Congress and state legislatures to allow abandoned military lands that contain significant wildlife habitat to become national or state wildlife refuges.

According to a General Accounting Office study, activities considered harmful to wildlife occur in nearly 60% of the nation's wildlife refuges. A 2002 study by the National Wildlife Refuge Association found that invasive species are wreaking havoc on many of the nation's wildlife refuges. Too much hunting and fishing (allowed on nearly two-thirds of the refuges) and use of powerboats and off-road vehicles

can take their toll on wildlife populations in heavily used refuges.

In 1997, the U.S. Congress passed the National Refuge System Improvement Act. It calls for insuring that the biological diversity and integrity and environmental health of the system are maintained. It also directs the USFWS to consider the effects of surrounding

Jeremy Woodhouse/Peter Arnold, Inc.

Figure 11-20
Natural capital protection: in 1903, U.S. President Theodore Roosevelt helped protect the brown pelican and several other bird species from extinction by establishing the nation's first wildlife refuge at Pelican Island, Florida.

areas on refuges and to develop research and management partnerships with other agencies, organizations, and neighboring landowners.

Gene Banks, Botanical Gardens, and Wildlife Farms (Science, Economics, and Stewardship)

Gene banks and botanical gardens and using farms to raise threatened species can help prevent extinction, but these options lack funding and storage space.

Gene or *seed banks* preserve genetic information and endangered plant species by storing their seeds in refrigerated, low-humidity environments. More than 100 seed banks around the world collectively hold about 3 million samples.

Scientists urge the establishment of many more such banks, especially in developing countries. But some species cannot be preserved in gene banks. Also, the banks are expensive to operate, and their accidental destruction by fire, power outages, or other means would destroy the seeds they store.

The world's 1,600 *botanical gardens* and *arboreta* contain living plants, representing almost one-third of the world's known plant species. These facilities help educate an estimated 150 million visitors each year about the need for plant conservation. But these sanctuaries have too little storage capacity and too little funding to preserve most of the world's rare and threatened plants. They contain only about 3% of the world's rare and threatened plant species.

Raising individuals on *farms* for commercial sale can take the pressure off some endangered and threatened species. Farms in Florida raise alligators for their meat and hides. *Butterfly farms* flourish in Papua New Guinea, where many butterfly species are threatened by development activities.

Zoos and Aquariums (Science, Economics, and Stewardship)

Zoos and aquariums can help protect endangered animal species, but there is too little funding and storage space.

Zoos, aquariums, game parks, and animal research centers are being used to preserve some individuals of critically endangered animal species, with the long-term goal of reintroducing them into protected wild habitats.

Two techniques for preserving endangered terrestrial species are egg pulling and captive breeding. *Egg pulling* involves collecting wild eggs laid by critically endangered bird species and then hatching them in zoos or research centers. In *captive breeding,* some or all of the wild individuals of a critically endangered species are captured for breeding in captivity, with the aim of reintroducing the offspring into the wild. Cap-

tive breeding has been used to save the peregrine falcon and the California condor (Figure 11-3).

Other techniques for increasing the populations of captive species include artificial insemination, surgical implantation of eggs of one species into a surrogate mother of another species (embryo transfer), use of incubators, and cross-fostering (in which the young of a rare species are raised by parents of a similar species). Scientists also use computer databases of the family lineages of species in zoos and DNA analysis to match individuals for mating—a computer dating service for zoo animals—and to prevent genetic erosion through inbreeding.

Proponents urge zoos and wildlife managers to collect and freeze cells of endangered species for possible cloning. In 2005, there was talk of trying to use DNA samples in birdshells in museums to bring back the Carolina parakeet that has been extinct since 1892.

? THINKING ABOUT BRINGING BACK EXTINCT SPECIES Do you favor using DNA samples to bring back extinct species? If so, what would be your favorite candidate for a comeback? Would the passenger pigeon (Core Case Study, p. 222) be on your list? Name a species that you would not like to see brought back.

The ultimate goal of captive breeding programs is to build up populations to a level where they can be reintroduced into the wild. But after more than two decades of captive breeding efforts, only a handful of endangered species have been returned to the wild. Examples shown in Figure 11-3 include the black-footed ferret, California condor, and golden lion tamarin. Most reintroductions fail because of lack of suitable habitat, inability of individuals bred in captivity to survive in the wild, or renewed overhunting or capture of some returned species.

Lack of space and money limits efforts to maintain populations of endangered species in zoos and research centers. The captive population of each species must number 100–500 individuals to avoid extinction through accident, disease, or loss of genetic diversity through inbreeding. Recent genetic research indicates that 10,000 or more individuals are needed for an endangered species to maintain its capacity for biological evolution.

Zoos and research centers contain only about 3% of the world's rare and threatened plant species. The major conservation role of these facilities will be to help educate the public about the ecological importance of the species they display and the need to protect their habitat.

Public aquariums that exhibit unusual and attractive fish and some marine animals such as seals and dolphins also help educate the public about the need to protect such species. In the United States, more than 35 million people visit aquariums each year. Many scientists and members of the public praise the Monterey

Bay Aquarium in Monterey, California (USA), for its educational and aquatic research efforts. Mostly because of limited funds public aquariums have not served as effective gene banks for endangered marine species, especially marine mammals that need large volumes of water.

Instead of seeing zoos and aquariums as sanctuaries, some critics claim that most of them imprison once-wild animals. They also contend that zoos and aquariums can foster the notion that we do not need to preserve large numbers of wild species in their natural habitats. Proponents counter that these facilities play an important role in educating the public about wildlife and the need to protect biodiversity.

Other people criticize zoos and aquariums for putting on shows in which animals wear clothes, ride bicycles, or perform tricks. They see such exhibitions as fostering the idea that the animals exist primarily to entertain us and, in the process, raise money for their keepers. **Question:** What is your stand on this issue? Explain.

Regardless of their benefits and drawbacks, zoos, aquariums, and botanical gardens are not biologically or economically feasible solutions for most of the world's current endangered species and the much larger number of species expected to be threatened over the next few decades.

RECONCILIATION ECOLOGY

Sharing the World with Other Species (Science and Stewardship)

Reconciliation ecology involves finding ways to share the places we dominate with other species.

In 2003, ecologist Michael L. Rosenzweig wrote a book entitled *Win–Win Ecology: How Earth's Species Can Survive in the Midst of Human Enterprise.* Rosenzweig strongly supports Edward O. Wilson's eight-point program to help sustain the earth's biodiversity (p. 219). He also supports the species protection strategies discussed in this chapter.

But he contends that, in the long run, these approaches will fail for two reasons. *First,* current fully protected reserves are devoted to saving only about 5% of the world's terrestrial area, excluding polar and other uninhabitable areas. To Rosenzweig, the real challenge is to help sustain wild species in the human-dominated portion of nature that makes up 95% of the planet's terrestrial ecological "cake."

Second, setting aside funds and refuges and passing laws to protect endangered and threatened species are essentially desperate attempts to save species that are in deep trouble. They can help a few species, but the real challenge is to keep more species away from the brink of extinction.

Rosenzweig suggests that we develop a new form of conservation biology, called **reconciliation ecology.** This science focuses on establishing and maintaining new habitats to conserve species diversity in places where people live, work, or play. In other words, we need to learn how to share the spaces we dominate with other species.

Implementing Reconciliation Ecology (Science and Stewardship)

Some people are finding creative ways to practice reconciliation ecology in their neighborhoods and cities.

Practicing reconciliation ecology begins by looking at the habitats we prefer. Given a choice, most people prefer a grassy and fairly open habitat with a few scattered trees and many people prefer to live near a stream, lake, river, or ocean. We also love flowers.

The problem is that most species do not like what we like or cannot survive in the habitats we prefer. No wonder so few of them live with us.

So what do we do? Reconciliation ecology goes far beyond efforts to attract birds to backyards. For example, we can protect vital insect pollinators such as native butterflies and bees that are especially vulnerable to insecticides and loss of critical habitat. Neighborhoods and cities could agree to reduce or eliminate the use of pesticides on their lawns, fields, golf courses, and parks. Neighbors could also work together in planting gardens of flowering plants as a source of food for pollinating insect species. And neighborhoods and farmers can build devices from wood and plastic straws that provide holes to serve as nesting sites for pollinating bees. Maintaining habitats for an insect-eating bat species (Case Study, p. 227) could keep down mosquitoes and other pesky insects in a neighborhood.

Some monoculture grass yards could be replaced with diverse yards using plant species adapted to local climates that are selected to attract certain species. This would help keep down insect pests, save water, and require less use of noisy and polluting lawnmowers.

Communities could have contests and awards for people who design the most biodiverse and species-friendly yards and gardens. Signs could describe the type of ecosystem being mimicked and the species being protected as a way to educate and encourage experiments by other people. Some creative person might be able to design more biologically diverse golf courses and cemeteries. People have already worked together to help preserve bluebirds within human-dominated habitats (Case Study, p. 246).

In Berlin, Germany, people have planted gardens on many large rooftops. These can be designed to support a variety of species by varying the depth and type of soil and their exposure to sun. Such roofs also save energy by providing insulation, help cool cities, and

Using Reconciliation Ecology to Protect Bluebirds (Science and Stewardship)

CASE STUDY

Populations of bluebirds in much of the eastern United States are declining, for two reasons.

First, these birds nest in tree holes of a certain size. In the past, dead and dying trees provided plenty of these holes. Today, timber companies often cut down all of the trees, and many homeowners manicure their property by removing dead and dying trees.

Second, two aggressive, abundant, and nonnative bird species—starlings and house sparrows—like to nest in tree holes and take them away from bluebirds. To make matters worse, starlings eat blueberries

Figure 11-B Sustaining natural capital: male Eastern bluebird on a nesting box.

that bluebirds need to survive during the winter.

People have come up with a creative way to help save the bluebird. They have designed nest boxes with holes large enough to accommodate bluebirds but too small for starlings (Figure 11-B). They also made the bluebird boxes deep enough to make them unattractive nesting sites for the sparrows.

In 1979, the North American Bluebird Society was founded to spread the word and encourage people to use the bluebird boxes on their properties and to keep house cats away from nesting bluebirds. Now bluebird numbers are building back up.

Critical Thinking

Can you come up with a reconciliation ecology project to help protect a threatened bird or other species in your neighborhood or school?

---

conserve water by reducing evapotranspiration. Reconciliation ecology proponents call for a global campaign to use the roofs of the world to help sustain biodiversity. *Green Career:* Rooftop garden designer

San Francisco's Golden Gate Park is a large oasis of gardens and trees in the midst of a major city. It is a good example of reconciliation ecology because it was designed and planted by humans who transformed it from a system of sand dunes.

The Department of Defense controls about 10 million hectares (25 million acres) of land in the United States. Biologists propose using some of this land for developing and testing reconciliation ecology projects. Some college campuses and schools might also serve as reconciliation ecology laboratories. How about yours? *Green Career:* Reconciliation ecology specialist

RESEARCH FRONTIER Determining where and how reconciliation ecology can work best

Individuals can also help prevent the premature extinction of wild species. In Thailand, Pilai Poonswad decided to do something about poachers taking hornbills—large, beautiful, and rare birds—from a rain forest. She visited the poachers in their villages and showed them why the birds are worth more alive than dead. Today, some ex-poachers are earning money by taking ecotourists into the forest to see these magnificent birds. Because of their vested financial interest in preserving the hornbills, they help protect the birds from poachers. Figure 11-21 lists some things you can do to help prevent the premature extinction of species by practicing stewardship or good earthkeeping.

What Can You Do?

Protecting Species

- Do not buy furs, ivory products, and other materials made from endangered or threatened animal species.

- Do not buy wood and paper products produced by cutting remaining old-growth forests in the tropics.

- Do not buy birds, snakes, turtles, tropical fish, and other animals that are taken from the wild.

- Do not buy orchids, cacti, and other plants that are taken from the wild.

- Spread the word. Talk to your friends and relatives about this problem and what they can do about it.

Figure 11-21 Individuals matter: ways to help premature extinction of species. QUESTIONS: *Which two of these actions do you think are the most important? Which of these actions do you plan to take?*

Revisiting Passenger Pigeons and Sustainability

The disappearance of the passenger pigeon (Core Case Study, p. 222) in a short time was a blatant example of the effects of uninformed and uncaring human activities. We have learned and done a lot since then to help protect some species from premature extinction due to our activities.

Despite these efforts there is overwhelming evidence that we are in the midst of wiping out as much as half of the world's species within your lifetime. Part of the problem is ecological ignorance, but most of it is not having the political and ethical will to act on what we know.

Acting to prevent the premature extinction of species and to preserve their habitats (Chapter 10) is a key to sustainability. It helps preserve the earth's biodiversity and not disrupt species interactions that help control population sizes and the energy flow and matter cycling in ecosystems. In the next chapter, you will learn about how we can help sustain the species and aquatic life zones that make up the world's aquatic biodiversity.

Protecting biodiversity is no longer simply a matter of passing and enforcing endangered species laws and setting aside parks and preserves. It will also require slowing climate change that will affect such protected habitats and species and reducing the size and ecological footprints of the human species.

We know what to do. Perhaps we will act in time.

EDWARD O. WILSON

CRITICAL THINKING

1. Discuss your gut-level reaction to the following statement: "Eventually, all species become extinct. Thus, it does not really matter that the passenger pigeon (Core Case Study, p. 222) is extinct, and that the whooping crane and the world's remaining tiger species are endangered mostly because of human activities." Be honest about your reaction, and give arguments for your position.

2. Do you accept the ethical position that each *species* has the inherent right to survive without human interference, regardless of whether it serves any useful purpose for humans? Explain. Would you extend this right to the *Anopheles* mosquito, which transmits malaria, to infectious bacteria, and to individual tigers that have killed people? Explain.

3. Explain why you agree or disagree with **(a)** using animals for research, **(b)** keeping animals captive in a zoo or aquarium, and **(c)** killing surplus animals produced by a captive-breeding program in a zoo when no suitable habitat is available for their release.

4. What would you do if **(a)** your yard and house were invaded by fire ants, **(b)** you found bats flying around your yard at night, and **(c)** deer invaded your yard and ate your shrubs, flowers, and vegetables?

5. Which of the following statements best describes your feelings toward wildlife:
  (a) As long as it stays in its space, wildlife is okay.
  (b) As long as I do not need its space, wildlife is okay.
  (c) I have the right to use wildlife habitat to meet my own needs.
  (d) When you have seen one redwood tree, elephant, or some other form of wildlife, you have seen them all, so lock up a few of each species in a zoo or wildlife park and do not worry about protecting the rest.
  (e) Wildlife should be protected. Reflect on what your answer reveals about your environmental worldview (p. 22).

6. List your three favorite species. Why are they your favorites? Reflect on what your choice of favorite species tells you about your attitudes toward most wildlife.

7. Environmental groups in a heavily forested state want to restrict logging in some areas to save the habitat of an endangered squirrel. Timber company officials argue that the survival of one type of squirrel is not as important as the well being of the many families who will be affected if the restriction causes the company to lay off hundreds of workers. If you had the power to decide this issue, what would you do and why? Can you come up with a compromise?

8. In 2006, Lester R. Brown estimated that protecting the earth's biodiversity would cost about $31 billion a year. As the world's most affluent country, suppose that the United States agreed to put up half of this money, at an average annual cost of $52 per American. Would you support doing this? Explain. What other part or parts of the federal budget would you decrease to come up with these funds?

9. Congratulations! You are in charge of preventing the premature extinction of the world's existing species from human activities. What three things would you do to accomplish this goal?

10. List two questions that you would like to have answered as a result of reading this chapter.

PROJECTS

1. Make a log of your own consumption of all products for a single day. Relate your level and types of consumption to the decline of wildlife species and the increased destruction, degradation, and fragmentation of wildlife habitats in the United States (or the country where you live) and in tropical forests.

2. Identify examples of habitat destruction or degradation in your community that have had harmful effects on the populations of various wild plant and animal species. Develop a management plan for rehabilitating these habitats and species.

3. Choose a particular animal or plant species that interests you and use the library or the Internet to find out **(a)** its numbers and distribution, **(b)** whether it is threatened with extinction, **(c)** the major future threats to its survival, **(d)** actions that are being taken to help sustain this species, and **(e)** a type of reconciliation ecology that might be useful in sustaining this species. You might want to use the world's largest database of species developed by the World Wildlife Fund. It can be accessed at **http://worldwildlife.org/wildfinder**.

4. Work with your classmates to develop an experiment in reconciliation ecology for your campus.

5. Make a concept map of this chapter's major ideas, using the section heads, subheads, and key terms (in boldface). Look on the website for this book for information about making concept maps.

LEARNING ONLINE

The website for this book contains study aids and many ideas for further reading and research. They include a chapter summary, review questions for the entire chapter, flash cards for key terms and concepts, a multiple-choice practice quiz, interesting Internet sites, references, information about green careers, and a guide for accessing thousands of InfoTrac® College Edition articles. Log into

www.thomsonedu.com/biology/miller

Then choose Chapter 11, and select a learning resource. For access to animations, additional quizzes, chapter outlines and summaries, register and log into

at **www.thomsonedu.com** using the access code card in the front of your book.

Active Graphing

Log into ThomsonNow at www.thomsonedu.com to explore the graphing exercise for this chapter.

12 Sustaining Aquatic Biodiversity

CORE CASE STUDY
A Biological Roller Coaster Ride in Lake Victoria

Lake Victoria, a large, shallow lake in East Africa (Figure 12-1, left), has been in ecological trouble for more than two decades.

Until the early 1980s, the lake had 500 species of fish found nowhere else. About 80% of them were small fishes known as cichlids (pronounced "SIK-lids"), which feed mostly on detritus, algae, and zooplankton. Since 1980 some 200 of the cichlid species have become extinct, and some of those that remain are in trouble.

Four factors caused this dramatic loss of aquatic biodiversity. *First,* there was a large increase in the population of the Nile perch (Figure 12-1, middle). This large predatory fish was deliberately introduced into the lake during the 1950s and 1960s to stimulate exports of the fish to several European countries, despite warnings by biologists that this huge fish with a big appetite would reduce or eliminate many defenseless native fish species. The population of this large and prolific fish exploded, devoured the cichlids and by 1986 had wiped out over 200 of these species.

Introducing the perch had other social and ecological effects. The mechanized fishing industry increased poverty and malnutrition by putting many small-scale fishers and fish vendors out of business. And local forests were depleted for firewood because of the perch's oily flesh that must be preserved by smoking instead of sun drying.

Second, in the 1980s, the lake began experiencing frequent algal blooms because of nutrient runoff from surrounding farms, deforested land, untreated sewage, and declines in the populations of the algae-eating cichlids.

Third, since 1987 the nutrient-rich lake has been invaded by the water hyacinth (Figure 12-1, right). This rapidly growing plant carpeted large areas of the lake, blocked sunlight, deprived fish and plankton of oxygen, and reduced the diversity of important aquatic plant species. Scientists reduced this problem at strategic locations by mechanical removal and by introducing two weevils for biological control of the hyacinth.

Fourth, the Nile perch population is decreasing because it severely reduced its own food supply of smaller fishes—an example of one of the four principles of sustainability in action—and also shows signs of being overfished. This may allow a gradual increase in the populations of some of the remaining cichlids.

This ecological story about the dynamics of large aquatic systems illustrates that we can never do just one thing when we intrude into a poorly understood ecosystem. There are always unintended consequences.

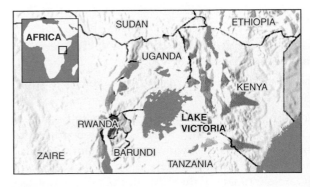

Figure 12-1 Natural capital degradation: the Nile perch (middle) is a fine food fish that can weigh more than 91 kilograms (200 pounds). However, this deliberately introduced fish has played a key role in a major loss of biodiversity in East Africa's Lake Victoria (left), which provides fish for 30 million people. The right photo shows how invasive water hyacinths, supported by nutrient runoff, blocked a ferry terminal on the Kenyan part of Lake Victoria in 1997.

The coastal zone may be the single most important portion of our planet. The loss of its biodiversity may have repercussions far beyond our worst fears.

G. Carleton Ray

This chapter looks at threats to aquatic diversity and how we can sustain more of this vital economic and ecological resource. It addresses the following questions:

- What do we know about aquatic biodiversity, and what is its economic and ecological importance?

- How are human activities affecting aquatic biodiversity?

- How can we protect and sustain marine biodiversity?

- How can we manage and sustain the world's marine fisheries?

- How can we protect, sustain, and restore wetlands?

- How can we protect, sustain, and restore lakes, rivers, and freshwater fisheries?

AQUATIC BIODIVERSITY

What Do We Know about Aquatic Biodiversity? (Science)

We know fairly little about the biodiversity of the world's marine and freshwater systems.

Although we live on a water planet (Figure 6-2, p. 127), we have explored only about 5% of the earth's global ocean and know fairly little about its biodiversity and how it works. We also have limited knowledge about freshwater biodiversity.

However, scientists have established three general patterns of marine biodiversity. *First,* the greatest marine biodiversity occurs in coral reefs, estuaries, and the deep-ocean floor. *Second,* biodiversity is higher near coasts than in the open sea because of the greater variety of producers and habitats in coastal areas. *Third,* biodiversity is higher in the bottom region of the ocean than in the surface region because of the greater variety of habitats and food sources on the ocean bottom.

Scientific investigation of poorly understood marine and freshwater aquatic systems is a *research frontier* whose study could result in immense ecological and economic benefits. Good news: Researchers are developing new techniques for studying and monitoring the ocean (Spotlight, at right).

F̵ RESEARCH FRONTIER Exploring marine and freshwater ecosystems, their species, and species interactions

SPOTLIGHT

High-Tech Ocean Exploration (Science)

Ocean scientists can now study and monitor the ocean without leaving their laboratories. Small underwater gliders that are 2 meters (6 feet) long and weigh about 45 kilograms (100 pounds) are now moving through the world's oceans and sending data to marine researchers.

These high-tech gliders run silently on very little power for month-long missions and can move up and down by sucking in and shooting out water to change their buoyancy. And they can be programmed to surface and beam data back to land-based labs via satellite.

These reusable gliders cost about $25,000 each, compared to up to $15,000 per day needed to run a large oceangoing research vessel. Currently, at least 15 ocean research laboratories in the United States each deploy up to 20 gliders in oceans throughout the world.

In 2005, scientists used an underwater glider to monitor pollutants discharged from the mouth of New York's Hudson River for two weeks. They used a harmless dye to find out how the pollutants interacted with the Atlantic Ocean.

Some envision using this technology to monitor the movement and power of waves to identify promising sites where wave power could be used to produce electricity. The gliders can also send back information on populations of commercially important fish and monitor the effectiveness of offshore aquaculture cages used to produce fish.

Critical Thinking

List two ways in which this new way of studying the ocean could benefit you.

Values of Aquatic Biodiversity (Science and Economics)

The world's marine and freshwater systems provide important ecological and economic services.

We should care about aquatic biodiversity because it helps keep us alive and supports our economies. Marine systems provide a variety of important ecological and economic services (Figure 6-4, p. 129). A very conservative estimate of the value of their ecological services is $21 trillion a year—about twice that of the world's terrestrial ecosystems, including croplands.

At least 3.5 billion people—more than half of the world's population—depend on the seas for their primary source of food. According to the United Nations, this number could double to 7 billion in 2025. Chemicals from several types of algae, sea anemones (Figure 7-9b, p. 154), sponges, and mollusks have antibiotic and anticancer properties. Anticancer chemicals have also been extracted from porcupine fish, puffer fish, and shark liver. We use chemicals from seaweeds and octopuses to treat hypertension and coral material to reconstruct our bones. These are only a few of many examples of the oceans' ecological and economic services.

Freshwater systems, which occupy only 1% of the earth's surface, also provide important ecological and economic services (Figure 6-14, p. 136). A very conservative estimate of the value of their ecological services is $1.7 trillion a year.

HUMAN IMPACTS ON AQUATIC BIODIVERSITY

Loss and Degradation of Aquatic Habitat: Our Large Aquatic Footprints (Science)

Human activities have destroyed or degraded a large proportion of the world's coastal wetlands, coral reefs, mangroves, and ocean bottom, and disrupted many of the world's freshwater ecosystems.

As with terrestrial biodiversity, human impacts on aquatic biodiversity can be summarized using the HIPPO acronym (p. 229). H stands for habitat loss and degradation, the greatest threat to the biodiversity of the world's oceans and freshwater systems. Some 90% of fish living in the ocean spawn in coral reefs, mangrove swamps, coastal wetlands, or rivers—areas that are under intense pressure from human activities.

According to the 2005 Millennium Ecosystem Assessment, for example, approximately 20% of the world's diverse coral reefs (Figure 6-1, p. 126, and Figure 6-3, p. 127) have been destroyed (up from 11% in 2000) and another 20% have been damaged, mostly by human activities (Figure 6-12, p. 135). And up to 58% of the world's coral reefs may be severely damaged or destroyed by 2050. Some 15% of the world seagrass beds have disappeared since 1995. And kelp beds (Figure 8-1, left, p. 161) are dying at an alarming rate, including 75% of those in Southern California alone.

Another threat to marine habitat is rising sea levels. During the past 100 years. average sea levels have risen by 10–25 centimeters (4–10 inches), and scientists estimate they will rise another 9–88 centimeters (4–35 inches) during this century, mostly from projected global warming. This would destroy more coral reefs, swamp some low-lying islands in the Pacific, and cover many highly productive coastal wetlands.

The 2005 Millennium Ecosystem Assessment estimated that since 1989, we have removed more than a third of the world's ecologically important mangrove forests (Figure 6-3, p. 127, and Figure 6-8, p. 131) to make way for shrimp farms and other uses. More will be flooded and lost as global warming causes further rises in sea levels.

More than half of the world's coastal wetlands, which serve as key nurseries for commercially important fish and shellfish, have disappeared—primarily the victims of human development.

Many sea-bottom habitats are being degraded and destroyed by dredging operations and trawler fishing boats, which, like giant submerged bulldozers, drag huge nets weighted down with heavy chains and steel plates over ocean bottoms to harvest bottom fish and shellfish (Figure 12-2). Each year thousands of trawlers scrape and disturb an area of ocean bottom about 150 times larger than the area of forests clear-cut each year.

Peter J. Auster/National Undersea Research Center

Peter J. Auster/National Undersea Research Center

Figure 12-2 Natural capital degradation: area of ocean bottom before (left) and after (right) a trawler net, acting like a gigantic plow, scraped it. Such ocean floor communities could take decades or centuries to recover. According to marine scientist Elliot Norse, "Bottom trawling is probably the largest human-caused disturbance to the biosphere." Trawler fishers disagree and claim that ocean bottom life recovers after trawling.

In 2004, some 1,134 scientists signed a statement urging the United Nations to declare a moratorium on bottom trawling in the open ocean—calling it the most destructive fishing practice on the high seas. Another effort involves a private sector buyout of trawler boats. This has been done in the U.S. state of California to help protect the Monterey Bay National Marine Sanctuary.

Habitat disruption is also a problem in freshwater aquatic zones. The 2005 Millennium Ecosystem Assessment reported that the amount of water held behind dams is currently three to six times the amount that flows in natural rivers. And we now take twice as much water each year from rivers and lakes (mostly for agriculture) as we have been withdrawing in the past. This destroys aquatic habitats and water flows and disrupts freshwater biodiversity.

Invasive Species: Aliens in the Water (Science)

Harmful invasive species are an increasing threat to marine and freshwater biodiversity.

Another problem is the deliberate or accidental introduction of hundreds of harmful invasive species (Figure 11-11, p. 234) into coastal waters, wetlands, and lakes throughout the world—the I in HIPPO. These bioinvaders can displace or cause the extinction of native species and disrupt ecosystem functions, as happened to Lake Victoria (Core Case Study, p. 249). Bioinvaders are blamed for about two-thirds of fish extinctions in the United States between 1900 and 2000 and cost the country an average of about $16 million *per hour.*

? THINKING ABOUT NILE PERCH AND LAKE VICTORIA
Would most of the now extinct cichlid fish species in Lake Victoria (Core Case Study, p. 249) still exist today if the Nile perch had not been introduced or might other factors come into play? Explain.

Many of these invaders arrive in the ballast water stored in tanks in large cargo ships to keep them stable. These ships take in ballast water—along with whatever microorganisms and tiny species it contains—in one harbor and dump it in another. Some hitch rides on trillions of pieces of pieces of floating plastic.

Let us take a look at two aquatic invader species. The *Asian swamp eel* has invaded the waterways of south Florida (USA), probably from the dumping of a home aquarium. This rapidly reproducing eel eats almost anything—including many prized fish species—by sucking them in like a vacuum cleaner. It can elude cold weather, drought, fires, and predators (including humans with nets) by burrowing into mud banks. It is also resistant to waterborne poisons because it can breathe air, and it can wriggle across dry land to invade new waterways, ditches, canals, and marshes.

Eventually, this eel could take over much of the waterways of the southeastern United States as far north as Chesapeake Bay.

The *purple loosestrife* (Figure 11-11, p. 234) is a perennial plant that grows in wetlands in parts of Europe. Since the 1880s, it has been imported and used in gardens as an ornamental plant in North America, Australia, and parts of Africa and South America, and it has spread rapidly. A single plant can produce more than 2.5 million seeds a year that are spread by water, in mud, and by becoming attached to wildlife, livestock, people, and tire treads. It reduces wetland biodiversity by displacing native vegetation and reducing habitat for some forms of wetland wildlife. Some conservationists call the spread of this plant the "purple plague."

Hopeful news. Some U.S. states have recently introduced two natural predators of loosestrife from Europe: a weevil species and a leaf-eating beetle. It will take some time to determine the effectiveness of this biological control approach and to be sure the introduced species themselves do not become pests.

Prevention and rapid action are the best ways to deal with invasive species. And there is no time to lose once an invader has been detected. In 1984, Monaco's oceanographic museum dumped a sprig of tropical seaweed (*Caluerpa taxifolia*) into the Mediterranean Sea. When it was discovered, 3 years later, its growth was not much larger than a bath mat. Today, the plant carpets 12,000 hectares (30,000 acres) of the Mediterranean Sea.

Population Growth and Pollution (Science and Economics)

Almost half of the world's people live on or near the world's coastal zones and 80% of ocean water pollution comes from land-based human activities.

In 2006, about 45% of the world's population lived along or near the coastal zone—the first P in HIPPO. By 2010, the UNEP projects that 80% of the world's people will be living in such areas, mostly in gigantic coastal cities. This coastal population growth will add to the already intense pressure on the world's coastal zones.

In 2004, the UNEP estimated that 80% of all ocean pollution comes from land-based coastal activities—the second P in HIPPO. Humans have doubled the flow of nitrogen, mostly from nitrate fertilizers, into the oceans since 1860, and the 2005 Millennium Ecosystem Assessment estimates this flow will increase by another two-thirds by 2050. These inputs of nitrogen (and phosphorus) result in eutrophication of marine and freshwater systems, which can lead to algae blooms (Figure 6-16, right, p. 138) and fish die-offs.

Figure 12-3 Threatened natural capital: before this discarded piece of plastic was removed, this Hawaiian monk seal was slowly starving to death. Each year plastic items dumped from ships and left as litter on beaches threaten the lives of millions of marine mammals, turtles, and seabirds that ingest, become entangled in, or are poisoned by such debris.

Similar pressures are growing in freshwater systems, as more and more people seek homes and places for recreation near lakes and streams. The result is massive inputs of sediment and other wastes from land into lakes (Figure 12-1, right) and streams.

⟲ ? THINKING ABOUT THE POPULATION, POLLUTION, AND LAKE VICTORIA Even without the Nile perch, do you think that many of Lake Victoria's cichlid species would be extinct because of overfertilization from nutrient runoff and the resulting loss of sunlight from the growth of water hyacinths (Figure 12-1, right)? Explain.

Toxic pollutants from industrial and urban areas can kill some forms of aquatic life by poisoning them. And each year plastic items dumped from ships and left as litter on beaches kill up to 1 million seabirds and 100,000 turtles. Such pollutants threaten the lives of millions of marine mammals (Figure 12-3) and countless fish that ingest, become entangled in, or are poisoned by them. And global warming is threatening to disrupt ocean life by increasing its acidity.

Overfishing and Extinction: Gone Fishing, Fish Gone (Science and Economics)

About 75% of the world's commercially valuable marine fish species are overfished or fished near their sustainable limits.

Overfishing is not new. Archaeological evidence indicates that for thousands of years humans living in some coastal areas overharvested fish, shellfish, seals, turtles, whales, and other marine mammals.

What is new today is that modern industrialized fishing fleets can overfish most of the oceans and de-

plete marine life rapidly—the O in HIPPO. Today fish are hunted throughout the world's oceans by a global fleet of millions of fishing boats—some of them longer than a football field. Modern industrial fishing can cause 80% depletion of a target fish species in only 10–15 years. (Spotlight, p. 254.)

The human demand for seafood is outgrowing the sustainable yield of most oceanic fisheries. According to ocean experts at the 2006 Third Global Conference on Oceans, Coasts, and Islands, about 75% of the world's 200 commercially valuable marine fish species are being fished at or beyond their sustainable capacity. In most cases, overfishing leads to *commercial extinction*, which occurs when it is no longer profitable to continue fishing the affected species.

Overfishing is usually only a temporary depletion of fish stocks, as long as depleted areas and fisheries are allowed to recover. But this is changing as industrialized fishing fleets vacuum up more and more of the world's available fish and shellfish. In 1992, for example, Canada's 500-year-old Newfoundland cod fishery collapsed, putting 40,000 fishers and fish processors out of work. By 2006, it had not recovered, despite a total ban on fishing.

According to a 2003 study by a Canadian–German team of scientists, 90% of the large, open-ocean fish such as tuna, swordfish, and marlin have disappeared since 1950. For example, Atlantic stocks of the heavily fished bluefin tuna have dropped by 94%. This is not surprising because a single large bluefin tuna used for sushi in Tokyo's restaurants can bring as much as $100,000. In addition to the economic loss, many of these large fish are important keystone species that help maintain ecosystem function and biodiversity.

Smaller fish are the next targets, as the fishing industry has begun working its way down marine food webs by shifting from large, desirable species to smaller, less-desirable ones. This practice can reduce the breeding stock needed for recovery of depleted species, unravel food webs, and disrupt marine ecosystems. If we keep vacuuming the seas, McDonald's may begin serving barnacle and jellyfish burgers instead of fish sandwiches.

Most fishing boats hunt and capture one or a few commercially valuable species. However, their gigantic nets and incredibly long lines of hooks also catch nontarget species, called *bycatch*. More than one-fourth of the world's annual fish catch consists of such species that are thrown overboard dead or dying. In addition to wasting potential sources of food, this can deplete the populations of bycatch species that play important ecological roles in oceanic food webs.

To sum up: *Many species are overfished, big fish are becoming scarce, smaller fish are next, we throw away*

SPOTLIGHT

Industrial fishing fleets dominate the world's marine fishing industry. They use global satellite positioning equipment, sonar, huge nets and long fishing lines, spotter planes, and huge refrigerated factory ships that can process and freeze their catches. These fleets help meet the growing demand for seafood. But critics say that these highly efficient fleets are *vacuuming the seas*.

Figure 12-A shows the major methods used for the commercial harvesting of various marine fish and shellfish. Until the mid-1980s, fishing fleets from developed counties dominated the ocean catch. Today, most of these fleets come from developing countries, especially in Asia.

Let us look at a few of these methods. *Trawler fishing* is used to catch fish and shellfish—especially shrimp, cod, flounder, and scallops—that live on or near the ocean floor. It involves dragging a funnel-shaped net held open at the neck along the ocean bottom; it is weighted down with chains or metal plates. This scrapes up almost everything that lies on the ocean floor and often destroys bottom habitats—somewhat like clear-cutting the ocean floor (Figure 12-2). Newer trawling nets are large enough to

swallow 12 jumbo jet planes and even larger ones are on the way!

The large mesh of the net allows most small fish to escape but can capture and kill other species such as seals and endangered and threatened sea turtles. Only the large fish are kept. Most of the fish and other aquatic species—called *bycatch*—are thrown back into the ocean dead or dying.

Another method, *purse-seine fishing*, involves catching surface-dwelling species such as tuna, mackerel, anchovies, and herring, which tend to feed in schools near the surface or in shallow areas. After locating a school the fishing vessel surrounds it with a large net called a purse seine. Then they close the net to trap the fish. Nets used to capture yellowfin tuna in the eastern tropical Pacific Ocean have killed large numbers of dolphins that swim on the surface above schools of tuna.

Fishing vessels also use *longlining*. It involves putting out lines up to 130 kilometers (80 miles) long, hung with thousands of baited hooks. The depth of the lines can be adjusted to catch open-ocean fish species such as swordfish, tuna, and sharks or bottom fishes such as halibut and cod. Longlines also hook large numbers of endangered sea turtles, sea-feeding albatross birds, pilot whales, sharks, and dolphins.

With *drift-net fishing*, fish are caught by huge drifting nets that can hang as much as 2,500 meters (1,600 feet) below the surface and be up to 240 kilometers (80 miles) long. This method can lead to overfishing of the desired species and may trap and kill large quantities of unwanted fish and marine mammals (such as dolphins, porpoises, and seals), marine turtles, and seabirds.

Since 1992, a UN ban on the use of drift nets longer than 2.5 kilometers (1.6 miles) in international waters has sharply reduced use of this technique. But longer nets continue to be used because compliance is voluntary and it is difficult to monitor fishing fleets over vast ocean areas. Also, the decrease in drift nets has led to increased use of longlines, which often have similar effects on marine wildlife.

Critical Thinking

What do you think would be the best way to institute and enforce controls over the most environmentally harmful of these fish harvesting methods? How might doing this benefit you?

Figure 12-A (facing page) Natural capital degradation: major commercial fishing methods used to harvest various marine species. These methods have become so effective that many fish species have become commercially extinct.

30% of the fish we catch, and we needlessly kill sea mammals and birds.

Fish species are also threatened with biological extinction, mostly from overfishing, water pollution, wetlands destruction, and excessive removal of water from rivers and lakes. According to the International Union for the Conservation of Nature and Natural Resources (IUCN), 34% of the world's known marine fish species and 71% of the world's freshwater fish species face biological extinction within your lifetime. Indeed, *marine and freshwater fish are threatened with extinction by human activities more than any other group of species* (Figure 11-5, p. 225).

F *RESEARCH FRONTIER* Learning more about how human activities affect aquatic biodiversity

Why Is It Difficult to Protect Aquatic Biodiversity?

Rapidly increasing human impacts, the invisibility of problems, citizen unawareness, and lack of legal jurisdiction hinder protection of aquatic biodiversity.

Protecting aquatic biodiversity, especially marine biodiversity, is difficult for several reasons. *First*, the human ecological footprint is expanding so rapidly into aquatic areas that it is difficult to monitor the impacts.

Second, much of the damage to the oceans is not visible to most people. *Third*, many people incorrectly view the seas as an inexhaustible resource that can absorb an almost infinite amount of waste and pollution.

Finally, most of the world's ocean area lies outside the legal jurisdiction of any country. Thus, it is an

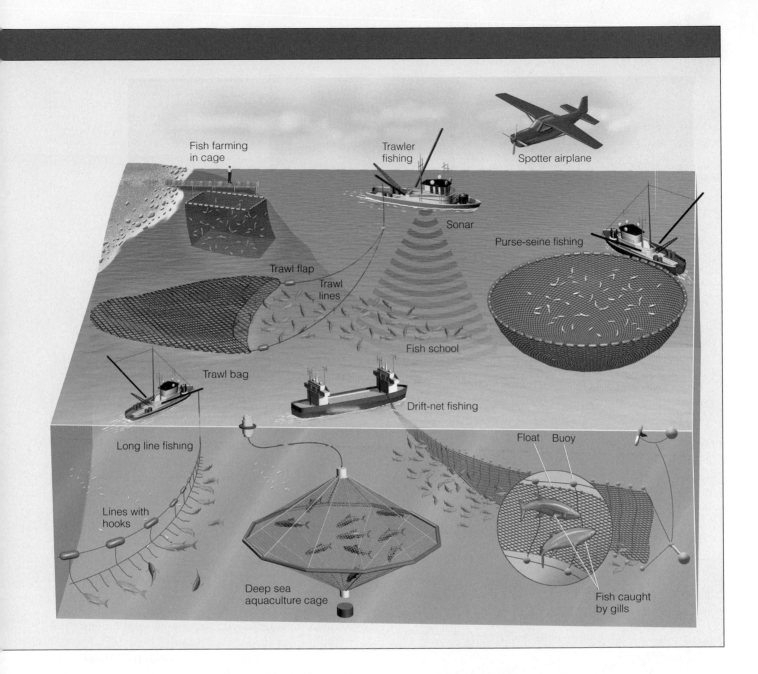

Fish farming
in cage

Trawler
fishing

Spotter airplane

Sonar

Purse-seine fishing

Trawl flap

Trawl
lines

Fish school

Trawl bag

Drift-net fishing

Long line fishing

Float   Buoy

Lines with
hooks

Deep sea
aquaculture cage

Fish caught
by gills

open-access resource, subject to overexploitation because of the tragedy of the commons (p. 12).

PROTECTING AND SUSTAINING MARINE BIODIVERSITY

Protecting Endangered and Threatened Marine Species: Legal and Economic Approaches (Science, Economics, and Politics)

Laws, international treaties, and education can help reduce the premature extinction of marine species.

National and international laws and treaties to help protect marine species include the 1975 Convention on International Trade in Endangered Species (CITES),

the 1979 Global Treaty on Migratory Species, the U.S. Marine Mammal Protection Act of 1972, the U.S. Endangered Species Act of 1973, the U.S. Whale Conservation and Protection Act of 1976, and the 1995 International Convention on Biological Diversity.

The U.S. Endangered Species Act (p. 240) has been used to identify and protect endangered and threatened marine mammal species such as whales, seals, sea lions, turtles, and the Florida manatee (Case Study p. 256).

Six of the world's seven major sea turtle species (Figure 12-4, p. 256)—some of which have been around for 100 million years—are threatened or endangered because of human activities. They include developments on turtle nesting beaches, overharvesting of their eggs for food, increased use of their shells

CASE STUDY

The Florida Manatee and Water Hyacinths (Science)

The Florida manatee, or sea cow (Figure 12-B), is often called a gentle giant because of its slow movement and peaceful nature and its average length of 3 meters (10 feet) and weight of 450 kilograms (1,000 pounds).

Each day these herbivores eat large quantities of aquatic plants found in Florida's salt marshes, mangroves, and bays and in some of its freshwater shallow rivers and canals. Like Lake Victoria (Core Case Study, p. 249), many of these nutrient-rich waters are clogged with growths of the water hyacinth (Figure 12-1, right and Figure 11-11, p. 234), supported by runoff of fertilizers and outputs from sewage treatment plants.

With its huge appetite, the manatee can help keep waters clear of this invasive species. But there is a catch. The Florida manatee is a highly threatened species, with only about 3,000 individuals left. They

Phillip Colla/BruceColeman USA

die from habitat loss, cold weather, entanglement in fishing lines and nets, and when they are hit by speeding boats or slashed by their propellers.

Their road to extinction is also hastened by susceptibility to stress from cold and their low reproductive rate. They have been protected as endangered, and more recently threatened, species by various federal and state acts since 1967. But

Figure 12-B Natural capital degradation: the West Indian manatee, or sea cow, could help control invasions of Florida's nutrient-rich waters by the water hyacinth (Figure 11-11. p. 234), but it is threatened with extinction.

even with such protection, they face a high risk of disappearing from the earth.

Critical Thinking

What three things would you do to help protect the Florida manatee from premature extinction? Use the Internet to find out whether these strategies have been implemented and how successful they have been.

© David B. Fleetham/Tom Stack & Associates

Figure 12-4 Threatened natural capital: endangered green sea turtle.

for making jewelry and their flippers for leather, and unintentional capture and drowning by commercial fishing boats—especially shrimp trawlers and those using long lines of hooks.

Since 1989 the U.S. government has required offshore shrimp trawlers to use turtle exclusion devices (TEDs) that help keep turtles from being caught in their nets. And, in 2004, the United States banned longline swordfish fishing off the Pacific coast to help save dwindling sea turtle populations. These are a few examples of how treaties and laws can help threatened and endangered species.

Other ways to protect these species involve using economic tools. For example, according to a 2004 World Wide Fund for Nature study, sea turtles are worth more to local communities alive than they are dead. The report estimates that sea turtle tourism brings in almost three times as much money as the sale of turtle products such as meat, leather, and eggs. The problem is that individuals seeking to make a quick gain take the turtles before their surrounding communities can realize the longer-term economic benefits and protect them.

Some people are using economic tools to protect, sustain, and restore aquatic systems. One example is an application of *reconciliation ecology* by a restaurant owner (Individuals Matter, at right).

Case Study: Commercial Whaling (Science, Politics, and Ethics)

After many of the world's whale species were overharvested, commercial whaling was banned in 1970, but the ban may be overturned.

Cetaceans are an order of mostly marine mammals ranging in size from the 0.9-meter (3-foot) porpoise to the giant 15- to 30-meter (50- to 100-foot) blue whale. They are divided into two major groups: *toothed whales* and *baleen whales* (Figure 12-5, p. 258).

Toothed whales, such as the porpoise, sperm whale, and killer whale (orca), bite and chew their food and feed mostly on squid, octopus, and other marine animals. *Baleen whales,* such as the blue, gray, humpback, and finback, are filter feeders. They use plates made of baleen, or whalebone, that hang down from their upper jaw to filter plankton from the seawater, especially tiny shrimplike krill (Figure 3-18, p. 65).

? *THINKING ABOUT WHALES* Why are baleen whales more abundant than toothed whales? (Hint: Think food chains.)

Whales are fairly easy to kill because of their large size and their need to come to the surface to breathe. Mass slaughter became efficient with the use of radar and airplanes to locate them, fast ships, harpoon guns, and inflation lances that pump dead whales full of air and make them float.

Whale harvesting, mostly in international waters, has followed the classic pattern of a tragedy of the commons, with whalers killing an estimated 1.5 million whales between 1925 and 1975. This overharvest-ing reduced the populations of 8 of the 11 major species to the point at which it no longer paid to hunt and kill them (*commercial extinction*). It also drove some commercially prized species such as the giant blue whale to the brink of biological extinction (Spotlight, p. 259) and has threatened other species.

In 1946, the International Convention for the Regulation of Whaling established the International Whaling Commission (IWC). Its mission was to regulate the whaling industry by setting annual quotas to prevent overharvesting and commercial extinction.

This did not work well for two reasons. *First,* IWC quotas often were based on inadequate data or ignored by whaling countries. *Second,* without powers of enforcement the IWC was not able to stop the decline of most commercially hunted whale species.

In 1970, the United States stopped all commercial whaling and banned all imports of whale products. Under pressure from conservationists, the U.S. government, and governments of many nonwhaling countries in the IWC, the IWC has imposed a moratorium on commercial whaling since 1986. It worked. The estimated number of whales killed commercially worldwide dropped from 42,480 in 1970 to about 1,300 in 2005.

Despite the ban, Japan, Norway, and Iceland kill about 1,300 whales of certain species each year (Figure 12-6, p. 258) for scientific purposes. Critics see these whale hunts as poorly disguised commercial whaling because the whale meat is sold to restaurants with each whale worth up to $30,000 wholesale. In 2005, Japan more than doubled its whaling catch, allegedly for scientific purposes, from 440 minke whales to 850

INDIVIDUALS MATTER

Creating an Artificial Coral Reef in Israel (Science, Economics, and Stewardship)

Let us zoom to the city of Eliat, Israel, at the northern tip of the Red Sea. There we find a magnificent coral reef at the water's edge, which is a major tourist attraction. To help protect the reef from excessive development and destructive tourism, Israel set aside part of the reef as a nature reserve.

But tourism, industry, and inadequate sewage treatment have destroyed most of the rest of the reef. Enter Reuven Yosef, a pioneer in reconciliation ecology, who has developed an underwater restaurant called the Red Sea Star Restaurant. Take an elevator down two floors and walk into a restaurant surrounded with windows looking out into a beautiful coral reef.

This reef was created from broken pieces of coral. When divers find broken pieces of coral in the nearby reserve they bring them to Yosef's coral hospital.

Most pieces of broken coral soon become infected and die. But researchers have learned how to treat the coral fragments with antibiotics and store them while they are healing in large tanks of fresh seawater.

After several months of healing, divers bring the fragments to the watery area outside the Red Sea Star Restaurant's windows. There they are wired to panels of iron mesh cloth. The coral grow and cover the iron matrix. Then fish and other creatures show up.

Using his creativity and working with nature, Yosef has helped create a small coral reef and provided a beautiful place for restaurant customers to see the reef without having to be divers or snorkelers.

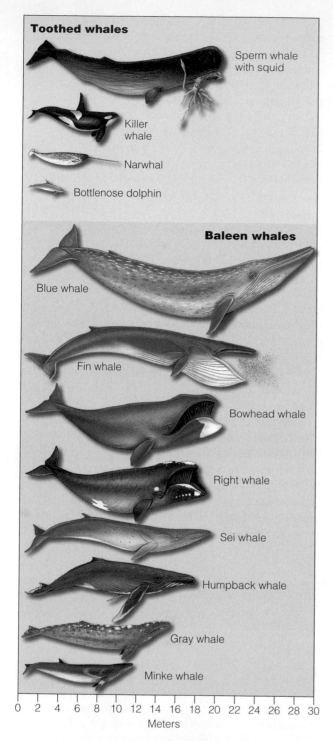

Toothed whales

Sperm whale with squid

Killer whale

Narwhal

Bottlenose dolphin

Baleen whales

Blue whale

Fin whale

Bowhead whale

Right whale

Sei whale

Humpback whale

Gray whale

Minke whale

| 0 | 2 | 4 | 6 | 8 | 10 | 12 | 14 | 16 | 18 | 20 | 22 | 24 | 26 | 28 | 30 |

Meters

Figure 12-5 Natural capital: examples of cetaceans, which can be classified as either toothed whales or baleen whales.

minke whales and 10 fin whales, and began harvesting humpback whales in 2006.

Japan, Norway, Iceland, Russia, and a growing number of small tropical island countries—which Japan brought into the IWC to support its position—hope to overthrow the IWC ban on commercial whaling and reverse the international ban on buying and selling whale products. They argue that commercial whaling should be allowed because it has been a traditional part of the economies and cultures of their countries. They also contend that the ban is based on emotion, not on updated scientific estimates of whale populations.

The moratorium on commercial whaling has led to a sharp rebound in the estimated populations of sperm, pilot, and minke whales. Proponents of whaling see no scientific reason for not resuming controlled and sustainable hunting of these and other whale species with populations of at least 1 million.

They argue that proposed hunting levels are too low to deplete stocks again. And Japan and other pro-whaling nations have agreed to a plan to allow independent observers on whaling vessels. They may also agree to a system that uses DNA tests of whale meat in markets to determine whether or not it came from whales killed according to IWC rules.

Most conservationists disagree. Some argue that whales are peaceful, intelligent, sensitive, and highly social mammals that pose no threat to humans and that should be protected for ethical reasons. Others question IWC estimates of the allegedly recovered whale species, noting the inaccuracy of past estimates of whale populations. Also, many conservationists fear that opening the door to any commercial whaling may eventually lead to widespread harvests of most whale species by weakening current international disapproval and legal sanctions against commercial whaling.

People in the whale watching business also argue that whales are worth more alive than dead. Each year

Tony Martin/WWI/Peter Arnold, Inc.

Figure 12-6 Natural capital degradation: Norwegian whalers harpooning a sperm whale. Norway, Japan, and Iceland kill about 1,300 whales a year, allegedly for scientific purposes. They also believe that increased but sustainable commercial whaling should be allowed for sperm, minke, and pilot whales whose stocks have built back to large numbers.

Code Blue: Near Extinction of the Blue Whale (Science, Economics, and Politics)

The endangered blue whale (Figure 12-5) is the world's largest animal. Fully grown, it is longer than three train boxcars and weighs more than 25 elephants. The adult has a heart the size of a Volkswagen Beetle car, and some of its arteries are so big that a child could swim through them.

Blue whales spend about 8 months a year in Antarctic waters. There they find an abundant supply of krill (Figure 3-18, p. 65), which they filter from seawater by the trillions daily. During the winter, they migrate to warmer waters where their young are born.

Before commercial whaling began an estimated 200,000 blue whales roamed the Antarctic Ocean. Today the species has been hunted to near biological extinction for its oil, meat, and bone. There are probably fewer than 10,000 of these whales left.

A combination of prolonged overharvesting and certain natural characteristics of blue whales caused their decline. Their huge size made them easy to spot. They were caught in large numbers because they grouped together in their Antarctic feeding grounds. They also take 25 years to mature sexually and have only one offspring every 2–5 years. This low reproductive rate makes it difficult for the species to recover once its population falls beneath a certain threshold.

Blue whales have not been hunted commercially since 1964 and have been classified as an endangered species since 1975. Despite this protection, some marine biologists fear that too few blue whales remain for the species to recover and avoid extinction. Others believe that with continued protection they will make a slow comeback.

Critical Thinking

What difference does it make if the blue whale becomes prematurely extinct mostly because of human activities? Explain.

---

about 10 million people go on whale watching boats in the world's $1 billion a year whale watching business.

Proponents of resuming whaling say that people in other countries have no right to tell Japanese, Norwegians, and people in other whaling countries not to eat whales, just because some people like whales. This would be the same as people in India who consider cows sacred telling Americans and Europeans that they should not be allowed to eat beef.

X *HOW WOULD YOU VOTE?* Should carefully controlled commercial whaling be resumed for species with populations of 1 million or more? Cast your vote online at www.thomsonedu.com/biology/miller.

Marine Sanctuaries (Science and Politics)

Fully protected marine reserves make up less than 0.3% of the world's ocean area.

By international law, a country's offshore fishing zone extends to 370 kilometers (200 statute miles) from its shores. Foreign fishing vessels can take certain quotas of fish within such zones, called *exclusive economic zones*, but only with a government's permission. Ocean areas beyond the legal jurisdiction of any country are known as the *high seas*. But laws and treaties pertaining to the high seas are difficult to monitor and enforce, especially when nations such as the United States have failed to ratify the international Law of the Sea Treaty.

Through this international law, the world's coastal nations have jurisdiction over 36% of the ocean surface and 90% of the world's fish stocks. But instead of using this law to protect their fishing grounds, many governments have promoted overfishing, subsidized new fishing fleets, and failed to establish and enforce stricter regulation of fish catches in their coastal waters.

There are attempts to protect marine biodiversity and sustain fisheries by establishing marine sanctuaries. Since 1986, the IUCN has helped establish a global system of *marine protected areas* (MPAs)—areas of ocean partially protected from human activities. There are 1,300 MPAs, almost 200 of them in U.S. waters. However, nearly all MPAs allow dredging, trawler fishing, and other resource extraction activities.

Marine reserves are areas where no extraction and alteration of any living or nonliving resource is allowed. These fully protected areas are designed to preserve entire aquatic ecosystems and to help rebuild commercial and recreational fisheries. More than 20 coastal nations, including the United States, have established marine reserves (see photo 10 in the Detailed Contents) that vary widely in size and their degree of protection.

Marine reserves work and they work fast. Scientific studies show that within fully protected marine reserves, fish populations double, fish size grows by almost a third, fish reproduction triples, and species diversity increases by almost one-fourth. Furthermore, this improvement occurs within 2–4 years after strict protection begins and lasts for decades. Studies also show that reserves benefit nearby fisheries because fish move out of the reserves and currents carry away fish larvae produced inside reserves.

According to the 2005 Millennium Ecosystem Assessment, less than 0.3% of the world's ocean area consists of fully protected marine reserves waters. In other words, *we have failed to strictly protect 99.7% of the world's ocean area from human exploitation.*

In 2004, a joint study by the World Wildlife Fund International and Great Britain's Royal Society for Protection of Birds called for fully protecting at least 30% of the world's oceans as marine reserves to help prevent overfishing and retire depleted commercial fisheries. The estimated cost of establishing and managing a global network of marine reserves would be $12–14 billion a year and would create more than 1 million jobs.

Good news. In 2006, U.S. President George W. Bush created the world's largest protected marine area. He established permanent government protection for 360,000 square kilometers (140,000 square miles) of federal waters surrounding a 1,900-kilometer- (1,200-mile-) long chain of remote and small northwestern Hawaiian islands and atolls.

The resulting Northwestern Hawaiian Marine National Monument is nearly as large as the state of California, larger than 46 of the 50 U.S. states, and more than seven times larger than the nation's 13 national marine sanctuaries combined. This area is home for more than 7,000 species, at least a fourth of them found nowhere else on the earth.

Marine scientists are evaluating how big reserves must be, how many we need, where they should be established, and how they can be connected to take advantage of ocean currents. Most owners of commercial fishing fleets and individuals who fish for recreation oppose establishing such reserves because it restricts where they can fish. But this is beginning to change. Fishers strongly opposed the establishment of a reserve for snapper off the coast of New England in the United States. Now they are strong supporters because the reserve has helped increase the local population of snapper 40-fold.

E **RESEARCH FRONTIER** Determining characteristics of marine reserves that will maximize their effectiveness

? **THINKING ABOUT MARINE RESERVES** Do you support setting aside 30% of the world's oceans as fully protected marine reserves? Explain. How would this affect your life?

Another approach involves using *zoning rules* to protect and manage marine resources for entire bodies of water. In 1975, Australia's Great Barrier Reef Marine Park was the first example of applying large-scale zon-

ing as part of its management plan. Today this reef is the world's largest protected marine reserve, with fishing and shipping banned on one-third of the reef. Nevertheless, in 2005, one of Australia's experts on marine studies warned that the reef could be wiped out by coral bleaching and other effects of global warming on its corals by 2050.

In 2000, New Zealand zoned all of its waters up to 320 kilometers (200 statue miles) off its coastline. A 2004 study by the Pew Oceans Commission recommended greatly increased use of such zoning in U.S. waters, but so far the federal government has not done this.

Integrated Coastal Management (Science and Politics)

Some communities work together to develop integrated plans for managing their coastal areas.

Integrated coastal management is a community-based effort to develop and use coastal resources more sustainably. The overall aim is for fishers scientists, conservationists, citizens, business interests, developers, and politicians competing for the use of coastal resources to identify shared problems and goals. Then they attempt to develop workable, cost-effective, and adaptable solutions that preserve biodiversity and environmental quality while meeting economic and social needs. This requires all participants to seek reasonable short-term trade-offs that can lead to long-term ecological and economic benefits.

Australia's huge Great Barrier Reef Marine Park is managed this way, and more than 100 integrated coastal management programs are being developed throughout the world. In the United States, 90 coastal counties are working to establish coastal management systems, but fewer than 20 of these plans have been implemented.

Revamping U.S. Ocean Policy (Science and Politics)

Two recent studies called for an overhaul of U.S. ocean policy and management.

In 2003 and 2004, the Pew Oceans Commission and the U.S. Commission on Ocean Policy carried out the first broad assessments of U.S. ocean policy in more than 30 years. The two reports differed in some details and recommendations. But both agreed that the coastal waters of the United States are in deep trouble and that laws protecting them need fundamental reforms.

Here are some of their major recommendations:

- Develop a unified national ocean policy such as a National Ocean Policy Act and a national oceans council of advisers within the Executive Office of the President

- Double the federal budget for ocean research
- Centralize and streamline the current fragmented management of the oceans into a high-level National Oceans Agency
- Set up a network of marine reserves, linked by protected corridors, to help protect fish breeding and nursery grounds
- Manage coastal development to minimize damage to coastal habitats and water quality
- Reorient fisheries management to protect and sustain ecosystem functions and productivity rather than relying mostly on catch limits for individual species
- Mount a program to increase public awareness of the economic and ecological values of the oceans

So far, progress in meeting these goals has been slow. In 2006, members of the U.S. commission on ocean policy gave the government a D$^+$ on their efforts so far.

? *THINKING ABOUT OCEAN POLICY* Do you support providing greatly increased government funding to mount a crash program to improve ocean policy in the United States or in the country where you live? Explain. List three ways doing this might benefit you.

MANAGING AND SUSTAINING MARINE FISHERIES

Sustainable Management of Marine Fisheries (Science and Economics)

There are a number of ways to manage marine fisheries more sustainably and protect marine biodiversity.

Figure 12-7 lists some measures for managing global fisheries more sustainably and protecting marine biodiversity.

Another way to reduce overfishing and improve fishery management is to make better estimates of fish populations. The traditional approach has used a *maximum sustained yield* (MSY) model to project the maximum number of fish that can be harvested annually from a fish stock without causing a population drop.

However, the MSY concept has not worked very well because of the difficulty in estimating the populations and growth rates of fish stocks. Also, harvesting a particular species at its estimated maximum sustainable level can affect the populations of other target and nontarget fish species and other marine organisms—those pesky connections again.

Solutions

Managing Fisheries

Fishery Regulations

Set catch limits well below the maximum sustainable yield

Improve monitoring and enforcement of regulations

Economic Approaches

Sharply reduce or eliminate fishing subsidies

Charge fees for harvesting fish and shellfish from publicly owned offshore waters

Certify sustainable fisheries

Protected Areas

Establish no-fishing areas

Establish more marine protected areas

Rely more on integrated coastal management

Consumer Information

Label sustainably harvested fish

Publicize overfished and threatened species

Bycatch

Use wide-meshed nets to allow escape of smaller fish

Use net escape devices for seabirds and sea turtles

Ban throwing edible and marketable fish back into the sea

Aquaculture

Restrict coastal locations for fish farms

Control pollution more strictly

Depend more on herbivorous fish species

Nonnative Invasions

Kill organisms in ship ballast water

Filter organisms from ship ballast water

Dump ballast water far at sea and replace with deep-sea water

Figure 12-7 Solutions: ways to manage fisheries more sustainably and protect marine biodiversity. QUESTION: *Which four of these solutions do you think are the most important?*

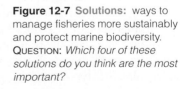

In recent years, some fishery biologists and managers have begun using the *optimum sustained yield* (OSY) concept. It attempts to take into account interactions with other species and to provide more room for error. But this concept still depends on the poorly understood biology of fish and changing ocean conditions.

F RESEARCH FRONTIER Studying fish and their habitats to make better estimates of optimum sustained yields for fisheries

Another approach is *multispecies management* of a number of interacting species, which takes into account their competitive and predator–prey interactions. Such models are still in the development and testing stage.

A more ambitious approach is to develop complex computer models for managing multispecies fisheries in *large marine systems*. However, it is a political challenge to get groups of nations to cooperate in planning and managing them.

A basic problem is the uncertainties built into using any of these approaches. As a result, many fishery scientists and environmentalists are increasingly interested in using the *precautionary principle* for managing fisheries and large marine systems. This means sharply reducing fish harvests and closing some overfished areas until they recover and we have more information about what levels of fishing can be sustained.

Regulating Fish Harvests: Cooperation Works (Economics, Politics, and Stewardship)

Some fishing communities regulate fish harvests on their own and others work with the government to regulate them.

Traditionally, many coastal fishing communities have developed allotment and enforcement systems that have sustained their fisheries, jobs, and communities for hundreds and sometimes thousands of years. An example is Norway's Lofoten fishery, one of the world's largest cod fisheries. For 100 years it has been self-regulated, with no government quota regulations and no participation by the Norwegian government. Cooperation can work.

However, the influx of large modern fishing boats and fleets has weakened the ability of many coastal communities to regulate and sustain local fisheries. Many community management systems have been replaced by *comanagement,* in which coastal communities and the government work together to manage fisheries. In this approach, a central government typically sets quotas for various species, divides the quotas among communities, and may limit fishing seasons and regulate the type of fishing gear that can be used to harvest a particular species.

Each community then allocates and enforces its quota among its members based on its own rules. Often communities focus on managing inshore fisheries and the central government manages the offshore fisheries. When it works, community-based comanagement illustrates that the tragedy of the commons is not inevitable. Supplement 16 on p. S51 discusses the use of the marketplace to control access to fisheries.

PROTECTING, SUSTAINING, AND RESTORING WETLANDS

Wetlands Protection in the United States (Economics and Politics)

Requiring government permits for filling or destroying U.S. wetlands has slowed their loss, but attempts to weaken this protection continue.

Coastal and inland wetlands are important reservoirs of aquatic biodiversity that provide vital ecological and economic services. At the 2005 World Summit on Sustainable Development, a study by over 1,300 scientists reported that the ecological services provided by an intact wetland in Canada is worth an estimated $6,000 per hectare ($2,400 per acre). By comparison, the wetland is worth only one-third of this amount if it is destroyed and used for intensive agriculture.

Despite their ecological value, the U. S. has lost more than half of its coastal and inland wetlands since 1900. Some other countries have lost even more of their wetlands. For example, Italy has lost 95% of its coastal wetlands. The U. S. state of Louisiana has more than 40% of the nation's saltwater marshes. Since the 1940s, the state has lost an area of these coastal wetlands equal to the size of the U.S. state of Rhode Island, mostly because the state's wetlands have been sinking. This occurs because extensive dams and levees on rivers flowing through the state to the sea interfere with the deposition of sediments that would replace those washed out to sea.

To make matters worse, most of the current coastal wetlands in Louisiana and other coastal states will probably be under water during your lifetime because of rising sea levels caused by global warming. What will happen to the aquatic diversity, including commercially important fish and shellfish, supported by coastal wetlands and the millions of migratory ducks and other birds that spend the winter in such wetlands? More bad news: Warmer water may support algal blooms and lead to huge fish kills.

? THINKING ABOUT WETLAND LOSSES List three aspects of your lifestyle that help cause loss of coastal wetlands.

In the United States, a federal permit is required to fill or to deposit dredged material into wetlands occupying more than 1.2 hectares (3 acres). According to the U.S. Fish and Wildlife Service, this law helped cut the average annual wetland loss by 80% since 1969.

However, there are continuing attempts to weaken wetlands protection by using unscientific criteria to classify areas as wetlands. Also, only about 6% of remaining inland wetlands are under federal protection, and federal, state, and local wetland protection is weak.

The stated goal of current U.S. federal policy is zero net loss in the function and value of coastal and inland wetlands. A policy known as *mitigation banking* allows destruction of existing wetlands as long as an equal area of the same type of wetland is created or restored

Some wetland restoration projects have been successful (Figure 12-8 and Individuals Matter, at right). However, a 2001 study by the National Academy of Sciences found that at least half of the attempts to create new wetlands fail to replace lost ones and most of the created wetlands do not provide the ecological functions of natural wetlands. The study also found that wetland creation projects often fail to meet the standards set for them and are not adequately monitored.

F RESEARCH FRONTIER Evaluating ecological services provided by wetlands, human impacts on wetlands, and how to preserve and restore wetlands

Increasingly, U.S. developers that destroy existing wetlands are hiring companies that specialize in creating and restoring wetlands in other areas to help make up for the loss of wetlands functions. Or they buy credits from private wetlands banks that have earned credits by restoring wetlands. More than 500 wetlands restoration banks operating in the United States have sold nearly $300 million in credits to developers. Wetlands restoration is becoming a big business.

INDIVIDUALS MATTER

Restoring a Wetland

Humans have drained, filled in, or covered over swamps, marshes, and other wetlands for centuries. They have done this to create rice fields and land to grow other crops, create land for urban development and highways, reduce disease such as malaria caused by mosquitoes, and extract minerals, oil, and natural gas.

Some people have begun to question such practices as we learn more about the ecological and economic importance of coastal and inland wetlands. Can we turn back the clock to restore or rehabilitate lost marshes?

California rancher Jim Callender decided to try. In 1982, he bought 20 hectares (50 acres) of a Sacramento Valley rice field that had been a marsh until the early 1970s. To grow rice, the previous owner had destroyed the marsh by bulldozing, draining, leveling, uprooting the native plants, and spraying with chemicals to kill the snails and other food of the waterfowl.

Callender and his friends set out to restore the marshland. They hollowed out low areas, built up islands, replanted bulrushes and other plants that once were there, reintroduced smartweed and other plants needed by birds, and planted fast-growing Peking willows. After years of care, hand planting, and annual seeding with a mixture of watergrass, smartweed, and rice, the marsh is once again used by migratory waterfowl.

Jim Callender and others have shown that at least part of the continent's degraded or destroyed wetlands can be reclaimed with scientific knowledge and hard work. Such restoration is important, but to most ecologists the real challenge is to protect remaining wetlands from harm in the first place.

U.S. Fish and Wildlife Service

U.S. Fish and Wildlife Service

Figure 12-8 Natural capital restoration: wetland restoration in the midwestern United States before (left) and after (right).

Legally protect existing wetlands

Steer development away from existing wetlands

Use mitigation banking only as a last resort

Require creation and evaluation of a new wetland before destroying an existing wetland

Restore degraded wetlands

Try to prevent and control invasions by nonnative species

Figure 12-9 Solutions: ways to help sustain the world's wetlands. QUESTION: *Which two of these solutions do you think are the most important?*

Ecologists urge using mitigation banking only as a last resort. They also call for creating and evaluating a new wetland *before* any existing wetland can be destroyed. This example of applying the precautionary principle is the reverse of current policy.

[?] *THINKING ABOUT WETLANDS MITIGATION* Should a new wetland be created and evaluated before anyone is allowed to destroy the wetland it is supposed to replace? Explain.

Figure 12-9 lists ways to help sustain wetlands in the United States and elsewhere. Restoring degraded wetlands is important, but the most important strategy is to help prevent wetland loss in the first place. Many developers, farmers, and resource extractors vigorously oppose these remedies.

Case Study: Restoring the Florida Everglades (Science and Politics)

The world's largest ecological restoration project involves trying to undo some of the damage inflicted on Florida's Everglades by human activities.

South Florida's Everglades (USA) was once a 100-kilometer-wide (60-mile-wide), knee-deep sheet of water flowing slowly south from Lake Okeechobee to Florida Bay (Figure 12-10). As this shallow body of water—known as the "River of Grass"—trickled south it created a vast network of wetlands with a variety of wildlife habitats.

Since 1948, a massive water control project has provided south Florida's rapidly growing population with a reliable water supply and flood protection. But is has also contributed to widespread degradation of the original Everglades ecosystem.

Much of the original Everglades has been drained, diverted, paved over, ravaged by nutrient pollution from agriculture, and invaded by a number of plant species. As a result, the Everglades is now less than half its original size. Much of it has also dried out, leaving large areas vulnerable to summer wildfires. And much of its biodiversity has been lost because of reduced water flows, invasive species, and habitat loss and fragmentation from urbanization.

Between 1962 and 1971, the U.S. Army Corps of Engineers transformed the wandering 166-kilometer- (103-mile-) long Kissimmee River (Figure 12-10) into a straight 84-kilometer (56-mile) concrete-lined canal flowing into Lake Okeechobee. The canal provided flood control by speeding the flow of water but it drained large wetlands north of Lake Okeechobee, which farmers then turned into cow pastures.

Below Lake Okeechobee, farmers planted and fertilized vast agricultural fields of sugarcane and vegetables. Historically, the Everglades has been a nutrient-poor aquatic system, with low phosphorus levels. But runoff of phosphorus from fertilizers has greatly increased phosphorus levels. This large nutrient input has stimulated the growth of invasive plants such as cattails, which have taken over and displaced native saw grass, choked waterways, and disrupted food webs in a vast area of the Everglades.

To help preserve the lower end of the system, in 1947, the U.S. government established Everglades National Park, which contains about a fifth of the remaining Everglades. But this did not work—as conservationists had predicted—because the massive water distribution and land development project to the north cut off much of the water flow needed to sustain the park's wildlife.

As a result, 90% of the park's wading birds have vanished, and populations of other vertebrates, from deer to turtles, are down 75–95%. Florida Bay, south of the Everglades is a shallow estuary with many tiny islands or keys. Large volumes of freshwater that once flowed through the park into Florida Bay have been diverted for crops and cities, causing the bay to become saltier and warmer. This and increased nutrient input from crop fields and cities have stimulated the growth of large algae blooms that sometimes cover 40% of the bay. This has threatened the coral reefs and the diving, fishing, and tourism industries of the bay and the Florida Keys—another example of unintended consequences.

By the 1970s, state and federal officials recognized that this huge plumbing project was reducing wildlife—a major source of tourism income for

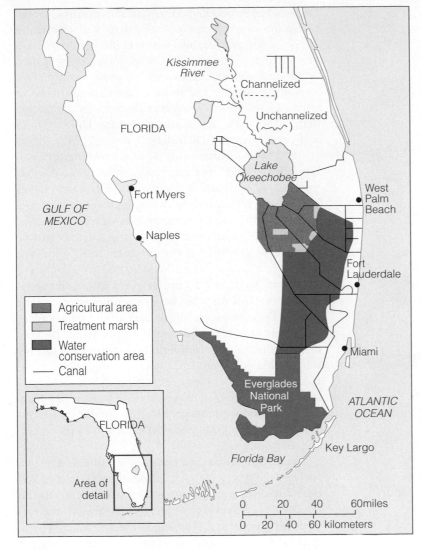

Figure 12-10 The world's largest ecological restoration project is an attempt to undo and redo an engineering project that has been destroying Florida's Everglades (USA) and threatening water supplies for south Florida's rapidly growing population.

create artificial marshes to filter agricultural runoff before it reaches Everglades National Park. *Fourth,* create 18 large reservoirs and underground water storage areas to ensure an adequate water supply for south Florida's current and projected population and the lower Everglades. *Fifth,* build new canals, reservoirs, and huge pumping systems to capture 80% of the water currently flowing out to sea and return it to the Everglades.

Will this huge ecological restoration project work? No one knows. It depends not only on the abilities of scientists and engineers but also on prolonged political and economic support from citizens, the powerful sugarcane and agricultural industries, and elected state and federal officials.

The carefully negotiated plan is beginning to unravel. In 2003, the politically powerful sugarcane growers persuaded the Florida legislature to increase the amount of phosphorus they could discharge and extend the deadline for reducing such discharge from 2006 to 2016. Overall funding for the project has fallen short of the projected needs and federal and state agencies are far behind on almost every component of the project.

According to critics, the main goal of the Everglades restoration plan is to provide water for urban and agricultural development with ecological restoration as a secondary goal. Also, the plan does not specify how much of the water rerouted toward south and central Florida will go to the parched park instead of to increased industrial, agricultural, and urban development. In 2002, a National Academy of Sciences panel said that the plan would probably not clear up Florida Bay's nutrient enrichment problems.

The need to make expensive and politically controversial efforts to undo some of the damage to the Everglades caused by 120 years of agricultural and urban development is another example of failure to heed two fundamental lessons from nature: Prevention is the cheapest and best way to go, and when we intervene in nature, unintended and often harmful consequences always occur.

? THINKING ABOUT EVERGLADES RESTORATION Do you support carrying out the proposed plan for restoring the Florida Everglades? Explain. Should the plan include much greater efforts for ecological restoration? Explain

Florida—and the water supply for the 6 million residents of south Florida. After more than 20 years of political haggling, in 1990 Florida's state government and the federal government agreed on the world's largest ecological restoration project, known as the Comprehensive Everglades Restoration Plan (CERP). The U.S. Army Corps of Engineers will carry out this 30-year, $8.4 billion federal and state plan to partially restore the Everglades by revamping south Florida's plumbing system.

The project has several ambitious goals. *First,* restore the curving flow of more than half of the Kissimmee River. *Second,* remove 400 kilometers (250 miles) of canals and levees blocking water flow south of Lake Okeechobee. *Third,* buy 240 square kilometers (93 square miles) of farmland and allow it to flood to

PROTECTING, SUSTAINING, AND RESTORING LAKES AND RIVERS

Managing Lakes (Science and Economics)

Lakes are difficult to manage and are vulnerable to planned or unplanned introductions of nonnative species.

Sustaining and restoring the biodiversity and ecological services provided by freshwater lakes and rivers is a complex and challenging task, as shown by the changes in Lake Victoria (Core Case Study, p. 249) from the deliberate introduction of Nile perch (Figure 12-1) and the Case Study that follows.

 THINKING ABOUT LAKE VICTORIA AND SUSTAINABILITY
What three things would you do to help improve the ecological and economic sustainability of Lake Victoria?

Case Study: Can the Great Lakes Survive Repeated Invasions by Alien Species? (Science and Economics)

For decades, invasions by nonnative species have caused major ecological and economic damage to North America's Great Lakes.

Invasions by nonnative species is a major threat to the biodiversity and ecological functioning of lakes, as illustrated by what has happened to the Great Lakes, which are located between the United States and Canada.

Collectively, the Great Lakes are the world's largest body of fresh water. Since the 1920s, they have been invaded by at least 162 nonnative species and the number keeps rising. Many of the alien invaders arrive on the hulls or in bilge water discharges of oceangoing ships that have been entering the Great Lakes through the St. Lawrence seaway for over 40 years.

One of the biggest threats, the *sea lamprey*, reached the western lakes through the Welland Canal as early as 1920. This parasite attaches itself to almost any kind of fish and kills the victim by sucking out its blood (Figure 11-11, p. 234). Over the years it has depleted populations of many important sport fish species such as lake trout. The United States and Canada keep the lamprey population down by applying a chemical that kills their larvae in their spawning streams—at a cost of about $15 million a year.

In 1986, larvae of the *zebra mussel* (Figure 11-11, p. 234) arrived in ballast water discharged from a European ship near Detroit, Michigan. This thumbnail-sized mollusk reproduces rapidly and has no known natural enemies in the Great Lakes. As a result, it has displaced other mussel species and depleted the food supply for some other Great Lakes species. The

mussels have also clogged irrigation pipes, shut down water intake pipes for power plants and city water supplies, fouled beaches, jammed ship rudders, and grown in huge masses on boat hulls, piers, pipes, rocks, and almost any exposed aquatic surface. This mussel has also spread to freshwater communities in parts of southern Canada and 18 states in the United States. Currently, the mussels cost the United States and Canada about $140 million a year.

Sometimes nature aids us in controlling an invasive alien species. For example, populations of zebra mussels are declining in some parts of the Great Lakes because a native sponge growing on their shells is preventing them from opening up their shells to breathe. However, it is not clear whether the sponges will be effective in controlling the invasive mussels in the long run.

Zebra mussels may not be good for us and some fish species but they can benefit a number of aquatic plants. By consuming algae and other microorganisms, the mussels increase water clarity, which permits deeper penetration of sunlight and more photosynthesis. This allows some native plants to thrive and could return the plant composition of Lake Erie (and presumably other lakes) closer to what it was 100 years ago. Because the plants provide food and increase dissolved oxygen, their comeback may benefit certain aquatic animals.

In 1991, a larger and potentially more destructive species, the *quagga mussel*, invaded the Great Lakes, probably discharged in the ballast water of a Russian freighter. It can survive at greater depths and tolerate more extreme temperatures than the zebra mussel can. There is concern that it may spread by river transport and eventually colonize areas such as Chesapeake Bay and waterways in parts of Florida.

The *Asian carp* may be the next invader. These highly prolific fish, which can quickly grow as long as 1.2 meters (4 feet) and weigh up to 50 kilograms (110 pounds), have no natural predators in the Great Lakes. In less than a decade, this hearty fish with a voracious appetite has dominated sections of the Mississippi River and its tributaries and is spreading toward the Great Lakes. The only barriers are a few kilometers of waterway and a little-tested underwater electric barrier spanning a canal near Chicago, Illinois (USA).

Managing River Basins (Science, Economics, and Politics)

Dams can provide many human benefits but can disrupt some of the ecological services that rivers provide.

Rivers and streams provide important ecological and economic services (Figure 12-11). But these services

- Deliver nutrients to sea to help sustain coastal fisheries

- Deposit silt that maintains deltas

- Purify water

- Renew and renourish wetlands

- Provide habitats for wildlife

Figure 12-11 Natural capital: important ecological services provided by rivers. Currently, the services are given little or no monetary value when the costs and benefits of dam and reservoir projects are assessed. According to environmental economists, attaching even crudely estimated monetary values to these ecosystem services would help sustain them. QUESTIONS: *Which two of these services do you think are the most important? Which two of these services do you think we are most likely to decline?*

can be disrupted by overfishing, pollution, dams, and water withdrawal for irrigation.

An example of such disruption is the Columbia River that runs through parts of southwestern Canada and the northwestern United States. It has 119 dams, 19 of which are major generators of inexpensive hydroelectric power. It also supplies water for several major urban areas and for irrigating large areas of agricultural land.

This has benefited many people but has sharply reduced populations of wild salmon. These migratory fish hatch in the upper reaches of streams and rivers, migrate to the ocean where they spend most of their adult lives, and then swim upstream to return to the place where they were hatched to spawn and die. Dams interrupt their life cycle.

Since the dams were built, the Columbia River's wild Pacific salmon population has dropped by 94% and nine Pacific Northwest salmon species are listed as endangered or threatened. Since 1980, the U.S. federal government has spent over $3 billion in efforts to save the salmon but none have been effective.

Conservationists, Native American tribes, and commercial salmon fishers want the government to remove four small hydroelectric dams on the lower Snake River in the U.S. state of Washington to restore salmon spawning habitat. Farmers, barge operators, and aluminum workers argue that removing the dams would hurt local economies by reducing irrigation water, eliminating cheap transportation of commodities by ship in the affected areas, and reducing the supply of cheap electricity for industries and consumers.

X *HOW WOULD YOU VOTE?* Should federal efforts to rebuild wild salmon populations in the Columbia River Basin be abandoned? Cast your vote online at www.thomsonedu.com/biology/miller.

In addition to such large-scale efforts, some people have worked to restore salmon populations in specific streams (Individuals Matter, below).

Managing and Sustaining Freshwater Fisheries (Science)

We can help sustain freshwater fisheries by building and protecting populations of desirable species, preventing overfishing, and decreasing populations of less desirable species.

Sustainable management of freshwater fish involves encouraging populations of commercial and sport fish species, preventing such species from being overfished,

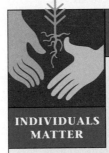

INDIVIDUALS MATTER

The Man Who Planted Trees to Restore a Stream

In 1980, heart problems forced John Beal, an engineer with the Boeing Company, to take some time off. To improve his health he began taking daily walks. His strolls took him by a small stream called Hamm Creek that flows from the southwest hills of Seattle, Washington (USA), into the Duwamish River that empties into Puget Sound.

He remembered when the stream was a spawning ground for salmon and evergreen trees lined its banks. Now the polluted stream had no fish and the trees were gone.

He decided to restore Hamm Creek. He persuaded companies to stop polluting the creek and hauled out many truckloads of garbage. Then he began a 15-year project of planting thousands of trees along the stream's banks. He also restored natural waterfalls and ponds and salmon spawning beds.

At first he worked alone, but word spread and other people joined him. TV news reports and newspaper articles about the restoration project brought more volunteers.

The creek's water now runs clear, its vegetation has been restored, and salmon have returned to spawn. Beal's reward is the personal satisfaction he feels about having made a difference for Hamm Creek and his community. His dedication to making the world a better place is an outstanding example of *stewardship* based on the idea that *all sustainability is local.*

and reducing or eliminating populations of less desirable species. Ways to do this include regulating the time and length of fishing seasons and the number and size of fish that can be taken.

Other techniques include building reservoirs and farm ponds and stocking them with fish, fertilizing nutrient-poor lakes and ponds, and protecting and creating fish spawning sites. In addition, fishery managers can protect fish habitats from sediment buildup and other forms of pollution, prevent excessive growth of aquatic plants from large inputs of plant nutrients, and build small dams to control water flow.

Improving habitats, breeding genetically resistant fish varieties, and using antibiotics and disinfectants can control predators, parasites, and diseases. Hatcheries can be used to restock ponds, lakes, and streams with prized species such as trout and salmon, and entire river basins can be managed to protect valued species such as salmon.

Protecting Wild and Scenic Rivers in the United States (Science and Politics)

A federal law helps protect a tiny fraction of U.S. wild and scenic rivers from dams and other forms of development.

In 1968, the U.S. Congress passed the National Wild and Scenic Rivers Act. It established the National Wild and Scenic Rivers System to protect rivers and river segments with outstanding scenic, recreational, geological, wildlife, historical, or cultural values.

Congress established a three-tiered classification scheme. *Wild rivers* are rivers or segments of rivers that are relatively inaccessible (except by trail) and untamed and that are not permitted to be widened, straightened, dredged, filled, or dammed.

Scenic rivers are free of dams, mostly undeveloped, accessible in only a few places by roads, and of great scenic value. *Recreational rivers* are rivers or sections of rivers that are readily accessible by road or railroad and that may have some dams or development along their shores.

Only 2% of U.S. rivers remain free-flowing and only 0.2% of the country's total river length are protected by the Wild and Scenic Rivers System. In contrast, dams and reservoirs are found on 17% of the country's total river length.

Conservationists urge Congress to add 1,500 additional river segments to the system, a goal vigorously opposed by some local communities and anti-environmental groups. Achieving this goal would protect about 2% of the country's river systems. Even this modest increase would still leave 98% of the country's river length for human activities.

↩ Revisiting Lake Victoria and Sustainability

This chapter began with a look at how human intrusion upset the ecological processes of Africa's Lake Victoria (Core Case Study, p. 249).

Lake Victoria and other cases examined in this chapter illustrate the significant human impacts on commercially valuable marine fish populations and the effects of invasive species on marine and freshwater aquatic biodiversity. We have seen that threats to aquatic biodiversity are growing and are even greater than threats to terrestrial biodiversity.

We also explored ways to manage the world's oceans, fisheries, wetlands, lakes, and rivers more sustainably by emphasizing the four principles of sustainability. This involves reducing inputs of sediments and excess nutrients, which cloud water and lessen the input of solar energy and which can also upset the natural cycling of nutrients in aquatic systems. In addition, emphasis should be placed on preserving the biodiversity of aquatic systems and not upsetting natural interactions among aquatic species that help prevent excessive population growth of species, as happened in Lake Victoria.

If we do not change our relationships with the world's aquatic systems, we risk depriving future generations of their biological and economic wealth and their wonder. Without healthy aquatic systems, we cannot have a healthy planet. There is an urgent need to confront these critical issues.

To promote conservation, fishers and officials need to view fish as a part of a larger ecological system, rather than simply as a commodity to extract.

ANNE PLATT McGINN

CRITICAL THINKING

↩ **1.** Explain how introducing the Nile perch into Lake Victoria (Core Case Study, p. 249) violated all four principles of sustainability.

↩ **2.** What difference does it make that the introduction of the Nile perch into Lake Victoria caused the extinction of more than 200 cichlid fish species? Explain.

3. What do you think are the three greatest threats to aquatic biodiversity? Why are aquatic species overall more vulnerable to premature extinction from human activities than terrestrial species?

4. Why is marine biodiversity higher **(a)** near coasts than in the open sea and **(b)** on the ocean's bottom than at its surface?

5. Why is it more difficult to identify and protect endangered marine species than to protect endangered species on land?

6. How can continued overfishing of marine species affect your lifestyle? What three things would you do to help prevent overfishing?

7. Should fishers harvesting fish from a country's publicly owned waters be required to pay the government (taxpayers) fees for the fish they catch? Explain. If your livelihood depended on commercial fishing, would you be for or against such fees?

8. Do you think the plan for restoring Florida's Everglades will succeed? Give three reasons why or why not.

9. Congratulations! You are in charge of protecting the world's aquatic biodiversity. List the three most important points of your policy to accomplish this goal.

10. List two questions that you would like to have answered as a result of reading this chapter.

PROJECTS

1. Survey the condition of a nearby wetland, coastal area, river, or stream and research its history. Has its condition improved or deteriorated during the last 10 years? What local, state, or national efforts are being used to protect this aquatic system? Develop a plan for protecting it.

2. Work with your classmates to develop an experiment in aquatic reconciliation ecology for your campus or local community.

3. Make a concept map of this chapter's major ideas, using the section heads, subheads, and key terms (in boldface). Look on the website for this book about how to prepare concept maps.

LEARNING ONLINE

The website for this book contains study aids and many ideas for further reading and research. They include a chapter summary, review questions for the entire chapter, flash cards for key terms and concepts, a multiple-choice practice quiz, interesting Internet sites, references, information about green careers, and a guide for accessing thousands of InfoTrac® College Edition articles. Log into

www.thomsonedu.com/biology/miller

Then choose Chapter 12, and select a learning resource. For access to animations, additional quizzes, chapter outlines and summaries, register and log into

at **www.thomsonedu.com** using the access code card in the front of your book.

Supplements

MEASUREMENT UNITS, PRECISION, AND ACCURACY (SCIENCE)
CHAPTER 2

LENGTH
Metric
1 kilometer (km) = 1,000 meters (m)
1 meter (m) = 100 centimeters (cm)
1 meter (m) = 1,000 millimeters (mm)
1 centimeter (cm) = 0.01 meter (m)
1 millimeter (mm) = 0.001 meter (m)

English
1 foot (ft) = 12 inches (in)
1 yard (yd) = 3 feet (ft)
1 mile (mi) = 5,280 feet (ft)
1 nautical mile = 1.15 miles

Metric–English
1 kilometer (km) = 0.621 mile (mi)
1 meter (m) = 39.4 inches (in)
1 inch (in) = 2.54 centimeters (cm)
1 foot (ft) = 0.305 meter (m)
1 yard (yd) = 0.914 meter (m)
1 nautical mile = 1.85 kilometers (km)

AREA
Metric
1 square kilometer (km^2) = 1,000,000 square meters (m^2)
1 square meter (m^2) = 1,000,000 square millimeters (mm^2)
1 hectare (ha) = 10,000 square meters (m^2)
1 hectare (ha) = 0.01 square kilometer (km^2)

English
1 square foot (ft^2) = 144 square inches (in^2)
1 square yard (yd^2) = 9 square feet (ft^2)
1 square mile (mi^2) = 27,880,000 square feet (ft^2)
1 acre (ac) = 43,560 square feet (ft^2)

Metric–English
1 hectare (ha) = 2.471 acres (ac)
1 square kilometer (km^2) = 0.386 square mile (mi^2)
1 square meter (m^2) = 1.196 square yards (yd^2)
1 square meter (m^2) = 10.76 square feet (ft^2)
1 square centimeter (cm^2) = 0.155 square inch (in^2)

VOLUME
Metric
1 cubic kilometer (km^3) = 1,000,000,000 cubic meters (m^3)
1 cubic meter (m^3) = 1,000,000 cubic centimeters (cm^3)
1 liter (L) = 1,000 milliliters (mL) = 1,000 cubic centimeters (cm^3)
1 milliliter (mL) = 0.001 liter (L)
1 milliliter (mL) = 1 cubic centimeter (cm^3)

English
1 gallon (gal) = 4 quarts (qt)
1 quart (qt) = 2 pints (pt)

Metric–English
1 liter (L) = 0.265 gallon (gal)
1 liter (L) = 1.06 quarts (qt)
1 liter (L) = 0.0353 cubic foot (ft^3)
1 cubic meter (m^3) = 35.3 cubic feet (ft^3)
1 cubic meter (m^3) = 1.30 cubic yards (yd^3)
1 cubic kilometer (km^3) = 0.24 cubic mile (mi^3)
1 barrel (bbl) = 159 liters (L)
1 barrel (bbl) = 42 U.S. gallons (gal)

MASS
Metric
1 kilogram (kg) = 1,000 grams (g)
1 gram (g) = 1,000 milligrams (mg)
1 gram (g) = 1,000,000 micrograms (μg)
1 milligram (mg) = 0.001 gram (g)
1 microgram (μg) = 0.000001 gram (g)
1 metric ton (mt) = 1,000 kilograms (kg)

English
1 ton (t) = 2,000 pounds (lb)
1 pound (lb) = 16 ounces (oz)

Metric–English
1 metric ton (mt) = 2,200 pounds (lb) = 1.1 tons (t)
1 kilogram (kg) = 2.20 pounds (lb)
1 pound (lb) = 454 grams (g)
1 gram (g) = 0.035 ounce (oz)

ENERGY AND POWER
Metric
1 kilojoule (kJ) = 1,000 joules (J)
1 kilocalorie (kcal) = 1,000 calories (cal)
1 calorie (cal) = 4.184 joules (J)

Metric–English
1 kilojoule (kJ) = 0.949 British thermal unit (Btu)
1 kilojoule (kJ) = 0.000278 kilowatt-hour (kW-h)
1 kilocalorie (kcal) = 3.97 British thermal units (Btu)
1 kilocalorie (kcal) = 0.00116 kilowatt-hour (kW-h)
1 kilowatt-hour (kW-h) = 860 kilocalories (kcal)
1 kilowatt-hour (kW-h) = 3,400 British thermal units (Btu)
1 quad (Q) = 1,050,000,000,000,000 kilojoules (kJ)
1 quad (Q) = 293,000,000,000 kilowatt-hours (kW-h)

TEMPERATURE CONVERSIONS
Fahrenheit ($^\circ$F) to Celsius ($^\circ$C):
   $^\circ$C = ($^\circ$F − 32.0) ÷ 1.80
Celsius ($^\circ$C) to Fahrenheit ($^\circ$F):
   $^\circ$F = ($^\circ$C × 1.80) + 32.0

Uncertainty, Accuracy, and Precision in Scientific Measurements

Scientists check their measuring instruments and repeat their measurements to reduce uncertainty.

How do we know whether a scientific measurement is correct? All scientific observations and measurements have some degree of *uncertainty* because people and measuring devices are not perfect.

However, scientists take great pains to reduce the errors in observations and measurements by using standard procedures and testing (calibrating) measuring devices. They also repeat their measurements several times, and then find the average value of these measurements.

In determining the uncertainty involved in a measurement it is important to distinguish between accuracy and precision. *Accuracy* is how well a measurement conforms to what is accepted as the correct value for the measured quantity, based on careful measurements by many people over a long time. *Precision* is a measure of *repro-*

Good accuracy and good precision

Poor accuracy and poor precision

Poor accuracy and good precision

Figure 1 The distinction between accuracy and precision. In scientific measurements, a measuring device that has not been calibrated to determine its accuracy may give precise or reproducible results that are not accurate.

ducibility, or how closely a series of measurements of the same quantity agree with *one another.*

The dartboard analogy shown in Figure 1 shows the difference between precision and accuracy. *Accuracy* depends on how close the darts are to the bull's-eye. *Precision* depends on how close the darts are to each other. Note that good precision is necessary for accuracy but does not guarantee it. Three closely spaced darts may be far from the bull's-eye.

SUPPLEMENT 2

GRAPHING (SCIENCE)

In science, graphs convey information that can be summarized numerically. This information, called *data,* is collected in experiments, surveys, historical studies, and other information gathering activities. This textbook and accompanying web-based Active Graphing exercises use three types of graphs—line graphs, bar charts, and pie charts.

Line graphs usually represent data that falls in some sort of sequence such as a series of measurements over time or distance. In most such cases, units of time or distance lie on the horizontal *x-axis* (see Figure 1). The possible measurements of some variable, such as temperature, that changes over time or distance usually lie on the vertical *y-axis*. The curving line on the graph represents the measurements taken at certain time or distance intervals.

A good example of this application is Figure 4-13, p. 94. Note that two sets of data are represented by two different lines on this graph for purposes of comparing the data sets.

Another important use of the line graph is to show experimental results such as changes in a dependent variable in response to changes in an independent variable. For example, we measure changes in toxicity of a chemical (dependent variable) with increases in the dosage of the chemical (independent variable).

The *bar chart* is used to compare measurements for one or more variables across categories. For instance, we could compare the amount of forested land in four regions of the U.S., as in Figure 2.

See also the Active Graphing exercise for Chapter 10, which includes the data in Figure 2 along with similar data sets for other years. This enables us to compare not only the regional forest coverage but also how that coverage has changed over time across the regions.

In these examples, the categories are laid out on the x-axis, while the range of measurements for the variable under consideration lies along the y-axis. This is usually the case, although the information on the axes can be reversed. See for example Figures 9-9 (p. 179) and 9-11 (p. 180) depicting age structure diagrams, in which the bars are placed horizontally. This is another way to compare two data sets (in this case, for males and females) across categories.

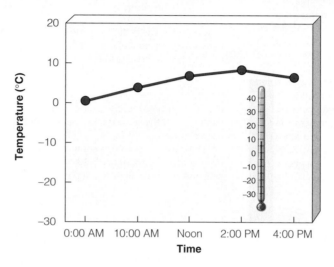

Figure 1 Temperature changes on a winter day (in Centigrade degrees).

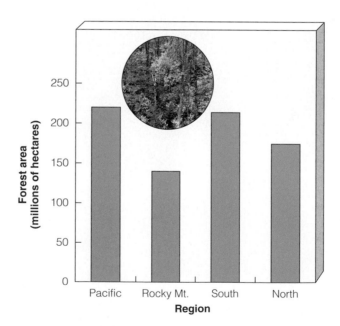

Figure 2 Forest area in four regions of the U.S. in 2002 (in millions of hectares).

Like bar charts, *pie charts* illustrate numerical values for more than one category. But in addition to that, they can also show each category's proportion of the total of all measurements. For example, Figure 3 shows how much each major energy source contributes to the world's total amount of energy used. Usually, the categories are ordered on the chart from largest to smallest, for ease of comparison, although this is not always the case.

The Active Graphing exercises available for various chapters on the website for this textbook will help you to apply this information. Register and log into

using the access code card in the front of your book. Choose a chapter with an Active Graphing exercise, click on the exercise, and begin learning more about graphing.

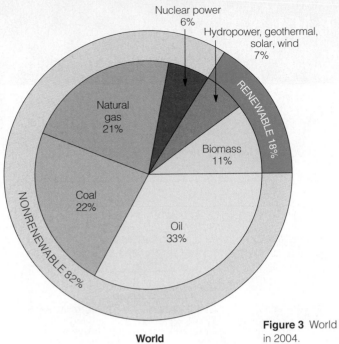

World

Figure 3 World energy use by source in 2004.

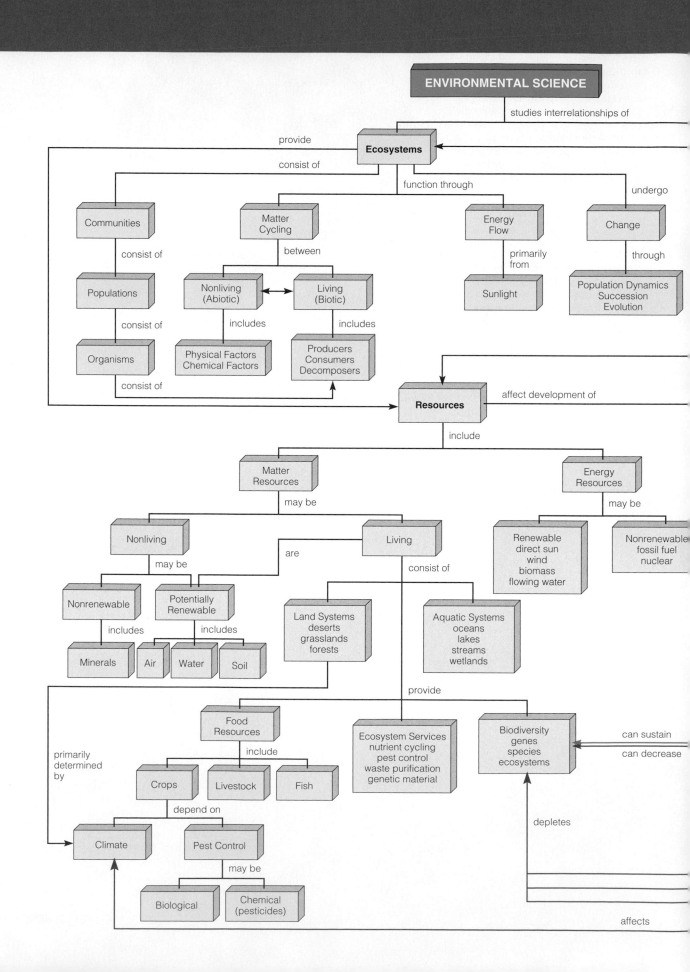

ENVIRONMENTAL SCIENCE: CONCEPT MAP OVERVIEW (SCIENCE)

Developed by **Jane Heinze-Fry** with assistance from G. Tyler Miller, Jr.
(For assistance in creating your own concept maps, see the website for this book.)

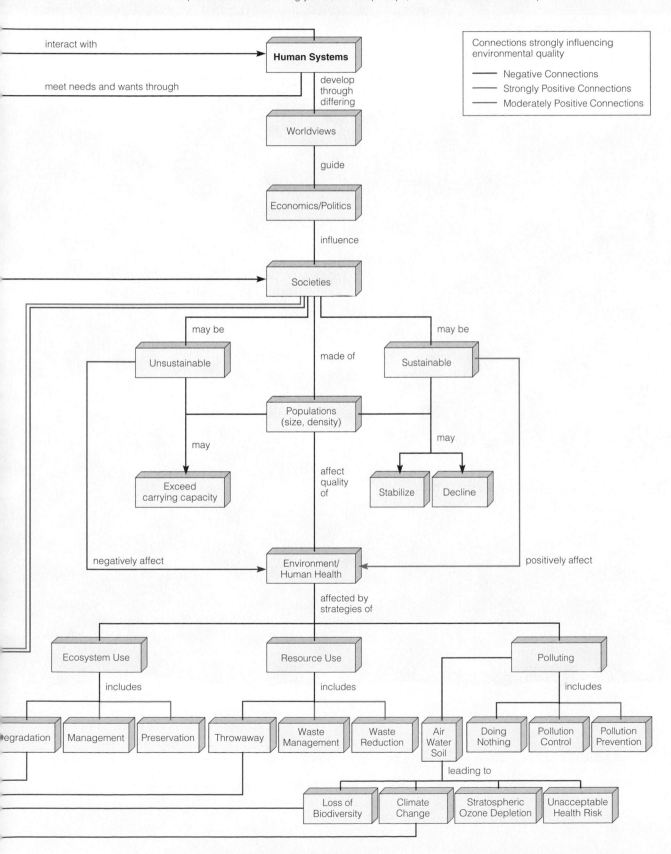

BIODIVERSITY AND ECOLOGICAL FOOTPRINTS (SCIENCE)
CHAPTERS 1, 3, 5, 9, 10

NASA Goddard Space Flight Center Image by Reto Stöckli (land surface, shallow water, clouds). Enhancements by Robert Simmon (ocean color, compositing, 3D globes, animation)

Figure 1 Natural capital: composite satellite view of the earth showing its major terrestrial and aquatic features.
QUESTION: *Where do you live on this map?*

Figure 2 **Natural capital:** global map of plant biodiversity. QUESTION: *How high is the plant diversity where you live?* (Used by permission from Kier, et al. 2005. "Global Patterns of Plant Diversity and Floristic Knowledge." *Journal of Biogeography,* vol. 32, Issue 6, pp. 921–1106 and Blackwell Publishing)

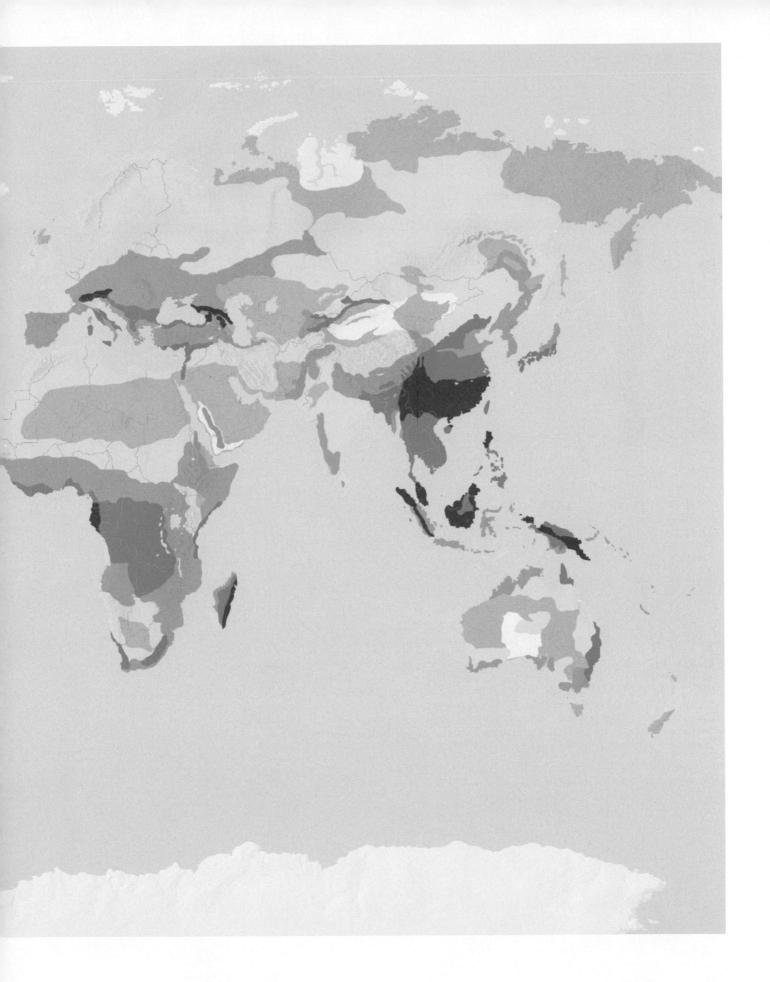

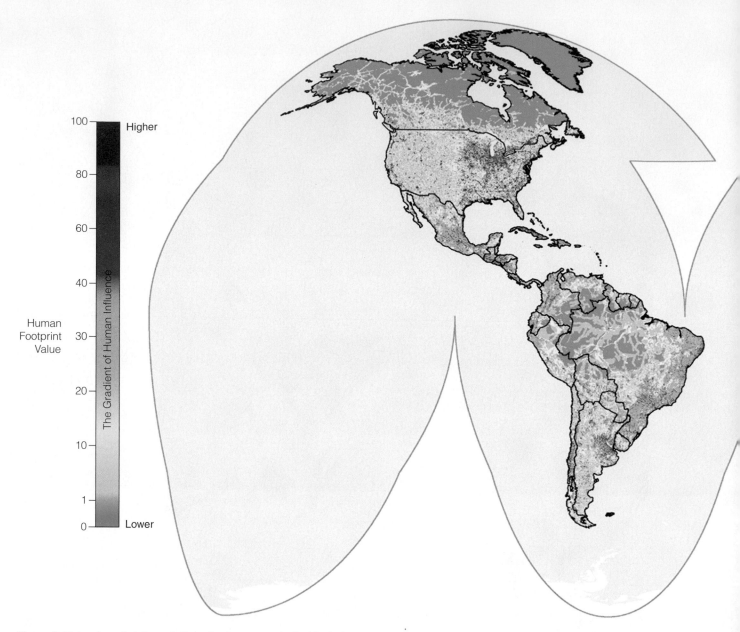

Figure 3 **Natural capital degradation:** the human ecological footprint on the earth's land surface—in effect the sum of all ecological footprints (Figure 1-7, p. 13) of the human population. Colors represent the percentage of each area influenced by human activities. Excluding Antarctica and Greenland, human activities have directly affected to some degree about 83% of the earth's land surface and 98% of the area where it is possible to grow rice, wheat, or maize. QUESTION: *How large is the human ecological footprint in the general area where you live?* (Data from Wildlife Conservation Society and the Center for International Science Earth Information Network at Columbia University)

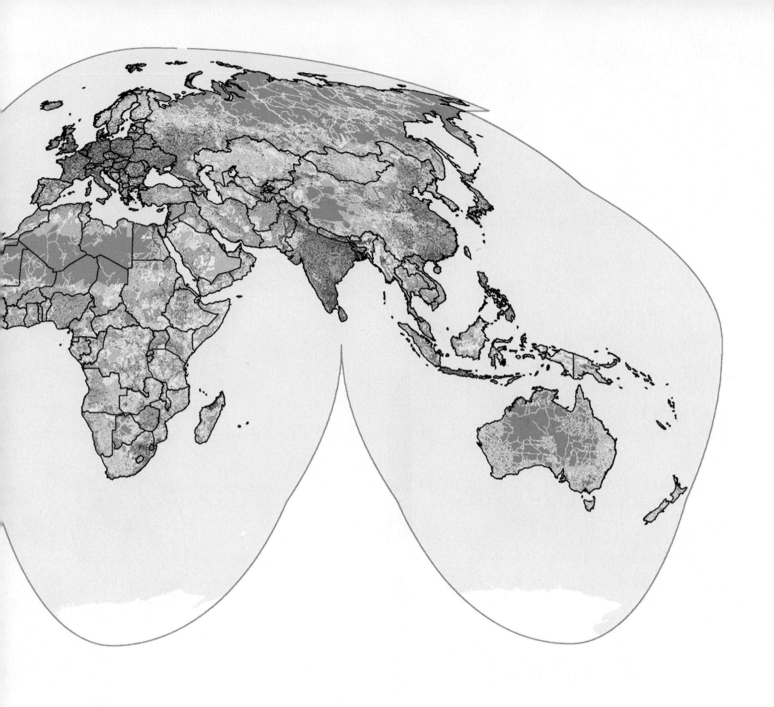

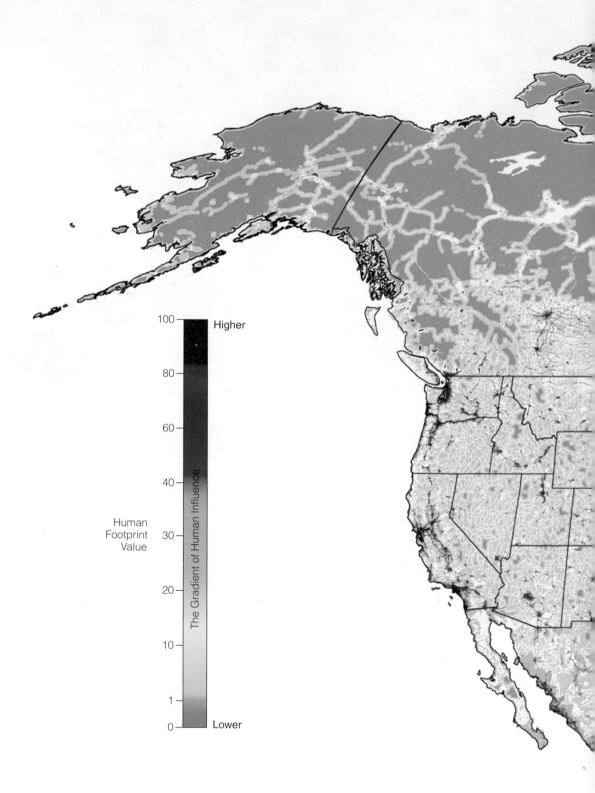

Figure 4 **Natural capital degradation:** the human ecological footprint in North America. Colors represent the percentage of each area influenced by human activities. This is an expanded portion of Figure 3 showing the human footprint on the earth's entire land surface. QUESTION: *How large is the human ecological footprint in the general area where you live?* (Data from Wildlife Conservation Society and the Center for International Earth Science Information Network at Columbia University)

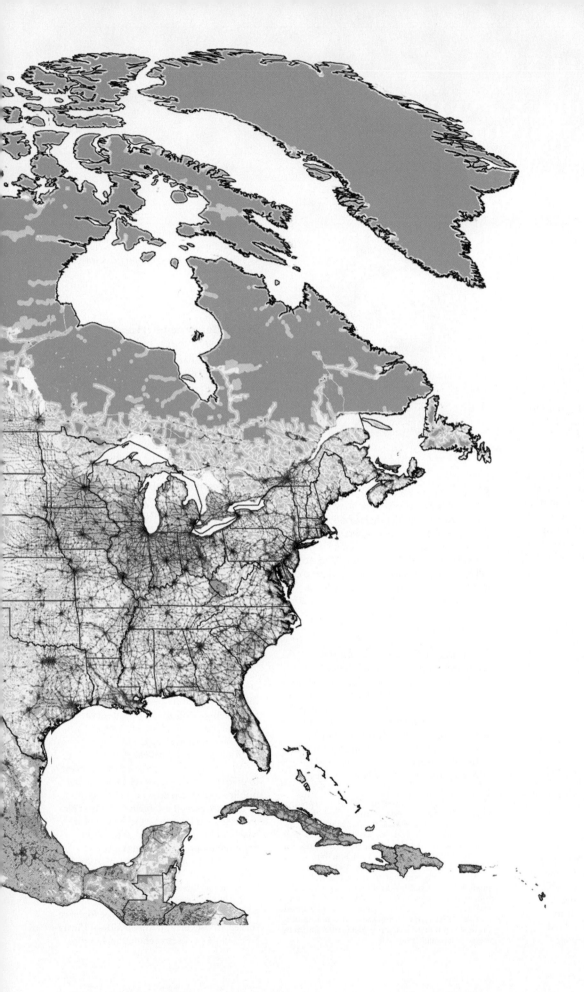

The Frontier Era (1607–1890)

For almost three centuries, the U.S. government encouraged settlers to spread across the continent.

During the frontier era, European settlers spread across the land by clearing forests for cropland and settlements and displaced the Native Americans that had lived on the land fairly sustainably for thousands of years.

The U.S. government accelerated this settling of the continent and use of its resources by transferring vast areas of public land to private interests. Between 1850 and 1890, more than half of the country's public land was given away or sold cheaply by the government to railroad, timber, and mining companies, land developers, states, schools, universities, and homesteaders to encourage settlement. This era came to an end when the government declared the frontier officially closed in 1890.

Early Conservationists (1832–70)

Several conservationists warned that the country was degrading its resource base, but few listened.

Between 1832 and 1870, some people became alarmed at the scope of resource depletion and degradation in the United States. They urged the government to preserve part of the unspoiled wilderness on public lands owned jointly by all people (but managed by the government) and protect it as a legacy to future generations.

Two of these early conservationists were Henry David Thoreau (1817–1862) and George Perkins Marsh (1812–1939). Thoreau (Figure 1) was alarmed at the loss of numerous wild species from his native eastern Massachusetts. To gain a better understanding of nature, he built a cabin in the woods on Walden Pond near Concord, Massachusetts, lived

Figure 1 Henry David Thoreau (1817–1862) was an American writer and naturalist who kept journals about his excursions into wild nature throughout parts of the northeastern United States and Canada and at Walden Pond in Massachusetts. He sought self-sufficiency, a simple lifestyle, and a harmonious coexistence with nature.

there alone for 2 years, and wrote *Life in the Woods*, an environmental classic.*

In 1864, George Perkins Marsh, a scientist and member of Congress from Vermont, published *Man and Nature*, which helped legislators and citizens see the need for resource conservation. Marsh questioned the idea that the country's resources were inexhaustible. He also used scientific studies and case studies to show how the rise and

---

*I can identify with Thoreau. I spent 15 years living in the deep woods studying and thinking about how nature works and writing early editions of the book you are reading. I lived in a school bus with an attached greenhouse. I used it as a scientific laboratory for evaluating things such as passive and active solar energy technologies for heating the bus and water, waste disposal (composting toilets), natural geothermal cooling (earth tubes), ways to save energy and water, and biological control of pests. It was great fun and I learned a lot. In 1990, I came out of the woods to find out. more about how to live more sustainably in urban areas, where most people live.

fall of past civilizations were linked to the use and misuse of their resource base. Some of his resource conservation principles are still used today.

What Happened between 1870 and 1930?

The government and newly formed private groups tried to protect more of the nation's natural resources and improve public health.

Between 1870 and 1930, a number of actions increased the role of the federal government and private citizens in resource conservation and public health (Figure 2). The *Forest Reserve Act of 1891* was a turning point in establishing the responsibility of the federal government for protecting public lands from resource exploitation.

In 1892, nature preservationist and activist John Muir (Figure 3) founded the Sierra Club. He became the leader of the *preservationist movement* that called for protecting large areas of wilderness on public lands from human exploitation, except for low-impact recreational activities such as hiking and camping. This idea was not enacted into law until 1964. Muir also proposed and lobbied for creation of a national park system on public lands.

Mostly because of political opposition, effective protection of forests and wildlife did not begin until Theodore Roosevelt (Figure 4, p. S18), an ardent conservationist, became president. His term of office, 1901–1909, has been called the country's *Golden Age of Conservation*.

While in office he persuaded Congress to give the president power to designate public land as federal wildlife refuges. In 1903, Roosevelt established the first federal refuge at Pelican Island off the east coast of Florida for preservation of the endangered brown pelican (Figure 11-20, p. 243), and he added 35 more reserves by 1904. He also more than tripled the size of the national forest reserves.

In 1905, Congress created the U.S. Forest Service to manage and protect the forest reserves. Roosevelt appointed Gifford Pinchot (1865–1946) as its first chief. Pinchot pio-

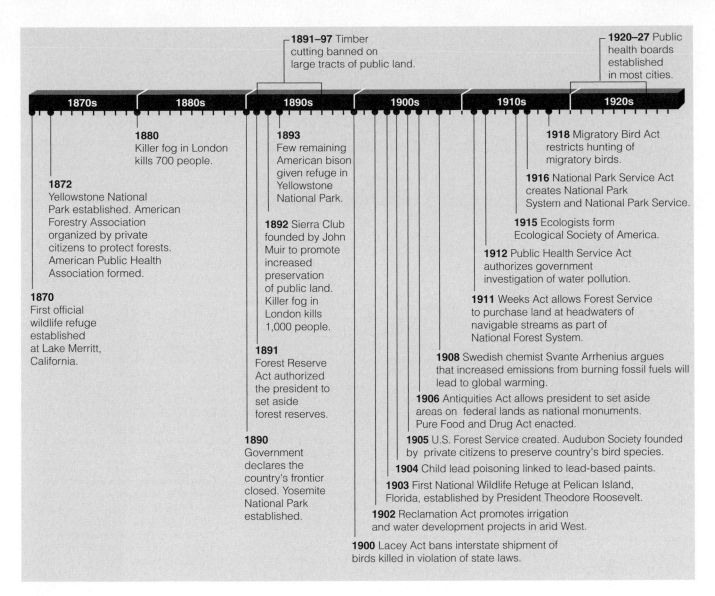

1891–97 Timber cutting banned on large tracts of public land.

1920–27 Public health boards established in most cities.

1870s | 1880s | 1890s | 1900s | 1910s | 1920s

1880 Killer fog in London kills 700 people.

1872 Yellowstone National Park established. American Forestry Association organized by private citizens to protect forests. American Public Health Association formed.

1870 First official wildlife refuge established at Lake Merritt, California.

1893 Few remaining American bison given refuge in Yellowstone National Park.

1892 Sierra Club founded by John Muir to promote increased preservation of public land. Killer fog in London kills 1,000 people.

1891 Forest Reserve Act authorized the president to set aside forest reserves.

1890 Government declares the country's frontier closed. Yosemite National Park established.

1918 Migratory Bird Act restricts hunting of migratory birds.

1916 National Park Service Act creates National Park System and National Park Service.

1915 Ecologists form Ecological Society of America.

1912 Public Health Service Act authorizes government investigation of water pollution.

1911 Weeks Act allows Forest Service to purchase land at headwaters of navigable streams as part of National Forest System.

1908 Swedish chemist Svante Arrhenius argues that increased emissions from burning fossil fuels will lead to global warming.

1906 Antiquities Act allows president to set aside areas on federal lands as national monuments. Pure Food and Drug Act enacted.

1905 U.S. Forest Service created. Audubon Society founded by private citizens to preserve country's bird species.

1904 Child lead poisoning linked to lead-based paints.

1903 First National Wildlife Refuge at Pelican Island, Florida, established by President Theodore Roosevelt.

1902 Reclamation Act promotes irrigation and water development projects in arid West.

1900 Lacey Act bans interstate shipment of birds killed in violation of state laws.

1870–1930

Figure 2 Examples of the increased role of the federal government in resource conservation and public health and the establishment of key private environmental groups, 1870–1930. QUESTION: *Which three of these events do you think are the most important?*

neered scientific management of forest resources on public lands. In 1906, Congress passed the *Antiquities Act,* which allows the president to protect areas of scientific or historical interest on federal lands as national monuments. Roosevelt used this act to protect the Grand Canyon and other areas that would later become national parks.

Congress became upset with Roosevelt in 1907, because by then he had added vast tracts to the forest reserves and passed a law banning further executive withdrawals of public forests. On the day before the bill became law, Roosevelt defiantly reserved another 6.5 million hectares (16 million acres). Most environmental historians view Roosevelt (a Republican) as the country's best environmental president.

Figure 3 John Muir (1838–1914) was a geologist, explorer, and naturalist. He spent 6 years studying, writing journals, and making sketches in the wilderness of California's Yosemite Valley and then went on to explore wilderness areas in Utah, Nevada, the Northwest, and Alaska. He was largely responsible for establishing Yosemite National Park in 1890. He also founded the Sierra Club and spent 22 years lobbying actively for conservation laws.

Figure 4 Theodore (Teddy) Roosevelt (1858–1919) was a writer, explorer, naturalist, avid birdwatcher, and twenty-sixth president of the United States. He was the first national political figure to bring the issues of conservation to the attention of the American public. According to many historians, he has contributed more than any other president to natural resource conservation in the United States.

Early in the twentieth century, the U.S. conservation movement split into two factions over how public lands should be used. The *wise-use,* or *conservationist,* school, led by Roosevelt and Pinchot, believed all public lands should be managed wisely and scientifically to provide needed resources. The *preservationist* school, led by Muir wanted wilderness areas on public lands to be left untouched. This controversy over use of public lands continues today.

In 1916, Congress passed the *National Park Service Act.* It declared that parks are to be maintained in a manner that leaves them unimpaired for future generations. The Act also established the National Park Service (within the Department of the Interior) to manage the system. Under its first head, Stephen T. Mather (1867–1930), the dominant park policy was to encourage tourist visits by allowing private concessionaires to operate facilities within the parks.

After World War I, the country entered a new era of economic growth and expansion. During the Harding, Coolidge, and Hoover administrations, the federal government promoted increased resource removal from public lands at low prices to stimulate economic growth.

President Herbert Hoover (a Republican) went even further and proposed that the federal government return all remaining federal lands to the states or sell them to private interests for economic development. But the Great Depression (1929–1941) made owning such lands unattractive to state governments and private investors. The depression was bad news for the country. But some say that without it we might have little if any of the public lands that make up about one-third of the country's land left today.

What Happened between 1930 and 1960?

To help get the United States out of a major economic depression, the government bought land and hired many workers to restore the country's degraded environment and build dams to supply electricity and water.

A second wave of national resource conservation and improvements in public health began in the early 1930s (Figure 5) as President Franklin D. Roosevelt (1882–1945) strove to bring the country out of the Great Depression. He persuaded Congress to enact federal government programs to provide jobs and help restore the country's degraded environment.

During this period the government purchased large tracts of land from cash-poor landowners, and established the *Civilian Conservation Corps* (CCC) in 1933. It put 2 million unemployed people to work planting trees and developing and maintaining parks and recreation areas. The CCC also restored silted waterways and built levees and dams for flood control.

The government built and operated many large dams in the Tennessee Valley and in the arid western states, including Hoover Dam on the Colorado River. The goals were to provide jobs, flood control, cheap irrigation water, and cheap electricity for industry.

In 1935, Congress enacted the Soil Conservation Act. It established the *Soil Erosion Service* as part of the Department of Agriculture to correct the enormous erosion problems that had ruined many farms in the Great Plains states during the depression (see Case Study on p. S23). Its name was later changed to the *Soil Conservation Service,* now called the *Natural Resources Conservation Service.* Many environmental historians praise Roosevelt (a Democrat) for his efforts to get the country out of a major economic depression and restore past environmental protection.

Federal resource conservation and public health policy during the 1940s and 1950s changed little, mostly because of preoccupation with World War II (1941–1945) and economic recovery after the war.

Between 1930 and 1960, improvements in public health included establishment of public health boards and agencies at the municipal, state, and federal levels; increased public education about health issues; introduction of vaccination programs; and a sharp reduction in waterborne infectious disease mostly because of improved sanitation and garbage collection.

What Happened during the 1960s?

The modern environmental movement began as many citizens urged the government to improve environmental quality.

A number of milestones in American environmental history occurred during the 1960s (Figure 6). In 1962, biologist Rachel Carson (1907–1964) published *Silent Spring,* which documented the pollution of air, water, and wildlife from pesticides such as DDT. This influential book helped broaden the concept of resource conservation to include preservation of the *quality* of the air, water, soil, and wildlife.

Many environmental historians mark Carson's wake-up call as the beginning of the modern *environmental movement* in the United States. It flourished when a growing number of citizens organized to demand that political leaders enact laws and develop policies to curtail pollution, clean up polluted environments, and protect unspoiled areas from environmental degradation.

In 1964, Congress passed the *Wilderness Act,* inspired by the vision of John Muir more than 80 years earlier. It authorized the government to protect undeveloped tracts of public land as part of the National Wilderness System, unless Congress later decides they are needed for the national good. Land in this system is to be used only for nondestructive forms of recreation such as hiking and camping (see pp. 215–217).

Between 1965 and 1970, the emerging science of *ecology* received widespread media attention. At the same time, the popular writings of biologists such as Paul Ehrlich, Barry Commoner, and Garrett Hardin awakened people to the interlocking relationships among population growth, resource use, and pollution. (Figure 1-14, p. 20).

During that period, a number of events increased public awareness of pollution (Figure 6). The public also became aware that pollution and loss of habitat were endangering well-known wildlife species such as the North American bald eagle, grizzly bear, whooping crane, and peregrine falcon.

During the 1969 U.S. Apollo mission to the moon, astronauts photographed the earth from space. This allowed people to see the earth as a tiny blue and white planet in the black void of space (Figure 4-1, p. 82) and led to the development of the *spaceship-earth environmental worldview.* It reminded us that we live on a planetary spaceship (Terra I) that we should not harm because it is the only home we have.

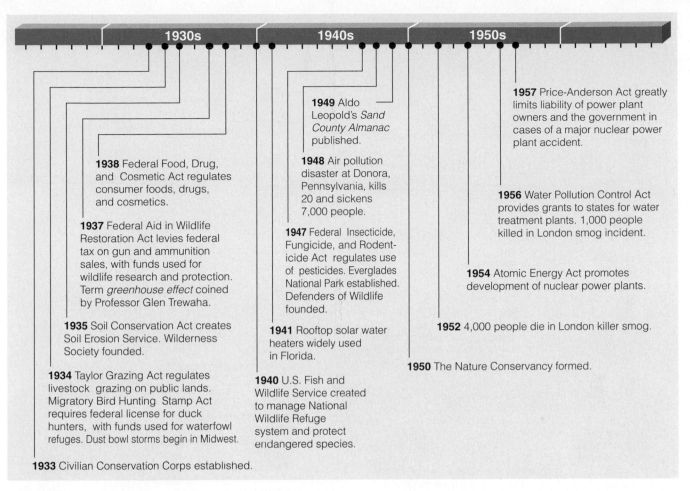

1957 Price-Anderson Act greatly limits liability of power plant owners and the government in cases of a major nuclear power plant accident.

1949 Aldo Leopold's *Sand County Almanac* published.

1956 Water Pollution Control Act provides grants to states for water treatment plants. 1,000 people killed in London smog incident.

1948 Air pollution disaster at Donora, Pennsylvania, kills 20 and sickens 7,000 people.

1938 Federal Food, Drug, and Cosmetic Act regulates consumer foods, drugs, and cosmetics.

1947 Federal Insecticide, Fungicide, and Rodent-icide Act regulates use of pesticides. Everglades National Park established. Defenders of Wildlife founded.

1954 Atomic Energy Act promotes development of nuclear power plants.

1937 Federal Aid in Wildlife Restoration Act levies federal tax on gun and ammunition sales, with funds used for wildlife research and protection. Term *greenhouse effect* coined by Professor Glen Trewaha.

1941 Rooftop solar water heaters widely used in Florida.

1952 4,000 people die in London killer smog.

1935 Soil Conservation Act creates Soil Erosion Service. Wilderness Society founded.

1950 The Nature Conservancy formed.

1934 Taylor Grazing Act regulates livestock grazing on public lands. Migratory Bird Hunting Stamp Act requires federal license for duck hunters, with funds used for waterfowl refuges. Dust bowl storms begin in Midwest.

1940 U.S. Fish and Wildlife Service created to manage National Wildlife Refuge system and protect endangered species.

1933 Civilian Conservation Corps established.

1930–1960

Figure 5 Some important conservation and environmental events, 1930–1960. QUESTION: *Which three of these events do you think are the most important?*

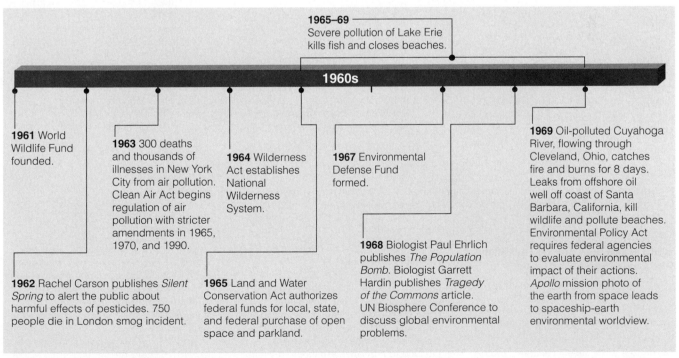

1965–69 Severe pollution of Lake Erie kills fish and closes beaches.

1961 World Wildlife Fund founded.

1963 300 deaths and thousands of illnesses in New York City from air pollution. Clean Air Act begins regulation of air pollution with stricter amendments in 1965, 1970, and 1990.

1964 Wilderness Act establishes National Wilderness System.

1967 Environmental Defense Fund formed.

1969 Oil-polluted Cuyahoga River, flowing through Cleveland, Ohio, catches fire and burns for 8 days. Leaks from offshore oil well off coast of Santa Barbara, California, kill wildlife and pollute beaches. Environmental Policy Act requires federal agencies to evaluate environmental impact of their actions. *Apollo* mission photo of the earth from space leads to spaceship-earth environmental worldview.

1962 Rachel Carson publishes *Silent Spring* to alert the public about harmful effects of pesticides. 750 people die in London smog incident.

1965 Land and Water Conservation Act authorizes federal funds for local, state, and federal purchase of open space and parkland.

1968 Biologist Paul Ehrlich publishes *The Population Bomb*. Biologist Garrett Hardin publishes *Tragedy of the Commons* article. UN Biosphere Conference to discuss global environmental problems.

1960s

Figure 6 Some important environmental events during the 1960s. QUESTION: *Which three of these events do you think are the most important?*

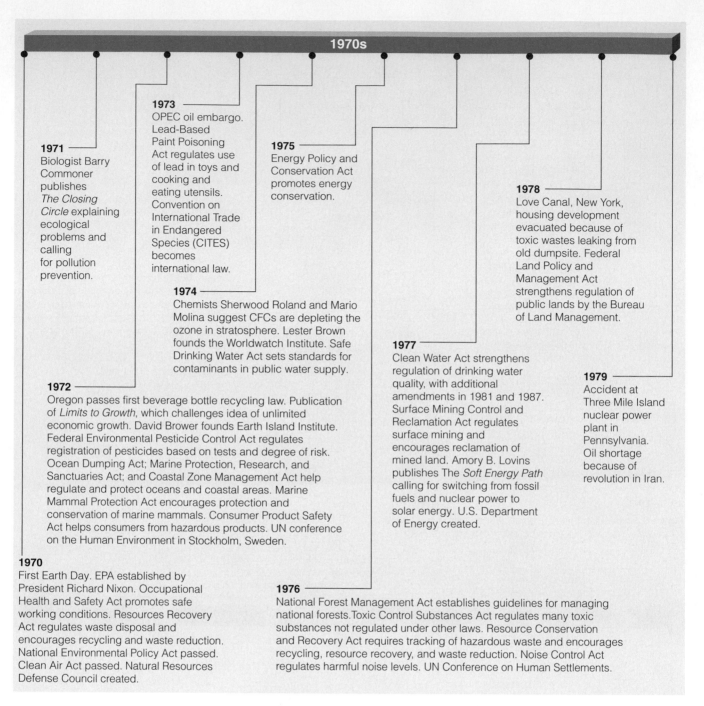

1970s

1971
Biologist Barry Commoner publishes *The Closing Circle* explaining ecological problems and calling for pollution prevention.

1973
OPEC oil embargo. Lead-Based Paint Poisoning Act regulates use of lead in toys and cooking and eating utensils. Convention on International Trade in Endangered Species (CITES) becomes international law.

1975
Energy Policy and Conservation Act promotes energy conservation.

1978
Love Canal, New York, housing development evacuated because of toxic wastes leaking from old dumpsite. Federal Land Policy and Management Act strengthens regulation of public lands by the Bureau of Land Management.

1974
Chemists Sherwood Roland and Mario Molina suggest CFCs are depleting the ozone in stratosphere. Lester Brown founds the Worldwatch Institute. Safe Drinking Water Act sets standards for contaminants in public water supply.

1977
Clean Water Act strengthens regulation of drinking water quality, with additional amendments in 1981 and 1987. Surface Mining Control and Reclamation Act regulates surface mining and encourages reclamation of mined land. Amory B. Lovins publishes The *Soft Energy Path* calling for switching from fossil fuels and nuclear power to solar energy. U.S. Department of Energy created.

1979
Accident at Three Mile Island nuclear power plant in Pennsylvania. Oil shortage because of revolution in Iran.

1972
Oregon passes first beverage bottle recycling law. Publication of *Limits to Growth*, which challenges idea of unlimited economic growth. David Brower founds Earth Island Institute. Federal Environmental Pesticide Control Act regulates registration of pesticides based on tests and degree of risk. Ocean Dumping Act; Marine Protection, Research, and Sanctuaries Act; and Coastal Zone Management Act help regulate and protect oceans and coastal areas. Marine Mammal Protection Act encourages protection and conservation of marine mammals. Consumer Product Safety Act helps consumers from hazardous products. UN conference on the Human Environment in Stockholm, Sweden.

1970
First Earth Day. EPA established by President Richard Nixon. Occupational Health and Safety Act promotes safe working conditions. Resources Recovery Act regulates waste disposal and encourages recycling and waste reduction. National Environmental Policy Act passed. Clean Air Act passed. Natural Resources Defense Council created.

1976
National Forest Management Act establishes guidelines for managing national forests. Toxic Control Substances Act regulates many toxic substances not regulated under other laws. Resource Conservation and Recovery Act requires tracking of hazardous waste and encourages recycling, resource recovery, and waste reduction. Noise Control Act regulates harmful noise levels. UN Conference on Human Settlements.

1970s

Figure 7 Some important environmental events during the 1970s, sometimes called the *environmental decade.*
QUESTION: *Which three of these events do you think are the most important?*

What Happened during the 1970s? The Environmental Decade

Increased awareness and public concern led Congress to pass a number of laws to improve environmental quality and conserve more of the nation's natural resources.

During the 1970s, media attention, public concern about environmental problems, scientific research, and action to address environmental concerns grew rapidly. This period is sometimes called the *first decade of the environment* (Figure 7).

The first annual *Earth Day* was held on April 20, 1970. During this event, proposed by Senator Gaylord Nelson (1916–2005), some 20 million people in more than 2,000 communities took to the streets to heighten awareness and to demand improvements in environmental quality.

Republican President Richard Nixon (1913–1994) responded to the rapidly growing environmental movement. He established the *Environmental Protection Agency* (EPA) in 1970 and supported passage of the *Endangered Species Act of 1973*. This greatly strengthened the role of the federal government in protecting endangered species and their habitats.

In 1978, the *Federal Land Policy and Management Act* gave the *Bureau of Land Management* (BLM) its first real authority to manage the public land under its control, 85% of which is in 12 western states. This law angered a number of western interests whose use of these public lands was restricted for the first time.

In response, a coalition of ranchers, miners, loggers, developers, farmers, some elected officials, and others launched a political campaign known as the *sagebrush rebellion*. It had two major goals. *First,* sharply reduce government regulation of the use of public lands. *Second,* remove most public lands in the western United States from federal ownership and management and turn them over to the states. Then the plan was to persuade state legislatures to sell or lease the resource-rich lands at low prices to ranching, mining, energy, timber, land development, and other private interests. This represented a return to President Hoover's plan to get rid of all public land that was thwarted by the Great Depression.

Jimmy Carter (a Democrat), president between 1977 and 1981, was very responsive to environmental concerns. He persuaded Congress to create the *Department of Energy* to develop a long-range energy strategy to reduce the country's heavy dependence on imported oil. He appointed respected environmental leaders to key positions in environmental and resource agencies and consulted with environmental interests on environmental and resource policy matters.

In 1980, Carter helped create a *Superfund* as part of the *Comprehensive Environment Response, Compensation, and Liability Act* to clean up abandoned hazardous waste sites, including the Love Canal in Niagara Falls, New York.

Carter also used the *Antiquities Act of 1906* to triple the amount of land in the National Wilderness System and double the area in the National Park System (primarily by adding vast tracts in Alaska). He used the Antiquities Act to protect more public land, in all 50 states, than any president before him.

What Happened during the 1980s? Environmental Backlash

An anti-environmental movement formed to weaken or do away with many of the environmental laws passed in the 1960s and 1970s and to destroy the political effectiveness of the U.S. environmental movement.

Figure 8 summarizes some key environmental events during the 1980s that shaped U.S. environmental policy. During this decade, farmers and ranchers and leaders of the oil, coal, automobile, mining, and timber industries strongly opposed many of the environmental laws and regulations developed in the 1960s and 1970s. They organized and

funded a strong *anti-environmental movement* that persists today.

In 1981, Ronald Reagan (a Republican, 1911–2004), a self-declared *sagebrush rebel* and advocate of less federal control, became president. During his 8 years in office, he angered environmentalists by appointing to key federal positions people who opposed most existing environmental and public land-use laws and policies.

Reagan greatly increased private energy and mineral development and timber cutting on public lands. He also drastically cut federal funding for research on energy conservation and renewable energy resources and eliminated tax incentives for residential solar energy and energy conservation enacted during the Carter administration. In addition, he lowered automobile gas mileage standards and relaxed federal air and water quality pollution standards.

Although Reagan was immensely popular, many people strongly opposed his environmental and resource policies. This resulted in strong opposition in Congress, public outrage, and legal challenges by environmental and conservation organizations, whose memberships soared during this period.

In 1988, an industry-backed, anti-environmental coalition called the *wise-use movement* was formed. Its major goals were to

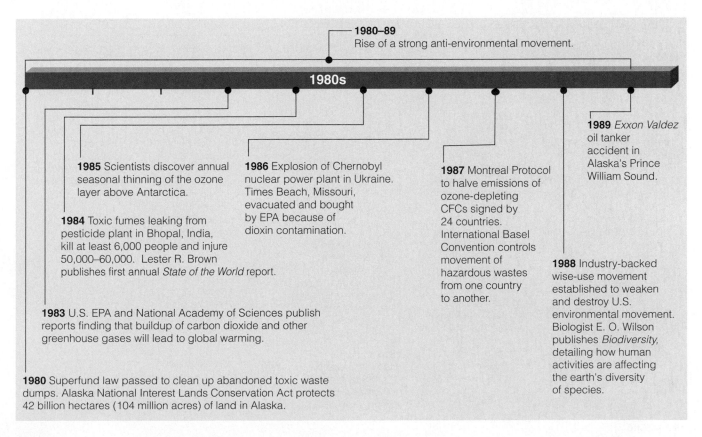

1980s

Figure 8 Some important environmental events during the 1980s. QUESTION: *Which three of these events do you think are the most important?*

weaken or repeal most of the country's environmental laws and regulations and destroy the effectiveness of the environmental movement in the United States. Politically powerful coal, oil, mining, automobile, timber, and ranching interests helped back this movement.

Upon his election in 1989, George H. W. Bush (a Republican) promised to be "the environmental president." But he received criticism from environmentalists for not providing leadership on such key environmental issues as population growth, global warming, and loss of biodiversity. He also continued support of exploitation of valuable resources on public lands at giveaway prices. In addition, he allowed some environmental laws to be undercut by the political influence of industry, mining, ranching, and real estate development interests. They argued that environmental laws had gone too far and were hindering economic growth.

What Happened from 1990 to 2006?

Since 1990, members of the environmental movement have spent most of their time and money trying to keep anti-environmentalists from weakening or eliminating most environmental laws passed in the 1960s and 1970s.

Figure 9 summarizes some key environmental events that took place between 1990 and 2006. In 1993, Bill Clinton (a Democrat) became president and promised to provide national and global environmental leadership. During his 8 years in office, he appointed respected environmental leaders to key positions in environmental and resource agencies and consulted with environmental interests about environmental policy, as Carter did.

He also vetoed most of the anti-environmental bills (or other bills passed with anti-environmental riders attached) passed by a Republican-dominated Congress between 1995 and 2000. He announced regulations requiring sport utility vehicles (SUVs) to meet the same air pollution emission standards as cars. Clinton also used executive orders to make forest health the primary priority in managing national forests and to declare many roadless areas in national forests off limits to the building of roads and to logging.

In addition, he used the Antiquities Act of 1906 to protect various parcels of public land in the West from development and resource exploitation by declaring them national monuments. He protected more public land as national monuments in the lower 48 states than any other president, including Teddy Roosevelt and Jimmy Carter. However, environmental leaders criticized Clinton for failing to push hard enough on key environmental issues such as global

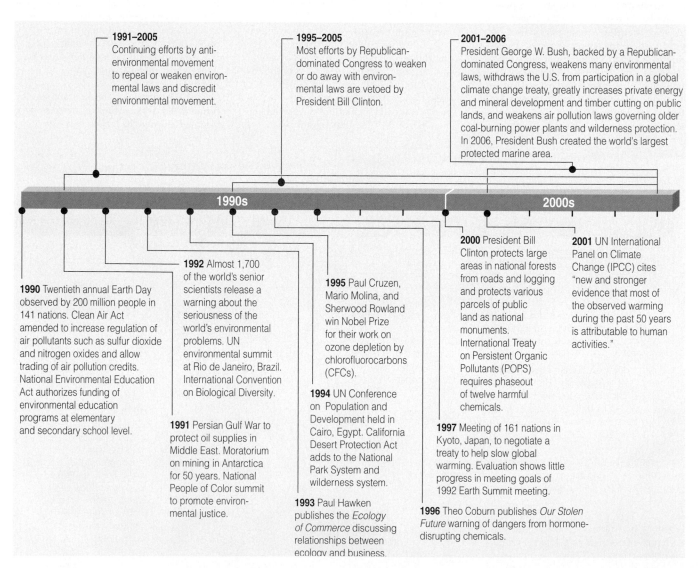

1991–2005 Continuing efforts by anti-environmental movement to repeal or weaken environmental laws and discredit environmental movement.

1995–2005 Most efforts by Republican-dominated Congress to weaken or do away with environmental laws are vetoed by President Bill Clinton.

2001–2006 President George W. Bush, backed by a Republican-dominated Congress, weakens many environmental laws, withdraws the U.S. from participation in a global climate change treaty, greatly increases private energy and mineral development and timber cutting on public lands, and weakens air pollution laws governing older coal-burning power plants and wilderness protection. In 2006, President Bush created the world's largest protected marine area.

1990s   2000s

1990 Twentieth annual Earth Day observed by 200 million people in 141 nations. Clean Air Act amended to increase regulation of air pollutants such as sulfur dioxide and nitrogen oxides and allow trading of air pollution credits. National Environmental Education Act authorizes funding of environmental education programs at elementary and secondary school level.

1991 Persian Gulf War to protect oil supplies in Middle East. Moratorium on mining in Antarctica for 50 years. National People of Color summit to promote environmental justice.

1992 Almost 1,700 of the world's senior scientists release a warning about the seriousness of the world's environmental problems. UN environmental summit at Rio de Janeiro, Brazil. International Convention on Biological Diversity.

1993 Paul Hawken publishes the *Ecology of Commerce* discussing relationships between ecology and business.

1994 UN Conference on Population and Development held in Cairo, Egypt. California Desert Protection Act adds to the National Park System and wilderness system.

1995 Paul Cruzen, Mario Molina, and Sherwood Rowland win Nobel Prize for their work on ozone depletion by chlorofluorocarbons (CFCs).

1996 Theo Coburn publishes *Our Stolen Future* warning of dangers from hormone-disrupting chemicals.

1997 Meeting of 161 nations in Kyoto, Japan, to negotiate a treaty to help slow global warming. Evaluation shows little progress in meeting goals of 1992 Earth Summit meeting.

2000 President Bill Clinton protects large areas in national forests from roads and logging and protects various parcels of public land as national monuments. International Treaty on Persistent Organic Pollutants (POPS) requires phaseout of twelve harmful chemicals.

2001 UN International Panel on Climate Change (IPCC) cites "new and stronger evidence that most of the observed warming during the past 50 years is attributable to human activities."

1990–2006

Figure 9 Some important environmental events, 1990–2006. QUESTION: *Which three of these events do you think are the most important?*

warming and global and national biodiversity protection.

Between 1990 and 2006, the anti-environmental movement gained strength. This occurred because of continuing political and economic support from corporate backers, who argued that environmental laws were hindering economic growth, and because federal elections gave Republicans (many of whom were generally unsympathetic to environmental concerns) a majority in Congress.

Since 1990, leaders and supporters of the environmental movement have had to spend much of their time and funds fighting efforts to discredit the movement and weaken or eliminate most environmental laws passed during the 1960s and 1970s. They also have had to counter claims by anti-environmental groups that problems such as global warming and ozone depletion are hoaxes or not very serious and that environmental laws and regulations were allegedly hindering economic growth.

During the 1990s, many small and mostly local grassroots environmental organizations sprang up to deal with environmental threats in their local communities. Interest in environmental issues increased on many college campuses and environmental studies programs at colleges and universities expanded. In addition, awareness of important, complex environmental issues, such as sustainability, population growth, biodiversity protection, and threats from global warming, increased.

In 2001, George W. Bush (a Republican) became president. Like Reagan in the 1980s, he appointed to key federal positions people who opposed or wanted to weaken many existing environmental and public land-use laws and policies because they allegedly threatened economic growth. Also like Reagan, he did not consult with environmental groups and leaders in developing environmental policies, and he greatly increased private energy and mineral development and timber cutting on public lands.

Bush also opposed increasing automobile gas mileage standards as a way to save energy and reduce dependence on oil imports, and he supported relaxation of various federal air and water quality standards. Like Reagan, he developed an energy policy that emphasized use of fossil fuels and nuclear power with much less support for reducing energy waste and relying more on renewable energy resources.

In addition, he withdrew the United States from participation in the international Kyoto treaty, designed to help reduce carbon dioxide emissions that can promote global warming. He also repealed or tried to weaken most of the pro-environmental measures established by Clinton.

In 2003, leaders of a dozen major environmental organizations charged that Bush, backed by a Republican-dominated Congress, was well on the way to compiling the worst environmental record of any president in the history of the country. By 2006, many of the country's environmental and public land-use laws and regulations had been seriously weakened.

A few moderate Republican members of Congress have urged their party to return to its environmental roots, put down during Teddy Roosevelt's presidency, and shed its anti-environmental approach to legislation. Most Democrats agree and assert that the environmental problems we face are much too serious to be held hostage by political squabbling. They call for cooperation, not confrontation. These Democrats and Republicans urge elected officials, regardless of party, to enter into a new pact in which the United States becomes the world leader in making this the *environmental century*. This would help sustain the country's rich heritage of natural capital and provide economic development, jobs, and profits in rapidly growing businesses such as solar and wind energy, energy efficient vehicles and buildings, ecological restoration, and pollution prevention. As of 2006, elected officials had ignored such calls.

Case Study: The U.S. Dust Bowl: An Environmental Lesson from Nature

In the 1930s, a large area of cropland in the midwestern United States had to be abandoned because of severe soil erosion caused by a combination of poor cultivation practices and prolonged drought.

In the 1930s, Americans learned a harsh environmental lesson when much of the topsoil in several dry and windy midwestern states was lost through a combination of poor cultivation practices and prolonged drought. This threatened to turn much of the U.S. plains into a vast desert.

Before settlers began grazing livestock and planting crops there in the 1870s, the deep and tangled root systems of native prairie grasses anchored the fertile topsoil firmly in place. But plowing the prairie tore up these roots, and the crops that settlers planted annually in their place had less extensive root systems.

After each harvest, the land was plowed and left bare for several months, exposing it to high winds. Overgrazing by livestock in some areas also destroyed large expanses of grass, denuding the ground.

The stage was set for severe wind erosion and crop failures; all that was needed was a long drought. It came between 1926 and 1937 when the annual precipitation dropped by almost two-thirds. In the 1930s, dust clouds created by hot, dry windstorms blowing across the barren exposed soil darkened the sky at midday in some areas (Figure 10). Rabbits and birds choked to death on the dust.

During May 1934, a cloud of topsoil blown off the Great Plains traveled some 2,400 kilometers (1,500 miles) and blanketed most of the eastern United States with dust. Laundry hung out to dry by women in the state of Georgia quickly became covered with dust blown in from the Midwest. Journalists gave the most eroded part of the Great Plains a new name: the *Dust Bowl* (Figure 11, *p.* S23).

During the "dirty thirties," large areas of cropland were stripped of topsoil and severely eroded. This triggered one of the largest internal migrations in U.S. history.

NOAA George E. Marsh Album

Figure 10 Natural capital degradation: dust storm of eroded soil approaching Stratford, Texas (USA) in 1935.

Figure 11 Natural capital degradation: the *Dust Bowl* of the Great Plains, where a combination of extreme drought and poor soil conservation practices led to severe wind erosion of topsoil in the 1930s.

Thousands of farm families from the states of Oklahoma, Texas, Kansas, and Colorado abandoned their dust-choked farms and dead livestock and migrated to California or to the industrial cities of the Midwest and East. Most found no jobs because the country was in the midst of the Great Depression.

In May 1934, Hugh Bennett of the U.S. Department of Agriculture (USDA) went before a congressional hearing in Washington to plead for new programs to protect the country's topsoil. Lawmakers took action when Great Plains dust began seeping into the hearing room. As Hugh Bennett put it, "This nation and civilization is founded upon nine inches of topsoil. And when that is gone there will no longer be any nation or any civilization."

In 1935, the United States passed the *Soil Erosion Act*, which established the Soil Conservation Service (SCS) as part of the USDA. With Bennett as its first head, the SCS (now called the Natural Resources Conservation Service) began promoting sound soil conservation practices, first in the Great Plains states and later elsewhere. Soil conservation districts were formed throughout the country, and farmers and ranchers were given technical assistance in setting up soil conservation programs.

In 1985, the U.S. Congress created the Conservation Reserve Program (CRP) that paid farmers to cover highly erodible land with vegetation under 10-year contracts. Since 1982, this program has cut soil erosion on cropland by about 40% and has provided a model to the world for soil conservation.

Critical Thinking

What is the ecological lesson from the Dust Bowl? Has the United States or the country where you live learned this lesson?

NORSE GREENLAND, SUMERIAN, AND ICELANDIC CIVILIZATIONS (SCIENCE AND HISTORY)
CHAPTERS 2, 10

Downfall of the Norse Greenland Civilization

Norse people who settled in Greenland during the tenth century survived for 450 years before destroying the vegetation and soil that supported them.

Greenland is a vast, mostly ice-covered island about three times the size of the U.S. state of Texas. During the tenth century, Viking explorers settled a small, flat portion of this island that was covered with vegetation and located near the water.

In his 2005 book, *Collapse: How Societies Choose to Fail or Succeed*, biogeographer Jared Diamond describes how this 450-year-old Norse settlement in Greenland collapsed in the 1400s from a combination of colder weather in the 1300s and abuse of its soil resources.

Diamond suggests the Norse made three major errors. First, they cut most of the trees and shrubs to clear fields, make lumber, and gather firewood. Without that vegetation, cold winds dried and eroded the already thin soil. The second error was overgrazing, which meant the depletion of remaining vegetation and trampling of the fragile soil.

Finally, when wood used for lumber was depleted, the Norse removed chunks of their turf and used it to build thick walls in their houses to keep out cold winds. Because they removed the turf faster than it could be regenerated, there was less land for grazing so livestock numbers fell. As a result, their food supply and civilization collapsed. Archeological evidence suggests the last residents starved or froze to death.

After about 500 years nature healed the ecological wounds the Norse caused and Greenland's meadows recovered. In the twentieth century, Danes who settled in Greenland reintroduced livestock. Today, over 56,000 people make their living there by mining, fishing, growing crops, and grazing livestock.

But there is evidence that Greenland's green areas—about 1% of its total land area—are again being overused and strained to their limits. Now Greenlanders have the scientific knowledge to avoid the tragedy of the commons by reducing livestock numbers to a sustainable level, not cutting trees faster than they are replenished, and practicing soil conservation. Time will tell whether they will work together to avoid the environmental tragedy they face.

Downfall of the Sumerian Civilization

The once-great Sumerian civilization collapsed mostly because long-term irrigation led to salt buildup in its soils and declining food productivity.

During the fourth century BC, a highly advanced urban and literate Sumerian civilization began emerging on the flood plains of the lower reaches of the Tigris and Euphrates Rivers in parts of what is now Iraq. This civilization developed science and mathematics, and to grow food, built a well-engineered irrigation system, which used dams to divert water from the Euphrates River through a network of gravity-fed canals.

The irrigated cropland produced a food surplus and allowed Sumerians to develop the world's first cities and written language (the cunneiform script). But the Sumerians also learned the painful lesson that long-term irrigation can lead to salt buildup in soils and sharp declines in food production.

Poor underground drainage slowly raised the water table to the surface and evaporation of the water left behind salts that sharply reduced crop productivity—a form of environmental degradation we now call *soil salinization*. As wheat yields declined, the Sumerians slowed the salinization by shifting to more salt-tolerant barley. But as salt concentrations continued to increase, barley yields declined and food production was undermined.

Around 2000 BC, this once-great civilization disappeared as a result of such environmental degradation, economic decline, and invasion by Semitic peoples.

Iceland's Environmental Struggles and Triumphs

Early settlers ecologically devastated Iceland, but the people learned from their mistakes and now have one of the world's most environmentally sustainable countries and a prosperous economy.

Iceland is a Northern European island country slightly smaller than the U.S. state of Kentucky. This volcanic island is located in the North Atlantic Ocean just south of the Arctic Circle between Greenland and Norway, Ireland, and Scotland. Glaciers cover about 10% of the country and it is subject to earthquakes and volcanic activity.

Immigrants from Scandinavia, Ireland, and Scotland began settling the country during the late ninth and tenth centuries AD. Since these settlements began, most of the country's trees and other vegetation have been destroyed and about half of its original soils have eroded into the sea. As a result, Iceland suffers more ecological degradation than any other European country.

The early settlers saw what appeared to be a country with deep and fertile soils, dense forests, and highland grasslands similar to those in their native countries. They did not realize that the soils built up by ash from volcanic eruptions were replenished very slowly and were highly susceptible to water and wind erosion when protective vegetation was removed for growing crops and grazing livestock. Within a few decades, the settlers degraded much of this natural capital that had taken thousands of years to build up.

When the settlers realized what was happening, they took corrective action by trying to save remaining trees and not raising ecologically destructive pigs and goats. The farmers joined together to slow soil erosion and preserve their grasslands to avoid what Garrett Hardin later called the tragedy of the commons (p. 12). They estimated how many sheep the communal highland grasslands could sustain and divided the allotted quotas among themselves.

Icelanders also learned how to tap into their country's abundance of fish, geothermal power from its numerous hot springs and heated rock formations, and hydroelectric power from its many rivers. Renewable hydropower and geothermal energy provide about 95% of its electricity and geothermal energy is used to heat 80% of its buildings and to grow most of its fruits and vegetables in greenhouses.

In terms of per capita income, Iceland is one of the world's ten richest countries, and in 2005 it had the world's thirteenth highest Environmental Performance Index of 133 countries. By 2050, Iceland plans to become the world's first country to run its entire economy on renewable hydropower, geothermal energy, and wind and to use these sources to produce hydrogen for running all of its motor vehicles and ships.

Critical Thinking

List two ecological lessons for us from these three stories.

S u p p l e m e n t 7

SOME BASIC CHEMISTRY (SCIENCE)
CHAPTERS 2, 3

The Periodic Table

The chemical elements can be arranged in a table based on their similar chemical properties.

Chemists have developed a way to classify the elements according to their chemical behavior, in what is called the *periodic table of elements* (Figure 1). Each horizontal row in the table is called a *period*. Each vertical column lists elements with similar chemical properties and is called a *group*.

The partial periodic table in Figure 1 shows how the elements can be classified as *metals, nonmetals,* and *metalloids*. Most of the elements found to the left and at the bottom of the table are *metals,* which usually conduct electricity and heat, and are shiny. Examples are sodium (Na), calcium (Ca), aluminum (Al), iron (Fe), lead (Pb), and mercury (Hg).

Atoms of metals tend to lose one or more of their electrons to form positively charged ions such as Na^+, Ca^{2+}, and Al^{3+}. For example, an atom of the metallic element sodium (Na, atomic number 11) with 11 positively charged protons and 11 negatively charged electrons can lose one of its electrons. It then becomes a sodium ion with a positive charge of 1 (Na^+) because it now has 11 positive charges (protons) but only 10 negative charges (electrons).

Nonmetals, found in the upper right of the table, do not conduct electricity very well and usually are not shiny. Examples are hydrogen (H), carbon (C), nitrogen (N), oxygen (O), phosphorus (P), sulfur (S), chlorine (Cl), and fluorine (F).

Atoms of some nonmetals such as chlorine, oxygen, and sulfur tend to gain one or more electrons lost by metallic atoms to form negatively charged ions such as O^{2-},

S^{2-}, and Cl^-. For example, an atom of the nonmetallic element chlorine (Cl, with an atomic number of 17) can gain an electron and become a chlorine ion. The ion has a negative charge of 1 (Cl^-) because it has 17 positively charged protons and 18 negatively charged electrons. Atoms of nonmetals can also combine with one another to form molecules in which they share one or more pairs of their electrons. Hydrogen, a nonmetal, is placed by itself above the center of the table because it does not fit very well into any of the groups.

The elements arranged in a diagonal staircase pattern between the metals and nonmetals have a mixture of metallic and nonmetallic properties and are called *metalloids*.

Figure 1 also identifies the elements required as *nutrients* (black squares) for all or some forms of life and elements that are

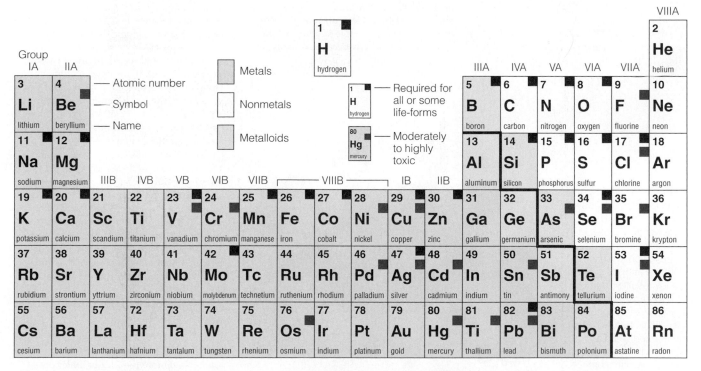

Figure 1 Abbreviated periodic table of elements. Elements in the same vertical column, called a *group*, have similar chemical properties. To simplify matters at this introductory level, only 72 of the 115 known elements are shown.

moderately or highly toxic to all or most forms of life (red squares). Six nonmetallic elements—carbon (C), oxygen (O), hydrogen (H), nitrogen (N), sulfur (S), and phosphorus (P)—make up about 99% of the atoms of all living things.

? *THINKING ABOUT THE PERIODIC TABLE* Use the periodic table to identify by name and symbol two elements that should have similar chemical properties to those of (a) Ca, (b) potassium, (c) S, (d) lead.

Ionic and Covalent Bonds

The forces of attraction between oppositely charged ions hold some compounds together; others are held together when their atoms share one or more pairs of electrons.

Sodium chloride (NaCl) consists of a three-dimensional network of oppositely charged *ions* (Na^+ and Cl^-) held together by the forces of attraction between opposite charges (Figure 2). The strong forces of attraction between such oppositely charged ions are called *ionic bonds*. Because ionic compounds consist of ions formed from atoms of metallic (positive ions) and nonmetallic (negative ions) elements (Figure 1), they can be described as *metal–nonmetal compounds*.

Sodium chloride and many other ionic compounds tend to dissolve in water and break apart into their individual ions (Figure 3).

$$NaCl \longrightarrow Na^+ + Cl^-$$

sodium chloride in water sodium ion + chloride ion

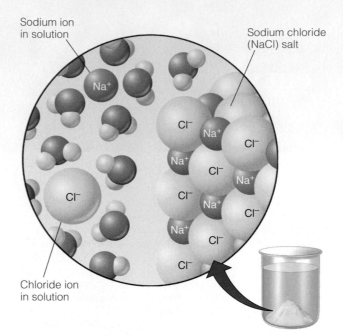

Sodium ion in solution

Sodium chloride (NaCl) salt

Figure 3 How a salt dissolves in water.

Chloride ion in solution

Water, a *covalent compound,* consists of molecules made up of uncharged atoms of hydrogen (H) and oxygen (O). Each water molecule consists of two hydrogen atoms chemically bonded to an oxygen atom, yielding H_2O molecules. The bonds between the atoms in such molecules are called *covalent bonds* and form when the atoms in the molecule share one or more pairs of their electrons. Because they are formed from atoms of nonmetallic elements (Figure 1), covalent compounds can be described as *nonmetal–nonmetal compounds.* Figure 4 shows the chemical formulas and shapes of the molecules that are the building blocks for several common *covalent compounds.*

Hydrogen Bonds

Weak forces of attraction can occur between molecules of covalent compounds such as water.

Ionic and covalent bonds form between the ions or atoms *within* a compound. There are also weaker forces of attraction *between* the molecules of covalent compounds (such as water) resulting from an unequal sharing of electrons by two atoms.

For example, an oxygen atom has a much greater attraction for electrons than does a hydrogen atom. Thus, in a water molecule the electrons shared between the oxygen atom and its two hydrogen atoms are pulled closer to the oxygen atom, but not actually transferred to the oxygen atom. As a result, the oxygen atom in a water molecule has a slightly negative partial charge and its two hydrogen atoms have a slightly positive partial charge (Figure 5).

The slightly positive hydrogen atoms in one water molecule are then attracted to the slightly negative oxygen atoms in another water molecule. These forces of attraction *between* water molecules are called *hydrogen bonds* (Figure 5). Hydrogen bonds also form between other covalent molecules or portions of such molecules containing hydrogen and nonmetallic atoms with a strong ability to attract electrons.

Macromolecules: The Building Blocks of Life

Four types of complex organic compounds are the building blocks of life.

Larger and more complex organic compounds, called *polymers,* consist of a number of basic structural or molecular units (*monomers*) linked by chemical bonds, some-

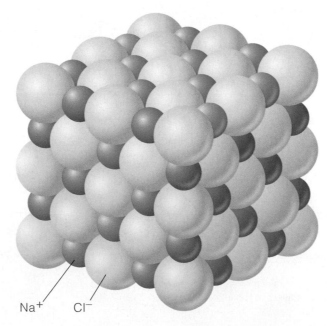

Figure 2 A solid crystal of an ionic compound such as sodium chloride consists of a three-dimensional array of opposite charged ions held together by *ionic bonds* resulting from the strong forces of attraction between opposite electrical charges. They are formed when an electron is transferred from a metallic atom such as sodium (Na) to a nonmetallic element such as chlorine (Cl).

Na^+ Cl^-

what like rail cars linked in a freight train. Four types of macromolecules—complex carbohydrates, proteins, nucleic acids, and lipids—are molecular building blocks of life.

Complex carbohydrates, consist of two or more monomers of *simple sugars* (such as glucose, Figure 6, p. S30) linked together. One example are the starches that plants use to store energy and also provide energy for animals that feed on plants. Another is cellulose, the earth's most abundant organic compound, that is found in the cell walls of bark, leaves, stems, and roots.

Proteins are large polymer molecules formed by linking together long chains of monomers called *amino acids* (Figure 7, p. S30). Living organisms use about 20 different amino acid molecules to build a variety of proteins, which play different roles. Some help store energy. Some are components of the *immune system* that helps protect the body against disease and harmful substances by forming antibodies that make invading agents harmless. Others are *hormones* that are used as chemical messengers in the bloodstream of animals to turn various bodily functions on or off. In animals, proteins are also components of hair, skin, muscle, and tendons. In addition, some proteins act as *enzymes* that catalyze or speed up certain chemical reactions.

Nucleic acids are large polymer molecules made by linking hundreds to thousands of four types of monomers called *nucleotides.* Two nucleic acids—DNA (**d**eoxyribo**n**ucleic **a**cid) and RNA (**r**ibo**n**ucleic **a**cid)—participate in the building of proteins and carry hereditary information used to pass traits from parent to offspring. Each nucleotide consists of a *phosphate group,* a *sugar molecule* containing five carbon atoms (deoxyribose in DNA molecules and ribose in RNA molecules), and one of four different *nucleotide bases* (represented by A, G, C, and T, the first letter in each of their names, or A, G, C, and U in RNA) (Figure 8, p. S30). In the cells of living organisms, these nucleotide units combine in different numbers and sequences to form *nucleic acids* such as various types of RNA and DNA (Figure 9, p. S31).

Hydrogen bonds formed between parts of the four nucleotides in DNA hold two DNA strands together like a spiral staircase, forming a double helix (Figure 9, p. S31). DNA molecules can unwind and replicate themselves.

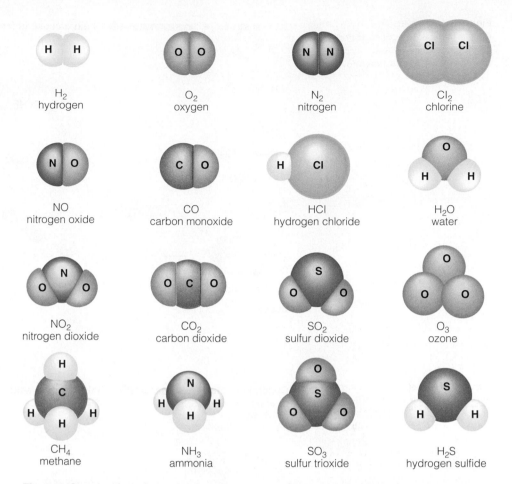

Figure 4 Chemical formulas and shapes for some *covalent compounds* formed when atoms of one or more nonmetallic elements combine with one another by sharing one or more pairs of electrons. The bonds between the atoms in such molecules are called *covalent bonds.*

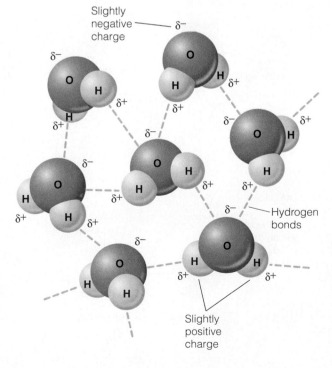

Figure 5 *Hydrogen bond:* slightly unequal sharing of electrons in the water molecule creates a molecule with a slightly negatively charged and a slightly positively charged end. Because of this electrical polarity, the hydrogen atoms of one water molecule are attracted to the oxygen atoms in other water molecules. These fairly weak forces of attraction *between* molecules (represented by the dashed lines) are called *hydrogen bonds.*

Figure 6 Straight-chain and ring structural formulas of glucose, a simple sugar that can be used to build long chains of complex carbohydrates such as starch and cellulose.

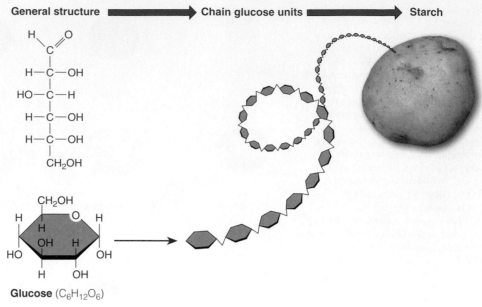

Glucose ($C_6H_{12}O_6$)

Figure 7 General structural formula and a specific structural formula of one of the 20 different amino acid molecules that can be linked together in chains to form proteins that fold up into more complex shapes.

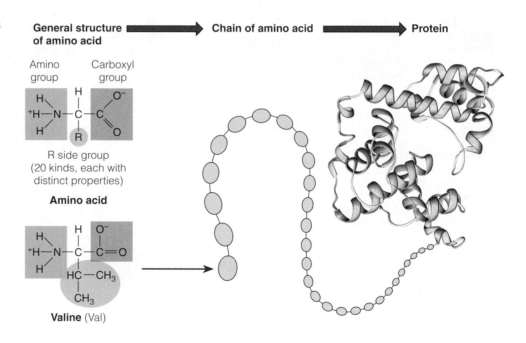

Amino group

Carboxyl group

R side group (20 kinds, each with distinct properties)

Amino acid

Valine (Val)

Figure 8 Generalized structure of the nucleotide molecules linked in various numbers and sequences to form large nucleic acid molecules such as various types of DNA (deoxyribonucleic acid) and RNA (ribonucleic acid). In DNA, the 5-carbon sugar in each nucleotide is deoxyribose; in RNA it is ribose. The four basic nucleotides used to make various forms of DNA molecules differ in the types of nucleotide bases they contain—guanine (G), cytosine (C), adenine (A), and thymine (T). (Uracil, U, occurs instead of thymine in RNA.)

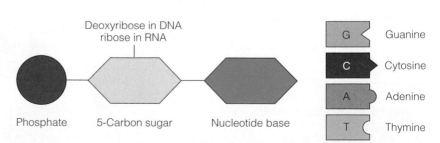

Deoxyribose in DNA ribose in RNA

Phosphate   5-Carbon sugar   Nucleotide base

G — Guanine
C — Cytosine
A — Adenine
T — Thymine

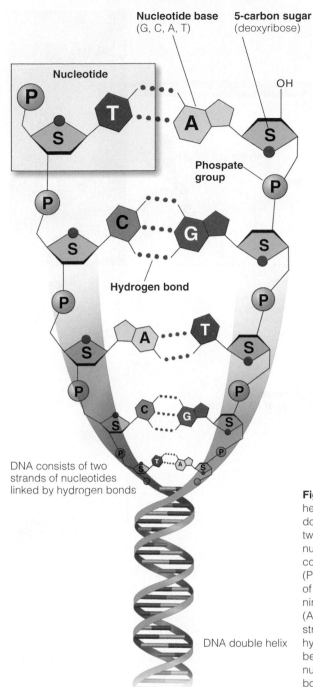

Nucleotide base (G, C, A, T)

5-carbon sugar (deoxyribose)

Nucleotide

Phospate group

Hydrogen bond

DNA consists of two strands of nucleotides linked by hydrogen bonds

DNA double helix

Fatty acid (lipid)

Fat molecule (triglyceride)

Fatty tissue (adipose cells)

Figure 10 Structural formula of fatty acid that is one form of lipid (left). Fatty acids are converted into more complex fat molecules (molecule) that are stored in adipose cells (right).

Figure 9 Portion of the double helix of a DNA molecule. The double helix is composed of two spiral (helical) strands of nucleotides. Each nucleotide contains a unit of phosphate (P), deoxyribose (S), and one of four nucleotide bases: guanine (G), cytosine (C), adenine (A), and thymine (T). The two strands are held together by hydrogen bonds formed between various pairs of the nucleotide bases. Guanine (G) bonds with cytosine (C), and adenine (A) with thymine (T).

The total weight of the DNA needed to reproduce the world's 6.6 billion people is only about 50 milligrams—the weight of a small match. If the DNA coiled in your body were unwound, it would stretch about 960 million kilometers (600 million miles)—more than six times the distance between the sun and the earth.

The different molecules of DNA that make up the millions of species found on the earth are like a vast and diverse genetic library. Each species is a unique book in that library. The *genome* of a species is made up of the entire sequence of DNA "letters" or base pairs that combine to "spell out" the chromosomes in typical members of each species. In 2002, scientists were able to map out the genome for the human species by analyzing the 3.1 billion base sequences in human DNA.

Lipids, a fourth building block of life, are a chemically diverse group of large organic compounds that do not dissolve in water. Examples are *fats and oils* for storing energy (Figure 10), *waxes* for structure, and *steroids* for producing hormones.

Figure 11 (p. S32) shows the relative sizes of simple and complex molecules, cells, and multicelled organisms.

Energy Storage and Release in Cells

Energy released by chemical reactions in cells is stored in ATP molecules for release as needed.

Chemical reactions occurring in photosynthesis (Spotlight, p. 59) release energy that is absorbed by adenosine diphosate (ADP) molecules and stored as chemical energy in adenosine triphosphate (ATP) molecules (Figure 12, left, p. S32). When cellular processes require energy, ATP molecules release it to form ADP molecules (Figure 12, right, p. S32).

Balancing Chemical Equations

Chemists use a shorthand system to represent chemical reactions and insure that no atoms are created or destroyed in a chemical reaction as required by the law of conservation of matter.

In keeping with the law of conservation of matter (p. 39), chemical equations are used as an accounting system to verify that no atoms are created or destroyed in a chemical reaction. As a consequence, each side of a chemical equation must have the same number of atoms of each element involved. Ensuring that this condition is met leads to what chemists call a *balanced chemical equation*. The equation for the burning of carbon

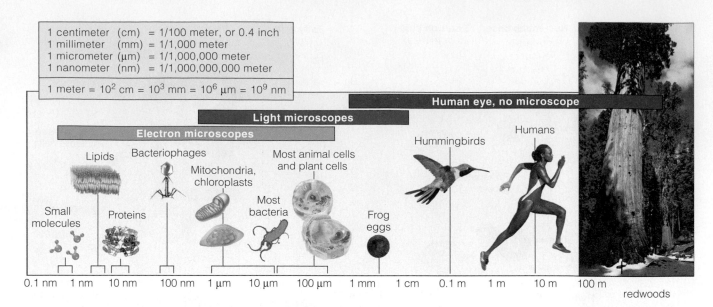

Figure 11 Natural capital: relative size of simple molecules, complex molecules, cells, and multi-cellular organisms. This scale is exponential, not linear. Each unit of measure is ten times larger than the unit preceding it. (Used by permission from Cecie Starr and Ralph Taggart, *Biology*, 11th ed, Belmont, Calif.: Thomson Brooks/Cole, © 2006)

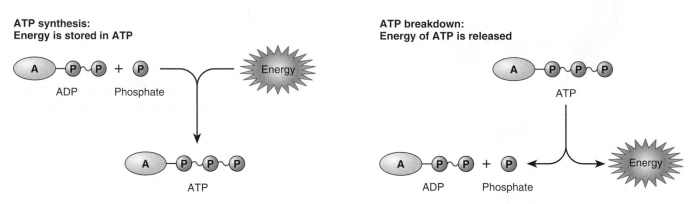

Figure 12 Energy storage and release in cells.

$(C + O_2 \longrightarrow CO_2)$ is balanced because one atom of carbon and two atoms of oxygen are on both sides of the equation.

Consider the following chemical reaction: When electricity passes through water (H_2O), the latter can be broken down into hydrogen (H_2) and oxygen (O_2), as represented by the following equation:

$$H_2O \longrightarrow H_2 + O_2$$
2 H atoms   2 H atoms   2 O atoms
1 O atom

This equation is unbalanced because one atom of oxygen is on the left side of the equation but two atoms are on the right side.

We cannot change the subscripts of any of the formulas to balance this equation because that would change the arrangements of the atoms, leading to different substances. Instead, we must use different

numbers of the molecules involved to balance the equation. For example, we could use two water molecules:

$$2 H_2O \longrightarrow H_2 + O_2$$
4 H atoms   2 H atoms   2 O atoms
2 O atoms

This equation is still unbalanced. Although the numbers of oxygen atoms on both sides of the equation are now equal, the numbers of hydrogen atoms are not.

We can correct this problem by having the reaction produce two hydrogen molecules:

$$2 H_2O \longrightarrow 2 H_2 + O_2$$
4 H atoms   4 H atoms   2 O atoms
2 O atoms

Now the equation is balanced, and the law of conservation of matter has been ob-

served. For every two molecules of water through which we pass electricity, two hydrogen molecules and one oxygen molecule are produced.

If scientists and engineers can find economical ways to decompose water by using electricity, heat, or solar energy, this reaction may be used as a way to produce hydrogen gas (H_2) for use as a fuel to help replace oil during this century. The hydrogen would be used in *fuel cells* where it would combine with oxygen gas to produce water and energy for heating houses and water and propelling motor vehicles and planes. Bringing about such a *hydrogen revolution* would reduce the world's dependence on dwindling supplies of oil, eliminate most forms of air pollution because the major emission from a fuel cell is water vapor, and help slow global warming by not emitting the carbon dioxide gas that is released when any carbon-containing fuel is burned (provided electricity produced by wind turbines or solar cells is used to decompose water).

Try to balance the chemical equation for the reaction of nitrogen gas (N_2) with hydrogen gas (H_2) to form ammonia gas (NH_3).

Nanotechnology: Building Materials from the Bottom Up

Scientists and engineers are learning how to build materials from the chemical elements.

Nanotechnology uses atoms and molecules to build materials from the bottom up using the elements in the periodic table as its raw materials. A *nanometer* is one billionth of a meter—like comparing the size of a marble to the size of the earth. A comma has the width of about half a million nanometers. A nanometer is about the length a man's beard grows during the time it takes him to lift a razor to his face. A DNA molecule (Figure 9) is about 2.5 nanometers wide.

At the nanoscale below about 100 nanometers, the properties of materials change dramatically. At this scale, materials can exhibit new properties that do not exist at the microscale of thousandths of a meter and the macroscale of everyday items (Figure 11).

For example, at the macroscale, carbon such as pencil lead is soft and fairly weak. But at the nanoscale, carbon nanoparticles can be 10 times stronger and 6 times lighter than steel. At the macroscale zinc oxide (ZnO) can be rubbed on the skin as a white paste to protect against the sun's harmful UV rays; at the nanoscale it becomes transparent and is being used as invisible coatings to protect the skin and fabrics from UV damage. Because silver can kill harmful bacteria, silver nanocrystals are being incorporated into bandages for wounds. And, at a size of 20–30 nanometers, particles of aluminum can explode, explaining why some are experimenting with adding it to rocket fuel.

Researchers hope to incorporate nanoparticles of hydroxyapatite, with the same chemical structure as tooth enamel, into toothpaste to put coatings on teeth that prevent bacteria from penetrating. Nanotech coatings now being used on cotton fabrics form an impenetrable barrier that causes liquids to bead and roll off. Such stain-resistant fabrics used to make clothing, rugs, and furniture upholstery could eliminate the need to use harmful chemicals for removing stains.

Self-cleaning window glass coated with a layer of nanoscale titanium dioxide (TiO_2) particles are now available. As the particles interact with UV rays from the sun, dirt on the surface of the glass loosens and washes off when it rains. Similar products can be used for self-cleaning sinks and toilet bowls.

Typically, the manufacture of solar cells requires a multimillion-dollar fabrication facility. Scientists are developing nanosolar cells made by mixing up a hundred dollars' worth of starter chemicals. The resulting nanosolar cells can be painted on window glass or walls to turn an entire building into a solar-energy generator. If the manufacturing process can be developed, this could change the world's energy supplies by applying one of the four principles of sustainability on a global scale.

Scientists are working on ways to replace the silicon in computer chips with carbon-based nanomaterials that greatly increase the processing power of computers. Biological engineers are working on nanoscale devices to deliver drugs and to penetrate cancer cells and deliver nanomolecules that kill the cancer cells from the inside. Researchers also hope to develop nanoscale crystals that can change color when they detect parts per trillion amounts of harmful substances such chemical and biological warfare agents and food pathogens. For example, a color change in food packaging could alert a consumer when a food is contaminated or has begun to spoil. The list of possibilities could go on.

By 2006, more than 720 products containing nanoscale particles were commercially available and thousands more are in the pipeline. Examples are found in cosmetics, sunscreens, fabrics, pesticides, and food additives.

So far, these products are unregulated and unlabeled. This concerns many health and environmental scientists because the tiny size of nanoparticles can allow them to penetrate the natural defenses of the body against invasions by foreign and potentially harmful chemicals and pathogens.

Engineered nanoscale particles have a larger surface area that can make them more chemically active. This property allows them to perform many useful functions, but it may also makes them more toxic than conventional-size particles. This means that a chemical that is harmless at the macroscale may be hazardous at the nanoscale when it is inhaled, ingested, or absorbed through the skin.

We know little about such effects and risks at a time when the use of a variety of untested and unregulated nanoparticles is increasing exponentially. A few toxicological studies are sending up red flags.

- In 2004, Eva Olberdorster, an environmental toxicologist at Southern Methodist University, found that fish swimming in water loaded with carbon buckyballs experienced brain damage within 48 hours.

- In 2005, NASA researchers found that injecting commercially available carbon nanotubes into rats caused significant lung damage.

- A 2005 study by researchers at the U.S. National Institute of Occupational Safety and Health found substantial damage to the heart and aortic arteries of mice exposed to carbon nanotubes.

- In 2005, researchers at New York's University of Rochester found increased blood clotting in rabbits inhaling carbon buckyballs.

On the other hand, a 2006 study by the Centre for Drug Delivery Research at the University of London's School of Pharmacy found that when mice were injected with water-soluble carbon nanotubes the tubes were excreted intact in urine. And a 2006 study found that buckyballs could be made less toxic by attaching chemicals known as hydroxyl groups. The question is whether this alters their beneficial properties.

In 2004, the British Royal Society and Royal Academy of Engineering recommended that we avoid the environmental release of nanoparticles and nanotubes as much as possible until more is known about their potential harmful impacts. They recommended as a precautionary measure that factories and research laboratories treat manufactured nanoparticles and nanotubes as if they were hazardous to their workers and to the general public. *Green Career:* Nanotechnology

---

? THINKING ABOUT NANOTECHNOLOGY
Do the benefits of nanotechnology outweigh its potentially harmful effects? Explain. What three things would you do to reduce its potentially harmful effects?

F RESEARCH FRONTIER Learning more about nanotechnology and how to reduce its potentially harmful effects

Biologists classify species into different *kingdoms,* on the basis of similarities and differences in their nutrition, cell structure, appearance, and developmental features.

In this book, the earth's organisms are classified into six kingdoms: *eubacteria, archaebacteria, protists, fungi, plants,* and *animals* (Figure 1).

Eubacteria consist of single-celled prokaryotic bacteria (Figure 2-6, bottom, p. 37) not including archaebacteria. Examples include various cyanobacteria and bacteria such as *staphylococcus* and *streptococcus.*

Archaebacteria are single-celled bacteria that are closer to eukaryotic cells

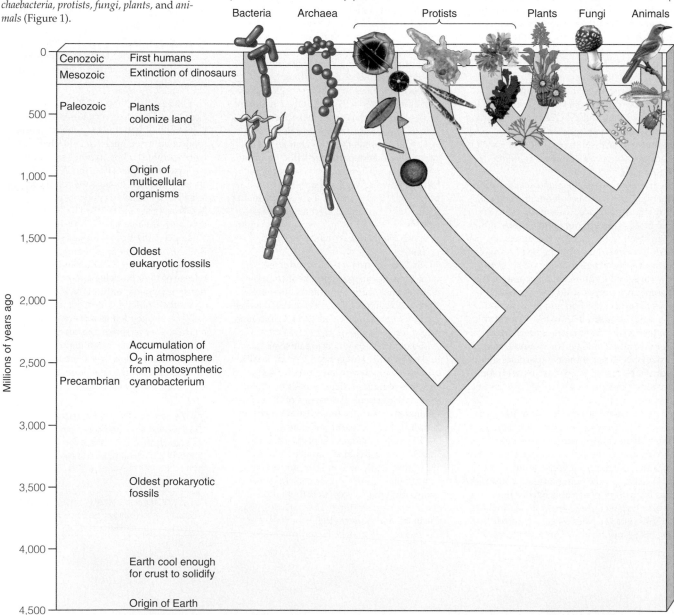

Prokaryotes

Bacteria   Archaea

Eukaryotes

Protists   Plants   Fungi   Animals

Millions of years ago

0	Cenozoic	First humans
	Mesozoic	Extinction of dinosaurs
500	Paleozoic	Plants colonize land
1,000		Origin of multicellular organisms
1,500		Oldest eukaryotic fossils
2,000		
2,500	Precambrian	Accumulation of O$_2$ in atmosphere from photosynthetic cyanobacterium
3,000		
3,500		Oldest prokaryotic fossils
4,000		Earth cool enough for crust to solidify
4,500		Origin of Earth

Figure 1 Natural capital: overview of the evolution of life on the earth into six major kingdoms of species as a result of natural selection (*p.* 85). This view sees the development of life as an ever-branching, never-crossing tree of species diversity, sometimes called the tree of life. More modern views of evolution picture life as a complex web in which species can interact to produce new species through natural selection, and more rapidly through the crossbreeding of closely related species (hybridization) and the transfer of genes between unrelated species (gene transfer, p. 86).

(Figure 2-6, top, p. 37) than to eubacteria. Examples include methanogens, which live in oxygen-free sediments of lakes and swamps and in animal guts; halophiles, which live in extremely salty water; and thermophiles, which live in hot springs, hydrothermal vents, and acidic soil. These organisms live in extreme environments.

Protists are mostly single-celled eukaryotic organisms such as diatoms, dinoflagellates, amoebas, golden brown and yellow-green algae, and protozoans. Some protists cause human diseases such as malaria and sleeping sickness.

Fungi are mostly many-celled, sometimes microscopic, eukaryotic organisms such as mushrooms, molds, mildews, and yeasts. Many fungi are decomposers. Other fungi kill various plants and cause huge losses of crops and valuable trees.

Plants are mostly many-celled eukaryotic organisms such as red, brown, and green algae and mosses, ferns, and flowering plants (whose flowers produce seeds that perpetuate the species). Some plants such as corn and marigolds are *annuals*, meaning that they complete their life cycles in one growing season. Others are *perennials*, which can live for more than 2 years, such as roses, grapes, elms, and magnolias.

Animals are also many-celled, eukaryotic organisms. Most have no backbones and hence are called *invertebrates*. Invertebrates include sponges, jellyfish, worms, arthropods (insects, shrimp, and spiders), mollusks (snails, clams, and octopuses), and echinoderms (sea urchins and sea stars). *Vertebrates* (animals with backbones and a brain protected by skull bones) include fishes (e.g., sharks and tuna), amphibians (e.g., frogs and salamanders), reptiles (e.g., crocodiles and snakes), birds (e.g., eagles and robins), and mammals (e.g., bats, elephants, whales, and humans).

Within each kingdom, biologists have created subcategories based on anatomical, physiological, and behavioral characteristics. Kingdoms are divided into *phyla*, which are divided into subgroups called *classes*. Classes are subdivided into *orders*, which are further divided into *families*. Families consist of *genera* (singular, *genus*), and each genus contains one or more species. Note that the word *species* is both singular and plural. Figure 2 shows this detailed taxonomic classification for the current human species.

Most people call a species by its common name, such as robin or grizzly bear.

Biologists use scientific names (derived from Latin) consisting of two parts (printed in italics, or underlined) to describe a species. The first word is the capitalized name (or abbreviation) for the genus to which the organism belongs. It is followed by a lowercase name that distinguishes the species from other members of the same genus. For example, the scientific name of the robin is *Turdus migratorius* (Latin for "migratory thrush") and the grizzly bear goes by the scientific name *Ursus horribilis* (Latin for "horrible bear").

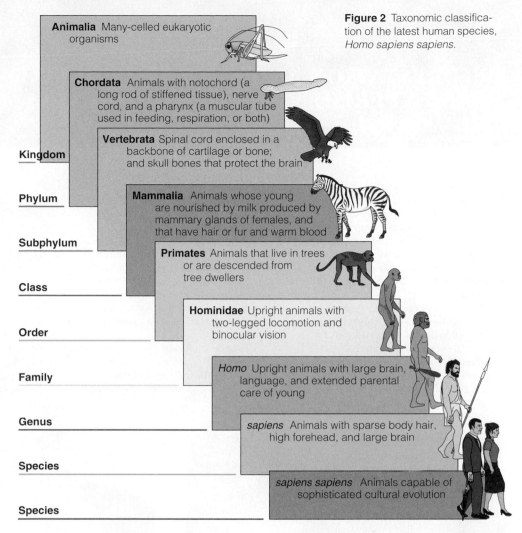

Figure 2 Taxonomic classification of the latest human species, *Homo sapiens sapiens*.

Kingdom — **Animalia** Many-celled eukaryotic organisms

Phylum — **Chordata** Animals with notochord (a long rod of stiffened tissue), nerve cord, and a pharynx (a muscular tube used in feeding, respiration, or both)

Subphylum — **Vertebrata** Spinal cord enclosed in a backbone of cartilage or bone; and skull bones that protect the brain

Class — **Mammalia** Animals whose young are nourished by milk produced by mammary glands of females, and that have hair or fur and warm blood

Order — **Primates** Animals that live in trees or are descended from tree dwellers

Family — **Hominidae** Upright animals with two-legged locomotion and binocular vision

Genus — **Homo** Upright animals with large brain, language, and extended parental care of young

Species — *sapiens* Animals with sparse body hair, high forehead, and large brain

Species — *sapiens sapiens* Animals capable of sophisticated cultural evolution

SUPPLEMENT 9

DEFORESTATION AND NUTRIENT CYCLING IN AN EXPERIMENTAL FOREST (SCIENCE)
CHAPTERS 3, 10

In the 1960s, F. H. Bormann of Yale University, Gene Likens of Cornell University, and their colleagues began carrying out a controlled experiment. The goal was to compare the loss of water and nutrients from an uncut forest ecosystem (the *control system*) with one that was stripped of its trees (the *experimental system*).

They built V-shaped concrete catchment dams across the creeks at the bottoms of several valleys in the Hubbard-Brook Experimental Forest in New Hampshire (Figure 1, left). The dams were anchored on impenetrable bedrock so all surface water leaving each forested valley ecosystem had to flow across the dams, where scientists could measure its volume and dissolved nutrient content.

The first project measured the amounts of water that entered and left an undisturbed (control) forest and the amount of dissolved nutrients in this inflow and outflow. These baseline data showed that an undisturbed mature forest ecosystem is very efficient at retaining chemical nutrients.

The next experiment disturbed the system and observed any changes that occurred. One winter the investigators cut down all trees and shrubs in one valley, left them where they fell, and sprayed with herbicides to prevent regrowth (Figure 1, right). Then they compared the inflow and outflow of water and nutrients in this modified experimental valley with those in the control valley for three years.

With no plants to absorb and transpire water from the soil, water runoff in the deforested valley increased by 30–40%. As this excess water ran rapidly over the surface of the ground, it eroded soil and carried nutrients out of the ecosystem. Overall, the loss of minerals from the cut forest was six to eight times that in a nearby undisturbed forest.

Figure 1 Controlled field experiment on the effects of deforestation on the loss of water and nutrients from a forest ecosystem. V–notched dams were built into the impenetrable bedrock at the bottoms of several forested valleys (left) so that all water and minerals flowing from each valley could be collected and measured for volume and mineral content. Baseline data were collected on several forested valleys that acted as the control group. Then all of the trees in one valley (the experimental group) were cut (right) and the flow of water and minerals from this experimental valley were measured for three years.

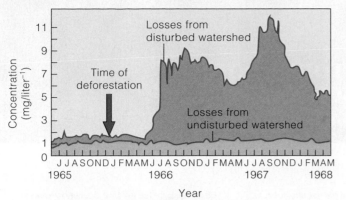

Figure 2 Loss of nitrate ions (NO_3^-) from a deforested watershed in the Hubbard-Brook Experimental Forest in New Hampshire. The concentration of nitrate ions in runoff from the deforested experimental watershed was many times greater than in a nearby unlogged watershed used as a control. (Data from F. H. Bormann and Gene Likens)

For example, chemical analysis of the water flowing through the dams showed a 60-fold rise in the concentration of nitrate ions (NO_3^-) (Figure 2). So much nitrogen as nitrates (NO_3^-) was lost from the experimental valley that the water flowing out of it was unsafe to drink, and the overfertilized stream below this valley became covered with populations of cyanobacteria and algae. After a few years, however, vegetation grew back, and nitrate levels returned to normal.

Critical Thinking

What ecological lesson can we learn from this experiment about more sustainable use of the world's forests?

What Is Weather?

Weather is the result of the atmospheric conditions in a particular area.

Weather is an area's short-term atmospheric conditions—typically those occurring over hours or days. Examples of atmospheric conditions include temperature, pressure, moisture content, precipitation, sunshine, cloud cover, and wind direction and speed.

Meteorologists use equipment on weather balloons, aircraft, ships, and satellites, as well as radar and stationary sensors, to obtain data on weather variables. They then feed these data into computer models to draw weather maps. Other computer models project the weather for the next several days by calculating the probabilities that air masses, winds, and other factors will move and change in certain ways.

Much of the weather you experience results from interactions between the leading edges of moving masses of warm and cold air. Weather changes as one air mass replaces or meets another. The most dramatic changes in weather occur along a **front,** the boundary between two air masses with different temperatures and densities.

A **warm front** is the boundary between an advancing warm air mass and the cooler one it is replacing (Figure 1, top). Because warm air is less dense (weighs less per unit of volume) than cool air, an advancing warm front rises up over a mass of cool air.

As the warm front rises, its moisture begins condensing into droplets, forming layers of clouds at different altitudes. Gradually the clouds thicken, descend to a lower altitude, and often release their moisture as rainfall. A moist warm front can bring days of cloudy skies and drizzle.

A **cold front** (Figure 1, bottom) is the leading edge of an advancing mass of cold air. Because cold air is denser than warm air, an advancing cold front stays close to the ground and wedges underneath less dense warmer air. An approaching cold front produces rapidly moving, towering clouds called *thunderheads.*

As a cold front passes through, we may experience high surface winds and thunderstorms. After it leaves the area, we usually have cooler temperatures and a clear sky.

Near the top of the troposphere, hurricane-force winds circle the earth. These powerful winds, called *jet streams,* follow rising and falling paths that have a strong influence on weather patterns.

Highs and Lows: Effects of Atmospheric Pressure

Weather is affected by up-and-down movements of air masses in conjunction with high and low atmospheric pressure.

Weather is also affected by changes in atmospheric pressure. *Air pressure* results from zillions of tiny molecules of gases (mostly nitrogen and oxygen) in the atmosphere zipping around at incredible speeds and hitting and bouncing off anything they encounter.

Atmospheric pressure is greater near the earth's surface because the molecules in the atmosphere are squeezed together under the weight of the air above them. An air mass with high pressure, called a **high,** contains cool, dense air that descends toward the earth's surface and becomes warmer. Fair weather follows as long as this high-pressure air mass remains over the area.

In contrast, a low-pressure air mass, called a **low,** produces cloudy and sometimes stormy weather. Because of its low pressure and low density, the center of a low rises, and its warm air expands and cools. When the temperature drops below a certain level where condensation takes place, called the *dew point,* moisture in the air condenses and forms clouds.

If the droplets in the clouds coalesce into large and heavy drops, then precipitation occurs. The condensation of water vapor into water drops usually requires that the air contain suspended tiny particles of material such as dust, smoke, sea salts, or volcanic ash. These so-called *condensation nuclei* provide surfaces on which the droplets of water can form and coalesce.

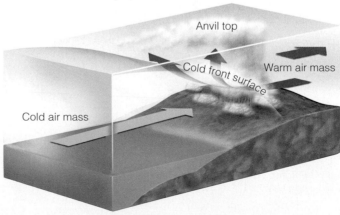

Figure 1 Natural capital: a *warm front* (top) arises when an advancing mass of warm air meets and rises up over a retreating mass of denser cool air. A *cold front* (bottom) forms when a mass of cold air wedges beneath a retreating mass of less dense warm air.

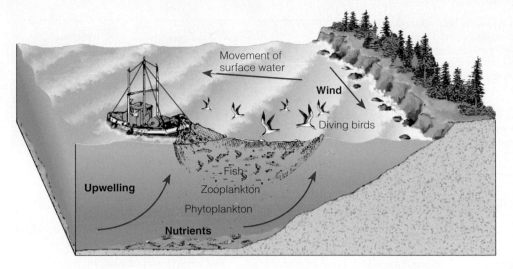

Figure 2 Natural capital: *shore upwelling* which occurs when deep, cool, nutrient-rich waters are drawn up to replace surface water moved away from a steep coast by wind flowing along the coast toward the equator.

Upwellings, El Niño and La Niña: Effects of Major Wind Shifts

El Niño occurs when a change in the direction of tropical winds warms coastal surface water, suppresses upwellings, and temporarily alters much of the earth's climate. La Niña is the reverse of this effect.

An **upwelling,** or upward movement of ocean water, can mix ocean water. An upwelling brings cool and nutrient-rich water from the bottom of the ocean to the surface where it supports large populations of phytoplankton, zooplankton, fish, and fish-eating seabirds.

Figure 5-2 (p. 101) shows the oceans' major upwelling zones. Upwellings far from shore occur when surface currents move apart and draw water up from deeper layers. Strong upwellings are also found along the steep western coasts of some continents when winds blowing along the coasts push surface water away from the land and draw water up from the ocean bottom (Figure 2).

Every few years in the Pacific Ocean, normal shore upwellings (Figure 3, left) are affected by changes in climate patterns called the *El Niño–Southern Oscillation*, or ENSO (Figure 3, right). In an ENSO, often called *El Niño,* prevailing tropical trade winds blowing westward weaken or reverse direction. This warms up surface water along the South and North American coasts, which suppresses the normal upwellings of cold, nutrient-rich water (Figure 3, right). The decrease in nutrients reduces primary productivity and causes a sharp decline in the populations of some fish species.

A strong ENSO can alter the weather of at least two-thirds of the globe (Figure 4, p. S40)—especially in lands along the Pacific and Indian Oceans.

La Niña, the reverse of El Niño, cools some coastal surface waters, and brings back upwellings. Typically, La Niña means more Atlantic Ocean hurricanes, colder winters in Canada and the northeastern United States, and warmer and drier winters in the southeastern and southwestern United States. It also usually leads to wetter winters in the Pacific Northwest, torrential rains in Southeast Asia, lower wheat yields in Argentina, and more wildfires in Florida.

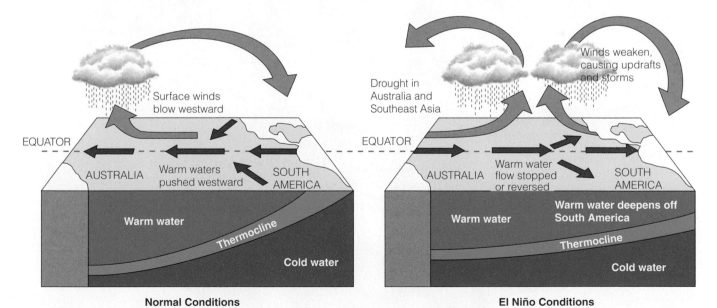

Figure 3 Normal trade winds blowing westward cause shore upwellings of cold, nutrient-rich bottom water in the tropical Pacific Ocean near the coast of Peru (left). A zone of gradual temperature change called the *thermocline* separates the warm and cold water. Every few years a shift in trade winds known as the *El Niño–Southern Oscillation* (ENSO) disrupts this pattern. Westward-blowing trade winds weaken or reverse direction, which depresses the coastal upwellings and warms the surface waters off South America (right). When an ENSO lasts 12 months or longer, it severely disrupts populations of plankton, fish, and seabirds in upwelling areas and can alter weather conditions over much of the globe.

Figure 4 Typical global weather effects of an El Niño–Southern Oscillation. During the 1996–98 ENSO, huge waves battered the coast in the U.S. state of California, and torrential rains caused widespread flooding and mudslides. In Peru, floods and mudslides killed hundreds of people, left about 250,000 people homeless, and ruined harvests. Drought in Brazil, Indonesia, and Australia led to massive wildfires in tinder-dry forests. India and parts of Africa also experienced severe drought. A catastrophic ice storm hit Canada and the northeastern United States, but the southeastern United States had fewer hurricanes. (Data from United Nations Food and Agriculture Organization)

El Niño

☐ Drought

■ Unusually high rainfall

■ Unusually warm periods

Tornadoes and Tropical Cyclones

Tornadoes and tropical storms are weather extremes that can cause tremendous damage but can sometimes have beneficial ecological effects.

Sometimes we experience *weather extremes.* Two examples are violent storms called *tornadoes* (which form over land) and *tropical cyclones* (which form over warm ocean waters and sometimes pass over coastal land).

Tornadoes or *twisters* are swirling funnel-shaped clouds that form over land. They can destroy houses and cause other serious damage in areas where they touch down on the earth's surface. The United States is the world's most tornado-prone country, followed by Australia.

Tornadoes in the plains of the midwestern United States usually occur when a large, dry, cold-air front moving southward from Canada runs into a large mass of humid air moving northward from the Gulf of Mexico. Most tornadoes occur in the spring and sum-

mer when fronts of cold air from the north penetrate deeply into the midwestern plains.

As the large warm-air mass moves rapidly over the more dense cold-air mass, it rises swiftly and forms strong vertical convection currents that suck air upward, as shown in Figure 5. Scientists hypothesize

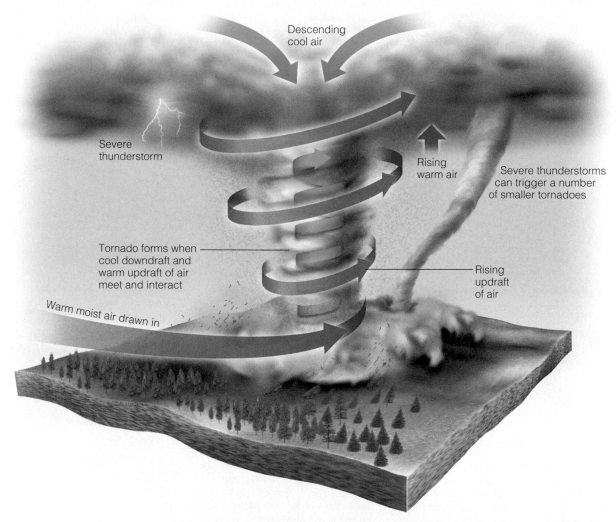

Descending cool air

Severe thunderstorm

Rising warm air

Severe thunderstorms can trigger a number of smaller tornadoes

Tornado forms when cool downdraft and warm updraft of air meet and interact

Rising updraft of air

Warm moist air drawn in

Figure 5 Formation of a *tornado* or *twister.* Although twisters can form at any time of the year, the most active tornado season in the United States is usually March through August. Meteorologists cannot tell us with great accuracy when and where most tornadoes will form.

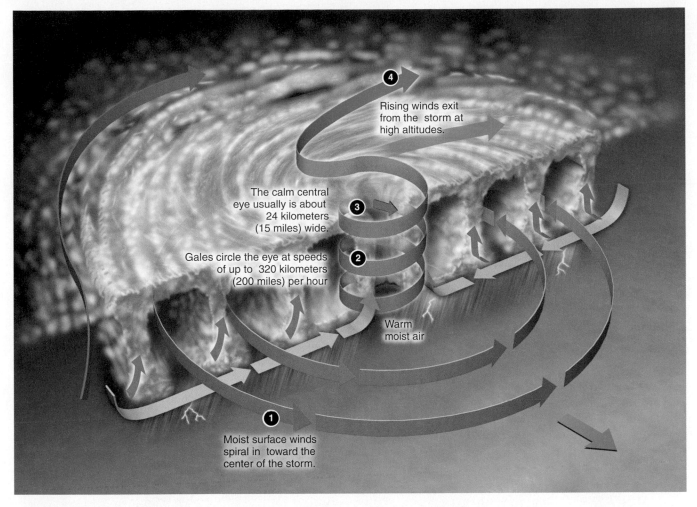

The calm central eye usually is about 24 kilometers (15 miles) wide.

Gales circle the eye at speeds of up to 320 kilometers (200 miles) per hour

Rising winds exit from the storm at high altitudes.

Warm moist air

Moist surface winds spiral in toward the center of the storm.

Figure 6 Formation of a *tropical cyclone.* Those forming in the Atlantic Ocean usually are called *hurricanes;* those forming in the Pacific Ocean usually are called *typhoons.*

that the rising vortex of air starts spinning because the air near the ground in the funnel is moving more slowly than the air above. This difference causes the air ahead of the advancing front to roll or spin in a vertically rising air mass or vortex.

Tropical cyclones are spawned by the formation of low-pressure cells of air over warm tropical seas. Figure 6 shows the formation and structure of a tropical cyclone. *Hurricanes* are tropical cyclones that form in the Atlantic Ocean; those forming in the Pacific Ocean usually are called *typhoons.* Tropical cyclones take a long time to form and gain strength. As a result, meteorologists can track their paths and wind speeds and warn people in areas likely to be hit by these violent storms.

For a tropical cyclone to form, the temperature of ocean water has to be at least 27 °C (80 °F) to a depth of 46 meters (150 feet). A tropical cyclone forms when areas of low pressure over the warm ocean draw in air from surrounding higher-pressure areas. The earth's rotation makes these winds spiral counterclockwise in the northern hemisphere and clockwise in the southern hemi-

sphere. Moist air warmed by the heat of the ocean rises in a vortex through the center of the storm until it becomes a tropical cyclone (Figure 6).

The intensities of tropical cyclones are rated in different categories based on their sustained wind speeds. *Category 1:* 119–153 kilometers per hour (74–95 miles per hour); *Category 2:* 154–177 kilometer per hour (96–110 miles per hour); *Category 3:* 178–209

kilometers per hour (111–130 miles per hour); *Category 4:* 210–249 kilometers per hour (131–155 miles per hour); *Category 5:* greater than 249 kilometers per hour (155 miles per hour). The longer a tropical cyclone stays over warm waters, the stronger it gets. Significant hurricane-force winds can extend 64–161 kilometers (40–100 miles) from the center or eye of a tropical cyclone

Figure 7 shows the change in the average surface temperature of the global ocean between 1880 and 2004. Note the rise in this temperature since 1980.

Figure 7 Change in global ocean temperature from its average baseline temperature from 1971 to 2000. (Data from National Oceanic and Atmospheric Administration)

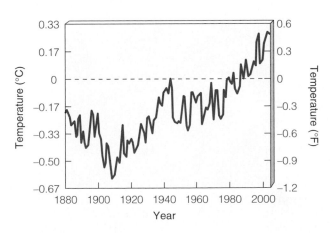

These higher temperatures help explain why the average intensity of tropical cyclones has increased since 1990. For example, between 1990 and 2006, there were 250 Category 4 and 5 tropical cyclones (120 of them in the West Pacific) compared to 161 between 1975 and 1989. With the number of people living along the world's coasts increasing, the danger to lives and property has risen dramatically.

Scientists have not been able to correlate the number of tropical cyclones with global warming of the troposphere and the world's oceans. However, they have found a statistical correlation between global warming of the troposphere and the global ocean (Figure 7) and the size and intensity of tropical cyclones. If this is correct, the size and intensity of tropical cyclones are expected to increase as the troposphere warms during this century.

Hurricanes and typhoons can kill and injure people and damage property and agricultural production. Sometimes, however, the long-term ecological and economic benefits of a tropical cyclone can exceed its short-term harmful effects.

For example, in parts of the U.S. state of Texas along the Gulf of Mexico, coastal bays and marshes normally are closed off from freshwater and saltwater inflows. In August 1999, Hurricane Brett struck this coastal area. According to marine biologists, it flushed out excess nutrients from land runoff and swept dead sea grasses and rotting vegetation from the coastal bays and marshes. It also carved out 12 channels through the barrier islands along the coast, allowing huge quantities of fresh seawater to flood the bays and marshes.

This flushing out of the bays and marshes reduced brown tides consisting of explosive growths of algae feeding on excess nutrients. It also increased growth of sea grasses, which serve as nurseries for shrimp, crabs, and fish and provide food for millions of ducks wintering in Texas bays. Production of commercially important species of shellfish and fish also increased.

EARTHQUAKES, TSUNAMIS, AND VOLCANIC ERUPTIONS (SCIENCE)
CHAPTER 5

Earthquakes

Earthquakes occur when a part of the earth's crust suddenly fractures, shifts to relieve stress, and releases energy as shock waves.

Stress in the earth's crust can cause solid rock to deform until it suddenly fractures and shifts along the fracture, producing a fault. The faulting or a later abrupt movement on an existing fault causes an **earthquake** (Figure 1).

Relief of the earth's internal stress releases energy as shock waves, which move outward from the earthquake's focus like ripples in a pool of water. Scientists measure the severity of an earthquake by the *magnitude* of its shock waves. The magnitude is a measure of the amount of energy released in the earthquake, as indicated by the amplitude (size) of the vibrations when they reach a recording instrument (seismograph).

Scientists use the *Richter scale,* on which each unit has amplitude 10 times greater than the next smaller unit. Thus a magnitude 5.0 earthquake is 10 times greater than a magnitude 4.0 earthquake, and a magnitude 6.0 quake is 100 times greater than a magnitude 4.0 quake. Seismologists rate earthquakes as *insignificant* (less than 4.0 on the Richter scale), *minor* (4.0–4.9), *damaging* (5.0–5.9), *destructive* (6.0–6.9), *major* (7.0–7.9), and *great* (over 8.0).

Earthquakes often have *aftershocks* that gradually decrease in frequency over a period of as long as several months. Some also are preceded by *foreshocks* that occur from seconds to weeks before the main shock.

The *primary effects of earthquakes* include shaking and sometimes a permanent vertical or horizontal displacement of the ground. These effects may have serious consequences for people and for buildings, bridges, freeway overpasses, dams, and pipelines. An earthquake is a very large rock-and-roll event.

Secondary effects of earthquakes include rockslides, urban fires, and flooding caused by *subsidence* (sinking) of land. One way to reduce the loss of life and property damage from earthquakes is to examine historical records and make geologic measurements

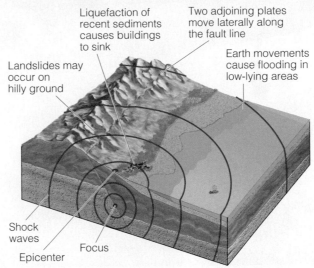

Landslides may occur on hilly ground

Liquefaction of recent sediments causes buildings to sink

Two adjoining plates move laterally along the fault line

Earth movements cause flooding in low-lying areas

Shock waves

Epicenter

Focus

Figure 1 Major features and effects of an *earthquake.*

to locate active fault zones. We can also map high-risk areas, establish building codes that regulate the placement and design of buildings in such areas, and increase research geared toward predicting when and where earthquakes will occur. Then people can decide how high the risk might be and whether they want to accept that risk and live in areas subject to earthquakes.

Engineers know how to make homes, large buildings, bridges, and freeways more earthquake resistant. But this can be expensive, especially the reinforcement of existing structures.

Tsunamis: Earthquakes and Huge Waves

Most tsunamis are a series of huge waves created when a large undersea earthquake causes a sudden up or down movement of the ocean floor.

A **tsunami** (from a Japanese word meaning "harbor wave") is a series of large waves generated when part of the ocean floor suddenly rises or drops (Figure 2). Most large tsunamis are caused when thrust faults (Figure 2) in the ocean floor move up or down as a result of a large underwater earthquake or a landslide caused by such an

earthquake. Such earthquakes often occur offshore at subduction zones where a tectonic plate slips under a continental plate.

Tsunamis are often called tidal waves, although they have nothing to do with tides. They travel very far and as fast as 890 kilometers (550 miles) per hour—the speed of a jet plane. At this speed, a tsunami could travel across the Pacific Ocean in less than a day.

In deep water the waves are very far apart—sometimes hundreds of kilometers—and their crests are not very high. When the tsunami reaches shallow water and approaches a coast it slows down and its wave crests squeeze closer together and their heights grow rapidly. It can hit a coast as a series of towering walls of water that can level buildings.

Tsunamis can be detected to provide some degree of early warning by using a network of ocean buoys. A pressure recorder on the ocean floor measures changes in water pressure as the waves of a tsunami pass over it. These data are relayed to a weather buoy, which then transmits the data via satellite to tsunami emergency warning centers.

Between 1900 and late 2004, an estimated 278,000 people in the Pacific Ocean

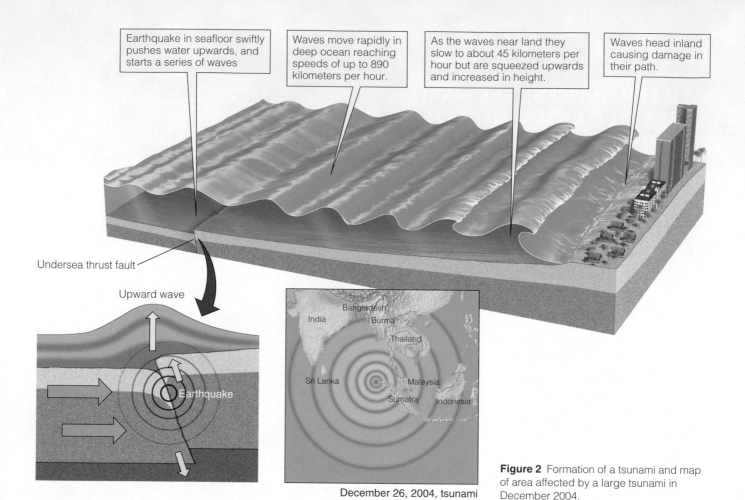

Earthquake in seafloor swiftly pushes water upwards, and starts a series of waves

Waves move rapidly in deep ocean reaching speeds of up to 890 kilometers per hour.

As the waves near land they slow to about 45 kilometers per hour but are squeezed upwards and increased in height.

Waves head inland causing damage in their path.

Undersea thrust fault

Upward wave

Earthquake

December 26, 2004, tsunami

Figure 2 Formation of a tsunami and map of area affected by a large tsunami in December 2004.

regions had been killed by tsunamis. The largest loss of life occurred in December 2004 when a large tsunami killed 228,000 people (168,000 of them in Indonesia) and devastated many coastal areas of Asia (Figure 3 and map in Figure 2).

Studies in February 2005 by the UN Environment Programme pointed to the role that healthy coral reefs (Figure 6-1, p. 126) and mangrove forests (Figure 6-8, p. 131) played in reducing the death toll and destruction from the 2004 tsunami. For

example, intact mangrove forests in parts of Thailand helped protect buildings and people from the force of the huge waves. In contrast, the extensive damage and high death toll from the 2004 tsunami in India's Tamus state has been attributed in part to

Figure 3 In December 2004, a major earthquake on the seafloor of the Pacific Ocean created a large tsunami that killed 168,000 people in Indonesia. These photos show the Banda Aceh Shore Gleebruk in Indonesia on June 23, 2004, before (left) the tsunami and on December 28, 2004, after it was struck by the tsunami (right).

the widespread clearing of a third of the coastal area's mangrove forest in recent decades. And in Sri Lanka, some of the greatest damage occurred where illegal coral mining and reef damage had caused severe beach erosion.

Volcanoes

Some volcanoes erupt quietly with oozing flows of molten rock. Others erupt explosively and spew large chunks of lava rock, ash, and harmful gases into the atmosphere.

An active **volcano** occurs where magma (molten rock) reaches the earth's surface through a central vent or a long crack (*fissure*; Figure 4). Volcanic activity can release *ejecta* (debris ranging from large chunks of lava rock to glowing hot ash), liquid lava, and gases (such as water vapor, carbon dioxide, and sulfur dioxide) into the environment.

Volcanic activity is concentrated for the most part in the same areas as seismic activity. Some volcanoes erupt explosively and eject large quantities of gases and particulate matter (soot and mineral ash) high into the troposphere. Most of this soot and ash soon falls back to the earth's surface. A gas such as sulfur dioxide remains in the atmosphere, however, where it is converted to tiny droplets of sulfuric acid. This acid may remain above the clouds and not be washed out by rain for as long as 3 years. The tiny droplets reflect some of the sun's energy and can cool the atmosphere for 1–4 years.

Other volcanoes erupt more quietly. They involve primarily lava flows, which can cover roads and villages and ignite brush, trees, and homes.

We tend to think of volcanic activity as an undesirable event, but it does provide some benefits. For example, it creates outstanding scenery in the form of majestic mountains, some lakes (such as Crater

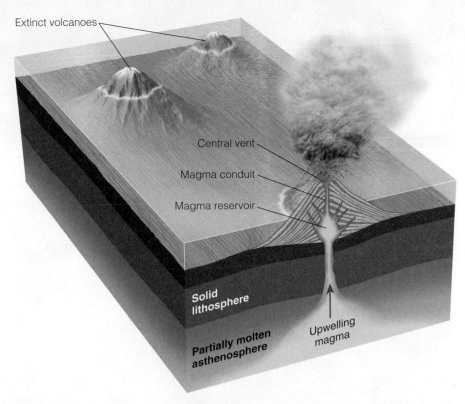

Figure 4 A *volcano* erupts when molten magma in the partially molten asthenosphere rises in a plume through the lithosphere to erupt on the surface as lava that can spill over or be ejected into the atmosphere. Chains of islands can be created by eruptions of volcanoes that then become inactive.

Lake in Oregon; Figure 6-16, left, p. 138), and other landforms. Perhaps the most important benefit of volcanism is the highly fertile soils produced by the weathering of lava.

We can reduce the loss of human life and sometimes property damage caused by volcanic eruptions in several ways. For example, we can use historical records and geologic measurements to identify high-risk areas so that people can try to avoid living in them. We can also develop effective evacuation plans and measurements that warn us when volcanoes are likely to erupt.

Scientists continue to study the phenomena that precede an eruption. Examples include tilting or swelling of the cone, changes in magnetic and thermal properties of the volcano, changes in gas composition, and increased seismic activity.

WOLF AND MOOSE INTERACTIONS ON ISLE ROYALE, MICHIGAN (USA) (SCIENCE)
CHAPTERS 7, 8

Isle Royale, Michigan (USA), is an isolated island in Lake Superior. For more than four decades, wildlife biologists, led by Rolf Peterson, have been studying the relationship between the moose and wolf populations on this island.

In the early 1900s, a small herd of moose wandered across the frozen ice of Lake Superior to this island. With an abundance of food and no predators, the moose population exploded. In 1928, a wildlife biologist visiting the island correctly predicted that the large moose population would crash as a result of stripping the island of most of its plant food resources.

In 1949, timber wolves (probably a single pair) wandered across the frozen lake from Canada and discovered abundant moose prey on the island. They stayed and slowly grew in numbers.

Since 1958, wildlife biologists have been tracking the populations of the two species and found that they appear to be interacting in an oscillating predator–prey cycle (Figure 1). See Supplement 17 on pp. S52–59 for an article describing how scientists can track the movement of animals such as wolves.

The simple explanation is that between 1958 and 1970 there were not enough wolves to control the moose population so the number of moose increased. Then as the wolf population increased and preyed on moose the number of moose decreased.

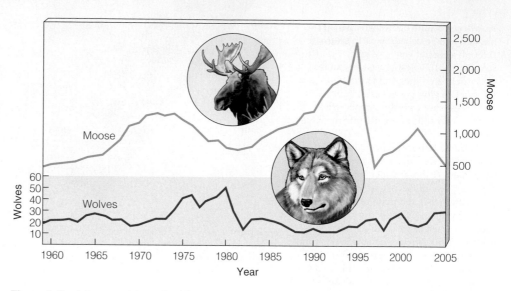

Figure 1 Predator–prey interactions between moose and wolf populations on Isle Royale, Michigan, from 1960 to 2005. (Data from Rolf O. Peterson)

The decline in the moose population then led to a decrease in the wolf population because of fewer prey for the wolves. In turn, this allowed the moose population to increase and started the predator–prey cycle again.

But things are not that simple. Researchers have identified four other factors that affect this predator–prey interaction. *First,* warmer than normal *summers* increase tick populations that weaken moose and make them more vulnerable to their wolf predators. *Second,* warmer than normal *winters* reduce snow cover and allow moose

to more readily escape capture by the wolves. *Third,* a canine virus introduced to wolves by dogs that migrated to the island may have weakened and killed some of the wolves. *Fourth,* the wolves may have a low reproduction rate because of a lack of genetic variability from inbreeding. Studying nature is fascinating!

Critical Thinking

What ecological lesson can we learn from studying the moose–wolf interaction on Isle Royale?

EFFECTS OF GENETIC VARIATIONS ON POPULATION SIZE (SCIENCE)
CHAPTER 8

Genetic Diversity and Population Size: Small Isolated Populations Are Vulnerable

Variations in genetic diversity can affect the survival of small, isolated populations.

In most large populations genetic diversity is fairly constant and the loss or addition of some individuals has little effect on the total gene pool. However, several genetic factors can play a role in the loss of genetic diversity and the survival of small, isolated populations. One called the *founder effect* can occur when a few individuals in a population colonize a new habitat that is geographically isolated from other members of the population (Figure 4-10, p. 92). In such cases, limited genetic diversity or variability may threaten the survival of the colonizing population.

Another factor is a *demographic bottleneck*. It occurs when only a few individuals in a population survive a catastrophe such as a fire or hurricane. Lack of genetic diversity may limit the ability of these individuals to rebuild the population.

A third factor is *genetic drift*. It involves random changes in the gene frequencies in a population that can lead to unequal reproductive success. For example, some individuals may breed more than others do and their genes may eventually dominate the gene pool of the population. This change in gene frequency could help or hinder the survival of the population. The founder effect is one cause of genetic drift.

A fourth factor is *inbreeding*. It occurs when individuals in a small population mate with one another. This can increase the frequency of defective genes within a population and affect its long-term survival.

Conservation biologists use the concepts of founder effects, demographic bottleneck, genetic drift, inbreeding, and island biogeography (Case Study, p. 146) to estimate the the *minimum viable population size* of rare and endangered species: the number of individuals such populations need for long-term survival.

Metapopulations: Exchanging Genes Now and Then

Separate subpopulations of mobile species can exchange genes regularly or occasionally if there are suitable corridors or migration routes.

Populations often live in areas where resources are found in patches. The individuals of a species that live in a habitat patch are called a **subpopulation.** Often areas of unsuitable habitat separate the subpopulations of a species located in suitable habitat patches. Most subpopulations are small and are thus vulnerable to being wiped out by diseases, invasions by predators, and local catastrophes such as fires, floods, or extreme weather. The smaller a subpopulation, the more likely it is to become locally extinct by such chance events.

This threat can be reduced if some individuals can move back and forth between different subpopulation patches. A set of subpopulations interconnected by occasional movement of individuals between them is called a *metapopulation*.

Some subpopulations where birth rates are higher than death rates produce excess individuals that can migrate to other local populations. Other subpopulations where death rates are greater than birth rates can accept individuals from other populations. Conservation biologists can map out the locations of metapopulations and use this information to provide corridors and migration routes to enhance the overall population size, genetic diversity, and survival of related local populations.

SHADE-GROWN COFFEE AND TROPICAL FORESTS (SCIENCE)
CHAPTER 10

If you are a coffee lover, your choice of coffee can help protect or destroy tropical forests. Traditionally, coffee beans are produced by small tree plants that grow under the shade of taller trees (Figure 1, right) in cool, mountain areas of the tropics.

This began changing a few decades ago with the development of new varieties of coffee plants that can be grown on plantations in full sunlight (Figure 1, left). Yields are higher because the plants get more solar energy for photosynthesis and more plants can be grown per unit of land area.

Currently, almost half the world's coffee is grown in unshaded plantations and the percentage is increasing rapidly. This is good business for coffee producers.

But there are ecological downsides to this shift to full-sun coffee plantations. Biodiversity decreases as tropical forests are cleared to provide land for the plantations. Indeed, 13 of the world's 34 ecologically endangered hotspots (Figure 10-26, p. 216) are located in traditional coffee-growing areas.

Typically, full-sun coffee plantations have half as many bird species and up to 90% fewer individual birds. Generally these plantations require more pesticides than do shade-grown coffee plantations, where birds and insects living in the forest canopy provide natural pest control. Without the nutrients forests add to the soil, the plantations also need more chemical fertilizers. And without the protective cover of forests that reduces evaporation, many of the plantations have to be irrigated.

Critical Thinking

To protest the increased use of environmentally damaging unshaded plantations, some conservationists urge coffee drinkers to buy organic shade-grown coffee. Do you support this idea? Explain. If so, do you buy only organic shade-grown coffee?

Figure 1 **Natural capital degradation:** an increasing amount of the world's coffee is being grown on unshaded plantations in countries such as Costa Rica (left) instead of in the shade of trees such as that grown in Colombia (right). The plantations increase productivity but decrease biodiversity and cause more pollution from runoff of fertilizers and pesticides than do shade-grown coffee operations.

REDUCING THE HARMFUL EFFECTS OF INSECTS AND PATHOGENS ON FORESTS (SCIENCE)
CHAPTER 10

We can reduce tree damage from diseases and insects by inspecting imported timber, removing diseased and infected trees, and using chemicals and natural predators to help control insect pests.

Accidental or deliberate introductions of foreign diseases and insects are a major threat to trees in the United States and elsewhere. In 1900, one of every four trees in the eastern deciduous forests of the United States was an American chestnut—a fast-growing and rot-free tree. Within 40 years, a fungal blight had wiped out more than 3.5 billion of these majestic trees. The blight arrived unknowingly on imported chestnut seedlings that were planted over much of the United States.

Figure 1 shows some other nonnative species of pests that are causing serious damage to certain tree species in parts of the United States. There are several ways

Sudden oak death   White pine blister rust   Pine shoot beetle   Beech bark disease   Hemlock woolly adelgid

Figure 1 Natural capital degradation: some of the nonnative insect species and disease organisms that have invaded U.S. forests and are causing billions of dollars in damages and tree loss. The light green and orange colors show areas where green or red overlap with yellow. (Data from U.S. Forest Service)

to reduce the harmful impacts of tree diseases and of insects on forests. One is to ban imported timber that might introduce harmful new pathogens or insect pests; another is to remove or clear-cut infected trees.

We can also develop tree species that are genetically resistant to common tree diseases. For example, since the late 1980s scientists have been working to genetically engineer a blight-resistant American chestnut.

Another approach is to control insect pests by applying conventional pesticides. Scientists also use biological control (bugs that eat harmful bugs) combined with very small amounts of conventional pesticides.

Critical Thinking

If insect pest began destroying a beautiful shade tree in your yard, what would you do?

SUPPLEMENT 16

USING THE MARKETPLACE TO CONTROL ACCESS TO FISHERIES (SCIENCE AND ECONOMICS)
CHAPTER 12

Some countries use a market-based system called *individual transfer rights* (ITRs) to help control access to fisheries. The government gives each fishing vessel owner a specified percentage of the total allowable catch (TAC) for a fishery in a given year.

Owners are permitted to buy, sell, or lease their fishing rights like private property. The ITR market-based system was introduced in New Zealand in 1986 (where it is now used for 93 fish species) and in Iceland in 1990. In these countries, there has been some reduction in overfishing and the overall fishing fleet size and an end to government fishing subsidies that encourage overfishing. But enforcement has been difficult, some fishers illegally exceed their quotas, and the wasteful bycatch has not been reduced.

In 1995, the U.S. introduced tradable quotas to regulate Alaska's halibut fishery, which had declined so much that the fishing season had been cut to only 48 hours a year. The number of fishers declined as some re-tired by selling their quotas. Halibut prices and fisher income rose, and with less fishing pressure the halibut population recovered. By 2005, the season was 258 days long.

Critics have identified four problems with the ITR approach and have made suggestions for its improvement. *First,* in effect it transfers ownership of publicly owned fisheries to private commercial fishers but still makes the public responsible for the costs of enforcing and managing the system. *Remedy:* Collect fees (not to exceed 5% of the value of the catch) from quota holders to pay for the costs of government enforcement and management of the ITR system.

Second, it can squeeze out small fishing vessels and companies because they do not have the capital to buy ITRs from others. For example, 5 years after the ITR system was implemented in New Zealand, three companies controlled half the ITRs. *Remedy:* Do not allow any fisher or fishing company to accumulate more than a fifth of the total rights to a fishery.

Third, the ITR system can increase poaching and sales of illegally caught fish on the black market, as has happened to some extent in New Zealand. Some of these black market sales come from small-scale fishers who receive no quota or too small a quota to make a living. Some also come from larger-scale fishers who deliberately exceed their quotas. *Remedy:* Require strict record keeping and have well-trained observers on all fishing vessels with ITRs.

Fourth, total fishing catches (TACS) are often set too high to prevent overfishing. *Remedy:* Set a limit of 50–90% of the estimated *optimal* sustainable yield and allow no fishing beyond that limit in any ITR fishery. Most fishing industry interests oppose these stricter rules for ITR systems.

Critical Thinking

Do you support or oppose widespread use of ITR systems to help control access to fisheries? Explain.

1 Title of the journal, which reports on science taking place in Arctic regions.

2 Volume number, issue number and date of the journal, and page numbers of the article.

3 Title of the article: a concise but specific description of the subject of study—one episode of long-range travel by a wolf hunting for food on the Arctic tundra.

4 Authors of the article: scientists working at the institutions listed in the footnotes below. Note #2 indicates that P. F. Frame is the *corresponding author*—the person to contact with questions or comments. His email address is provided.

5 Date on which a draft of the article was received by the journal editor, followed by date on which a revised draft was accepted for publication. Between these dates, the article was reviewed and critiqued by other scientists, a process called peer review. The authors revised the article to make it clearer, according to those reviews.

6 ABSTRACT: A brief description of the study containing all basic elements of this report. First sentence summarizes the *background* material. Second sentence encapsulates the *methods* used. The rest of the paragraph sums up the *results*. Authors introduce the main *subject* of the study—a female wolf (#388) with pups in a den—and refer to later *discussion* of possible explanations for her behavior.

7 Key words are listed to help researchers using computer databases. Searching the databases using these key words will yield a list of studies related to this one.

8 RÉSUMÉ: The French translation of the abstract and key words. Many researchers in this field are French Canadian. Some journals provide such translations in French or in other languages.

9 INTRODUCTION: Gives the background for this wolf study. This paragraph tells of known or suspected wolf behavior that is important for this study. Note that (a) major species mentioned are always accompanied by scientific names, and (b) statements of fact or *postulations* (claims or assumptions about what is likely to be true) are followed by references to studies that established those facts or supported the postulations.

10 This paragraph focuses directly on the wolf behaviors that were studied here.

11 This paragraph starts with a statement of the *hypothesis* being tested, one that originated in other studies and is supported by this one. The hypothesis is restated more succinctly in the last sentence of this paragraph. This is the *inquiry* part of the scientific process—asking questions and suggesting possible answers.

This journal article reports on the movements of a female wolf during the summer of 2002 in northwestern Canada. It also reports on a scientific process of inquiry, observation and interpretation to learn where, how and why the wolf traveled as she did. In some ways, this article reflects the story of "how to do science" told on pp. 29–32 of this textbook. These notes are intended to help you read and understand how scientists work and how they report on their work.

① ARCTIC

② VOL. 57, NO. 2 (JUNE 2004) P. 196–203

③ Long Foraging Movement of a Denning Tundra Wolf

④ Paul F. Frame,[1,2] David S. Hik,[1] H. Dean Cluff,[3] and Paul C. Paquet[4]

⑤ (Received 3 September 2003; accepted in revised form 16 January 2004)

---

⑥ ABSTRACT. Wolves (*Canis lupus*) on the Canadian barrens are intimately linked to migrating herds of barren-ground caribou (*Rangifer tarandus*). We deployed a Global Positioning System (GPS) radio collar on an adult female wolf to record her movements in response to changing caribou densities near her den during summer. This wolf and two other females were observed nursing a group of 11 pups. She traveled a minimum of 341 km during a 14-day excursion. The straight-line distance from the den to the farthest location was 103 km, and the overall minimum rate of travel was 3.1 km/h. The distance between the wolf and the radio-collared caribou decreased from 242 km one week before the excursion to 8 km four days into the excursion. We discuss several possible explanations for the long foraging bout.

⑦ Key words: wolf, GPS tracking, movements, *Canis lupus*, foraging, caribou, Northwest Territories

⑧ RÉSUMÉ. Les loups (*Canis lupus*) dans la toundra canadienne sont étroitement liés aux hardes de caribous des toundras (*Rangifer tarandus*). On a équipé une louve adulte d'un collier émetteur muni d'un système de positionnement mondial (GPS) afin d'enregistrer ses déplacements en réponse au changement de densité du caribou près de sa tanière durant l'été. On a observé cette louve ainsi que deux autres en train d'allaiter un groupe de 11 louveteaux. Elle a parcouru un minimum de 341 km durant une sortie de 14 jours. La distance en ligne droite de la tanière à l'endroit le plus éloigné était de 103 km, et la vitesse minimum durant tout le voyage était de 3,1 km/h. La distance entre la louve et le caribou muni du collier émetteur a diminué de 242 km une semaine avant la sortie à 8 km quatre jours après la sortie. On commente diverses explications possibles pour ce long épisode de recherche de nourriture.

Mots clés: loup, repérage GPS, déplacements, *Canis lupus*, recherche de nourriture, caribou, Territoires du Nord-Ouest

Traduit pour la revue *Arctic* par Nésida Loyer.

---

⑨ Introduction

Wolves (*Canis lupus*) that den on the central barrens of mainland Canada follow the seasonal movements of their main prey, migratory barren-ground caribou (*Rangifer tarandus*) (Kuyt, 1962; Kelsall, 1968; Walton et al., 2001). However, most wolves do not den near caribou calving grounds, but select sites farther south, closer to the tree line (Heard and Williams, 1992). Most caribou migrate beyond primary wolf denning areas by mid-June and do not return until mid-to-late July (Heard et al., 1996; Gunn et al., 2001). Conse-

quently, caribou density near dens is low for part of the summer.

During this period of spatial separation from the main caribou herds, wolves must either search near the homesite for scarce caribou or alternative prey (or both), travel to where prey are abundant, or use a combination of these strategies. **⑩**

Walton et al. (2001) postulated that the travel of tundra wolves outside their normal summer ranges is a response to low caribou availability rather than a pre-dispersal exploration like that observed in territorial wolves (Fritts and Mech, 1981; Messier, 1985). The authors postulated this because most such travel was directed toward caribou calving grounds. We report details of such a long-distance excursion by a breeding female tundra wolf wearing a GPS radio collar. We discuss the relationship of the excursion to movements of satellite-collared caribou (Gunn et al., 2001), supporting the hypothesis that tundra wolves make directional, rapid, long-distance movements in response to seasonal prey availability. **⑪**

[1] Department of Biological Sciences, University of Alberta, Edmonton, Alberta T6G 2E9, Canada
[2] Corresponding author: pframe@ualberta.ca
[3] Department of Resources, Wildlife, and Economic Development, North Slave Region, Government of the Northwest Territories, P.O. Box 2668, 3803 Bretzlaff Dr., Yellowknife, Northwest Territories X1A 2P9, Canada; Dean_Cluff@gov.nt.ca
[4] Faculty of Environmental Design, University of Calgary, Calgary, Alberta T2N 1N4, Canada; current address: P.O. Box 150, Meacham, Saskatchewan S0K 2V0, Canada

196

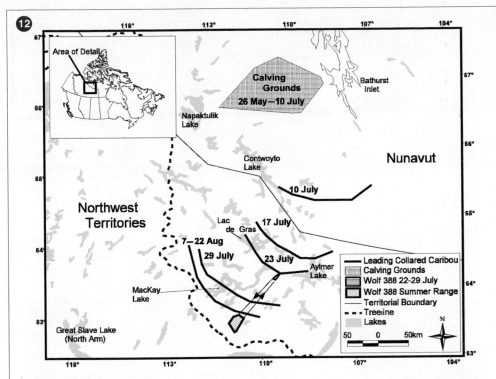

12 This map shows the study area and depicts wolf and caribou locations and movements during one summer. Some of this information is explained below.

13 STUDY AREA: This section sets the stage for the study, locating it precisely with latitude and longitude coordinates and describing the area (illustrated by the map in Figure 1).

14 Here begins the story of how prey (caribou) and predators (wolves) interact on the tundra. Authors describe movements of these nomadic animals throughout the year.

15 We focus on the denning season (summer) and learn how wolves locate their dens and travel according to the movements of caribou herds.

Figure 1. Map showing the movements of satellite radio-collared caribou with respect to female wolf 388's summer range and long foraging movement, in summer 2002.

13 Study Area

Our study took place in the northern boreal forest–low Arctic tundra transition zone (63° 30' N, 110° 00' W; Figure 1; Timoney et al., 1992). Permafrost in the area changes from discontinuous to continuous (Harris, 1986). Patches of spruce (*Picea mariana, P. glauca*) occur in the southern portion and give way to open tundra to the northeast. Eskers, kames, and other glacial deposits are scattered throughout the study area. Standing water and exposed bedrock are characteristic of the area.

14 Details of the Caribou-Wolf System

The Bathurst caribou herd uses this study area. Most caribou cows have begun migrating by late April, reaching calving grounds by June (Gunn et al., 2001; Figure 1). Calving peaks by 15 June (Gunn et al., 2001), and calves begin to travel with the herd by one week of age (Kelsall, 1968). The movement patterns of bulls are less known, but bulls frequent areas near calving grounds by mid-June (Heard et al., 1996; Gunn et al., 2001). In summer, Bathurst caribou cows generally travel south from their calving grounds and then, parallel to the tree line, to the northwest. The rut usually takes place at the tree line in October (Gunn et al., 2001). The winter range of the Bathurst herd varies among years, ranging through the taiga and along the tree line from south of Great Bear Lake to southeast of Great Slave Lake. Some caribou spend the winter on the tundra (Gunn et al., 2001; Thorpe et al., 2001).

15 In winter, wolves that prey on Bathurst caribou do not behave territorially. Instead, they follow the herd throughout its winter range (Walton et al., 2001; Musiani, 2003). However, during denning (May–

16 Other variables are considered—prey other than caribou and their relative abundance in 2002.

17 METHODS: There is no one scientific method. Procedures for each and every study must be explained carefully.

18 Authors explain when and how they tracked caribou and wolves, including tools used and the exact procedures followed.

19 This important subsection explains what data were calculated (average distance …) and how, including the software used and where it came from. (The calculations are listed in Table 1.) Note that the behavior measured (traveling) is carefully defined.

20 RESULTS: The heart of the report and the *observation* part of the scientific process. This section is organized parallel to the Methods section.

21 This subsection is broken down by periods of observation. Pre-excursion period covers the time between 388's capture and the start of her long-distance travel. The investigators used visual observations as well as telemetry (measurements taken using the global positioning system (GPS)) to gather data. They looked at how 388 cared for her pups, interacted with other adults, and moved about the den area.

Table 1. Daily distances from wolf 388 and the den to the nearest radio-collared caribou during a long excursion in summer 2002.

Date (2002)	Mean distance from caribou to wolf (km)	Daily distance from closest caribou to den
12 July	242	241
13 July	210	209
14 July	200	199
15 July	186	180
16 July	163	162
17 July	151	148
18 July	144	137
19 July[1]	126	124
20 July	103	130
21 July	73	130
22 July	40	110
23 July[2]	9	104
29 July[3]	16	43
30 July	32	43
31 July	28	44
1 August	29	46
2 August[4]	54	52
3 August	53	53
4 August	74	74
5 August	75	75
6 August	74	75
7 August	72	75
8 August	76	75
9 August	79	79

[1] Excursion starts.
[2] Wolf closest to collared caribou.
[3] Previous five days' caribou locations not available.
[4] Excursion ends.

August, parturition late May to mid-June), wolf movements are limited by the need to return food to the den. To maximize access to migrating caribou, many wolves select den sites closer to the tree line than to caribou calving grounds (Heard and Williams, 1992). Because of caribou movement patterns, tundra denning wolves are separated from the main caribou herds by several hundred kilometers at some time during summer (Williams, 1990:19; Figure 1; Table 1).

16 Muskoxen do not occur in the study area (Fournier and Gunn, 1998), and there are few moose there (H.D. Cluff, pers. obs.). Therefore, alternative prey for wolves includes waterfowl, other ground-nesting birds, their eggs, rodents, and hares (Kuyt, 1972; Williams, 1990:16; H.D. Cluff and P.F. Frame, unpubl. data). During 56 hours of den observations, we saw no ground squirrels or hares, only birds. It appears that the abundance of alternative prey was relatively low in 2002.

17 Methods

Wolf Monitoring

18 We captured female wolf 388 near her den on 22 June 2002, using a helicopter net-gun (Walton et al., 2001). She was fitted with a releasable GPS radio collar (Merrill et al., 1998) programmed to acquire locations at 30-minute intervals. The collar was electronically released (e.g., Mech and Gese, 1992) on 20 August 2002. From 27 June to 3 July 2002, we observed 388's den with a 78 mm spotting scope at a distance of 390 m.

Caribou Monitoring

In spring of 2002, ten female caribou were captured by helicopter net-gun and fitted with satellite radio collars, bringing the total number of collared Bathurst cows to 19. Eight of these spent the summer of 2002 south of Queen Maud Gulf, well east of normal Bathurst caribou range. Therefore, we used 11 caribou for this analysis. The collars provided one location per day during our study, except for five days from 24 to 28 July. Locations of satellite collars were obtained from Service Argos, Inc. (Landover, Maryland).

Data Analysis

19 Location data were analyzed by ArcView GIS software (Environmental Systems Research Institute Inc., Redlands, California). We calculated the average distance from the nearest collared caribou to the wolf and the den for each day of the study.

Wolf foraging bouts were calculated from the time 388 exited a buffer zone (500 m radius around the den) until she re-entered it. We considered her to be traveling when two consecutive locations were spatially separated by more than 100 m. Minimum distance traveled was the sum of distances between each location and the next during the excursion.

We compared pre- and post-excursion data using Analysis of Variance (ANOVA; Zar, 1999). We first tested for homogeneity of variances with Levene's test (Brown and Forsythe, 1974). No transformations of these data were required.

Results **20**

Wolf Monitoring

Pre-Excursion Period: **21** Wolf 388 was lactating when captured on 22 June. We observed her and two other females nursing a group of 11 pups between 27 June and 3 July. During our observations, the pack consisted of at least four adults (3 females and 1 male) and 11 pups. On 30 June, three pups were moved to a location 310 m from the other eight and cared for by an uncollared female. The male was not seen at the den after the evening of 30 June.

Before the excursion, telemetry indicated 18 foraging bouts. The mean distance traveled during these bouts was 25.29 km (± 4.5 SE, range 3.1–82.5 km). Mean greatest distance from the den on foraging

198   P.F. Frame, et al

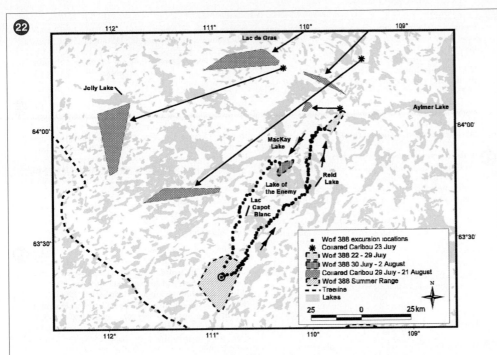

Figure 2. Details of a long foraging movement by female wolf 388 between 19 July and 2 August 2002. Also shown are locations and movements of three satellite radio-collared caribou from 23 July to 21 August 2002. On 23 July, the wolf was 8 km from a collared caribou. The farthest point from the den (103 km distant) was recorded on 27 July. Arrows indicate direction of travel.

22 The key in the lower right-hand corner of the map shows areas (shaded) within which the wolves and caribou moved, and the dotted trail of 388 during her excursion. From the results depicted on this map, the investigators tried to determine when and where 388 might have encountered caribou and how their locations affected her traveling behavior.

23 The wolf's excursion (her long trip away from the den area) is the focus of this study. These paragraphs present detailed measurements of daily movements during her two-week trip—how far she traveled, how far she was from collared caribou, her time spent traveling and resting, and her rate of speed. Authors use the phrase "minimum distance traveled" to acknowledge they couldn't track every step but were measuring samples of her movements. They knew that she went at least as far as they measured. This shows how scientists try to be exact when reporting results. Results of this study are depicted graphically in the map in Figure 2.

bouts was 7.1 km (± 0.9 SE, range 1.7–17.0 km). The average duration of foraging bouts for the period was 20.9 h (± 4.5 SE, range 1–71 h).

The average daily distance between the wolf and the nearest collared caribou decreased from 242 km on 12 July, one week before the excursion period, to 126 km on 19 July, the day the excursion began (Table 1).

Excursion Period: On 19 July at 2203, after spending 14 h at the den, 388 began moving to the northeast and did not return for 336 h (14 d; Figure 2). Whether she traveled alone or with other wolves is unknown. During the excursion, 476 (71%) of 672 possible locations were recorded. The wolf crossed the southeast end of Lac Capot Blanc on a small land bridge, where she paused for 4.5 h after traveling for 19.5 h (37.5 km). Following this rest, she traveled for 9 h (26.3 km) onto a peninsula in Reid Lake, where she spent 2 h before backtracking and stopping for 8 h just off the peninsula. Her next period of travel lasted 16.5 h (32.7 km), terminating in a pause of 9.5 h just 3.8 km from a concentration of locations at the far end of her excursion, where we presume she encountered caribou. The mean duration of these three movement periods was 15.7 h (± 2.5 SE), and that of the pauses, 7.3 h (± 1.5). The wolf required 72.5 h (3.0 d) to travel a minimum of 95 km from her den to this area near caribou (Figure 2). She remained there (35.5 km2) for 151.5 h (6.3 d) and then moved south to Lake of the Enemy, where she stayed (31.9 km^2) for 74 h (3.1 d) before returning to her den. Her greatest distance from the den, 103 km, was recorded 174.5 h (7.3 d) after the excursion

Foraging Movement of A Tundra Wolf **199**

24 Post-excursion measurements of 388's movements were made to compare with those of the pre-excursion period. In order to compare, scientists often use *means*, or averages, of a series of measurements—mean distances, mean duration, etc.

25 In the comparison, authors used statistical calculations (F and df) to determine that the differences between pre- and post-excursion measurements were *statistically insignificant*, or close enough to be considered essentially the same or similar.

26 As with wolf 388, the investigators measured the movements of caribou during the study period. The areas within which the caribou moved are shown in Figure 2 by shaded polygons mentioned in the second paragraph of this subsection.

27 This subsection summarizes how distances separating predators and prey varied during the study period.

28 DISCUSSION: This section is the *interpretation* part of the scientific process.

29 This subsection reviews observations from other studies and suggests that this study fits with patterns of those observations.

30 Authors discuss a prevailing *theory* (CBFT) which might explain why a wolf would travel far to meet her own energy needs while taking food caught closer to the den back to her pups. The results of this study seem to fit that pattern.

---

began, at 0433 on 27 July. She was 8 km from a collared caribou on 23 July, four days after the excursion began (Table 1).

The return trip began at 0403 on 2 August, 318 h (13.2 d) after leaving the den. She followed a relatively direct path for 18 h back to the den, a distance of 75 km.

The minimum distance traveled during the excursion was 339 km. The estimated overall minimum travel rate was 3.1 km/h, 2.6 km/h away from the den and 4.2 km/h on the return trip.

Post-Excursion Period: We saw three pups when recovering the collar on 20 August, but others may have been hiding in vegetation.

Telemetry recorded 13 foraging bouts in the post-excursion period. The mean distance traveled during these bouts was 18.3 km (+ 2.7 SE, range 1.2–47.7 km), and mean greatest distance from the den was 7.1 km (+ 0.7 SE, range 1.1–11.0 km). The mean duration of these post-excursion foraging bouts was 10.9 h (+ 2.4 SE, range 1–33 h).

When 388 reached her den on 2 August, the distance to the nearest collared caribou was 54 km. On 9 August, one week after she returned, the distance was 79 km (Table 1).

Pre- and Post-Excursion Comparison

We found no differences in the mean distance of foraging bouts before and after the excursion period (F = 1.5, df = 1, 29, p = 0.24). Likewise, the mean greatest distance from the den was similar pre- and post-excursion (F = 0.004, df = 1, 29, p = 0.95). However, the mean duration of 388's foraging bouts decreased by 10.0 h after her long excursion (F = 3.1, df = 1, 29, p = 0.09).

Caribou Monitoring

Summer Movements: On 10 July, 5 of 11 collared caribou were dispersed over a distance of 10 km, 140 km south of their calving grounds (Figure 1). On the same day, three caribou were still on the calving grounds, two were between the calving grounds and the leaders, and one was missing. One week later (17 July), the leading radio-collared cows were 100 km farther south (Figure 1). Two were within 5 km of each other in front of the rest, who were more dispersed. All radio-collared cows had left the calving grounds by this time. On 23 July, the leading radio-collared caribou had moved 35 km farther south, and all of them were more widely dispersed. The two cows closest to the leader were 26 km and 33 km away, with 37 km between them. On the next location (29 July), the most southerly caribou were 60 km

farther south. All of the caribou were now in the areas where they remained for the duration of the study (Figure 2).

A Minimum Convex Polygon (Mohr and Stumpf, 1966) around all caribou locations acquired during the study encompassed 85 119 km².

Relative to the Wolf Den: The distance from the nearest collared caribou to the den decreased from 241 km one week before the excursion to 124 km the day it began. The nearest a collared caribou came to the den was 43 km away, on 29 and 30 July. During the study, four collared caribou were located within 100 km of the den. Each of these four was closest to the wolf on at least one day during the period reported.

Discussion

Prey Abundance

Caribou are the single most important prey of tundra wolves (Clark, 1971; Kuyt, 1972; Stephenson and James, 1982; Williams, 1990). Caribou range over vast areas, and for part of the summer, they are scarce or absent in wolf home ranges (Heard et al., 1996). Both the long distance between radio-collared caribou and the den the week before the excursion and the increased time spent foraging by wolf 388 indicate that caribou availability near the den was low. Observations of the pups' being left alone for up to 18 h, presumably while adults were searching for food, provide additional support for low caribou availability locally. Mean foraging bout duration decreased by 10.0 h after the excursion, when collared caribou were closer to the den, suggesting an increase in caribou availability nearby.

Foraging Excursion

One aspect of central place foraging theory (CPFT) deals with the optimality of returning different-sized food loads from varying distances to dependents at a central place (i.e., the den) (Orians and Pearson, 1979). Carlson (1985) tested CPFT and found that the predator usually consumed prey captured far from the central place, while feeding prey captured nearby to dependants. Wolf 388 spent 7.2 days in one area near caribou before moving to a location 23 km back towards the den, where she spent an additional 3.1 days, likely hunting caribou. She began her return trip from this closer location, traveling directly to the den. While away, she may have made one or more successful kills and spent time meeting her own energetic needs before returning to the den. Alternatively, it may have taken several attempts to make a kill, which she then fed on before beginning her return trip. We do not know if she returned food to the pups, but such behavior would be supported by CPFT.

31 Other workers have reported wolves' making long round trips and referred to them as "extraterritorial" or "pre-dispersal" forays (Fritts and Mech, 1981; Messier, 1985; Ballard et al., 1997; Merrill and Mech, 2000). These movements are most often made by young wolves (1–3 years old), in areas where annual territories are maintained and prey are relatively sedentary (Fritts and Mech, 1981; Messier, 1985). The long excursion of 388 differs in that tundra wolves do not maintain annual territories (Walton et al., 2001), and the main prey migrate over vast areas (Gunn et al., 2001).

Another difference between 388's excursion and those reported earlier is that she is a mature, breeding female. No study of territorial wolves has reported reproductive adults making extraterritorial movements in summer (Fritts and Mech, 1981; Messier, 1985; Ballard et al., 1997; Merrill and Mech, 2001). However, Walton et al. (2001) also report that breeding female tundra wolves made excursions.

Direction of Movement

32 Possible explanations for the relatively direct route 388 took to the caribou include landscape influence and experience. Considering the timing of 388's trip and the locations of caribou, had the wolf moved northwest, she might have missed the caribou entirely, or the encounter might have been delayed.

A reasonable possibility is that the land directed 388's route. The barrens are crisscrossed with trails worn into the tundra over centuries by hundreds of thousands of caribou and other animals (Kelsall, 1968; Thorpe et al., 2001). At river crossings, lakes, or narrow peninsulas, trails converge and funnel towards and away from caribou calving grounds and summer range. Wolves use trails for travel (Paquet et al., 1996; Mech and Boitani, 2003; P. Frame, pers. observation). Thus, the landscape may direct an animal's movements and lead it to where cues, such as the odor of caribou on the wind or scent marks of other wolves, may lead it to caribou.

33 Another possibility is that 388 knew where to find caribou in summer. Sexually immature tundra wolves sometimes follow caribou to calving grounds (D. Heard, unpubl. data). Possibly, 388 had made such journeys in previous years and killed caribou. If this were the case, then in times of local prey scarcity she might travel to areas where she had hunted successfully before. Continued monitoring of tundra wolves may answer questions about how their food needs are met in times of low caribou abundance near dens.

34 Caribou often form large groups while moving south to the tree line (Kelsall, 1968). After a large aggregation of caribou moves through an area, its scent can linger for weeks (Thorpe et al., 2001:104). It is conceivable that 388 detected caribou scent on the wind, which was blowing from the northeast on 19–21 July (Environment Canada, 2003), at the same time her excursion began. Many factors, such as odor strength and wind direction and strength, make systematic study of scent detection in wolves difficult under field conditions (Harrington and Asa, 2003). However, humans are able to smell odors such as forest fires or oil refineries more than 100 km away. The olfactory capabilities of dogs, which are similar to wolves, are thought to be 100 to 1 million times that of humans (Harrington and Asa, 2003). Therefore, it is reasonable to think that under the right wind conditions, the scent of many caribou traveling together could be detected by wolves from great distances, thus triggering a long foraging bout.

Rate of Travel

35 Mech (1994) reported the rate of travel of Arctic wolves on barren ground was 8.7 km/h during regular travel and 10.0 km/h when returning to the den, a difference of 1.3 km/h. These rates are based on direct observation and exclude periods when wolves moved slowly or not at all. Our calculated travel rates are assumed to include periods of slow movement or no movement. However, the pattern we report is similar to that reported by Mech (1994), in that homeward travel was faster than regular travel by 1.6 km/h. The faster rate on return may be explained by the need to return food to the den. Pup survival can increase with the number of adults in a pack available to deliver food to pups (Harrington et al., 1983). Therefore, an increased rate of travel on homeward trips could improve a wolf's reproductive fitness by getting food to pups more quickly.

Fate of 388's Pups

36 Wolf 388 was caring for pups during den observations. The pups were estimated to be six weeks old, and were seen ranging as far as 800 m from the den. They received some regurgitated food from two of the females, but were unattended for long periods. The excursion started 16 days after our observations, and it is improbable that the pups could have traveled the distance that 388 moved. If the pups died, this would have removed parental responsibility, allowing the long movement.

Our observations and the locations of radio-collared caribou indicate that prey became scarce in

31 Here our authors note other possible explanations for wolves' excursions presented by other investigators, but this study does not seem to support those ideas.

32 Authors discuss possible reasons for why 388 traveled directly to where caribou were located. They take what they learned from earlier studies and apply it to this case, suggesting that the lay of the land played a role. Note that their description paints a clear picture of the landscape.

33 Authors suggest that 388 may have learned in traveling during previous summers where the caribou were. The last two sentences suggest ideas for future studies.

34 Or maybe 388 followed the scent of the caribou. Authors acknowledge difficulties of proving this, but they suggest another area where future studies might be done.

35 Authors suggest that results of this study support previous studies about how fast wolves travel to and from the den. In the last sentence, they speculate on how these observed patterns would fit into the theory of evolution.

36 Authors also speculate on the fate of 388's pups while she was traveling. This leads to . . .

37 Discussion of cooperative rearing of pups and, in turn, to speculation on how this study and what is known about cooperative rearing might fit into the animal's strategies for survival of the species. Again, the authors approach the broader theory of evolution and how it might explain some of their results.

38 And again, they suggest that this study points to several areas where further study will shed some light.

39 In conclusion, the authors suggest that their study supports the hypothesis being tested here. And they touch on the implications of increased human activity on the tundra predicted by their results.

40 ACKNOWLEDGEMENTS: Authors note the support of institutions, companies, and individuals. They thank their reviewers and list permits under which their research was carried on.

41 REFERENCES: List of all studies cited in the report. This may seem tedious, but is a vitally important part of scientific reporting. It is a record of the sources of information on which this study is based. It provides readers with a wealth of resources for further reading on this topic. Much of it will form the foundation of future scientific studies like this one.

the area of the den as summer progressed. Wolf 388 may have abandoned her pups to seek food for herself. However, she returned to the den after the excursion, where she was seen near pups. In fact, she foraged in a similar pattern before and after the excursion, suggesting that she again was providing for pups after her return to the den.

(37) A more likely possibility is that one or both of the other lactating females cared for the pups during 388's absence. The three females at this den were not seen with the pups at the same time. However, two weeks earlier, at a different den, we observed three females cooperatively caring for a group of six pups. At that den, the three lactating females were observed providing food for each other and trading places while nursing pups. Such a situation at the den of 388 could have created conditions that allowed one or more of the lactating females to range far from the den for a period, returning to her parental duties afterwards. However, the pups would have been weaned by eight weeks of age (Packard et al., 1992), so nonlactating adults could also have cared for them, as often happens in wolf packs (Packard et al., 1992; Mech et al., 1999).

Cooperative rearing of multiple litters by a pack could create opportunities for long-distance foraging movements by some reproductive wolves during summer periods of local food scarcity. We have recorded multiple lactating females at one or more tundra wolf dens per year since 1997. This reproductive strategy may be an adaptation to temporally and **(38)** spatially unpredictable food resources. All of these possibilities require further study, but emphasize both the adaptability of wolves living on the barrens and their dependence on caribou.

Long-range wolf movement in response to caribou **(39)** availability has been suggested by other researchers (Kuyt, 1972; Walton et al., 2001) and traditional ecological knowledge (Thorpe et al., 2001). Our report demonstrates the rapid and extreme response of wolves to caribou distribution and movements in summer. Increased human activity on the tundra (mining, road building, pipelines, ecotourism) may influence caribou movement patterns and change the interactions between wolves and caribou in the region. Continued monitoring of both species will help us to assess whether the association is being affected adversely by anthropogenic change.

(40) Acknowledgements

This research was supported by the Department of Resources, Wildlife, and Economic Development, Government of the Northwest Territories; the Department of Biological Sciences at the University of Alberta; the Natural Sciences and Engineering Research Council of Canada; the Department of Indian and Northern Affairs Canada; the Canadian Circumpolar Institute; and DeBeers Canada, Ltd. Lorna Ruechel assisted with den observations. A. Gunn provided caribou location data. We thank Dave Mech for the use of GPS collars. M. Nelson, A. Gunn, and three anonymous reviewers made helpful comments on earlier drafts of the manuscript. This work was done under Wildlife Research Permit – WL002948 issued by the Government of the Northwest Territories, Department of Resources, Wildlife, and Economic Development.

(41) References

BALLARD, W.B., AYRES, L.A., KRAUSMAN, P.R., REED, D.J., and FANCY, S.G. 1997. Ecology of wolves in relation to a migratory caribou herd in northwest Alaska. Wildlife Monographs 135. 47 p.

BROWN, M.B., and FORSYTHE, A.B. 1974. Robust tests for the equality of variances. Journal of the American Statistical Association 69:364–367.

CARLSON, A. 1985. Central place foraging in the red-backed shrike (*Lanius collurio* L.): Allocation of prey between forager and sedentary consumer. Animal Behaviour 33:664–666.

CLARK, K.R.F. 1971. Food habits and behavior of the tundra wolf on central Baffin Island. Ph.D. Thesis, University of Toronto, Ontario, Canada.

ENVIRONMENT CANADA. 2003. National climate data information archive. Available online: http://www.climate.weatheroffice.ec.gc.ca/Welcome_e.html

FOURNIER, B., and GUNN, A. 1998. Musk ox numbers and distribution in the NWT, 1997. File Report No. 121. Yellowknife: Department of Resources, Wildlife, and Economic Development, Government of the Northwest Territories. 55 p.

FRITTS, S.H., and MECH, L.D. 1981. Dynamics, movements, and feeding ecology of a newly protected wolf population in northwestern Minnesota. Wildlife Monographs 80. 79 p.

GUNN, A., DRAGON, J., and BOULANGER, J. 2001. Seasonal movements of satellite-collared caribou from the Bathurst herd. Final Report to the West Kitikmeot Slave Study Society, Yellowknife, NWT. 80 p. Available online: http://www.wkss.nt.ca/HTML/08_ProjectsReports/PDF/Seasonal MovementsFinal.pdf

HARRINGTON, F.H., and ASA, C.S. 2003. Wolf communication. In: Mech, L.D., and Boitani, L., eds. Wolves: Behavior, ecology, and conservation. Chicago: University of Chicago Press. 66–103.

HARRINGTON, F.H., MECH, L.D., and FRITTS, S.H. 1983. Pack size and wolf pup survival: Their relationship under varying ecological conditions. Behavioral Ecology and Sociobiology 13:19–26.

HARRIS, S.A. 1986. Permafrost distribution, zonation and stability along the eastern ranges of the cordillera of North America. Arctic 39(1):29–38.

HEARD, D.C., and WILLIAMS, T.M. 1992. Distribution of wolf dens on migratory caribou ranges in the Northwest

Territories, Canada. Canadian Journal of Zoology 70:1504–1510.

HEARD, D.C., WILLIAMS, T.M., and MELTON, D.A. 1996. The relationship between food intake and predation risk in migratory caribou and implication to caribou and wolf population dynamics. Rangifer Special Issue No. 2:37–44.

KELSALL, J.P. 1968. The migratory barren-ground caribou of Canada. Canadian Wildlife Service Monograph Series 3. Ottawa: Queen's Printer. 340 p.

KUYT, E. 1962. Movements of young wolves in the Northwest Territories of Canada. Journal of Mammalogy 43:270–271.

———. 1972. Food habits and ecology of wolves on barren-ground caribou range in the Northwest Territories. Canadian Wildlife Service Report Series 21. Ottawa: Information Canada. 36 p.

MECH, L.D. 1994. Regular and homeward travel speeds of Arctic wolves. Journal of Mammalogy 75:741–742.

MECH, L.D., and BOITANI, L. 2003. Wolf social ecology. In: Mech, L.D., and Boitani, L., eds. Wolves: Behavior, ecology, and conservation. Chicago: University of Chicago Press. 1–34.

MECH, L.D., and GESE, E.M. 1992. Field testing the Wildlink capture collar on wolves. Wildlife Society Bulletin 20:249–256.

MECH, L.D., WOLFE, P., and PACKARD, J.M. 1999. Regurgitative food transfer among wild wolves. Canadian Journal of Zoology 77:1192–1195.

MERRILL, S.B., and MECH, L.D. 2000. Details of extensive movements by Minnesota wolves (Canis lupus). American Midland Naturalist 144:428–433.

MERRILL, S.B., ADAMS, L.G., NELSON, M.E., and MECH, L.D. 1998. Testing releasable GPS radiocollars on wolves and white-tailed deer. Wildlife Society Bulletin 26:830–835.

MESSIER, F. 1985. Solitary living and extraterritorial movements of wolves in relation to social status and prey abundance. Canadian Journal of Zoology 63:239–245.

MOHR, C.O., and STUMPF, W.A. 1966. Comparison of methods for calculating areas of animal activity. Journal of Wildlife Management 30:293–304.

MUSIANI, M. 2003. Conservation biology and management of wolves and wolf-human conflicts in western North America. Ph.D. Thesis, University of Calgary, Calgary, Alberta, Canada.

ORIANS, G.H., and PEARSON, N.E. 1979. On the theory of central place foraging. In: Mitchell, R.D., and Stairs, G.F., eds. Analysis of ecological systems. Columbus: Ohio State University Press. 154–177.

PACKARD, J.M., MECH, L.D., and REAM, R.R. 1992. Weaning in an arctic wolf pack: Behavioral mechanisms. Canadian Journal of Zoology 70:1269–1275.

PAQUET, P.C., WIERZCHOWSKI, J., and CALLAGHAN, C. 1996. Summary report on the effects of human activity on gray wolves in the Bow River Valley, Banff National Park, Alberta. In: Green, J., Pacas, C., Bayley, S., and Cornwell, L., eds. A cumulative effects assessment and futures outlook for the Banff Bow Valley. Prepared for the Banff Bow Valley Study. Ottawa: Department of Canadian Heritage.

STEPHENSON, R.O., and JAMES, D. 1982. Wolf movements and food habits in northwest Alaska. In: Harrington, F.H., and Paquet, P.C., eds. Wolves of the world. New Jersey: Noyes Publications. 223–237.

THORPE, N., EYEGETOK, S., HAKONGAK, N., and QITIRMIUT ELDERS. 2001. The Tuktu and Nogak Project: A caribou chronicle. Final Report to the West Kitikmeot/Slave Study Society, Ikaluktuuttiak, NWT. 160 p.

TIMONEY, K.P., LA ROI, G.H., ZOLTAI, S.C., and ROBINSON, A.L. 1992. The high subarctic forest-tundra of northwestern Canada: Position, width, and vegetation gradients in relation to climate. Arctic 45(1):1–9.

WALTON, L.R., CLUFF, H.D., PAQUET, P.C., and RAMSAY, M.A. 2001. Movement patterns of barren-ground wolves in the central Canadian Arctic. Journal of Mammalogy 82:867–876.

WILLIAMS, T.M. 1990. Summer diet and behavior of wolves denning on barren-ground caribou range in the Northwest Territories, Canada. M.Sc. Thesis, University of Alberta, Edmonton, Alberta, Canada.

ZAR, J.H. 1999. Biostatistical analysis. 4th ed. New Jersey: Prentice Hall. 663 p.

GLOSSARY

abiotic Nonliving. Compare *biotic*.

acid See *acid solution*.

acidic solution Any water solution that has more hydrogen ions (H^+) than hydroxide ions (OH^-); any water solution with a pH less than 7. Compare *basic solution*, *neutral solution*.

adaptation Any genetically controlled structural, physiological, or behavioral characteristic that helps an organism survive and reproduce under a given set of environmental conditions. It usually results from a beneficial mutation. See *biological evolution*, *differential reproduction*, *mutation*, *natural selection*.

adaptive radiation Process in which numerous new species evolve to fill vacant and new ecological niches in changed environments, usually after a mass extinction. Typically, this process takes millions of years.

adaptive trait See *adaptation*.

aerobic respiration Complex process that occurs in the cells of most living organisms, in which nutrient organic molecules such as glucose ($C_6H_{12}O_6$) combine with oxygen (O_2) to produce carbon dioxide (CO_2), water (H_2O), and energy. Compare *photosynthesis*.

affluenza Unsustainable addiction to overconsumption and materialism exhibited in the lifestyles of affluent consumers in the United States and other developed countries.

age structure Percentage of the population (or number of people of each sex) at each age level in a population.

alien species See *nonnative species*.

alpha particle Positively charged matter, consisting of two neutrons and two protons, that is emitted as radioactivity from the nuclei of some radioisotopes. See also *beta particle*, *gamma ray*.

altitude See *elevation*.

anaerobic respiration Form of cellular respiration in which some decomposers get the energy they need through the breakdown of glucose (or other nutrients) in the absence of oxygen. Compare *aerobic respiration*.

ancient forest See *old-growth forest*.

annual Plant that grows, sets seed, and dies in one growing season. Compare *perennial*.

anthropocentric Human-centered.

aquatic Pertaining to water. Compare *terrestrial*.

aquatic life zone Marine and freshwater portions of the biosphere. Examples include freshwater life zones (such as lakes and streams) and ocean or marine life zones (such as estuaries, coastlines, coral reefs, and the deep ocean).

aquifer Porous, water-saturated layers of sand, gravel, or bedrock that can yield an economically significant amount of water.

arid Dry. A desert or other area with an arid climate has little precipitation.

artificial selection Process by which humans select one or more desirable genetic traits in the population of a plant or animal species and then use *selective breeding* to produce populations containing many individuals with the desired traits. Compare *genetic engineering*, *natural selection*.

asexual reproduction Reproduction in which a mother cell divides to produce two identical daughter cells that are clones of the mother cell. This type of reproduction is common in single-celled organisms. Compare *sexual reproduction*.

atmosphere Mass of air surrounding the earth. See *stratosphere*, *troposphere*.

atom Minute unit, made of subatomic particles, that is the basic building block of all chemical elements and thus all matter; the smallest unit of an element that can exist and still have the unique characteristics of that element. Compare *ion*, *molecule*.

atomic number Number of protons in the nucleus of an atom. Compare *mass number*.

autotroph See *producer*.

background extinction Normal extinction of various species as a result of changes in local environmental conditions. Compare *mass depletion*, *mass extinction*.

bacteria Prokaryotic, one-celled organisms. Some transmit diseases. Most act as decomposers and get the nutrients they need by breaking down complex organic compounds in the tissues of living or dead organisms into simpler inorganic nutrient compounds.

barrier islands Long, thin, low offshore islands of sediment that generally run parallel to the shore along some coasts.

basic solution Water solution with more hydroxide ions (OH^-) than hydrogen ions (H^+); water solution with a pH greater than 7. Compare *acidic solution*, *neutral solution*.

benthos Bottom-dwelling organisms. Compare *decomposer*, *nekton*, *plankton*.

beta particle Swiftly moving electron emitted by the nucleus of a radioactive isotope. See also *alpha particle*, *gamma ray*.

biocentric Life-centered. Compare *anthropocentric*.

biodegradable Capable of being broken down by decomposers.

biodegradable pollutant Material that can be broken down into simpler substances (elements and compounds) by bacteria or other decomposers. Paper and most organic wastes such as animal manure are biodegradable but can take decades to biodegrade in modern landfills. Compare *degradable pollutant*, *nondegradable pollutant*, *slowly degradable pollutant*.

biodiversity Variety of different species (*species diversity*), genetic variability among individuals within each species (*genetic diversity*), variety of ecosystems (*ecological diversity*), and functions such as energy flow and matter cycling needed for the survival of species and biological communities (*functional diversity*).

biogeochemical cycle Natural processes that recycle nutrients in various chemical forms from the nonliving environment to living organisms and then back to the non-living environment. Examples include the carbon, oxygen, nitrogen, phosphorus, sulfur, and hydrologic cycles.

biological community See *community*.

biological diversity See *biodiversity*.

biological evolution Change in the genetic makeup of a population of a species in successive generations. If continued long enough, it can lead to the formation of a new species. Note that populations—not individuals—evolve. See also *adaptation, differential reproduction, natural selection, theory of evolution*.

biomass Organic matter produced by plants and other photosynthetic producers; total dry weight of all living organisms that can be supported at each trophic level in a food chain or web; dry weight of all organic matter in plants and animals in an ecosystem; plant materials and animal wastes used as fuel.

biome Terrestrial regions inhabited by certain types of life, especially vegetation. Examples include various types of deserts, grasslands, and forests.

biosphere Zone of the earth where life is found. It consists of parts of the atmosphere (the troposphere), hydrosphere (mostly surface water and groundwater), and lithosphere (mostly soil and surface rocks and sediments on the bottoms of oceans and other bodies of water) where life is found.

biotic Living organisms. Compare *abiotic*.

biotic pollution Harmful ecological and economic effects from the presence of accidentally or deliberately introduced species into ecosystems.

biotic potential Maximum rate at which the population of a given species can increase when there are no limits on its rate of growth. See *environmental resistance*.

birth rate See *crude birth rate*.

broadleaf deciduous plants Plants such as oak and maple trees that survive drought and cold by shedding their leaves and becoming dormant. Compare *broadleaf evergreen plants, coniferous evergreen plants*.

broadleaf evergreen plants Plants that keep most of their broad leaves year-round. An example is the trees found in the canopies of tropical rain forests. Compare *broadleaf deciduous plants, coniferous evergreen plants*.

calorie Unit of energy; amount of energy needed to raise the temperature of 1 gram of water by 1C° (unit on Celsius temperature scale). See also *kilocalorie*.

carbon cycle Cyclic movement of carbon in different chemical forms from the environment to organisms and then back to the environment.

carnivore Animal that feeds on other animals. Compare *herbivore, omnivore*.

carrying capacity (K) Maximum population of a particular species that a given habitat can support over a given period.

cell Smallest living unit of an organism. Each cell is encased in an outer membrane or wall and contains genetic material (DNA) and other parts to perform its life function. Organisms such as bacteria consist of only one cell, but most organisms contain many cells.

chain reaction Multiple nuclear fissions, taking place within a certain mass of a fissionable isotope, that release an enormous amount of energy in a short time.

chemical One of the millions of different elements and compounds found naturally and synthesized by humans. See *compound, element*.

chemical change Interaction between chemicals in which the chemical composition of the elements or compounds involved changes. Compare *nuclear change, physical change*.

chemical evolution Formation of the earth and its early crust and atmosphere, evolution of the biological molecules necessary for life, and evolution of systems of chemical reactions needed to produce the first living cells. These processes are believed to have occurred about 1 billion years before biological evolution. Compare *biological evolution*.

chemical formula Shorthand way to show the number of atoms (or ions) in the basic structural unit of a compound. Examples include H_2O, NaCl, and $C_6H_{12}O_6$.

chemical reaction See *chemical change*.

chemosynthesis Process in which certain organisms (mostly specialized bacteria) extract inorganic compounds from their environment and convert them into organic nutrient compounds without the presence of sunlight. Compare *photosynthesis*.

chlorinated hydrocarbon Organic compound made up of atoms of carbon, hydrogen, and chlorine. Examples include DDT and PCBs.

chromosome Grouping of genes and associated proteins in plant and animal cells that carry certain types of genetic information. See *genes*.

clear-cutting Method of timber harvesting in which all trees in a forested area are

removed in a single cutting. Compare *selective cutting, strip cutting*.

climate Physical properties of the troposphere of an area based on analysis of its weather records over a long period (at least 30 years). The two main factors determining an area's climate are the *temperature*, with its seasonal variations, and the amount and distribution of *precipitation*. Compare *weather*.

climax community See *mature community*.

coastal wetland Land along a coastline, extending inland from an estuary that is covered with saltwater all or part of the year. Examples include marshes, bays, lagoons, tidal flats, and mangrove swamps. Compare *inland wetland*.

coastal zone Warm, nutrient-rich, shallow part of the ocean that extends from the high-tide mark on land to the edge of a shelf-like extension of continental land masses known as the continental shelf. Compare *open sea*.

coevolution Evolution in which two or more species interact and exert selective pressures on each other that can lead each species to undergo adaptations. See *evolution, natural selection*.

cold front Leading edge of an advancing mass of cold air. Compare *warm front*.

commensalism Interaction between organisms of different species in which one type of organism benefits and the other type is neither helped nor harmed to any great degree. Compare *mutualism*.

commercial extinction Depletion of the population of a wild species used as a resource to a level at which it is no longer profitable to harvest the species.

common-property resource Resource that people normally are free to use; each user can deplete or degrade the available supply. Most such resources are renewable and owned by no one. Examples include clean air, fish in parts of the ocean not under the control of a coastal country, migratory birds, gases of the lower atmosphere, and the ozone content of the upper atmosphere (stratosphere). See *tragedy of the commons*.

community Populations of all species living and interacting in an area at a particular time.

competition Two or more individual organisms of a single species (*intraspecific competition*) or two or more individuals of different species (*interspecific competition*) attempting to use the same scarce resources in the same ecosystem.

complex carbohydrates Two or more monomers of *simple sugars* (such as glucose) linked together.

compound Combination of atoms, or oppositely charged ions, of two or more elements held together by attractive forces called chemical bonds. Compare *element*.

concentration Amount of a chemical in a particular volume or weight of air, water, soil, or other medium.

coniferous evergreen plants Cone-bearing plants (such as spruces, pines, and firs) that keep some of their narrow, pointed leaves (needles) all year. Compare *broadleaf deciduous plants, broadleaf evergreen plants*.

coniferous trees Cone-bearing trees, mostly evergreens, that have needle-shaped or scale-like leaves. They produce wood known commercially as softwood. Compare *deciduous plants*.

consensus science See *sound science*.

conservation Sensible and careful use of natural resources by humans. People with this view are called *conservationists*.

conservation biology Multidisciplinary science created to deal with the crisis of maintaining the genes, species, communities, and ecosystems that make up earth's biological diversity. Its goals are to investigate human impacts on biodiversity and to develop practical approaches to preserving biodiversity.

conservationist Person concerned with using natural areas and wildlife in ways that sustain them for current and future generations of humans and other forms of life.

constancy Ability of a living system, such as a population, to maintain a certain size. Compare *inertia, resilience*.

consumer Organism that cannot synthesize the organic nutrients it needs and gets its organic nutrients by feeding on the tissues of producers or of other consumers; generally divided into *primary consumers* (herbivores), *secondary consumers* (carnivores), *tertiary (higher-level) consumers, omnivores*, and *detritivores* (decomposers and detritus feeders). In economics, one who uses economic goods. Compare *producer*.

controlled burning Deliberately set, carefully controlled surface fires that reduce flammable litter and decrease the chances of damaging crown fires. See *ground fire, surface fire*.

convergent plate boundary Area where the earth's lithospheric plates are pushed together. See *subduction zone*. Compare *divergent plate boundary, transform fault*.

coral reef Formation produced by massive colonies containing billions of tiny coral animals, called polyps, that secrete a stony substance (calcium carbonate) around themselves for protection. When the corals die, their empty outer skeletons form layers and cause the reef to grow. Coral reefs are found in the coastal zones of warm tropical and subtropical oceans.

core Inner zone of the earth. It consists of a solid inner core and a liquid outer core. Compare *crust, mantle*.

corrective feedback loop See *negative feedback loop*.

critical mass Amount of fissionable nuclei needed to sustain a nuclear fission chain reaction.

crown fire Extremely hot forest fire that burns ground vegetation and treetops. Compare *controlled burning, ground fire, surface fire*.

crude birth rate Annual number of live births per 1,000 people in the population of a geographic area at the midpoint of a given year. Compare *crude death rate*.

crude death rate Annual number of deaths per 1,000 people in the population of a geographic area at the midpoint of a given year. Compare *crude birth rate*.

crust Solid outer zone of the earth. It consists of oceanic crust and continental crust. Compare *core, mantle*.

DDT Dichlorodiphenyltrichloroethane, a chlorinated hydrocarbon that has been widely used as an insecticide but is now banned in some countries.

death rate See *crude death rate*.

debt-for-nature swap Agreement in which a certain amount of foreign debt is canceled in exchange for local currency investments that will improve natural resource management or protect certain areas in the debtor country from harmful development.

deciduous plants Trees, such as oaks and maples, and other plants that survive during dry seasons or cold seasons by shedding their leaves. Compare *coniferous trees, succulent plants*.

decomposer Organism that digests parts of dead organisms and cast-off fragments and wastes of living organisms by breaking down the complex organic molecules in those materials into simpler inorganic compounds and then absorbing the soluble nutrients. Producers return most of these chemicals to the soil and water for reuse. Decomposers consist of various bacteria and fungi. Compare *consumer, detritivore, producer*.

deductive reasoning Using logic to arrive at a specific conclusion based on a generalization or premise. Compare *inductive reasoning*.

deforestation Removal of trees from a forested area without adequate replanting.

degradable pollutant Potentially polluting chemical that is broken down completely or reduced to acceptable levels by natural physical, chemical, and biological processes. Compare *biodegradable pollutant, nondegradable pollutant, slowly degradable pollutant*.

demographic transition Hypothesis that countries, as they become industrialized, have declines in death rates followed by declines in birth rates.

desert Biome in which evaporation exceeds precipitation and the average amount of precipitation is less than 25 centimeters (10 inches) per year. Such areas have little vegetation or have widely spaced, mostly low vegetation. Compare *forest, grassland*.

detritivore Consumer organism that feeds on detritus, parts of dead organisms, and cast-off fragments and wastes of living organisms. The two principal types are *detritus feeders* and *decomposers*.

detritus Parts of dead organisms and cast-off fragments and wastes of living organisms.

detritus feeder Organism that extracts nutrients from fragments of dead organisms and their cast-off parts and organic wastes. Examples include earthworms, termites, and crabs. Compare *decomposer*.

deuterium (D; hydrogen-2) Isotope of the element hydrogen, with a nucleus containing one proton and one neutron and a mass number of 2.

developed country Country that is highly industrialized and has a high per capita GDP. Compare *developing country*.

developing country Country that has low to moderate industrialization and low to moderate per capita GDP. Most are located in Africa, Asia, and Latin America. Compare *developed country*.

dieback Sharp reduction in the population of a species when its numbers exceed the carrying capacity of its habitat. See *carrying capacity*.

differential reproduction Phenomenon in which individuals with adaptive genetic traits produce more living offspring than do individuals without such traits. See *natural selection*.

dissolved oxygen (DO) content Amount of oxygen gas (O_2) dissolved in a given volume of water at a particular temperature and pressure, often expressed as a concentration in parts of oxygen per million parts of water.

distribution Area over which a species can be found.

disturbance Event that disrupts an ecosystem or community. Examples of *natural disturbances* include fires, hurricanes, tornadoes, droughts, and floods. Examples of *human-caused disturbances* include deforestation, overgrazing, and plowing.

divergent plate boundary Area where the earth's lithospheric plates move apart in opposite directions. Compare *convergent plate boundary, transform fault.*

DNA (deoxyribonucleic acid) Large molecules in the cells of organisms that carry genetic information in living organisms.

domesticated species Wild species tamed or genetically altered by crossbreeding for use by humans for food (cattle, sheep, and food crops), pets (dogs and cats), or enjoyment (animals in zoos and plants in gardens). Compare *wild species.*

doubling time Time it takes (usually in years) for the quantity of something growing exponentially to double. It can be calculated by dividing the annual percentage growth rate into 70.

drought Condition in which an area does not get enough water because of lower-than-normal precipitation or higher-than-normal temperatures that increase evaporation.

durability Ability of earth's various systems, including human cultural systems and economies, to survive and adapt to changing environmental conditions indefinitely. This is another name for sustainability.

earthquake Shaking of the ground resulting from the fracturing and displacement of rock, which produces a fault, or from subsequent movement along the fault.

ecological diversity Variety of forests, deserts, grasslands, oceans, streams, lakes, and other biological communities interacting with one another and with their nonliving environment. See *biodiversity.* Compare *functional diversity, genetic diversity, species diversity.*

ecological efficiency Percentage of energy transferred from one trophic level to another in a food chain or web.

ecological footprint Amount of biologically productive land and water needed to supply a population with the renewable resources it uses and to absorb or dispose of the wastes from such resource use. It measures the average environmental impact of populations in different countries and areas. See *per capita ecological footprint*

ecological niche Total way of life or role of a species in an ecosystem. It includes all physical, chemical, and biological conditions that a species needs to live and reproduce in an ecosystem. See *fundamental niche, realized niche.*

ecological restoration Deliberate alteration of a degraded habitat or ecosystem to restore as much of its ecological structure and function as possible.

ecological succession Process in which communities of plant and animal species in a particular area are replaced over time by a series of different and often more complex communities. See *primary succession, secondary succession.*

ecologist Biological scientist who studies relationships between living organisms and their environment.

ecology Biological science that studies the relationships between living organisms and their environment; study of the structure and functions of nature.

economic development Improvement of human living standards by economic growth. Compare *economic growth, environmentally sustainable economic development.*

economic growth Increase in the capacity to provide people with goods and services; an increase in gross domestic product (GDP). Compare *economic development, environmentally sustainable economic development.* See *gross domestic product.*

ecosphere See *biosphere.*

ecosystem Community of different species interacting with one another and with the chemical and physical factors making up its nonliving environment.

ecosystem services Natural services or natural capital that support life on the earth and are essential to the quality of human life and the functioning of the world's economies. See *natural resources.*

electromagnetic radiation Forms of kinetic energy traveling as electromagnetic waves. Examples include radio waves, TV waves, microwaves, infrared radiation, visible light, ultraviolet radiation, X rays, and gamma rays. Compare *ionizing radiation, nonionizing radiation.*

electron (e) Tiny particle moving around outside the nucleus of an atom. Each electron has one unit of negative charge and almost no mass. Compare *neutron, proton.*

element Chemical, such as hydrogen (H), iron (Fe), sodium (Na), carbon (C), nitrogen (N), or oxygen (O), whose distinctly different atoms serve as the basic building blocks of all matter. Two or more elements combine to form the compounds that make up most of the world's matter. Compare *compound.*

elevation Distance above sea level. Compare *latitude.*

emigration Movement of people out of a specific geographic area. See *migration.* Compare *immigration.*

endangered species Wild species with so few individual survivors that the species could soon become extinct in all or most of its natural range. Compare *threatened species.*

endemic species Species that is found in only one area. Such species are especially vulnerable to extinction.

energy Capacity to do work by performing mechanical, physical, chemical, or electrical tasks or to cause a heat transfer between two objects at different temperatures.

energy quality Ability of a form of energy to do useful work. High-temperature heat and the chemical energy in fossil fuels and nuclear fuels are concentrated high-quality energy. Low-quality energy such as low-temperature heat is dispersed or diluted and cannot do much useful work. See *high-quality energy, low-quality energy.*

environment All external conditions and factors, living and nonliving (chemicals and energy), that affect any living organism or other specified system.

environmental degradation Depletion or destruction of a potentially renewable resource such as soil, grassland, forest, or wildlife that is used faster than it is naturally replenished. If such use continues, the resource becomes nonrenewable (on a human time scale) or nonexistent (extinct). See also *sustainable yield.*

environmental ethics Human beliefs about what is right or wrong with how we treat the environment.

environmentalism Social movement dedicated to protecting the earth's life-support systems for us and other species.

environmentalist Person who is concerned about the impact of people on environmental quality and believe that some human actions are degrading parts of the earth's life-support systems for humans and many other forms of life.

environmentally sustainable economic development Development that *encourages* forms of economic growth that meet the basic needs of the current generations of humans and other species without preventing future generations of humans and other species from meeting their basic needs and *discourages* environmentally harmful and unsustainable forms of economic growth. It is the economic component of an *environmentally sustainable society*. Compare *economic development, economic growth*.

environmentally sustainable society Society that meets the current and future basic needs of its people for basic resources in a just and equitable manner without compromising the ability of future generations of humans and other species from meeting their basic needs.

environmental movement Environmental movement that flourished in the United States in the 1960s when a growing number of citizens organized to demand that political leaders enact laws and develop policies to curtail pollution, clean up polluted environments, and protect unspoiled areas from environmental degradation.

environmental resistance All of the limiting factors that act together to limit the growth of a population. See *biotic potential, limiting factor*.

environmental revolution Cultural change involving halting population growth and altering lifestyles, political and economic systems, and the way we treat the environment so that we can help sustain the earth for ourselves and other species. It requires working with the rest of nature by learning more about how nature sustains itself. See *environmental wisdom worldview*.

environmental science Interdisciplinary study that uses information from the physical sciences and social sciences to learn how the earth works, how we interact with the earth, and how to deal with environmental problems.

environmental scientist Scientist who uses information from the physical sciences and social sciences to understand how the earth works, learn how humans interact with the earth, and develop solutions to environmental problems.

environmental wisdom worldview We are part of and totally dependent on nature and nature exists for all species, not just for us, and we should encourage earth-sustaining forms of economic growth and development and discourage earth-degrading forms. Our success depends on learning how the earth sustains itself and integrating such environmental wisdom into the ways we think and act. Compare *planetary management worldview, stewardship worldview*.

environmental worldview Set of assumptions and beliefs about how people think the world works, what they think their role in the world should be, and what they believe is right and wrong environmental behavior (environmental ethics).

EPA U.S. Environmental Protection Agency; responsible for managing federal efforts to control air and water pollution, radiation and pesticide hazards, environmental research, hazardous waste, and solid waste disposal.

epiphyte Plant that uses its roots to attach itself to branches high in trees, especially in tropical forests.

estuary Partially enclosed coastal area at the mouth of a river where its freshwater, carrying fertile silt and runoff from the land, mixes with salty seawater.

eukaryotic organism Classification of cell structure in which the cell is surrounded by a membrane and has a distinct nucleus and several other internal parts. Most organisms consist of eukaryotic cells. Compare *prokaryotic organism*.

euphotic zone Upper layer of a body of water through which sunlight can penetrate and support photosynthesis.

eutrophic lake Lake with a large or excessive supply of plant nutrients, mostly nitrates and phosphates. Compare *mesotrophic lake, oligotrophic lake*.

evaporation Conversion of a liquid to a gas.

even-aged management Method of forest management in which trees, sometimes of a single species in a given stand, are maintained at roughly the same age and size and are harvested all at once. Compare *uneven-aged management*.

evergreen plants Plants that keep some of their leaves or needles throughout the year. Examples include ferns and cone-bearing trees (conifers) such as firs, spruces, pines, redwoods, and sequoias. Compare *deciduous plants, succulent plants*.

evolution See *biological evolution*.

exhaustible resource See *nonrenewable resource*.

exotic species See *nonnative species*.

experiment Procedure a scientist uses to study a phenomenon under known conditions. Scientists conduct some experiments in the laboratory and others in nature. The resulting scientific data or facts must be verified or confirmed by repeated observations and measurements, ideally by several different investigators.

exponential growth Growth in which some quantity, such as population size or economic output, increases at a constant rate per unit of time. An example is the growth sequence 2, 4, 8, 16, 32, 64, and so on. When the increase in quantity over time is plotted, this type of growth yields a curve shaped like the letter J. Compare *linear growth*.

extinction Complete disappearance of a species from the earth. It happens when a species cannot adapt and successfully reproduce under new environmental conditions or when a species evolves into one or more new species. Compare *speciation*. See also *endangered species, mass depletion, mass extinction, threatened species*.

family planning Providing information, clinical services, and contraceptives to help people choose the number and spacing of children they want to have.

feedback loop Occurs when an output of matter, energy, or information is fed back into the system as an input and leads to changes in that system.

fermentation See *anaerobic respiration*.

fertility Number of births that occur to an individual woman.

first law of thermodynamics In any physical or chemical change, no detectable amount of energy is created or destroyed, but energy can be changed from one form to another; you cannot get more energy out of something than you put in; in terms of energy quantity, you cannot get something for nothing (there is no free lunch). This law does not apply to nuclear changes, in which energy can be produced from small amounts of matter. See *second law of thermodynamics*.

fissionable isotope Isotope that can split apart when hit by a neutron at the right speed and thus undergo nuclear fission. Examples include uranium-235 and plutonium-239.

flows See *throughputs*.

food chain Series of organisms in which each eats or decomposes the preceding one. Compare *food web*.

food web Complex network of many interconnected food chains and feeding relationships. Compare *food chain*.

forest Biome with enough average annual precipitation (at least 76 centimeters, or 30 inches) to support the growth of tree species and smaller forms of vegetation. Compare *desert, grassland*.

fossils Skeletons, bones, shells, body parts, leaves, seeds, or impressions of such items that provide recognizable evidence of organisms that lived long ago.

foundation species Species that plays a major role in shaping communities by creating and enhancing a habitat that benefits other species. Compare *indicator species, keystone species, native species, nonnative species*.

free-access resource See *common-property resource*.

freshwater life zones Aquatic systems where water with a dissolved salt concentration of less than 1% by volume accumulates on or flows through the surfaces of terrestrial biomes. Examples include *standing (lentic)* bodies of freshwater such as lakes, ponds, and inland wetlands and *flowing (lotic)* systems such as streams and rivers. Compare *biome*.

front Boundary between two air masses with different temperatures and densities. See *cold front, warm front*.

frontier environmental worldview View by European colonists settling North America in the 1600s that the continent had vast resources and was a wilderness to be conquered by settlers clearing and planting land.

frontier science Preliminary scientific data, hypotheses, and models that have not been widely tested and accepted. Compare *junk science, sound science*.

functional diversity Biological and chemical processes or functions such as energy flow and matter cycling needed for the survival of species and biological communities. See *biodiversity, ecological diversity, genetic diversity, species diversity*.

fundamental niche Full potential range of the physical, chemical, and biological factors a species can use if it does not face any competition from other species. See *ecological niche*. Compare *realized niche*.

gamma ray Form of ionizing electromagnetic radiation with a high energy content emitted by some radioisotopes. It readily penetrates body tissues. See also *alpha particle, beta particle*.

GDP See *gross domestic product*.

gene mutation See *mutation*.

gene pool Sum total of all genes found in the individuals of the population of a particular species.

generalist species Species with a broad ecological niche. They can live in many different places, eat a variety of foods, and tolerate a wide range of environmental conditions. Examples include flies, cockroaches, mice, rats, and humans. Compare *specialist species*.

genes Coded units of information about specific traits that are passed from parents to offspring during reproduction. They consist of segments of DNA molecules found in chromosomes.

gene splicing See *genetic engineering*.

genetic adaptation Changes in the genetic makeup of organisms of a species that allow the species to reproduce and gain a competitive advantage under changed environmental conditions. See *differential reproduction, evolution, mutation, natural selection*.

genetically modified organism (GMO) Organism whose genetic makeup has been altered by genetic engineering.

genetic diversity Variability in the genetic makeup among individuals within a single species. See *biodiversity*. Compare *ecological diversity, functional diversity, species diversity*.

genetic engineering Insertion of an alien gene into an organism to give it a beneficial genetic trait. Compare *artificial selection, natural selection*.

geographic isolation Separation of populations of a species for long times into different areas.

geology Study of the earth's dynamic history. Geologists study and analyze rocks and the features and processes of the earth's interior and surface.

global climate change Changes in any aspects of the earth's climate, including temperature, precipitation, and storm intensity and patterns.

global warming Warming of the earth's atmosphere because of increases in the concentrations of one or more greenhouse gases primarily as a result of human activities. See *greenhouse effect, greenhouse gases*. Compare *global climate change*.

grassland Biome found in regions where moderate annual average precipitation (25–76 centimeters, or 10–30 inches) is enough to support the growth of grass and small plants but not enough to support large stands of trees. Compare *desert, forest*.

greenhouse effect Natural effect that releases heat in the atmosphere (troposphere) near the earth's surface. Water vapor, carbon dioxide, ozone, and other gases in the lower atmosphere (troposphere) absorb some of the infrared radiation (heat) radiated by the earth's surface. Their molecules vibrate and transform the absorbed energy into longer-wavelength infrared radiation (heat) in the troposphere. If the atmospheric concentrations of these greenhouse gases increase and other natural processes do not remove them, the average temperature of the lower atmosphere will increase gradually. Compare *global warming*. See also *natural greenhouse effect*.

greenhouse gases Gases in the earth's lower atmosphere (troposphere) that cause the greenhouse effect. Examples include carbon dioxide, chlorofluorocarbons, ozone, methane, water vapor, and nitrous oxide.

gross domestic product (GDP) Annual market value of all goods and services produced by all firms and organizations, foreign and domestic, operating within a country. See *per capita GDP*.

gross primary productivity (GPP) Rate at which an ecosystem's producers capture and store a given amount of chemical energy as biomass in a given length of time. Compare *net primary productivity*.

ground fire Fire that burns decayed leaves or peat deep below the ground surface. Compare *crown fire, surface fire*.

groundwater Water that sinks into the soil and is stored in slowly flowing and slowly renewed underground reservoirs called aquifers; underground water in the zone of saturation, below the water table. Compare *runoff, surface water*.

habitat Place or type of place where an organism or population of organisms lives. Compare *ecological niche*.

habitat fragmentation Breakup of a habitat into smaller pieces, usually as a result of human activities.

half-life Time needed for one-half of the nuclei in a radioisotope to emit their radiation. Each radioisotope has a characteristic half-life, which may range from a few millionths of a second to several billion years. See *radioisotope*.

heat Total kinetic energy of all randomly moving atoms, ions, or molecules within a given substance, excluding the overall motion of the whole object. Heat always flows spontaneously from a hot sample of matter to a colder sample of matter. This is one way to state the second law of thermodynamics. Compare *temperature*.

herbivore Plant-eating organism. Examples include deer, sheep, grasshoppers, and zooplankton. Compare *carnivore, omnivore*.

heterotroph See *consumer*.

high Air mass with a high pressure. Compare *low*.

high-quality energy Energy that is concentrated and has great ability to perform useful work. Examples include high-temperature heat and the energy in electricity, coal, oil, gasoline, sunlight, and nuclei of uranium-235. Compare *low-quality energy*.

high-quality matter Matter that is concentrated and contains a high concentration of a useful resource. Compare *low-quality matter*.

high-throughput economy Situation in most advanced industrialized countries, in which ever-increasing economic growth is sustained by maximizing the rate at which matter and energy resources are used, with little emphasis on pollution prevention, recycling, reuse, reduction of unnecessary waste, and other forms of resource conservation. Compare *low-throughput economy, matter-recycling economy*.

high-waste economy See *high-throughput economy*.

host Plant or animal on which a parasite feeds.

human capital See *human resources*.

human resources People's physical and mental talents that provide labor, innovation, culture, and organization. Human capital also includes the cultural skills people pick up from their families and friends, the ability to be trustworthy, and the drive to achieve.

hunter–gatherers People who get their food by gathering edible wild plants and other materials and by hunting wild animals and fish.

hydrocarbon Organic compound of hydrogen and carbon atoms. The simplest hydrocarbon is methane (CH_4), the major component of natural gas.

hydrologic cycle Biogeochemical cycle that collects, purifies, and distributes the earth's fixed supply of water from the environment to living organisms and then back to the environment.

hydrosphere Earth's *liquid water* (oceans, lakes, other bodies of surface water, and underground water), *frozen water* (polar ice caps, floating ice caps, and ice in soil, known as permafrost), and *water vapor* in the atmosphere. See also *hydrologic cycle*.

immature community Community at an early stage of ecological succession. It usually has a low number of species and ecological niches and cannot capture and use energy and cycle critical nutrients as efficiently as more complex, mature communities. Compare *mature community*.

immigrant species See *nonnative species*.

immigration Migration of people into a country or area to take up permanent residence. See *migration*. Compare *emigration*.

indicator species Species that serve as early warnings that a community or ecosystem is being degraded. Compare *foundation species, keystone species, native species, nonnative species*.

inductive reasoning Using specific observations and measurements to arrive at a general conclusion or hypothesis. Compare *deductive reasoning*.

inertia Ability of a living system to resist being disturbed or altered.

infant mortality rate Number of babies out of every 1,000 born each year who die before their first birthday.

inland wetland Land away from the coast, such as a swamp, marsh, or bog, that is covered all or part of the time with freshwater. Compare *coastal wetland*.

inorganic compounds All compounds not classified as organic compounds. See *organic compounds*.

inorganic fertilizer See *commercial inorganic fertilizer*.

input Matter, energy, or information entering a system. Compare *output, throughput*.

input pollution control See *pollution prevention*.

instrumental value Value of an organism, species, ecosystem, or the earth's biodiversity based on its usefulness to humans. Compare *intrinsic value*.

interspecific competition Attempts by members of two or more species to use the same limited resources in an ecosystem. See *competition, intraspecific competition*.

intertidal zone The area of shoreline between low and high tides.

intraspecific competition Attempts by two or more organisms of a single species to use the same limited resources in an ecosystem. See *competition, interspecific competition*.

intrinsic rate of increase (r) Rate at which a population could grow if it had unlimited resources. Compare *environmental resistance*.

intrinsic value Value of an organism, species, ecosystem, or the earth's biodiversity based on its existence, regardless of whether it has any usefulness to humans. Compare *instrumental value*.

invasive species See *nonnative species*.

invertebrates Animals that have no backbones. Compare *vertebrates*.

ion Atom or group of atoms with one or more positive (+) or negative (−) electrical charges. Compare *atom, molecule*.

ionizing radiation Fast-moving alpha or beta particles or high-energy radiation (gamma rays) emitted by radioisotopes. They have enough energy to dislodge one or more electrons from atoms they hit, thereby forming charged ions in tissue that can react with and damage living tissue. Compare *nonionizing radiation*.

isotopes Two or more forms of a chemical element that have the same number of protons but different mass numbers because they have different numbers of neutrons in their nuclei.

J-shaped curve Curve with a shape similar to that of the letter J; can represent prolonged exponential growth. See *exponential growth*.

junk science Scientific results or hypotheses presented as sound science without having undergone the rigors of the peer review process. Compare *frontier science, sound science*.

keystone species Species that play roles affecting many other organisms in an ecosystem. Compare *foundation species, indicator species, native species, nonnative species*.

kilocalorie (kcal) Unit of energy equal to 1,000 calories. See *calorie*.

kinetic energy Energy that matter has because of its mass and speed or velocity. Compare *potential energy*.

K-selected species Species that produce a few, often fairly large offspring but invest a great deal of time and energy to ensure that most of those offspring reach reproductive age. Compare *r-selected species*.

K-strategists See *K-selected species*.

lake Large natural body of standing freshwater formed when water from precipitation, land runoff, or groundwater flow fills a depression in the earth created by glaciation, earth movement, volcanic activity, or a giant meteorite. See *eutrophic lake, mesotrophic lake, oligotrophic lake*.

land degradation Decrease in the ability of land to support crops, livestock, or wild species in the future as a result of natural or human-induced processes.

latitude Distance from the equator. Compare *altitude*.

law of conservation of energy See *first law of thermodynamics*.

law of conservation of matter In any physical or chemical change, matter is neither created nor destroyed but merely changed from one form to another; in physical and chemical changes, existing atoms are rearranged into different spatial patterns (physical changes) or different combinations (chemical changes).

law of tolerance The existence, abundance, and distribution of a species in an ecosystem are determined by whether the levels of one or more physical or chemical factors fall within the range tolerated by the species. See *threshold effect*.

LDC See *developing country*.

less developed country (LDC) See *developing country*.

life-centered environmental worldview Belief that we have an ethical responsibility to prevent degradation of the earth's ecosystems, biodiversity, and biosphere, and that there is *inherent* or *intrinsic value* of all forms of life, regardless of their potential or actual use to humans.

life expectancy Average number of years a newborn infant can be expected to live.

limiting factor Single factor that limits the growth, abundance, or distribution of the population of a species in an ecosystem. See *limiting factor principle.*

limiting factor principle Too much or too little of any abiotic factor can limit or prevent growth of a population of a species in an ecosystem, even if all other factors are at or near the optimal range of tolerance for the species.

linear growth Growth in which a quantity increases by some fixed amount during each unit of time. An example is growth that increases in the sequence 2, 4, 6, 8, 10, and so on. Compare *exponential growth.*

lipids Chemically diverse group of large organic compounds that do not dissolve in water. Examples are *fats and oils* for storing energy, *waxes* for structure, and *steroids* for producing hormones.

lithosphere Outer shell of the earth, composed of the crust and the rigid, outermost part of the mantle outside the asthenosphere; material found in the earth's plates. See *crust, mantle.*

logistic growth Pattern in which exponential population growth occurs when the population is small, and population growth decreases steadily with time as the population approaches the carrying capacity. See *S-shaped curve.*

low An air mass with a low pressure. Compare *high.*

low-quality energy Energy that is dispersed and has little ability to do useful work. An example is low-temperature heat. Compare *high-quality energy.*

low-quality matter Matter that is dilute or dispersed or contains a low concentration of a useful resource. Compare *high-quality matter.*

low-throughput economy Economy based on working with nature by recycling and reusing discarded matter; preventing pollution; conserving matter and energy resources by reducing unnecessary waste and use; not degrading renewable resources; building things that are easy to recycle, reuse, and repair; not allowing population size to exceed the carrying capacity of the environment; and preserving biodiversity and ecological integrity. Compare *high-throughput economy, matter-recycling economy.*

low-waste economy See *low-throughput economy.*

mangrove swamps Swamps found on the coastlines in warm tropical climates. They are dominated by mangrove trees, any of about 55 species of trees and shrubs that can live partly submerged in the salty environment of coastal swamps.

mantle Zone of the earth's interior between its core and its crust. Compare *core, crust.* See *lithosphere.*

mass The amount of material in an object.

mass depletion Widespread, often global period during which extinction rates are higher than normal but not high enough to classify as a mass extinction. Compare *background extinction, mass extinction.*

mass extinction A catastrophic, widespread, often global event in which major groups of species are wiped out over a short time compared with normal (background) extinctions. Compare *background extinction, mass depletion.*

mass number Sum of the number of neutrons (n) and the number of protons (p) in the nucleus of an atom. It gives the approximate mass of that atom. Compare *atomic number.*

matter Anything that has mass (the amount of material in an object) and takes up space. On the earth, where gravity is present, we weigh an object to determine its mass.

matter quality Measure of how useful a matter resource is, based on its availability and concentration. See *high-quality matter, low-quality matter.*

matter-recycling-and-reuse economy Economy that emphasizes recycling the maximum amount of all resources that can be recycled and reused. The goal is to allow economic growth to continue without depleting matter resources and without producing excessive pollution and environmental degradation. Compare *high-throughput economy, low-throughput economy.*

mature community Fairly stable, self-sustaining community in an advanced stage of ecological succession; usually has a diverse array of species and ecological niches; captures and uses energy and cycles critical chemicals more efficiently than simpler, immature communities. Compare *immature community.*

maximum sustainable yield See *sustainable yield.*

MDC See *developed country.*

mesotrophic lake Lake with a moderate supply of plant nutrients. Compare *eutrophic lake, oligotrophic lake.*

metabolism Ability of a living cell or organism to capture and transform matter and energy from its environment to supply its needs for survival, growth, and reproduction.

microorganisms Organisms such as bacteria that are so small that it takes a microscope to see them.

migration Movement of people into and out of a specific geographic area. See *immigration, emigration.*

mineral Any naturally occurring inorganic substance found in the earth's crust as a crystalline solid. See *mineral resource.*

mineral resource Concentration of naturally occurring solid, liquid, or gaseous material in or on the earth's crust in a form and amount such that extracting and converting it into useful materials or items is currently or potentially profitable. Mineral resources are classified as *metallic* (such as iron and tin ores) or *nonmetallic* (such as fossil fuels, sand, and salt).

minimum viable population (MVP) Estimate of the smallest number of individuals necessary to ensure the survival of a population in a region for a specified time period, typically ranging from decades to 100 years.

mixture Combination of one or more elements and compounds.

model Approximate representation or simulation of a system being studied.

molecule Combination of two or more atoms of the same chemical element (such as O_2) or different chemical elements (such as H_2O) held together by chemical bonds. Compare *atom, ion.*

more developed country (MDC) See *developed country.*

mutation Random change in DNA molecules making up genes that can alter anatomy, physiology, or behavior in offspring.

mutualism Type of species interaction in which both participating species generally benefit. Compare *commensalism.*

nanotechnology Using atoms and molecules to build materials from the bottom up using the elements in the periodic table as its raw materials.

native species Species that normally live and thrive in a particular ecosystem. Compare *foundation species, indicator species, keystone species, nonnative species.*

natural capital Natural resources and natural services that keep us and other species alive and support our economies

natural greenhouse effect Heat buildup in the troposphere because of the presence of certain gases, called greenhouse gases. Without this effect, the earth would be nearly as cold as Mars, and life as we know it could not exist. Compare *global warming*.

natural law See *scientific law*.

natural radioactive decay Nuclear change in which unstable nuclei of atoms spontaneously shoot out particles (usually alpha or beta particles) or energy (gamma rays) at a fixed rate.

natural rate of extinction See *background extinction*.

natural resources See *natural capital*.

natural selection Process by which a particular beneficial gene (or set of genes) is reproduced in succeeding generations more than other genes. The result of natural selection is a population that contains a greater proportion of organisms better adapted to certain environmental conditions. See *adaptation, biological evolution, differential reproduction, mutation*.

negative feedback loop Causes a system to change in the opposite direction. See *feedback loop*. Compare *positive feedback loop*.

nekton Strongly swimming organisms found in aquatic systems. Compare *benthos, plankton*.

net primary productivity (NPP) Rate at which all the plants in an ecosystem produce net useful chemical energy; equal to the difference between the rate at which the plants in an ecosystem produce useful chemical energy (gross primary productivity) and the rate at which they use some of that energy through cellular respiration. Compare *gross primary productivity*.

neutral solution Water solution containing an equal number of hydrogen ions (H^+) and hydroxide ions (OH^-); water solution with a pH of 7. Compare *acidic solution, basic solution*.

neutron (n) Elementary particle in the nuclei of all atoms (except hydrogen-1). It has a relative mass of 1 and no electric charge. Compare *electron, proton*.

niche See *ecological niche*.

nitrogen cycle Cyclic movement of nitrogen in different chemical forms from the environment to organisms and then back to the environment.

nitrogen fixation Conversion of atmospheric nitrogen gas into forms useful to plants by lightning, bacteria, and cyanobacteria; it is part of the nitrogen cycle.

nondegradable pollutant Material that is not broken down by natural processes. Examples include the toxic elements lead and mercury. Compare *biodegradable pollutant, degradable pollutant, slowly degradable pollutant*.

nonionizing radiation Forms of radiant energy such as radio waves, microwaves, infrared light, and ordinary light that do not have enough energy to cause ionization of atoms in living tissue. Compare *ionizing radiation*.

nonnative species Species that migrate into an ecosystem or are deliberately or accidentally introduced into an ecosystem by humans. Compare *native species*.

nonpersistent pollutant See *degradable pollutant*.

nonpoint source Large or dispersed land areas such as crop fields, streeets, and lawns that discharge pollutants into the environment over a large area. Compare *point source*.

nonrenewable resource Resource that exists in a fixed amount (stock) in the earth's crust and has the potential for renewal by geological, physical, and chemical processes taking place over hundreds of millions to billions of years. Examples include copper, aluminum, coal, and oil. We classify these resources as exhaustible because we are extracting and using them at a much faster rate than they are formed. Compare *renewable resource*.

nuclear change Process in which nuclei of certain isotopes spontaneously change, or are forced to change, into one or more different isotopes. The three principal types of nuclear change are natural radioactivity, nuclear fission, and nuclear fusion. Compare *chemical change, physical change*.

nuclear energy Energy released when atomic nuclei undergo a nuclear reaction such as the spontaneous emission of radioactivity, nuclear fission, or nuclear fusion.

nuclear fission Nuclear change in which the nuclei of certain isotopes with large mass numbers (such as uranium-235 and plutonium-239) are split apart into lighter nuclei when struck by a neutron. This process releases more neutrons and a large amount of energy. Compare *nuclear fusion*.

nuclear fusion Nuclear change in which two nuclei of isotopes of elements with a low mass number (such as hydrogen-2 and hydrogen-3) are forced together at extremely high temperatures until they fuse to form a heavier nucleus (such as helium-4). This process releases a large amount of energy. Compare *nuclear fission*.

nucleic acids Large polymer molecules made by linking hundreds to thousands of four types of monomers called *nucleotides*.

nucleus Extremely tiny center of an atom, making up most of the atom's mass. It contains one or more positively charged protons and one or more neutrons with no electrical charge (except for a hydrogen-1 atom, which has one proton and no neutrons in its nucleus).

nutrient Any food or element an organism must take in to live, grow, or reproduce.

nutrient cycle See *biogeochemical cycle*.

old-growth forest Virgin and old, second-growth forests containing trees that are often hundreds—sometimes thousands—of years old. Examples include forests of Douglas fir, western hemlock, giant sequoia, and coastal redwoods in the western United States. Compare *second-growth forest, tree plantation*.

oligotrophic lake Lake with a low supply of plant nutrients. Compare *eutrophic lake, mesotrophic lake*.

omnivore Animal that can use both plants and other animals as food sources. Examples include pigs, rats, cockroaches, and humans. Compare *carnivore, herbivore*.

open sea Part of an ocean that lies beyond the continental shelf. Compare *coastal zone*.

organic compounds Compounds containing carbon atoms combined with each other and with atoms of one or more other elements such as hydrogen, oxygen, nitrogen, sulfur, phosphorus, chlorine, and fluorine. All other compounds are called *inorganic compounds*.

organism Any form of life.

output Matter, energy, or information leaving a system. Compare *input, throughput*.

output pollution control See *pollution cleanup*.

ozone layer Layer of gaseous ozone (O_3) in the stratosphere that protects life on earth by filtering out most harmful ultraviolet radiation from the sun.

paradigm shifts Shifts in scientific thinking that occur when the majority of scientists in a field or related fields agree that a new explanation or theory is better than the old one.

parasite Consumer organism that lives on or in, and feeds on, a living plant or animal, known as the host, over an extended period. The parasite draws nourishment from and gradually weakens its host; it may or may not kill the host. See *parasitism*.

parasitism Interaction between species in which one organism, called the parasite, preys on another organism, called the host, by living on or in the host. See *host, parasite.*

per capita ecological footprint Amount of biologically productive land and water needed to supply each person or population with the renewable resources they use and to absorb or dispose of the wastes from such resource use. It measures the average environmental impact of individuals or populations in different countries and areas. Compare *ecological footprint.*

per capita GDP Annual gross domestic product (GDP) of a country divided by its total population at midyear. It gives the average slice of the economic pie per person. Used to be called per capita gross national product (GNP). See *gross domestic product.*

perennial Plant that can live for more than 2 years. Compare *annual.*

permafrost Perennially frozen layer of the soil that forms when the water there freezes. It is found in arctic tundra.

perpetual resource Essentially inexhaustible resource on a human time scale because it is renewed continuously. Solar energy is an example. Compare *nonrenewable resource, renewable resource.*

persistence How long a pollutant stays in the air, water, soil, or body. See *inertia.*

persistent pollutant See *slowly degradable pollutant.*

pH Numeric value that indicates the relative acidity or alkalinity of a substance on a scale of 0 to 14, with the neutral point at 7. Acid solutions have pH values lower than 7; basic or alkaline solutions have pH values greater than 7.

phosphorus cycle Cyclic movement of phosphorus in different chemical forms from the environment to organisms and then back to the environment.

photosynthesis Complex process that takes place in cells of green plants. Radiant energy from the sun is used to combine carbon dioxide (CO_2) and water (H_2O) to produce oxygen (O_2), carbohydrates (such as glucose, $C_6H_{12}O_6$), and other nutrient molecules. Compare *aerobic respiration, chemosynthesis.*

physical change Process that alters one or more physical properties of an element or a compound without changing its chemical composition. Examples include changing the size and shape of a sample of matter (crushing ice and cutting aluminum foil) and changing a sample of matter from one physical state to another (boiling and freezing water). Compare *chemical change, nuclear change.*

phytoplankton Small, drifting plants, mostly algae and bacteria, found in aquatic ecosystems. Compare *plankton, zooplankton.*

pioneer community First integrated set of plants, animals, and decomposers found in an area undergoing primary ecological succession. See *immature community, mature community.*

pioneer species First hardy species—often microbes, mosses, and lichens—that begin colonizing a site as the first stage of ecological succession. See *ecological succession, pioneer community.*

planetary management worldview We are separate from nature, nature exists mainly to meet our needs and increasing wants, and we can use our ingenuity and technology to manage the earth's life-support systems, mostly for our benefit. It assumes that economic growth is essentially unlimited. Compare *environmental wisdom worldview, stewardship worldview.*

plankton Small plant organisms (phytoplankton) and animal organisms (zooplankton) that float in aquatic ecosystems.

point source Single identifiable source that discharges pollutants into the environment. Examples include the smokestack of a power plant or an industrial plant, drainpipe of a meatpacking plant, chimney of a house, or exhaust pipe of an automobile. Compare *nonpoint source.*

pollutant Particular chemical or form of energy that can adversely affect the health, survival, or activities of humans or other living organisms. See *pollution.*

pollution An undesirable change in the physical, chemical, or biological characteristics of air, water, soil, or food that can adversely affect the health, survival, or activities of humans or other living organisms.

pollution cleanup Device or process that removes or reduces the level of a pollutant after it has been produced or has entered the environment. Examples include automobile emission control devices and sewage treatment plants. Compare *pollution prevention.*

pollution prevention Device or process that prevents a potential pollutant from forming or entering the environment or sharply reduces the amount entering the environment. Compare *pollution cleanup.*

population Group of individual organisms of the same species living in a particular area.

population change Increase or decrease in the size of a population. It is equal to (Births + Immigration) − (Deaths + Emigration).

population density Number of organisms in a particular population found in a specified area or volume.

population dispersion General pattern in which the members of a population are arranged throughout its habitat.

population distribution Variation of population density over a particular geographic area. For example, a country has a high population density in its urban areas and a much lower population density in rural areas.

population dynamics Major abiotic and biotic factors that tend to increase or decrease the population size and affect the age and sex composition of a species.

population size Number of individuals making up a population's gene pool.

positive feedback loop Causes a system to change further in the same direction. See *feedback loop.* Compare *negative feedback loop.*

potential energy Energy stored in an object because of its position or the position of its parts. Compare *kinetic energy.*

poverty Inability to meet basic needs for food, clothing, and shelter.

prairies See *grasslands.*

precautionary principle When there is some scientific uncertainty about potentially serious harm from chemicals or technologies, decision makers should act to prevent harm to humans and the environment. See *pollution prevention.*

precipitation Water in the form of rain, sleet, hail, and snow that falls from the atmosphere onto land and bodies of water.

predation Situation in which an organism of one species (the predator) captures and feeds on parts or all of an organism of another species (the prey).

predator Organism that captures and feeds on parts or all of an organism of another species (the prey).

predator–prey relationship Interaction between two organisms of different species in which one organism, called the *predator,* captures and feeds on parts or all of the other organism, called the *prey.*

prey Organism that is captured and serves as a source of food for an organism of another species (the predator).

primary consumer Organism that feeds on all or part of plants (herbivore) or on other producers. Compare *detritivore, omnivore, secondary consumer.*

primary productivity See *gross primary productivity, net primary productivity.*

primary succession Ecological succession in a bare area that has never been occupied by a community of organisms. See *ecological succession*. Compare *secondary succession*.

producer Organism that uses solar energy (green plants) or chemical energy (some bacteria) to manufacture the organic compounds it needs as nutrients from simple inorganic compounds obtained from its environment. Compare *consumer, decomposer*.

prokaryotic organism Classification of cell structure in which the cell contains no distinct nucleus or organelles enclosed by membranes. A prokaryotic cell is much simpler and usually much smaller than a eukaryotic cell. All bacteria are single-celled prokaryotic organisms. Compare *eukaryotic cell*.

proteins Large polymer molecules formed by linking together long chains of monomers called *amino acids*.

proton (p) Positively charged particle in the nuclei of all atoms. Each proton has a relative mass of 1 and a single positive charge. Compare *electron, neutron*.

pyramid of energy flow Diagram representing the flow of energy through each trophic level in a food chain or food web. With each energy transfer, only a small part (typically 10%) of the usable energy entering one trophic level is transferred to the organisms at the next trophic level.

radiation Fast-moving particles (particulate radiation) or waves of energy (electromagnetic radiation). See *alpha particle, beta particle, gamma ray*.

radioactive decay Change of a radioisotope to a different isotope by the emission of radioactivity.

radioactive isotope See *radioisotope*.

radioactive waste Waste products of nuclear power plants, research, medicine, weapon production, or other processes involving nuclear reactions. See *radioactivity*.

radioactivity Nuclear change in which unstable nuclei of atoms spontaneously shoot out "chunks" of mass, energy, or both at a fixed rate. The three principal types of radioactivity are gamma rays and fast-moving alpha particles and beta particles.

radioisotope Isotope of an atom that spontaneously emits one or more types of radioactivity (alpha particles, beta particles, gamma rays).

rain shadow effect Low precipitation on the far side (leeward side) of a mountain when prevailing winds flow up and over a high mountain or range of high mountains. This creates semiarid and arid conditions on the leeward side of a high mountain range.

range See *distribution*.

range of tolerance Range of chemical and physical conditions that must be maintained for populations of a particular species to stay alive and grow, develop, and function normally. See *law of tolerance*.

rare species Species that has naturally small numbers of individuals (often because of limited geographic ranges or low population densities) or that has been locally depleted by human activities.

realized niche Parts of the fundamental niche of a species that are actually used by that species. See *ecological niche*. Compare *fundamental niche*.

recombinant DNA DNA that has been altered to contain genes or portions of genes from organisms of different species.

reconciliation ecology Science of inventing, establishing, and maintaining new habitats to conserve species diversity in places where people live, work, or play.

recycling Collecting and reprocessing a resource so that it can be made into new products. An example is collecting aluminum cans, melting them down, and using the aluminum to make new cans or other aluminum products. Compare *reuse*.

reforestation Renewal of trees and other types of vegetation on land where trees have been removed; can be done naturally by seeds from nearby trees or artificially by planting seeds or seedlings.

reliable runoff Surface runoff of water that generally can be counted on as a stable source of water from year to year. See *runoff*.

renewable resource Resource that can be replenished rapidly (hours to several decades) through natural processes as long as it is not used up faster than it is replaced. Examples include trees in forests, grasses in grasslands, wild animals, fresh surface water in lakes and streams, most groundwater, fresh air, and fertile soil. If such a resource is used faster than it is replenished, it can be depleted and converted into a nonrenewable resource. Compare *nonrenewable resource, perpetual resource*. See also *environmental degradation*.

replacement-level fertility Number of children a couple must have to replace them. The average for a country or the world usually is slightly higher than 2 children per couple (2.1 in the United States and 2.5 in some developing countries) because some children die before reaching their reproductive years. See also *total fertility rate*.

reproduction Production of offspring by one or more parents.

reproductive isolation Long-term geographic separation of members of a particular sexually reproducing species.

reproductive potential See *biotic potential*.

resilience Ability of a living system to bounce back and repair damage after a disturbance that is not too drastic.

resource Anything obtained from the environment to meet human needs and wants. It can also be applied to other species.

resource partitioning Process of dividing up resources in an ecosystem so that species with similar needs (overlapping ecological niches) use the same scarce resources at different times, in different ways, or in different places. See *ecological niche, fundamental niche, realized niche*.

respiration See *aerobic respiration*.

restoration ecology Research and scientific study devoted to restoring, repairing, and reconstructing damaged ecosystems.

reuse Using a product over and over again in the same form. An example is collecting, washing, and refilling glass beverage bottles. Compare *recycling*.

r-selected species Species that reproduce early in their life span and produce large numbers of usually small and short-lived offspring in a short period. Compare *K-selected species*.

r-strategists See *r-selected species*.

rule of 70 Doubling time (in years) = 70/(percentage growth rate). See *doubling time, exponential growth*.

runoff Freshwater from precipitation and melting ice that flows on the earth's surface into nearby streams, lakes, wetlands, and reservoirs. See *reliable runoff, surface runoff, surface water*. Compare *groundwater*.

salinity Amount of various salts dissolved in a given volume of water.

scavenger Organism that feeds on dead organisms that were killed by other organisms or died naturally. Examples include vultures, flies, and crows. Compare *detritivore*.

science Attempts to discover order in nature and use that knowledge to make predictions about what should happen in nature. See *frontier science, scientific data, scientific hypothesis, scientific law, scientific methods, scientific model, scientific theory, sound science*.

scientific data Facts obtained by making observations and measurements. Compare *scientific hypothesis, scientific law, scientific methods, scientific model, scientific theory*.

scientific hypothesis Educated guess that attempts to explain a scientific law or certain scientific observations. Compare *scientific data, scientific law, scientific methods, scientific model, scientific theory.*

scientific law Description of what scientists find happening in nature repeatedly in the same way, without known exception. See *first law of thermodynamics, law of conservation of matter, second law of thermodynamics.* Compare *scientific data, scientific hypothesis, scientific methods, scientific model, scientific theory.*

scientific methods Ways scientists gather data and formulate and test scientific hypotheses, models, theories, and laws. See *scientific data, scientific hypothesis, scientific law, scientific model, scientific theory.*

scientific model Simulation of complex processes and systems. Many are mathematical models that are run and tested using computers.

scientific theory Well-tested and widely accepted scientific hypothesis. Compare *scientific data, scientific hypothesis, scientific law, scientific methods, scientific model.*

secondary consumer Organism that feeds only on primary consumers. Compare *detritivore, omnivore, primary consumer.*

secondary succession Ecological succession in an area in which natural vegetation has been removed or destroyed but the soil is not destroyed. See *ecological succession.* Compare *primary succession.*

second-growth forest Stands of trees resulting from secondary ecological succession. Compare *old-growth forest, tree farm.*

second law of energy See *second law of thermodynamics.*

second law of thermodynamics In any conversion of heat energy to useful work, some of the initial energy input is always degraded to lower-quality, more dispersed, less useful energy—usually low-temperature heat that flows into the environment; you cannot break even in terms of energy quality. See *first law of thermodynamics.*

selective cutting Cutting of intermediate-aged, mature, or diseased trees in an uneven-aged forest stand, either singly or in small groups. This encourages the growth of younger trees and maintains an uneven-aged stand. Compare *clear-cutting, strip-cutting.*

sexual reproduction Reproduction in organisms that produce offspring by combining sex cells or *gametes* (such as ovum and sperm) from both parents. It produces offspring that have combinations of traits from their parents. Compare *asexual reproduction.*

slowly degradable pollutant Material that is slowly broken down into simpler chemicals or reduced to acceptable levels by natural physical, chemical, and biological processes. Compare *biodegradable pollutant, degradable pollutant, nondegradable pollutant.*

social capital Positive force created when people with different views and values find common ground and work together to build understanding, trust, and informed shared visions of what their communities, states, nations, and the world could and should be. Compare *natural capital.*

soil Complex mixture of inorganic minerals (clay, silt, pebbles, and sand), decaying organic matter, water, air, and living organisms.

soil conservation Methods used to reduce soil erosion, prevent depletion of soil nutrients, and restore nutrients previously lost by erosion, leaching, and excessive crop harvesting.

soil erosion Movement of soil components, especially topsoil, from one place to another, usually by wind, flowing water, or both. This natural process can be greatly accelerated by human activities that remove vegetation from soil.

soil horizons Horizontal zones that make up a particular mature soil. Each horizon has a distinct texture and composition that vary with different types of soils. See *soil profile.*

soil profile Cross-sectional view of the horizons in a soil. See *soil horizon.*

solar capital Solar energy that warms the planet and supports photosynthesis, the process that plants use to provide food for themselves and for us and other animals. This direct input of solar energy also produces indirect forms of renewable solar energy such as wind and flowing water. Compare *natural capital.*

solar energy Direct radiant energy from the sun and a number of indirect forms of energy produced by the direct input of such radiant energy. Principal indirect forms of solar energy include wind, falling and flowing water (hydropower), and biomass (solar energy converted into chemical energy stored in the chemical bonds of organic compounds in trees and other plants).

sound science Concepts and ideas that are widely accepted by experts in a particular field of the natural or social sciences. These results of science are very reliable. Compare *frontier science, junk science.*

specialist species Species with a narrow ecological niche. They may be able to live in only one type of habitat, tolerate only a narrow range of climatic and other environmental conditions, or use only one type or a few types of food. Compare *generalist species.*

speciation Formation of two species from one species because of divergent natural selection in response to changes in environmental conditions; usually takes thousands of years. Compare *extinction.*

species Group of organisms that resemble one another in appearance, behavior, chemical makeup and processes, and genetic structure. Organisms that reproduce sexually are classified as members of the same species only if they can actually or potentially interbreed with one another and produce fertile offspring.

species diversity Number of different species and their relative abundances in a given area. See *biodiversity.* Compare *ecological diversity, genetic diversity.*

species equilibrium model Widely accepted model which says that the number of different species found on an island is determined by a balance between two factors: the rate at which new species immigrate to the island and the rate at which existing species become extinct on the island.

species evenness Abundance of individuals within each species contained in a community.

species richness Number of different species contained in a community.

S-shaped curve Leveling off of an exponential, J-shaped curve when a rapidly growing population exceeds the carrying capacity of its environment and ceases to grow.

stewardship worldview We can manage the earth for our benefit but we have an ethical responsibility to be caring and responsible managers, or *stewards,* of the earth. It calls for encouraging environmentally beneficial forms of economic growth and discouraging environmentally harmful forms. Compare *environmental wisdom worldview, planetary management worldview.*

stratosphere Second layer of the atmosphere, extending about 17–48 kilometers (11–30 miles) above the earth's surface. It contains small amounts of gaseous ozone (O_3), which filters out about 95% of the incoming harmful ultraviolet (UV) radiation emitted by the sun. Compare *troposphere.*

stream Flowing body of surface water. Examples are creeks and rivers.

strip-cutting Variation of clear-cutting in which a strip of trees is clear-cut along the contour of the land, with the corridor being narrow enough to allow natural regeneration within a few years. After regeneration, another strip is cut above the first, and so on. Compare *clear-cutting, selective cutting.*

subatomic particles Extremely small particles—electrons, protons, and neutrons—that make up the internal structure of atoms.

subduction zone Area in which oceanic lithosphere is carried downward (subducted) under an island arc or continent at a convergent plate boundary. A trench ordinarily forms at the boundary between the two converging plates. See convergent *plate boundary.*

subpopulation The individuals of a species that live in a habitat patch.

succession See *ecological succession, primary succession, secondary succession.*

succulent plants Plants, such as desert cacti, that survive in dry climates by having no leaves, thus reducing the loss of scarce water. They store water and use sunlight to produce the food they need in the thick, fleshy tissue of their green stems and branches. Compare *deciduous plants, evergreen plants.*

sulfur cycle Cyclic movement of sulfur in various chemical forms from the environment to organisms and then back to the environment.

surface fire Forest fire that burns only undergrowth and leaf litter on the forest floor. Compare *crown fire, ground fire.* See *controlled burning.*

surface runoff Water flowing off the land into bodies of surface water. See *reliable runoff.*

surface water Precipitation that does not infiltrate the ground or return to the atmosphere by evaporation or transpiration. See *runoff.* Compare *groundwater.*

survivorship curve Graph showing the number of survivors in different age groups for a particular species.

sustainability Ability of earth's various systems, including human cultural systems and economies, to survive and adapt to changing environmental conditions indefinitely.

sustainable development See *environmentally sustainable economic development.*

sustainable living Taking no more potentially renewable resources from the natural world than can be replenished naturally and not overloading the capacity of the

environment to cleanse and renew itself by natural processes.

sustainable society Society that manages its economy and population size without doing irreparable environmental harm by overloading the planet's ability to absorb environmental insults, replenish its resources, and sustain human and other forms of life over a specified period, usually hundreds to thousands of years. During this period, the society satisfies the needs of its people without depleting natural resources that would jeopardize the prospects of current and future generations of humans and other species.

sustainable yield (sustained yield) Highest rate at which a potentially renewable resource can be used indefinitely without reducing its available supply. See also *environmental degradation.*

synergistic interaction Interaction of two or more factors or processes causing the combined effect to be greater than the sum of their separate effects.

synergy See *synergistic interaction.*

system Set of components that function and interact in some regular and theoretically predictable manner.

temperature Measure of the average speed of motion of the atoms, ions, or molecules in a substance or combination of substances at a given moment. Compare *heat.*

terrestrial Pertaining to land. Compare *aquatic.*

tertiary (higher-level) consumers Animals that feed on animal-eating animals. They feed at high trophic levels in food chains and webs. Examples include hawks, lions, bass, and sharks. Compare *detritivore, primary consumer, secondary consumer.*

theory of evolution Widely accepted scientific idea that all life forms developed from earlier life forms. Although this theory conflicts with the creation stories of many religions, it is the way biologists explain how life has changed over the past 3.6–3.8 billion years and why it is so diverse today.

theory of island biogeography See *species equilibrium model.*

third and higher level consumers Carnivores that feed on other carnivores.

threatened species Wild species that is still abundant in its natural range but is likely to become endangered because of a decline in numbers. Compare *endangered species.*

threshold effect Harmful or fatal effect of a small change in environmental conditions that exceeds the limit of tolerance of an organism or population of a species. See *law of tolerance.*

throughput Rate of flow of matter, energy, or information through a system. Compare *input, output.*

throwaway society See *high-throughput economy.*

time delays Amount of time in a feedback loop between the input of a stimulus and the response to it. See *feedback loop, negative feedback loop, positive feedback loop.*

tolerance limits Minimum and maximum limits for physical conditions (such as temperature) and concentrations of chemical substances beyond which no members of a particular species can survive. See *law of tolerance.*

total fertility rate (TFR) Estimate of the average number of children who will be born alive to a woman during her lifetime if she passes through all her childbearing years (ages 15–44) conforming to age-specific fertility rates of a given year. More simply, it is an estimate of the average number of children a woman will have during her childbearing years.

tragedy of the commons Depletion or degradation of a potentially renewable resource to which people have free and unmanaged access. An example is the depletion of commercially desirable fish species in the open ocean beyond areas controlled by coastal countries. See *common-property resource.*

transform fault Area where the earth's lithospheric plates move in opposite but parallel directions along a fracture (fault) in the lithosphere. Compare *convergent plate boundary, divergent plate boundary.*

transgenic organisms See *genetically modified organisms (GMOs)*

transpiration Process in which water is absorbed by the root systems of plants, moves up through the plants, passes through pores (stomata) in their leaves or other parts, and evaporates into the atmosphere as water vapor.

tree farm See *tree plantation.*

tree plantation Site planted with one or only a few tree species in an even-aged stand. When the stand matures it is usually harvested by clear-cutting and then replanted. These farms normally raise rapidly growing tree species for fuelwood, timber, or pulpwood. See *even-aged management.* Compare *old-growth forest, second-growth forest, uneven-aged management.*

trophic level All organisms that are the same number of energy transfers away from the original source of energy (for example, sunlight) that enters an ecosystem. For example, all producers belong to the first trophic level, and all herbivores belong to the second trophic level in a food chain or a food web.

troposphere Innermost layer of the atmosphere. It contains about 75% of the mass of earth's air and extends about 17 kilometers (11 miles) above sea level. Compare *stratosphere*.

tsunami Series of large waves generated when part of the ocean floor suddenly rises or drops usually because of an earthquake.

ultraplankton Huge populations of extremely small photosynthetic bacteria that may be responsible for 70% of the primary productivity near the ocean surface.

uneven-aged management Method of forest management in which trees of different species in a given stand are maintained at many ages and sizes to permit continuous natural regeneration. Compare *even-aged management*.

upwelling Movement of nutrient-rich bottom water to the ocean's surface. It can occur far from shore but usually takes place along certain steep coastal areas where the surface layer of ocean water is pushed away from shore and replaced by cold, nutrient-rich bottom water.

utilitarian value See *instrumental value*.

vertebrates Animals that have backbones. Compare *invertebrates*.

volcano Vent or fissure in the earth's surface through which magma, liquid lava, and gases are released into the environment.

warm front Boundary between an advancing warm air mass and the cooler one it is replacing. Because warm air is less dense than cool air, an advancing warm front rises over a mass of cool air. Compare *cold front*.

water cycle See *hydrologic cycle*.

watershed Land area that delivers water, sediment, and dissolved substances via small streams to a major stream (river).

weather Short-term changes in the temperature, barometric pressure, humidity, precipitation, sunshine, cloud cover, wind direction and speed, and other conditions in the troposphere at a given place and time. Compare *climate*.

wetland Land that is covered all or part of the time with saltwater or freshwater, excluding streams, lakes, and the open ocean. See *coastal wetland, inland wetland*.

wilderness Area where the earth and its community of life have not been seriously disturbed by humans and where humans are only temporary visitors.

wildlife All free, undomesticated species. Sometimes the term is used to describe animals only.

wildlife resources Wildlife species that have actual or potential economic value to people.

wild species Species found in the natural environment. Compare *domesticated species*.

worldview How people think the world works and what they think their role in the world should be. See *environmental wisdom worldview, planetary management worldview, stewardship worldview*.

zooplankton Animal plankton; small floating herbivores that feed on plant plankton (phytoplankton). Compare *phytoplankton*.

INDEX

Note: Page numbers of **boldface** type indicate definitions of key terms. Page numbers followed by italicized *f, t,* or *b* indicate figures, tables, and boxes.

Overfishing, 253–54, 257–59
   industrialized fish harvesting methods, 254b
Overgrazing, 208, **209f**
Overpopulation, human, 171
Overshoot of carrying capacity, 164, 165f
Overturn in lakes, 137
Ozone (O_3), 54

Pangaea, 88f
Paper, alternative plants for production of, 203f
Paradigm shifts in scientific theories, **31**–32
Parasites, 153–54
Parasitism, 153–54
Parent material (C horizon), soil, 68f, 69
Passenger pigeon, extinction of, 222, 247
Pastures, **208**–9
Pathogens. *See also* Infectious disease
   tree, S49–S50
Peace, global human aging and, 182
Peer review, 4
Pension systems, 176
Per capita ecological footprint, **13**
Per capita GDP (gross domestic product), **10**
Periodic table of elements, S27f, S28
Permafrost, 54, **114**–15
Perpetual resource, **12**
Persistence
   inertia and, **158**
   of pollutants, **40**
Pest(s), insect, 50, 90b, S49–S50
Pesticides
   bioaccumulation and biomagnification of DDT, 237f
   threat of, to wild species, S18
   unintended consequences of using, 189b
Pets, market for exotic, 238–39
pH, 35, **36**
   scale, 36f
Pharmaceutical products, 148, 227
   from tropical forests, 205f
Phosphorus cycle, 76
   human effects on, 76, 77f
Photosynthesis, **58**, 59b
   CO_2 levels and, 73
Physical change in matter, **39**
Phytoplankton, 58
Pie charts, S4, S5f
Pimentel, David, 233
Pinchot, Gifford, S16
Pioneer (early successional) species, 156
Planetary management worldview, **22**–23
Plankton, **128**, 149
Plant(s). *See also* Tree(s)
   chaparral, 115f
   classification and naming, S35
   commensalisms among, 155
   desert, 108–10
   early successional, midsuccessional, and late successional, 156
   global biodiversity in, S10–S11f
   in evergreen coniferous forests, 120, 121f
   market for exotic, 238–39
   pharmaceuticals derived from tropical forests, 205f
   roots of, and mycorrhizal fungi, 154f

in temperate deciduous forests, 119, 120f
in temperate grasslands, 111, 112f, 113f
in terrestrial communities, 144f
in tropical rain forests, 117–19
Plant nutrients, effects of, on lakes, 137, 138f
Plasma, 38
Point sources of pollution, **15**
   of air pollution, 16f
Polar grasslands, 112f, 114–15
Pollutants, types and persistence of, 40
Pollution, **15**–16
   biotic, 233
   effects of, on aquatic biodiversity, 252–53
   laws to control (*see* Environmental law)
   point, and nonpoint sources of, 15
   prevention versus cleanup of, 16
   threat of, to wild species, 237
Pollution cleanup, **16**
Pollution prevention, 16
Polymers, S28
Polyps, coral, 126
Population(s), **53**. *See also* Population dynamics
   biotic potential of, 163
   conditions necessary for biological evolution of, 86
   growth (*see* Population growth)
   human (*see* Human population)
   size of (*see* Population size)
   sustainability and control of, 24f
Population change, **173**
   effects of population density on, 165
   factors governing, 162
   population change curves in nature, types of, 166
Population control as sustainability principle, **24**
Population crash (die-back), 164, 165f
Population density, effects on population change, **165**
Population distribution, 162
Population dynamics, **161**, 162–67
   carrying capacity and, 164–65
   case study of white-tailed deer, 166–67
   factors governing population-size changes, 162
   limits on population growth, 163
   population age structure, 162–63
   population change curves in nature, types of, 166
   population density, population change, and, 165
   population diebacks, 164, 165f
   population distribution, 162
   population growth, logistic and exponential, 163–64
   reproductive patterns, 167–69
   of southern sea otters, 161, 169
Population ecology, 161–70
   applied to humans (*see* Human population)
   case study of sea otters and, 161
   population dynamics, carrying capacity and, 162–67
   reproductive patterns, survival, and, 167–69
   sustainability and, 169

Population growth
   exponential and logistic, 163f
   human, 172–73 (*see also* Human population)
   intrinsic rate of increase (r), 163
   limits on, 57–58, 163
Population size. *See also* Population growth
   carrying capacity and, 163f
   effects of genetic variations on, S47
   effects of population density on, 165
   factors affecting human, 173–78, 182–85 (*see also* Human population)
   growth and decrease factors in, 162
   patterns of variation in, 166
   role of predation in, 166f
   top-down, and bottom-up controls on, 166f
Positive feedback loop, **33**
Potential energy, 42–**43**
Poverty, 6, **18**
   economic development and, 10–11
   environmental problems related to, 18
   global distribution of, 11f
   illegal smuggling of wild species linked to, 237–38
Prairie potholes, 140, 141b
Prairies, **111**, 112f, 113f. *See also* Grasslands
Precautionary approach
   in fisheries, 262
   to protection of invasive species, 236
Precipitation
   acid deposition/acid rain, 75
   as biome limiting factor, 58, 101, 107f
   climate and, 102
   monsoon, 105
   rain shadow effect and, 105f
Precision in scientific measurement, S3
Predation, **151**. *See also* Predators; Prey
   defenses against, 152–53
Predator–prey relationship, **151**
   snowshoe hare and Canadian lynx, 166f
   wolf and moose, S46
Predators, 151
   as keystone species, 143, 148, 161, 191
   killing of, by humans, 238
   relationship of, to prey, 151, 166f
   sharks as, 149b
   strategies of, 152
Preservationist movement, S16
Preservationist school of public lands management, S18
Prey, 151
   avoidance or defense against predators among, 152, 153f
Primary consumers, **60**
Primary succession, **155**, 156f
Primm, Stuart, 94, 226
Private ownership of land, species protection and, 241
Producers, **58**–61
   feeding relationships among consumers, decomposers, and, 56–57
   rate of productivity in, 64–67
Profundal zones, lake, 137
Prokaryotic cell, **37f**, S34f. *See also* Bacteria